COMPREHENSIVE CHEMICAL KINETICS

COMPREHENSIVE

CHEMICAL KINETICS

EDITED BY

C. H. BAMFORD

M.A., Ph.D., Sc.D. (Cantab.), F.R.I.C., F.R.S.

Campbell-Brown Professor of Industrial Chemistry,
University of Liverpool

AND

C. F. H. TIPPER

Ph.D. (Bristol), D.Sc. (Edinburgh)

Senior Lecturer in Physical Chemistry,
University of Liverpool

VOLUME 2

THE THEORY OF KINETICS

ELSEVIER PUBLISHING COMPANY

AMSTERDAM - LONDON - NEW YORK

1969

ELSEVIER PUBLISHING COMPANY
335 JAN VAN GALENSTRAAT
P.O. BOX 211, AMSTERDAM, THE NETHERLANDS

ELSEVIER PUBLISHING CO. LTD.
BARKING, ESSEX, ENGLAND

AMERICAN ELSEVIER PUBLISHING COMPANY, INC.
52 VANDERBILT AVENUE
NEW YORK, NEW YORK 10017

LIBRARY OF CONGRESS CARD NUMBER 68-29646.
STANDARD BOOK NUMBER 444-40674-3

WITH 77 ILLUSTRATIONS AND 13 TABLES.

PRINTED IN THE NETHERLANDS

COMPREHENSIVE CHEMICAL KINETICS

Contributors to Volume 2

I. D. CLARK   Physical Chemistry Laboratory,
University of Oxford,
Oxford, England

L. G. HARRISON   Department of Chemistry,
University of British Columbia,
Vancouver, B.C., Canada

V. N. KONDRATIEV   Institute of Chemical Physics,
Academy of Sciences of the U.S.S.R.,
Moscow, U.S.S.R.

Z. G. SZABÓ   Institute of Inorganic
and Analytical Chemistry,
L. Eötvös University,
Budapest, Hungary

R. P. WAYNE   Physical Chemistry Laboratory,
University of Oxford,
Oxford, England

Preface

The rates of chemical processes and their variation with conditions have been studied for many years, usually for the purpose of determining reaction mechanisms. Thus, the subject of chemical kinetics is a very extensive and important part of chemistry as a whole, and has acquired an enormous literature. Despite the number of books and reviews, in many cases it is by no means easy to find the required information on specific reactions or types of reaction or on more general topics in the field. It is the purpose of this series to provide a background reference work, which will enable such information to be obtained either directly, or from the original papers or reviews quoted.

The aim is to cover, in a reasonably critical way, the practice and theory of kinetics and the kinetics of inorganic and organic reactions in gaseous and condensed phases and at interfaces (excluding biochemical and electrochemical kinetics, however, unless very relevant) in more or less detail. The series will be divided into sections covering a relatively wide field; a section will consist of one or more volumes, each containing a number of articles written by experts in the various topics. Mechanisms will be thoroughly discussed and relevant non-kinetic data will be mentioned in this context. The methods of approach to the various topics will, of necessity, vary somewhat depending on the subject and the author(s) concerned.

It is obviously impossible to classify chemical reactions in a completely logical manner, and the editors have in general based their classification on types of chemical element, compound or reaction rather than on mechanisms, since views on the latter are subject to change. Some duplication is inevitable, but it is felt that this can be a help rather than a hindrance.

Volume 1 dealt with experimental methods and the processing of the data obtained. Having reached this stage it is necessary to decide whether the reaction is 'elementary' or complex, and, if the latter, what the mechanism is. The first two chapters of this volume deal with the general characteristics of complex but non-chain processes, and of chain reactions. In the other three, the theory of elementary reactions occurring in the gaseous phase and in condensed phases is discussed.

The Editors wish to express once again their sincere appreciation of the advice and support so readily given by the members of the Advisory Board.

Liverpool
May, 1969

C. H. BAMFORD
C. F. H. TIPPER

Contents

Chapter 4 (I. D. CLARK AND R. P. WAYNE)

The theory of elementary reactions in solution · · · · · · · · · · · · · · · 302

Chapter 5 (L. G. HARRISON)

The theory of solid phase kinetics 377

Chapter 1

Kinetic Characterization of Complex Reaction Systems

Z. G. SZABÓ

Introduction

The concept of molecularity is connected with and only with elementary chemical reactions. The molecularity of an elementary process means the number of particles taking part in the step leading to the reaction itself. In terms of the transition state theory, molecularity means the number of molecules, radicals or free atoms, giving rise to the activated complex. Molecularity, being a number expressing deeper chemical significance, is sometimes not clearly distinguished from the concept of the reaction order, the latter being an empirical factor of the chemical kinetics. There are cases, of course, when molecularity and reaction order may be the same.

The reaction order should be defined only in connection with chemical processes the rate expression of which can be given in the following form

$$w = kc_1^{n_1} c_2^{n_2} \dots \tag{1}$$

In eqn. (1) w is the rate of the reaction, k the rate coefficient and $c_1, c_2, \dots$ are concentrations. In this case the order of the reaction is the sum of the exponents of the concentrations, that is

$$n = n_1 + n_2 + \dots \tag{2}$$

where $n_1, n_2 \dots$ are orders referring to reactants $A_1, A_2 \dots$

The type of rate expression (1) may occur in connection with uni-, bi- and ter-molecular elementary reactions and some complex reactions. The kinetics of most complex reactions cannot be described in such a simple form by means of the reaction order; this is especially the case for reactions in the gas phase. These complex systems are reactions consisting of several steps. In such cases the introduction of the concept of reaction order has no practical importance. Our purpose in the present chapter, discussing systems with complex kinetics, is to examine for certain reaction types the mathematical form of the expression of the reaction rate and—if possible—the concentration change with time. We look for an expression—mostly in integrated form—which tests the applicability of a given mechanism to the experimental results at hand. In many cases the integrated expressions cannot be linearized and are so complicated that the applicability of the mechanism has to be checked by means of the differential form. Finally, in some cases we deal with the problems of the calculation of rate coefficients of elementary reactions. When possible, we also show the connection between the mechanism and the order of reaction.

Subsequently we shall discuss complex chemical reactions, with the exception of chain processes. An exact classification of these complex systems is not possible because of overlapping. The categories, described below, are essentially formal, especially with respect to III, V and VII. However, this systematization must be taken into consideration as it may play the role of a signpost in the analysis of experimental results, thus permitting a formal mechanism to be selected. The next step is to fill this formal mechanism with chemical content, to establish the chemistry itself. The classification, which also serves as a basis for the whole discussion, is described as follows.

According to the reaction order of the elementary processes involved in a complex reaction, we separately deal with (1) first order reactions, (2) reactions of the second and mixed order, *i.e.* first–second order reactions, (3) reactions of the third order, and (4) reactions of nth order, where n may be 1, 2, ... Depending on the order of the elementary processes we distinguish between the following categories:

(I) reversible reactions,
(II) parallel reactions,
(III) parallel reactions with a reversible reaction step (or steps),
(VI) competitive-consecutive reactions,
(VII) competitive-consecutive reactions with a reversible reaction step (or steps).

Systems involving two reaction steps of opposite direction and occurring with commensurable probability are reversible reactions. The connection with certain reversible reactions enables us to deal here with some heterogeneous, reversible and unidirectional, reactions as well, which are difficult to classify.

Processes in which the reaction partners may react in two or more independent ways are called parallel reactions. Parallel reactions containing one or more reversible reaction steps are dealt with in class III.

Reactions in which the product of the preceding reaction is the starting material for the following one are described in the group of consecutive reactions. To this group belong systems in which the starting substances are converted into the final products through intermediates. In group V we treat consecutive reactions with one or more reversible reaction steps.

The designation *competitive, consecutive reactions* refers to systems in which parallel and consecutive reactions both occur. Some very important reaction types are to be found among the representatives of this group. As a result of recent investigations the number of such reactions known has tremendously increased.

Finally, group VII includes competitive, consecutive reactions with one or more reversible reaction steps. The commonest and most complicated complex chemical reactions are treated here.

The reaction types are listed and classified in the following Tables. For the sake of simplicity we suppose that the reaction schemes also express the stoichiometry of the processes. The possible deviations are dealt with separately in the appropriate places.

1. First order reactions

1.I REVERSIBLE REACTIONS

1.I.1 $\quad A_1 \underset{k_2}{\overset{k_1}{\rightleftarrows}} A_2$

1.I.2 $\quad A_1 \overset{k_1}{\rightarrow} A_2 \quad$ (heterogeneous)

1.II PARALLEL REACTIONS

1.II.1 $\quad A_2 \overset{k_2}{\leftarrow} A_1 \overset{k_3}{\rightarrow} A_3$
$$\phantom{A_2 \overset{k_2}{\leftarrow} A_1} \downarrow k_4$$
$$\phantom{A_2 \overset{k_2}{\leftarrow} A} A_4$$

1.II.2 $\quad A_1 \overset{k_1}{\rightarrow} A_3 \overset{k_2}{\leftarrow} A_2$

1.III PARALLEL REACTIONS WITH REVERSIBLE REACTION STEP

No example considered.

1.IV CONSECUTIVE REACTIONS

1.IV.1 $\quad A_1 \overset{k_1}{\rightarrow} A_2 \overset{k_2}{\rightarrow} A_3$

1.IV.2 $\quad A_1 \overset{k_1}{\rightarrow} A_2 \overset{k_2}{\rightarrow} A_3 \overset{k_3}{\rightarrow} A_4$

1.IV.3 $\quad A_1 \overset{k_1}{\rightarrow} A_2 \overset{k_2}{\rightarrow} A_3 \ldots \overset{k_{n-1}}{\rightarrow} A_n \overset{k_n}{\rightarrow} A_{n+1}$

1.V CONSECUTIVE REACTIONS WITH REVERSIBLE REACTION STEP

1.V.1 $A_1 \underset{k_2}{\overset{k_1}{\rightleftarrows}} A_2 \overset{k_3}{\rightarrow} A_3$

1.V.2 $A_1 \overset{k_1}{\rightarrow} A_2 \underset{k_3}{\overset{k_2}{\rightleftarrows}} A_3$

1.V.3 $A_1 \underset{k_{21}}{\overset{k_{12}}{\rightleftarrows}} A_2 \overset{k_{23}}{\rightarrow} A_3 \overset{k_{34}}{\rightarrow} A_4$

1.V.4 $A_1 \underset{k_2}{\overset{k_1}{\rightleftarrows}} A_2 \underset{k_4}{\overset{k_3}{\rightleftarrows}} \ldots \underset{k_{2i-2}}{\overset{k_{2i-3}}{\rightleftarrows}} A_i \underset{k_{2i}}{\overset{k_{2i-1}}{\rightleftarrows}} \ldots \underset{k_{2n-2}}{\overset{k_{2n-3}}{\rightleftarrows}} A_n \underset{k_{2n}}{\overset{k_{2n-1}}{\rightleftarrows}} A_{n+1}$

1.V.5 $A_1 \underset{k_2}{\overset{k_1}{\rightleftarrows}} A_2 \underset{k_4}{\overset{k_3}{\rightleftarrows}} A_3$

1.VI COMPETITIVE-CONSECUTIVE REACTIONS

No example considered.

1.VII COMPETITIVE-CONSECUTIVE REACTIONS WITH REVERSIBLE REACTION STEP

1.VII.1

1.VII.2

1.VII.3 $A_1 \underset{k_2}{\overset{k_1}{\rightleftarrows}} A_2 \overset{k_3}{\leftarrow} A_3 \overset{k_4}{\rightarrow} A_4$

1.VII.4 $A_i \underset{k_{ji}}{\overset{k_{ij}}{\rightleftarrows}} A_j$ $(i = 1, 2, 3, \ldots, n; \; j = 1, 2, 3, \ldots, n; \; i \neq j)$

2. Second and mixed first–second order reactions

2.I REVERSIBLE REACTIONS

2.I.1 $2A_1 \underset{k_2}{\overset{k_1}{\rightleftarrows}} A_2 + A_3$

2.I.2 $A_1 \underset{k_2}{\overset{k_1}{\rightleftarrows}} A_2 + A_3$

2.I.3 $A_1 \overset{k_1}{\rightarrow} A_2 + A_3$ (heterogeneous)

2.I.4 $A_1 + A_2 \overset{k_1}{\rightarrow} A_3$ (heterogeneous)

2.I.5 $A_1 + A_2 \underset{k_2}{\overset{k_1}{\rightleftarrows}} A_3 + A_4$

2.II PARALLEL REACTIONS

2.II.1 $A_1 + A_2 \overset{k_1}{\rightarrow} A_3$
$A_1 \overset{k_2}{\rightarrow} A_4$

2.II.2 $A_1 \overset{k_1}{\rightarrow} A_2$
$2A_1 \overset{k_2}{\rightarrow} A_3$

2.II.3 $A_1 + A_2 \overset{k_1}{\rightarrow} A_4 + A_5$
$A_1 + A_3 \overset{k_2}{\rightarrow} A_5 + A_6$

References pp. 79–80

2.II.4 $A_1 \overset{k_1}{\rightarrow} A_3$

$A_1 + A_2 \overset{k_2}{\rightarrow} A_4$

$2A_1 \overset{k_3}{\rightarrow} A_5$

2.III PARALLEL REACTIONS WITH REVERSIBLE REACTION STEP

2.III.1 $A_1 + A_2 \underset{k_2}{\overset{k_1}{\rightleftarrows}} A_3 + A_4$

$A_1 + A_5 \underset{k_4}{\overset{k_3}{\rightleftarrows}} A_3 + A_6$

2.IV CONSECUTIVE REACTIONS

2.IV.1 $A_1 \overset{k_1}{\rightarrow} A_2$

$2A_2 \overset{k_2}{\rightarrow} A_3$

2.IV.2 $2A_1 \overset{k_1}{\rightarrow} A_2$

$A_2 \overset{k_2}{\rightarrow} A_3$

2.IV.3 $2A_1 \overset{k_1}{\rightarrow} A_2$

$2A_2 \overset{k_2}{\rightarrow} A_3$

2.V CONSECUTIVE REACTIONS WITH REVERSIBLE REACTION STEP

No example considered.

2.VI COMPETITIVE-CONSECUTIVE REACTIONS

2.VI.1 $A_1 \overset{k_1}{\rightarrow} A_2$

$A_1 + A_2 \overset{k_2}{\rightarrow} A_3$

2.VI.2 $A_1 \overset{k_1}{\rightarrow} A_2$

$A_1 + A_2 \overset{k_2}{\rightarrow} 2A_2 + A_3$

2.VI.3 $A_1 + A_2 \overset{k_1}{\rightarrow} A_3$

$A_2 + A_3 \overset{k_2}{\rightarrow} A_4$

2.VI.4₁ $A_1 + A_2 \overset{k_1}{\rightarrow} A_3 + A_4$

$A_1 + A_3 \overset{k_2}{\rightarrow} A_4 + A_5$

2.VI.5 $A_1 + A_2 \overset{k_1}{\rightarrow} A_3 + A_6$

$A_1 + A_3 \overset{k_2}{\rightarrow} A_4 + A_7$

$A_1 + A_2 \overset{k_3}{\rightarrow} A_5 + A_7$

$A_1 + A_5 \overset{k_4}{\rightarrow} A_4 + A_6$

2.VI.6 $A_1 + A_1 \overset{k_1}{\rightarrow} A_2$

$A_1 + A_2 \overset{k_2}{\rightarrow} A_3$

$A_1 + A_3 \overset{k_3}{\rightarrow} A_4$

$\vdots$

$A_1 + A_n \overset{k_n}{\rightarrow} A_{n+1}$

2.VI.7 $A_1 + A_1 \overset{k_1}{\rightarrow} A_2 + A_4$

$A_1 + A_2 \overset{k_2}{\rightarrow} A_3 + A_4$

2.VI.8 $A_1 \overset{k_1}{\rightarrow} A_2 \overset{k_2}{\underset{A_2}{\rightarrow}} A_4$

$\downarrow k_3$

A_3

2.VII COMPETITIVE-CONSECUTIVE REACTIONS WITH REVERSIBLE REACTION STEP

2.VII.1 $A_1 + A_2 \underset{k_2}{\overset{k_1}{\rightleftarrows}} A_3$

$A_1 + A_3 \overset{k_3}{\rightarrow} A_4$

2.VII.2 $A_1 + A_2 \overset{k_1}{\underset{k_2}{\rightleftarrows}} A_3$

$A_1 + A_4 \overset{k_3}{\rightarrow} A_5$

$A_3 + A_4 \overset{k_4}{\rightarrow} A_6$

3. Third order reactions

3.I REVERSIBLE REACTIONS

3.I.1 $A_1 + A_2 + A_3 \overset{k_1}{\underset{k_2}{\rightleftarrows}} A_4 + A_5 + A_6$

4. *n*th order reactions

4.I PARALLEL REACTIONS

4.I.1 $n_1 A_1 + n_2 A_2 \overset{k_1}{\rightarrow} A_3 + \ldots$

$n_1 A_1 + n_2 A_2 \overset{k_2}{\rightarrow} A_4 + \ldots$

$n_1 A_1 + n_2 A_2 \overset{k_3}{\rightarrow} A_5 + \ldots$

4.II CONSECUTIVE REACTIONS

4.II.1 $a_1 A_1 \overset{k_1}{\rightarrow} A_2$ (*m*th order)

$a_2 A_2 \overset{k_2}{\rightarrow} A_3$ (*n*th order)

1. First order reactions

1.I REVERSIBLE REACTIONS

1.I.1 $A_1 \overset{k_1}{\underset{k_2}{\rightleftarrows}} A_2$

This is the simplest case of first order reversible reactions. The rate expression is

$$d[A_1]/dt = -k_1[A_1] + k_2[A_2] \tag{1}$$

where k_1 is the rate coefficient of conversion of A_1, and k_2 is that of A_2.

Let the initial concentrations be $[A_1]_0 \neq 0$ and $[A_2]_0 \neq 0$. Then $[A_2]$ can be related at any time to $[A_1]$ by the stoichiometric equation. Thus

$$[A_2] = ([A_1]_0 + [A_2]_0) - [A_1] \tag{2}$$

After substitution and integration

$$\ln \frac{k_1[A_1] - k_2[A_2]}{k_1[A_1]_0 - k_2[A_2]_0} = -(k_1 + k_2)t \tag{3}$$

At equilibrium the rates of the two opposing reactions are equal, and thus

$$k_1[A_1]_e = k_2[A_2]_e \tag{4}$$

where the subscript e refers to equilibrium concentrations. The equilibrium constant is

$$K = \frac{k_1}{k_2} = \frac{[A_2]_e}{[A_1]_e} \tag{5}$$

Dividing the numerator and denominator of the logarithmic term of the integrated expression (3) by k_2 and substituting the equilibrium constant, we get

$$\ln \frac{K[A_1] - [A_2]}{K[A_1]_0 - [A_2]_0} = -(k_1 + k_2)t \tag{6}$$

From this and K, obtained by measuring the equilibrium concentrations, the sum of the rate coefficients can be determined. On the other hand, the ratio of the equilibrium concentrations gives the ratio of the rate coefficients. Thus the values of both k_1 and k_2 can be calculated.

If the initial concentrations are $[A_1]_0 \neq 0$ and $[A_2]_0 = 0$, the expressions are simpler. As $[A_2] = [A_1]_0 - [A_1]$, the integrated equation is

$$\ln \frac{(k_1 + k_2)[A_1] - k_2[A_1]_0}{k_1[A_1]_0} = -(k_1 + k_2)t \tag{7}$$

The ratio of the equilibrium and initial concentrations of substance A_1 are

$$[A_1]_e/[A_1]_0 = k_2/(k_1 + k_2) \tag{8}$$

Thus from the integrated form we obtain

$$\ln \frac{[A_1]_0 - [A_1]_e}{[A_1] - [A_1]_e} = (k_1 + k_2)t \tag{9}$$

As can be seen from the expression, the system approaches equilibrium as a first order process, the "rate coefficient" being the sum of the rate coefficients of the opposing reactions.

Experimental data for reactions consisting of only one forward and one reverse process can also be discussed using the treatment of Micka[1].

There are numerous reactions which can be described by the above scheme, *e.g.* isotopic exchange reactions, gas-phase *cis-trans* isomerizations, racemisations such as those of 2,2'-diamino-6,6'-dimethyldiphenyl and α- and β-glucoses.

The mutual conversion of A_1 and A_2 can give rise to somewhat complicated kinetics when the reaction takes place on a surface. In such a case, besides the chemical reaction the adsorption and desorption processes of the initial substance and the reaction product, respectively, must also be considered. The full scheme for the heterogeneous reaction is as follows

$$A_1 + S \underset{(k_{A_1})_d}{\overset{(k_{A_1})_a}{\rightleftarrows}} (A_1)_a \tag{10}$$

$$(A_1)_a \underset{k_2}{\overset{k_1}{\rightleftarrows}} (A_2)_a \tag{11}$$

$$(A_2)_a \underset{(k_{A_2})_a}{\overset{(k_{A_2})_d}{\rightleftarrows}} A_2 + S \tag{12}$$

where subscripts a and d refer to adsorption (or adsorbed state) and desorption, respectively, and S symbolises the free adsorption sites. Adsorption is regarded as a reaction, in which gas molecules and active sites on the surface take part.

Let us now consider the kinetic equations for the heterogeneous reaction, supposing that for the adsorbed substances A_1 and A_2 the concept of quasi-stationary concentrations is approximately valid. Three cases can be distinguished, depending on which particular process becomes rate controlling. The overall rate of reaction can be determined either by the adsorption of component A_1, or by the surface chemical reaction, or by the desorption of product A_2.

(*i*) *The rate determining step is the adsorption of component* A_1. Let us introduce the following notation: Q is the surface area of the catalyst, [S] is the concentration of free adsorption sites and $[S]_0$ the total number of sites. Let W_{A_1} and W_{A_2} be the net adsorption rates of substances A_1 and A_2, respectively (rate of adsorption –rate of desorption), and W the rate of chemical reaction on the surface (*i.e.*

rate of $A_1 \to A_2$ minus rate of $A_2 \to A_1$). Finally we define the following "equilibrium constants"

$$K_{A_1} = (k_{A_1})_a/(k_{A_1})_d, \qquad K = k_1/k_2,$$
$$K_{A_2} = (k_{A_2})_a/(k_{A_2})_d \tag{13}$$

From the assumption of quasi-stationarity it follows that

$$W = W_{A_1} = W_{A_2} \tag{14}$$

Substituting the rate expressions for the corresponding elementary processes, we get

$$Qk_1[A_1]_a - Qk_2[A_2]_a = Q(k_{A_1})_a[A_1][S] - Q(k_{A_1})_d[A_1]_a \tag{15}$$

and

$$Q(k_{A_1})_a[A_1][S] - Q(k_{A_1})_d[A_1]_a = Q(k_{A_2})_d[A_2]_a - (k_{A_2})_a[A_2][S] \tag{16}$$

After rearrangement and substitution

$$\frac{k_1}{(k_{A_1})_a} = \frac{[A_1][S] - \dfrac{1}{K_{A_1}}[A_1]_a}{[A_1]_a - \dfrac{1}{K}[A_2]_a} \tag{17}$$

and

$$\frac{(k_{A_2})_d}{(k_{A_1})_a} = \frac{[A_1][S] - \dfrac{1}{K_{A_1}}[A_1]_a}{[A_2]_a - K_{A_2}[A_2][S]} \tag{18}$$

On the basis of the statement concerning the rate determining step

$$(k_{A_1})_a \ll k_1, (k_{A_2})_d \tag{19}$$

Accordingly the right hand sides of both (17) and (18) are $\gg 1$. From this it follows that

$$[A_2]_a \approx K[A_1]_a; \qquad [A_2]_a \approx K_{A_2}[A_2][S]$$

and so

$$[A_1]_a \approx \frac{K_{A_2}}{K}[A_2][S] \tag{20}$$

As

$$[S] = [S]_0 - [A_1]_a - [A_2]_a \tag{21}$$

it follows from (20) and (21) that when the adsorption of substance A_1 is the rate determining process, the rate expression is given by

$$W = (k_{A_1})_a Q[S]_0 \frac{[A_1] - \frac{1}{K} \cdot \frac{K_{A_2}}{K_{A_1}} [A_2]}{1 + K_{A_2}[A_2] \left(1 + \frac{1}{K}\right)} \tag{22}$$

(*ii*) *The surface chemical reaction as rate determining step.* The extreme case when the surface chemical reaction itself is the rate determining process can be treated in a similar way. Then, considering equation (14) we can write equation (23); the other equation we need is equation (17), which is still valid.

$$\frac{k_1}{(k_{A_2})_d} = \frac{[A_2]_a - K_{A_2}[A_2][S]}{[A_1]_a - \frac{1}{K} \cdot [A_2]_a} \tag{23}$$

The surface chemical reaction being the rate determining process

$$k_1 \ll (k_{A_1})_a, (k_{A_2})_d \tag{24}$$

it follows that the right hand side of equations (17) and (23) are $\ll 1$. Furthermore, it follows that for the surface concentrations of substances A_1 and A_2 the following approximate equations are valid

$$[A_1]_a \approx K_{A_1}[A_1][S]; \qquad [A_2]_a \approx K_{A_2}[A_2][S] \tag{25}$$

The rate determining process being a surface reaction, we can, after considering equations (21) and (25), express the overall rate as follows

$$W = k_1 K_{A_1} Q[S]_0 \frac{[A_1] - \frac{1}{K} \cdot \frac{K_{A_2}}{K_{A_1}} [A_2]}{1 + K_{A_1}[A_1] + K_{A_2}[A_2]} \tag{26}$$

(*iii*) *Desorption of the product as rate determining process.* In this case we have

$$(k_{A_2})_d \ll k_1, (k_{A_1})_a \tag{27}$$

Accordingly it follows that the right hand side of equation (18) is $\ll 1$, while that of equation (23) is $\gg 1$. The conditions which fulfil the inequalities are

$$[A_1]_a \approx K_{A_1}[A_1][S];$$
$$[A_2]_a \approx K[A_1]_a \approx KK_{A_1}[A_1][S] \tag{28}$$

Thus from equations (21) and (28) the rate expression in this extreme case is the following

$$W = (k_{A_2})_d K K_{A_1} Q[S]_0 \frac{[A_1] - \dfrac{1}{K} \cdot \dfrac{K_{A_2}}{K_{A_1}} [A_2]}{1 + K_{A_1}[A_1](1 + K)} \tag{29}$$

Equations (22), (26) and (29) are the rate expressions for the heterogeneous $A_1 \rightleftarrows A_2$ process in three extreme cases: when the rate of the process is determined by the adsorption of substance A_1, by the chemical reaction and by the desorption of substance A_2, respectively. From these three expressions it is possible to derive mathematically the rate equations for further extreme cases, according to whether the terms of the expressions already given are commensurable or whether some of them can be neglected. We will not treat these cases in detail, but refer the reader to the literature[77]. The experimentally determined rate equations generally correspond to certain sub-cases. As different mechanisms may often lead to the same mathematical form of the rate equation, special care must be taken when analysing the experimental data. Conclusions can be drawn concerning the reaction mechanism, the nature of the rate determining process and the empirical rate constants on the basis of the empirical rate equation only in exceptional cases, due to this difficulty. Nevertheless, the compilations given in the literature can prove to be very useful for comparison of the experimental and theoretical rate equations of catalytic reactions.

1.1.2 $A_1 \xrightarrow{k_1} A_2$

The conversion of one substance into another by a first order reaction does not require any comment. However, if the process is a heterogeneous one, it is complicated by the fact that the initial material has to be adsorbed on the surface and the end-product to be desorbed from the catalyst. Thus the full reaction scheme is the following

$$A_1 + S \underset{(k_{A_1})_d}{\overset{(k_{A_1})_a}{\rightleftarrows}} (A_1)_a \tag{1}$$

$$(A_1)_a \xrightarrow{k_1} (A_2)_a \tag{2}$$

$$(A_2)_a \underset{(k_{A_2})_a}{\overset{(k_{A_2})_d}{\rightleftarrows}} A_2 + S \tag{3}$$

where, as before, a and d refer to adsorption (or adsorbed state) and desorption, respectively, and S symbolises the free adsorption sites.

The various rate expressions are derived in a similar way to that described in detail when treating the heterogeneous $A_1 \rightleftarrows A_2$ system.

Let us suppose that the condition of quasi-stationary concentrations is approximately correct for the adsorbed substances A_1 and A_2. It is easy to deduce the results in this case by substituting $K \gg 1$ into equations (22), (26) and (29) of Section 1.I.1 since $K (= k_1/k_2)$ is the ratio of the rate coefficients of two opposing surface reactions. The process $A_1 \rightarrow A_2$ is a special case of $A_1 \rightleftarrows A_2$, when $k_1 \gg k_2$.

Consequently, the rate equations for the catalytic reaction $A_1 \rightarrow A_2$ in three extreme cases are as follows.

(*i*) When the overall rate is determined by the adsorption of substance A_1

$$W = (k_{A_1})_a Q[S]_0 \cdot \frac{[A_1]}{1 + K_{A_2}[A_2]} \tag{4}$$

(*ii*) When the rate is determined by the surface chemical reaction

$$W = k_1 K_{A_1} Q[S]_0 \cdot \frac{[A_1]}{1 + K_{A_1}[A_1] + K_{A_2}[A_2]} \tag{5}$$

(*iii*) When the rate is determined by the desorption of substance A_2

$$W = (k_{A_2})_d Q[S]_0 \tag{6}$$

The above equations give the rate of reaction in three limiting cases and are called the kinetic equations of the process. In equations (4), (5) and (6) the various terms are not necessarily commensurable; one or the other can often be neglected. Thus we can get from all the three expressions to several other extreme cases.

1.II PARALLEL REACTIONS

1.II.1 $\quad A_2 \overset{k_2}{\leftarrow} A_1 \overset{k_3}{\rightarrow} A_3$
$$\quad\quad\quad\quad\quad \downarrow k_4$$
$$\quad\quad\quad\quad\quad A_4$$

According to this simple competitive scheme several chemical processes take place. The rate expression describing the consumption of substance A_1 is

$$-\frac{d[A_1]}{dt} = k_2[A_1] + k_3[A_1] + k_4[A_1] \tag{1}$$

Thus,

$$-\frac{d[A_1]}{dt} = k[A_1]$$

where

$$k = k_2 + k_3 + k_4 \tag{2}$$

Integrating equation (2) we obtain

$$[A_1] = [A_1]_0 e^{-kt} \tag{3}$$

As A_2, A_3 and A_4 are formed by first order processes, the change in their concentrations with time is described by the following equations (the initial conditions being $[A_1]_0 \neq 0$, $[A_2]_0 \neq 0$, $[A_3]_0 \neq 0$ and $[A_4]_0 \neq 0$)

$$[A_2] = [A_2]_0 + \frac{k_2[A_1]_0}{k}(1 - e^{-kt}) \tag{4}$$

$$[A_3] = [A_3]_0 + \frac{k_3[A_1]_0}{k}(1 - e^{-kt}) \tag{5}$$

$$[A_4] = [A_4]_0 + \frac{k_4[A_1]_0}{k}(1 - e^{-kt}) \tag{6}$$

If $[A_2]_0 = [A_3]_0 = [A_4]_0 = 0$, the above equations are simplified and obviously

$$\frac{[A_3]}{[A_2]} = \frac{k_3}{k_2}; \quad \frac{[A_4]}{[A_2]} = \frac{k_4}{k_2} \tag{7}$$

or $[A_2] : [A_3] : [A_4] = k_2 : k_3 : k_4$.

Thus, in this case, the ratio of the concentrations of products does not depend on the time, but only on the ratio of the corresponding rate coefficients.

The overall coefficient k can be determined from the dependence of the concentration of A_1 on time. It is enough to determine only two of the individual rate coefficients, as the third can be calculated from the expression $k = k_2 + k_3 + k_4$. The isomerization of α-pinene can serve as an example[2].

The general scheme for parallel first order reactions, *viz.*

$$A_1 \overset{k_i}{\underset{}{\rlap{\;\;k_j}\Longleftarrow}} \qquad\qquad (8)$$

can be treated similarly, *i.e.*

$$-\frac{d[A_1]}{dt} = \sum_j k_j[A_1] \qquad (9)$$

$$[A_i] = [A_i]_0 + \frac{k_i}{\sum_j k_j}[A_1]_0(1 - e^{-\Sigma_j k_j t}) \qquad (10)$$

The ratio of the concentrations of any two of the products is obviously given by the following equation

$$\frac{[A_i] - [A_i]_0}{[A_j] - [A_j]_0} = \frac{k_i}{k_j} \qquad (11)$$

I.II.2 $A_1 \overset{k_1}{\to} A_3 \overset{k_2}{\leftarrow} A_2$

The rate expressions are

$$-\frac{d[A_1]}{dt} = k_1[A_1] \qquad (1)$$

$$-\frac{d[A_2]}{dt} = k_2[A_2] \qquad (2)$$

$$\frac{d[A_3]}{dt} = k_1[A_1] + k_2[A_2] \qquad (3)$$

With the initial conditions, $[A_1] = [A_1]_0$, $[A_2] = [A_2]_0$, $[A_3] = 0$, the equations describing the time dependence of the concentrations of the substances are[2]

$$[A_1] = [A_1]_0 e^{-k_1 t} \qquad (4)$$

$$[A_2] = [A_2]_0 e^{-k_2 t} \qquad (5)$$

$$[A_3] = [A_1]_0 - [A_1]_0 e^{-k_1 t} + [A_2]_0 - [A_2]_0 e^{-k_2 t} \qquad (6)$$

Since $[A_1]_0 + [A_2]_0 = [A_3]_\infty$, the concentration of A_3 can be given in the following form as well

$$[A_3] = [A_3]_\infty - [A_1]_0 e^{-k_1 t} - [A_2]_0 e^{-k_2 t} \tag{7}$$

The rearranged and logarithmised form of equation (7) suggests that, on plotting $\log([A_3]_\infty - [A_3])$ against time, we get a straight line if $k_1 = k_2$, and if $k_1 \neq k_2$ we get a curve which transforms into a straight line at large values of t. Concerning the latter, it must be taken into consideration that, as the concentration of the more reactive substance diminishes, equation (7) becomes simpler and turns into the logarithmic expressions corresponding to equations (4) or (5), *viz.*

$$\log[A_1] = \log([A_3]_\infty - [A_3]) = \log[A_1]_0 - (k_1 t/2.303) \tag{8}$$

and

$$\log[A_2] = \log([A_3]_\infty - [A_3]) = \log[A_2]_0 - (k_3 t/2.303) \tag{9}$$

From the slope of the straight part of the plot it is possible to calculate the rate coefficient of the slower reaction and from the ordinate values corresponding to this straight part, one can calculate the concentration of the more slowly reacting substance at various times and, by extrapolation, at the beginning of the reaction. Thus, knowing $[A_3]_\infty$ and $[A_3]$, the concentration of the more reactive substance at various times can be calculated, $[A_3]_\infty = [A_1] + [A_2] + [A_3]$, and the rate coefficient of the more rapid reaction can be determined.

However, a separate analysis of reactions $A_1 \rightarrow A_3$ and $A_2 \rightarrow A_3$ is a preferable way of investigation.

1.III PARALLEL REACTIONS WITH REVERSIBLE REACTION STEP

No example considered.

1.IV CONSECUTIVE REACTIONS

1.IV.1 $A_1 \xrightarrow{k_1} A_2 \xrightarrow{k_2} A_3$

This scheme represents the simplest consecutive reaction system and has thus received considerable theoretical and practical attention[1-8].

The set of differential equations describing the change of concentrations with time is

$$d[A_1]/dt = -k_1[A_1] \tag{1}$$

$$d[A_2]/dt = k_1[A_1] - k_2[A_2] \tag{2}$$

$$d[A_3]/dt = k_2[A_2] \tag{3}$$

For the concentration of A_1 we immediately obtain by integration the usual expression for first order reactions

$$[A_1] = [A_1]_0 e^{-k_1 t} \tag{4}$$

Substituting (4) into (2) and solving the first order linear differential equation for the case when $[A_2]_0 \neq 0$, we get the time dependence of $[A_2]$

$$[A_2] = [A_2]_0 e^{-k_2 t} + \frac{k_1[A_1]_0}{k_2 - k_1}(e^{-k_1 t} - e^{-k_2 t}) \tag{5}$$

The expression for $[A_3]$ can be found in the simplest way by substituting the expressions for $[A_1]$ and $[A_2]$ into equation (6)

$$[A_1] + [A_2] + [A_3] = [A_1]_0 + [A_2]_0 + [A_3]_0 \tag{6}$$

Accordingly

$$[A_3] = [A_1]_0 \left\{ 1 - e^{-k_1 t} - \frac{k_1}{k_2 - k_1}(e^{-k_1 t} - e^{-k_2 t}) \right\}$$

$$+ [A_2]_0(1 - e^{-k_2 t}) + [A_3]_0 \tag{7}$$

The concentration of A_1 decreases exponentially with time. The change of $[A_3]$ is rather complicated, although it does increase continuously with time. On the other hand, the variation in the concentration of A_2 depends on $[A_1]_0$ and $[A_2]_0$ and on the numerical values of k_1 and k_2. From equation (2) it is clearly seen that the slope of the curve of $[A_2]$ *versus* t is positive when $k_1[A_1] > k_2[A_2]$, is zero when $k_1[A_1] = k_2[A_2]$, and is negative when $k_1[A_1] < k_2[A_2]$. Thus the concentration of A_2 may pass through a maximum. Starting from equation (5) the following expression is obtained for the time t_{max} needed to reach the maximum value

$$t_{max} = \frac{1}{k_2 - k_1} \ln \frac{k_2}{k_1} \left(1 + \frac{[A_2]_0}{[A_1]_0} - \frac{k_2}{k_1} \frac{[A_2]_0}{[A_1]_0} \right) \tag{8}$$

Thus the curve of $[A_2]$ *versus* t has a maximum if $k_1[A_1]_0 > k_2[A_2]_0$ and the concentration of A_2, after reaching a maximum, decreases asymptotically to zero. If $k_1[A_1]_0 < k_2[A_2]_0$, the concentration of A_2 decreases monotonously from the beginning to a final value of $[A_2] = 0$ (which is attained at $t = \infty$).

The situation is simpler if the initial concentrations are $[A_1]_0 \neq 0$, $[A_2]_0 = [A_3]_0 = 0$. Fig. 1 shows a typical case of the concentration dependences on time.

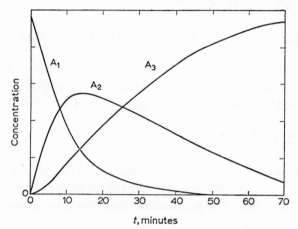

Fig. 1. Concentration–time curves for substances A_1, A_2 and A_3. $k_1 = 0.1$ min^{-1} and $k_2 = 0.05$ min^{-1}.

The time dependence of $[A_1]$, $[A_2]$ and $[A_3]$ are directly obtained from equations (4), (5) and (7). Introducing the dimensionless parameters and variables

$$\alpha_1 = \frac{[A_1]}{[A_1]_0}, \qquad \alpha_2 = \frac{[A_2]}{[A_1]_0}, \qquad \alpha_3 = \frac{[A_3]}{[A_1]_0},$$

$$\tau = k_1 t, \qquad \kappa = \frac{k_2}{k_1} \tag{9}$$

the expressions have an even more simple form[2,9,10]

$$\alpha_1 = e^{-\tau} \tag{10}$$

$$\alpha_2 = \frac{1}{\kappa-1}\left(e^{-\tau} - e^{-\kappa\tau}\right) \tag{11}$$

$$\alpha_3 = 1 + \frac{1}{1-\kappa}\left(\kappa e^{-\tau} - e^{-\kappa\tau}\right) \tag{12}$$

Besides α_1, α_2 and α_3 (the values of which may change during the reaction in the range from 0 to 1) it is useful to introduce another dimensionless quantity, α_4. This is defined by equation (13)

$$\alpha_4 = \alpha_2 + 2\alpha_3 \tag{13}$$

and accordingly it may change from 0 to 2. From equations (11), (12) and (13)

$$\alpha_4 = 2 - \frac{(1-2\kappa)}{(1-\kappa)} e^{-\tau} - \frac{1}{(1-\kappa)} e^{-\kappa\tau} \tag{14}$$

The value of α_4, as a rule, expresses the extent of the reaction better than do α_2 or α_3 alone. In consecutive reactions of compounds containing two identical functional groups, for example the hydrolysis of dihalides or of esters of diacids, α_4 is generally directly connected with a quantity measured experimentally. (For further details the reader is referred to the literature, and especially to ref. 2).

1.IV.2 $A_1 \xrightarrow{k_1} A_2 \xrightarrow{k_2} A_3 \xrightarrow{k_3} A_4$

This reaction scheme can be regarded as an extreme case of the general system of first order reversible, consecutive reactions, and can be derived from $A_1 \rightarrow A_2 \rightarrow A_3$ by assuming $n \rightarrow n+1$. The set of differential equations for the reaction rates is

$$dx_1/dt = k_1([A_1]_0 - x_1) \tag{1}$$

$$dx_2/dt = k_2(x_1 - x_2) \tag{2}$$

$$dx_3/dt = k_3(x_2 - x_3) \tag{3}$$

This set can be solved (Rakowski[8]) as illustrated in the discussion of the general scheme $A_1 \rightleftarrows A_2 \rightleftarrows \ldots A_i \rightleftarrows \ldots$ (see also Thiersch[5]).

1.IV.3 $A_1 \xrightarrow{k_1} A_2 \xrightarrow{k_2} A_3 \ldots \xrightarrow{k_{n-1}} A_n \xrightarrow{k_n} A_{n+1}$

The set of differential equations describing the change of concentrations with time is

$$d[A_1]/dt = -k_1[A_1] \tag{1}$$

$$d[A_2]/dt = k_1[A_1] - k_2[A_2] \tag{2}$$

$$d[A_3]/dt = k_2[A_2] - k_3[A_3] \tag{3}$$

$$\vdots$$

$$d[A_{n+1}]/dt = k_n[A_n] \tag{4}$$

and, of course, the concentration of A_1 as a function of time can be obtained directly

$$[A_1] = [A_1]_0 e^{-k_1 t} \tag{5}$$

By re-writing this system of equations in a more convenient symbolic form, where the differentiation with respect to time is given the symbol D, we get

$$(D+k_1)[A_1] = 0 \tag{6}$$

$$(D+k_2)[A_2] = k_1[A_1] \tag{7}$$

$$(D+k_3)[A_3] = k_2[A_2] \tag{8}$$

$$\vdots$$

$$(D+k_i)[A_i] = k_{i-1}[A_{i-1}] \tag{9}$$

After differentiation of equation (7) and substitution of the differential quotient $d[A_1]/dt$ by an expression obtained from equations (6) and (7), a secondary linear differential equation for $[A_2]$ is obtained

$$(D+k_1)(D+k_2)[A_2] = 0 \tag{10}$$

The solution of this is well-known

$$[A_2] = \alpha_{21} e^{-k_1 t} + \alpha_{22} e^{-k_2 t} \tag{11}$$

After similar steps, introducing successive differential equations and substitutions, the differential equation of the ith order is obtained

$$(D+k_1)(D+k_2)(D+k_3) \ldots (D+k_i)[A_i] = 0 \tag{12}$$

Its general solution (when all the rate coefficients are different) is

$$[A_i] = \alpha_{i1} e^{-k_1 t} + \alpha_{i2} e^{-k_2 t} + \ldots + \alpha_{ii} e^{-k_i t} \tag{13}$$

The integration constants (α) can be determined by considering the initial conditions[11]. For instance the constants in equation (11) are

$$\alpha_{21} = -\alpha_{22} = k_1[A_1]_0/(k_1 - k_2) \tag{14}$$

Using this latter equation, the change of concentration A_2 as a function of time can be given in the following form

$$[A_2] = [A_1]_0 \frac{k_1}{k_2 - k_1} (e^{-k_1 t} - e^{-k_2 t}) \tag{15}$$

The solutions are different if some of the rate coefficients are identical. If all the rate coefficients are the same, for example, the solution of equation (12) can be given in the following simple form (the initial concentrations being zero with the exception of that of A_1)

$$[A_i] = \alpha_{ii} t^{i-1} e^{-k_1 t} \tag{16}$$

Determining the constant (α_{ii}) by using the initial conditions, the expression for $[A_i]$ is[11]

$$[A_i] = [A_1]_0 \frac{(k_1 t)^{i-1}}{(i-1)!} e^{-k_1 t} \tag{17}$$

It has been pointed out by Erofeev[12] that a simple solution for the kinetics of a series of consecutive, unidirectional first order reactions can be obtained by introducing the so-called "kinetic determinants" which involve the rate coefficients. The suggested determinants are

$$C(k_1 \ldots k_i) = \begin{vmatrix} k_1^{-1} & 1 & k_1 \ldots k_1^{i-2} \\ k_2^{-1} & 1 & k_2 \ldots k_2^{i-2} \\ \cdot & \cdot & \cdot \quad \cdot \quad \cdot \\ k_i^{-1} & 1 & k_i \ldots k_i^{i-2} \end{vmatrix} = \frac{D(k_1 \ldots k_i)}{k_1 \ldots k_i} \tag{18}$$

$$D(k_1 \ldots k_i) = \begin{vmatrix} 1 & k_1 & k_1^2 \ldots k_1^{i-1} \\ 1 & k_2 & k_2^2 \ldots k_2^{i-1} \\ \cdot & \cdot & \cdot \quad \cdot \quad \cdot \\ 1 & k_i & k_i^2 \ldots k_i^{i-1} \end{vmatrix} \tag{19}$$

$$E(t, k_1 \ldots k_i) = \begin{vmatrix} e^{-k_1 t} & 1 & k_1 \ldots k_1^{i-2} \\ e^{-k_2 t} & 1 & k_2 \ldots k_2^{i-2} \\ \cdot & \cdot & \cdot \quad \cdot \quad \cdot \\ e^{-k_i t} & 1 & k_i \ldots k_i^{i-2} \end{vmatrix} \tag{20}$$

$$F(t, k_1 \ldots k_i) = \begin{vmatrix} k_1 e^{-k_1 t} & 1 & k_1 \ldots k_1^{i-2} \\ k_2 e^{-k_2 t} & 1 & k_2 \ldots k_2^{i-2} \\ \cdot & \cdot & \cdot \quad \cdot \quad \cdot \\ k_i e^{-k_i t} & 1 & k_i \ldots k_i^{i-2} \end{vmatrix} \tag{21}$$

$$G(t, k_1 \ldots k_n) = \begin{vmatrix} e^{-k_1 t} & k_1 \ldots k_1^{n-1} \\ e^{-k_2 t} & k_2 \ldots k_2^{n-1} \\ \cdot & \cdot \quad \cdot \quad \cdot \\ e^{-k_n t} & k_n \ldots k_n^{n-1} \end{vmatrix} \tag{22}$$

By means of these kinetic determinants, taking $[A_1]_0 \neq 0$ and $[A_2]_0 = [A_3]_0 = \ldots = [A_i]_0 \ldots = [A_{n+1}]_0 = 0$, the time dependence of the concentration of any intermediate product can be given, according to Erofeev, in the following form

$$[A_i] = \frac{[A_1]_0}{k_i} \frac{E(t, k_1 \ldots k_i)}{C(k_1 \ldots k_i)} \tag{23}$$

The equation expressing the connection between concentration of the final product and time is a special case of this expression. Using the kinetic determinants, we get for the end-product concentration

$$[A_{n+1}] = [A_1]_0 \left\{ 1 - \frac{G(t, k_1 \ldots k_n)}{D(k_1 \ldots k_n)} \right\} \tag{24}$$

By differentiating the expressions for the concentrations, rate expressions for the variation in concentration of an intermediate product and of the end-product formation can be derived

$$\frac{d[A_i]}{dt} = -\frac{[A_1]_0}{k_i} \cdot \frac{F(t, k_1 \ldots k_i)}{C(k_1 \ldots k_i)} \tag{25}$$

and

$$\frac{d[A_{n+1}]}{dt} = [A_1]_0 \cdot \frac{E(t, k_1 \ldots k_n)}{C(k_1 \ldots k_n)} \tag{26}$$

Let us examine the time dependence of $[A_i]$ and $[A_{n+1}]$ in the initial stage of the reaction. On expanding the exponential terms in the expressions for the concentrations, and considering only the first terms of the series, these time dependences are given by simple approximative equations. Thus, it can be stated that the amount of the ith intermediate product is proportional to $t^{(i-1)}$ and the amount of the end-product is proportional to t^n at the beginning of the reaction. In the initial stages of the reaction, therefore, the amounts of the intermediate products and of the end-product are all proportional to integral powers of the time, the power being given by the number of reaction steps preceding the formation of the substance investigated.

Erofeev's method makes it possible to examine the number of reaction steps preceding the formation of the substance in question. For this two methods are suggested. One consists in plotting the concentrations obtained experimentally, during the initial period of the reaction, in the coordinate system log [A]–log t. The initial slope of the curve obtained gives the required value directly. According to the other method, the functions $[A]/t^n$ are plotted against t for different positive values of n. We take that value of n for which the relation between $[A]/t^n$ and t

is the nearest to linearity in the initial period of the reaction. The n value obtained gives directly the number of consecutive steps which precede the formation of the product in question.

The formulae given can be applied also to the case when $[A_2]_0 \neq [A_3]_0 \neq \ldots \neq [A_{n+1}]_0 \neq 0$. In this case the suggested expressions are as follows

$$[A_i] = \frac{[A_1]_0}{k_i} \frac{E(t, k_1 \ldots k_i)}{C(k_1 \ldots k_i)} + \frac{[A_2]_0}{k_i} \frac{E(t, k_2 \ldots k_i)}{C(k_2 \ldots k_i)} + \ldots$$
$$+ \frac{[A_i]_0}{k_i} \frac{E(t, k_i)}{C(k_i)} \tag{27}$$

$$[A_{n+1}] = ([A_1]_0 + [A_2]_0 + \ldots + [A_{n+1}]_0)$$
$$- \left\{ [A_1]_0 \frac{G(t, k_1 \ldots k_n)}{D(k_1 \ldots k_n)} + [A_2]_0 \frac{G(t, k_2 \ldots k_n)}{D(k_2 \ldots k_n)} + \ldots \right.$$
$$\left. + [A_n]_0 \frac{G(t, k_n)}{D(k_n)} \right\} \tag{28}$$

$$\frac{d[A_i]}{dt} = - \left\{ \frac{[A_i]_0}{k_i} \frac{F(t, k_1 \ldots k_i)}{C(k_1 \ldots k_i)} + \frac{[A_2]_0}{k_i} \frac{F(t, k_2 \ldots k_i)}{C(k_2 \ldots k_i)} + \ldots \right.$$
$$\left. + \frac{[A_i]_0}{k_i} \frac{F(t, k_i)}{C(k_i)} \right\} \tag{29}$$

$$\frac{d[A_{n+1}]}{dt} = [A_1]_0 \frac{E(t, k_1 \ldots k_n)}{C(k_1 \ldots k_n)} + [A_2]_0 \frac{E(t, k_2 \ldots k_n)}{C(k_2 \ldots k_n)}$$
$$+ \ldots + [A_n]_0 \frac{E(t, k_n)}{C(k_n)} \tag{30}$$

These are general expressions describing the kinetics of a system of first order consecutive, unidirectional reactions, consisting of any number of steps, for the case of arbitrary initial concentrations (see also ref. 6).

1.V CONSECUTIVE REACTIONS WITH REVERSIBLE REACTION STEP

$$\textit{1.V.1} \qquad A_1 \underset{k_2}{\overset{k_1}{\rightleftarrows}} A_2 \overset{k_3}{\rightarrow} A_3$$

The set of differential equations describing the time dependences of the concentrations is

$$d[A_1]/dt = -k_1[A_1]+k_2[A_2] \tag{1}$$

$$d[A_2]/dt = k_1[A_1]-(k_2+k_3)[A_2] \tag{2}$$

$$d[A_3]/dt = k_3[A_2] \tag{3}$$

Dividing equation (2) by (1), introducing the parameters $y = [A_2]/[A_1]$, $k_1/k_2 = K_1$, $k_3/k_2 = K_3$ and finally $z = K_1-y$, the following equation is obtained

$$\frac{z\,dz}{z^2-(1+K_1+K_3)z+K_1K_3} = -d\ln[A_1] \tag{4}$$

The solution of equation (4), with z_1 and z_2 as the zero-locus of the nominator of the left hand side of the equation, is

$$\ln\left[1-\frac{y}{K_1-z_1}\right]-\left(\frac{z_2}{z_1}\right)\ln\left[1-\frac{y}{K_1-z_2}\right] = -\left(1-\frac{z_2}{z_1}\right)\ln\frac{[A_1]}{[A_1]_0} \tag{5}$$

Using the values of z_1 and z_2 (given in tables by Benson[7] for different values of K_1 and K_3) the maximum concentration ratio, $([A_2]/[A_1])_{max} = y_{max}$, can be determined. This is to be compared with the concentration ratio under stationary conditions

$$y_s = \left(\frac{[A_2]}{[A_1]}\right)_s = k_1/(k_2+k_3) = K_1/(1+K_3) \tag{6}$$

Benson[7] tabulates values of y_{max}/y_s for different values of K_1 and K_3 $(y_{max} = K_1-z_2, y_s = K_1/(1+K_3), y_{max}/y_s = [1-z_2/K_1](1+K_3))$. The table well illustrates that $([A_2]/[A_1])_{max}$ exceeds $([A_2]/[A_1])_s$, i.e. $y_{max}/y_s > 1$.

Sievert et al.[13] introduced into equations (1), (2) and (3) the dimensionless concentrations $\alpha = [A_1]/[A_1]_0$, $\beta = [A_2]/[A_1]_0$ and $\gamma = [A_3]/[A_1]_0$ together with the dimensionless time $\tau = k_1t$, and obtained the following expressions

$$\frac{d\alpha}{d\tau} = -\alpha+\kappa_1\beta \tag{7}$$

$$\frac{d\beta}{d\tau} = -\kappa_1\beta-\kappa_2\beta+\alpha \tag{8}$$

$$\frac{d\gamma}{d\tau} = \kappa_2\beta \tag{9}$$

where $\kappa_1 = k_2/k_1$ and $\kappa_2 = k_3/k_1$. Solving the above equations for α and β we find

$$\alpha = \frac{\lambda_2 - \kappa_2}{\lambda_2(\lambda_2 - \lambda_3)} e^{-\lambda_2 \tau} + \frac{\kappa_2 - \lambda_3}{\lambda_3(\lambda_2 - \lambda_3)} e^{-\lambda_3 \tau} \tag{10}$$

$$\beta = \frac{1}{\lambda_2 - \lambda_3} [e^{-\lambda_3 \tau} - e^{-\lambda_2 \tau}] \tag{11}$$

where

$$\lambda_2 = \frac{1 + \kappa_1 + \kappa_2 + [(1 + \kappa_1 + \kappa_2)^2 - 4\kappa_2]^{\frac{1}{2}}}{2} \tag{12}$$

and

$$\lambda_3 = \frac{1 + \kappa_1 + \kappa_2 - [(1 + \kappa_1 + \kappa_2)^2 - 4\kappa_2]^{\frac{1}{2}}}{2} \tag{13}$$

γ has been plotted as a function of τ at different values of κ.

McDaniel and Smoot[14] have discussed the kinetics of a reaction of the above type assuming a stationary state.

$$\textit{1.V.2} \quad A_1 \xrightarrow{k_1} A_2 \underset{k_3}{\overset{k_2}{\rightleftarrows}} A_3$$

The set of differential equations for the rates is

$$dx_1/dt = k_1([A_1]_0 - x_1) \tag{1}$$

$$dx_2/dt = k_2([A_2]_0 + x_1 - x_2) - k_3([A_3]_0 + x_2) \tag{2}$$

where $[A_1] = [A_1]_0 - x_1$, $[A_2] = [A_2]_0 + x_1 - x_2$ and $[A_3] = [A_3]_0 + x_2$. The integration was performed by Rakowski[8]. The treatment is similar to the one given in the discussion of the general scheme $A_1 \rightleftarrows A_2 \rightleftarrows A_3 \rightleftarrows \ldots$. Let us introduce the following parameters $\alpha_1 = [A_1]_\infty - [A_1]_0$, $\alpha_2 = [A_2]_\infty - [A_2]_0$ and $\alpha_3 = [A_3]_\infty - [A_3]_0$, where $[A]_0$ and $[A]_\infty$ are the values of corresponding concentrations at times $t = 0$ and $t = \infty$. The time dependence of the concentrations of A_1, A_2 and A_3 is given by equations (3) to (5)

$$[A_1] = [A_1]_0 e^{-k_1 t} \tag{3}$$

$$[A_2] = [A_2]_0 + \alpha_2 + \frac{1}{k_2 + k_3 - k_1} \{ [A_1]_0 (k_1 - k_3) e^{-k_1 t}$$
$$- [\alpha_2 k_2 - \alpha_3 (k_3 - k_1)] e^{-(k_2 + k_3)t} \} \tag{4}$$

$$[A_3] = [A_3]_0 + \alpha_3 - \frac{1}{k_2 + k_3 - k_1} \{ [A_1]_0 k_2 e^{-k_1 t}$$
$$- [\alpha_2 k_2 - \alpha_3 (k_3 - k_1)] e^{-(k_2 + k_3)t} \} \tag{5}$$

Expressions can obviously be derived from equations $(3)-(5)$ for the case when $[A_2]_0 = [A_3]_0 = 0$. In this case it should be noted that $k_2\alpha_2 = k_3\alpha_3$.

The concentration of A_2 passes through a maximum during the reaction. The time (t_{max}) needed to reach the maximum concentration can be given in the following form

$$t_{max} = \frac{\ln(k_1-k_3)-\ln k_2}{k_1-k_2-k_3} \tag{6}$$

The kinetics have been derived using dimensionless variables by Lowry and John[15], as well as by Sievert et al.[13]. The results have been applied to a "plug-flow", as well as to a continuous stirred tank, reactor.

1.V.3 $\quad A_1 \underset{k_{21}}{\overset{k_{12}}{\rightleftharpoons}} A_2 \overset{k_{23}}{\rightarrow} A_3 \overset{k_{34}}{\rightarrow} A_4$

This is a special case of the system $(A_1 \rightleftharpoons A_2 \rightleftharpoons \ldots A_n \rightleftharpoons \ldots)$ of first order consecutive reversible reactions. The treatment is similar, and so we refer here without any further details to what will be mentioned below for the general case (p. 28) (for details of the method see ref. 16). In the general case referred to, the set of differential equations is

$$d[A_i]/dt = \sum_{j \neq i}(k_{ji}[A]_j - k_{ij}[A_i]) \tag{1}$$

The general solution of such a set of homogeneous first order linear equations is

$$[A_i] = \sum_{k=1}^{n} a_{ik}e^{-b_k t} + \text{constant} \tag{2}$$

The integration constant can be evaluated from the initial conditions, but to determine the constants a_{ik} and b_k the general solution is re-substituted into the original system of differential equations. In this way we get a row of homogeneous algebraic equations

$$\sum_{j \neq i}[k_{ji}a_j - (k_{ij}+b)a_i] = 0 \tag{3}$$

The secular equation for determining the b values (remembering that in the present case the rate coefficients, with four exceptions, are equal to zero) is

$$\begin{vmatrix} b-k_{12} & k_{21} & 0 & 0 \\ k_{12} & b-k_{21}-k_{23} & 0 & 0 \\ 0 & k_{23} & b-k_{34} & 0 \\ 0 & 0 & k_{34} & b \end{vmatrix} = 0 \tag{4}$$

The solutions of (4) are $b_1 = 0$, $b_2 = k_{34}$, as well as b_3 and b_4, the roots of the quadratic equation

$$b^2 - b(k_{12} + k_{21} + k_{23}) + k_{12}k_{23} = 0 \tag{5}$$

Knowing the values of b those of a can be determined by substitution of the former into equation (3).

The variations of concentration of A_1 and A_2 with time are independent of the presence and concentration of A_3 and A_4, so the solution is the same as that deduced for the system $A_1 \rightleftarrows A_2 \rightarrow A_3$. A stationary state treatment is also possible if special relations exist between the k_{ij} rate coefficients, and in this case the concentration of A_2 can be written as

$$[A_2]/[A_1] = \frac{k_{12}}{k_{21} + k_{23}} \tag{6}$$

Then the rate equations for A_1 and A_3 are the following

$$\frac{d[A_1]}{dt} = -k_{12}\frac{k_{23}}{k_{21} + k_{23}}[A_1] \tag{7}$$

$$\frac{d[A_3]}{dt} = \frac{k_{12}k_{23}}{k_{21} + k_{23}}[A_1] - k_{34}[A_3] \tag{8}$$

Introducing appropriate parameters the system formally reduces to the case $A_1 \xrightarrow{k_1} A_2 \xrightarrow{k_2} A_3$. The expressions for the variations of $[A_1]$, $[A_3]$ and $[A_4]$ will be the same as those for the system consisting of two first order consecutive reactions.

Benson[7] discusses in detail the conditions for the occurrence of a stationary state in the given system and also the factors influencing the length of the "induction period".

$$1.V.4 \quad A_1 \underset{k_2}{\overset{k_1}{\rightleftarrows}} A_2 \underset{k_4}{\overset{k_3}{\rightleftarrows}} \dots \underset{k_{2i-2}}{\overset{k_{2i-3}}{\rightleftarrows}} A_i \underset{k_{2i}}{\overset{k_{2i-1}}{\rightleftarrows}} \dots \underset{k_{2n-2}}{\overset{k_{2n-3}}{\rightleftarrows}} A_n \underset{k_{2n}}{\overset{k_{2n-1}}{\rightleftarrows}} A_{n+1}$$

By the treatment of the above system, a general solution for the kinetics of a system involving n reversible consecutive reactions can be given. The concentration–time relationships for simpler systems (special cases of this scheme) can easily be obtained from the general solution if the values of the corresponding rate coefficients are put equal to zero.

The set of differential equations for the first order reversible consecutive reactions was integrated by Rakowski[8]. However, we will not follow his treatment. An exact, explicit mathematical solution for the general scheme, and for simpler ones as well (derivable from the former), is possible. The set of differential equations is

$$d[A_1]/dt = -k_1[A_1]+k_2[A_2]$$
$$d[A_2]/dt = k_1[A_1]-k_2[A_2]-k_3[A_2]+k_4[A_3]$$
$$\vdots$$
$$d[A_i]/dt = k_{2i-3}[A_{i-1}]-k_{2i-2}[A_i]-k_{2i-1}[A_i]+k_{2i}[A_{i+1}]$$
$$\vdots$$
$$d[A_{n+1}]/dt = k_{2n-1}[A_n]-k_{2n}[A_{n+1}]$$

(1)

The secular equation is

$$0 = \begin{vmatrix} k_1+\rho & -k_2 & & & & \\ -k_1 & +(k_2+k_3)+\rho & -k_4 & & & \\ & -k_3 & +(k_4+k_5)+\rho & -k_6 & & \\ & & & \ddots & & \\ & & & -k_{2n-3} & +(k_{2n-2}+k_{2n})+\rho & -k_{2n-1} \\ & & & & -k_{2n-1} & +k_{2n}+\rho \end{vmatrix}$$

and the treatment thereafter is the same as described for the scheme $A_i \rightleftarrows A_j$ (p. 34).

I.V.5 $A_1 \underset{k_2}{\overset{k_1}{\rightleftarrows}} A_2 \underset{k_4}{\overset{k_3}{\rightleftarrows}} A_3$

Several authors have dealt with the integration of the differential equations representing the kinetics of this system, and with the analysis of the equations describing the variation of concentrations of the components with time[8,15,17-21]. The differential equations were written by Vriens[22] in the following form

$$\frac{dx_1}{dt} = k_2 x_2 - k_1 x_1$$

(1)

$$\frac{dx_2}{dt} = k_1 x_1 - k_2 x_2 - k_3 x_2 + k_4 x_3$$

(2)

$$\frac{dx_3}{dt} = k_3 x_2 - k_4 x_3$$

(3)

$$x_1 + x_2 + x_3 = 1$$

where x_1, x_2 and x_3 are the mole fractions of substances A_1, A_2 and A_3 at time t.

From the above equations the following set of second order differential equations can be derived

$$\frac{d^2x_1}{dt^2} + \frac{dx_1}{dt}(\sum k) + x_1(\sum kk) = k_2 k_4 \tag{4}$$

$$\frac{d^2x_3}{dt^2} + \frac{dx_3}{dt}(\sum k) + x_3(\sum kk) = k_1 k_3 \tag{5}$$

where

$$\sum k = k_1 + k_2 + k_3 + k_4 \quad \text{and} \quad \sum kk = k_1 k_3 + k_2 k_4 + k_1 k_4$$

By corresponding substitutions all the rate coefficients except one can be eliminated from these equations. Introducing the following parameters

$$\frac{k_1}{k_2} = K_1, \quad \frac{k_3}{k_4} = K_2, \quad \frac{k_3}{k_1} = \kappa \quad \text{and} \quad k_1 t = \theta,$$

and if the initial concentrations are $x_{2_0} = x_{3_0} = 0$ and $x_{1_0} = 1$, then, when $\theta = 0$, $dx_1/d\theta = -1$ and $dx_3/d\theta = 0$. After integration of the set of differential equations one obtains

$$x_1 = C_1 e^{D_1\theta} + C_2 e^{D_2\theta} + \frac{\kappa}{K_1 K_2 E_2} \tag{6}$$

$$x_3 = C_3 e^{D_1\theta} + C_4 e^{D_2\theta} + \frac{\kappa}{E_2} \tag{7}$$

where

$$C_1 = (-1 - D_2 - \kappa/K_1 K_2 D_1)/(D_1 - D_2)$$
$$C_2 = (1 + D_1 - \kappa/K_1 K_2 D_2)/(D_1 - D_2)$$
$$C_3 = (\kappa/D_1)/(D_1 - D_2)$$
$$C_4 = (-\kappa/D_2)/(D_1 - D_2)$$
$$D_1 = (-E_1 + \sqrt{E_1^2 - 4E_2})^2$$
$$D_2 = (-E_1 - \sqrt{E_1^2 - E_2})/2$$
$$E_1 = 1 + (1/K_1) + \kappa + (\kappa/K_2)$$
$$E_2 = \kappa\{1 + (1/K_1 K_2 + 1/K_2)\} = D_1 D_2$$

By measuring the equilibrium concentrations of substances A_1, A_2 and A_3 the values of K_1 and K_2 can be determined experimentally. If x_2 passes through a maximum as t increases, then, to a first approximation, we can get information on the value of κ from

$$x_{2\,\text{max}} = \kappa e^{\kappa/(1-\kappa)}$$

Knowing K_1, K_2 and κ the $x_1(x_2, x_3) = f(\theta)$ expressions can be plotted from equations (6) or (7). θ can be calculated, and hence k_1 may be derived. From the equations defining K_1 and K_2, the other rate coefficients can be calculated.

1.VI COMPETITIVE-CONSECUTIVE REACTIONS

No example considered.

1.VII COMPETITIVE-CONSECUTIVE REACTIONS WITH REVERSIBLE REACTION STEP

1.VII.1

A number of authors[23-30] have dealt with the integration of the differential equations representing the kinetics of the system. Some treat it as a circulation reaction, being of importance especially with the isotopic-exchange and isomerization processes.

Integration of the differential equations describing the change with time of the concentrations of the components leads to the following expressions for the concentrations of A_1, A_2 and A_3, when the initial conditions are $[A_1] = [A_1]_0$, $[A_2] = [A_2]_0$ and $[A_3] = [A_3]_0$

$$[A_1]_0 - [A_1] = \frac{1}{\lambda_1 - \lambda_2} \left\{ \frac{\lambda_2 T'}{\theta_1}(e^{-\theta_1 t} - 1) - \frac{\lambda_1 T''}{\theta_2}(e^{-\theta_2 t} - 1) \right\} \tag{1}$$

$$[A_2]_0 - [A_2] = \frac{1}{\lambda_1 - \lambda_2} \left\{ -\frac{T'}{\theta_1}(e^{-\theta_1 t} - 1) + \frac{T''}{\theta_2}(e^{-\theta_2 t} - 1) \right\} \tag{2}$$

$$[A_3]_0 - [A_3] = \frac{1}{\lambda_1 - \lambda_2} \left\{ -(\lambda_2 - 1)\frac{T'}{\theta_1}(e^{-\theta_1 t} - 1) + (\lambda_1 - 1)\frac{T''}{\theta_2}(e^{-\theta_2 t} - 1) \right\} \tag{3}$$

where

$$\theta_{1(2)} = \frac{S_1+S_2}{2}(\pm)\sqrt{\frac{(S_1-S_2)^2}{4}+(k_{12}-k_{32})(k_{21}-k_{31})}$$

$$S_1 = k_{12}+k_{31}+k_{13}$$

$$S_2 = k_{21}+k_{23}+k_{32}$$

$$\lambda_1 = \frac{k_{21}-k_{31}}{S_2-\theta_1} \tag{4}$$

$$\lambda_2 = \frac{k_{21}-k_{31}}{S_2-\theta_2}$$

$$T' = T_1+\lambda_1 T_2$$

$$T'' = T_1+\lambda_2 T_2$$

$$T_1 = (k_{12}+k_{13})[A_1]_0-k_{21}[A_2]_0-k_{31}[A_3]_0$$

$$T_2 = -k_{12}[A_1]_0+(k_{21}+k_{23})[A_2]_0-k_{32}[A_3]_0$$

From equations (1), (2), (3) the equilibrium conditions for the system follow by putting $t = \infty$

$$\frac{[A_2]_e}{[A_1]_e} = \frac{k_{12}k_{32}+k_{12}k_{31}+k_{32}k_{13}}{k_{21}k_{32}+k_{21}k_{31}+k_{23}k_{31}} = K_1 \tag{5}$$

$$\frac{[A_3]_e}{[A_2]_e} = \frac{k_{12}k_{23}+k_{21}k_{13}+k_{23}k_{13}}{k_{12}k_{32}+k_{12}k_{31}+k_{32}k_{13}} = K_2 \tag{6}$$

$$\frac{[A_1]_e}{[A_3]_e} = \frac{k_{21}k_{32}+k_{21}k_{31}+k_{23}k_{31}}{k_{12}k_{23}+k_{21}k_{13}+k_{23}k_{13}} = K_3 = \frac{1}{K_1 K_2} \tag{7}$$

where subscript e refers to equilibrium concentrations. From these equilibrium conditions it follows that

$$k_{21}k_{32}k_{13} = k_{12}k_{23}k_{31} \tag{8}$$

The conversion of *endo*-cyclopentadiene/maleic anhydride adduct (N), labelled in its carbonyl group with ^{14}C, takes place according to the above mechanism when it is heated in hot decalin solution with an equimolar amount of maleic anhydride (MA). A rapid exchange between N and MA occurs and simultaneously N changes slowly into its *exo*-isomeric form[31].

1.VII.2

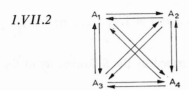

This scheme serves to describe a reaction system when any two of the four substances present can mutually change into each other by a first order reversible reaction. The system has no great practical importance; its theoretical importance derives from the fact that the kinetic expressions include solutions for simpler systems which are special cases of the above scheme.

A general solution of the problem was given by Matsen and Franklin[32].

1.VII.3 $A_1 \underset{k_2}{\overset{k_1}{\rightleftarrows}} A_2 \overset{k_3}{\leftarrow} A_3 \overset{k_4}{\rightarrow} A_4$

This scheme has been dealt with by Lemieux *et al.*[33].

The variation of $[A_3]$ and $[A_4]$ with time can be expressed in the form of equations (1) and (2) (see the treatment of scheme $A_1 \rightleftarrows A_2 \rightarrow A_3$, p. 24) the concentrations being given as mole fractions (α) of the total amount of substance (A_1, A_2, A_3 and A_4)

$$\alpha_3 = e^{-(k_3+k_4)t} \tag{1}$$

$$\alpha_4 = \frac{k_4}{k_3+k_4}\{1-e^{-(k_3+k_4)t}\} \tag{2}$$

Using these equations the values of k_3 and k_4 can be determined from the experimental results.

From the reaction scheme the rate expression for the change of $[A_2]$ is

$$d\alpha_2/dt = (k_3+k_4)\alpha_3 - k_4\alpha_3 - k_2\alpha_2 + k_1\alpha_1 \tag{3}$$

Substituting for α_3 and α_4 from equations (1) and (2), and considering that $\alpha_1 = 1-\alpha_2-\alpha_3-\alpha_4$ (by the definition of the concentrations), the following integrable rate equation is obtained from (3)

$$d\alpha_2/dt = (k_3-k_1)e^{-(k_3+k_4)t} - (k_1+k_2)\alpha_2 + k_1 - \frac{k_1 k_4}{k_3+k_4}\{1-e^{-(k_3+k_4)t}\} \tag{4}$$

$$I.VII.4 \qquad A_i \underset{k_{ji}}{\overset{k_{ij}}{\rightleftarrows}} A_j \qquad (i = 1, 2, 3, \ldots, n; \; j = 1, 2, 3, \ldots, n; \; i \neq j)$$

The kinetics of such a reaction can be described by the following set of linear homogeneous differential equations

$$\frac{d[A_1]}{dt} = -(k_{12}+k_{13}+ \ldots +k_{1n})[A_1]+k_{21}[A_2]+$$

$$+k_{31}[A_3]+ \ldots +k_{n1}[A_n]$$

$$\frac{d[A_2]}{dt} = k_{12}[A_1]-(k_{21}+k_{23}+ \ldots +k_{2n})[A_2]+k_{32}[A_3]+ \ldots +k_{n2}[A_n]$$

$$\vdots \tag{1}$$

$$\frac{d[A_n]}{dt} = k_{1n}[A_1]+k_{2n}[A_2]+k_{3n}[A_3]+ \ldots -(k_{n1}+k_{n2}+ \ldots +k_{nn-1})[A_n]$$

This always has a set of solutions

$$[A_i] = B_i e^{\rho t} \tag{2}$$

Beside the trivial solution $(B_1 = B_2 = \ldots B_n = 0)$ there are other solutions. After differentiation, substitution, rearrangement and simplifying, according to equation (3)

$$[(k_{12}+k_{13}+ \ldots +k_{1n})+\rho]B_1 -k_{12}B_2- \ldots -k_{n1}B_n = 0$$

$$-k_{12}B_1+[(k_{21}+k_{23}+ \ldots +k_{2n})+\rho]B_2- \ldots -k_{n2}B_n = 0 \tag{3}$$

$$\vdots$$

$$-k_{1n}B_1-k_{2n}B_2- \ldots +[(k_{n1}+k_{n2}+ \ldots +k_{nn-1})+\rho]B_n = 0$$

The condition of the solutions is that the $F(\rho)$ determinant of the system be equal to zero

$$F(\rho) \begin{vmatrix} (k_{12}+k_{13}+ \ldots +k_{1n})+\rho & -k_{21} \ldots & -k_{n1} \\ -k_{13} & (k_{21}+k_{23}+ \ldots +k_{2n})+\rho \ldots & -k_{n2} \\ \vdots & & \\ -k_{1n} & -k_{2n} \ldots (k_{n1}+k_{n2}+ \ldots +k_{nn-1})+\rho \end{vmatrix} = 0 \tag{4}$$

By means of equation (3) $B_1, B_2, \ldots B_n$ can be determined for any ρ root of the equation (e.g. $B_n = 1$). We have n values of B for each of the roots $\rho_1, \rho_2, \ldots \rho_n$.

The solution for ρ_k is

$$[A_1] = B_{1k}e^{\rho_k t}$$
$$[A_2] = B_{2k}e^{\rho_k t}$$
$$\vdots \qquad \vdots$$
$$[A_{n-1}] = B_{(n-1)k}e^{\rho_k t}$$
$$[A_n] = 1 \cdot e^{\rho_k t}$$

(5)

The general solution is a linear combination of these

$$[A_1] = \sum_{k=1}^{n} \lambda_k B_{1k}e^{\rho_k t}$$
$$[A_2] = \sum_{k=1}^{n} \lambda_k B_{2k}e^{\rho_k t}$$
$$\vdots$$
$$[A_n] = \sum_{k=1}^{n} \lambda_k B_{nk}e^{\rho_k t}$$

(6)

In equation (6) the λ_k values are determined from the initial conditions. This may be done by means of n linear equations.

2. Second and mixed first–second order reactions

2.I REVERSIBLE REACTIONS

$2.I.1 \qquad 2A_1 \underset{k_2}{\overset{k_1}{\rightleftarrows}} A_2 + A_3$

The kinetics of this system can be derived from the more general $A_1 + A_2 \rightleftarrows A_3 + A_4$ reversible second order reaction when $A_1 = A_2$. The latter scheme will be treated in detail later (p. 43), and the solutions described there can be simply applied to the present case.

If none of the concentrations of A_1, A_2 and A_3 are zero at $t = 0$, the rate expression is

$$d[A_2]/dt = d[A_3]/dt = -d[A_1]/2dt = k_1[A_1]^2 - k_2[A_2][A_3]$$

(1)

Assuming that $[A_1]_0 = [A_2]_0 = [A_3]_0$ at time $t = 0$, the expression for $[A_1]$ is the same as the reaction $A_1 + A_2 \rightleftarrows A_3 + A_4$, except that $k_2/4$ must be substituted for k_2.

References pp. 79–80

If the initial concentrations are $[A_2]_0 = [A_3]_0 = 0$ and $[A_1]_0 \neq 0$, the rate expression is

$$d[A_2]/dt = d[A_3]/dt = -d[A_1]/2dt = k_1[A_1]^2 - \frac{k_2}{4}([A_1]_0 - [A_1])^2 \quad (2)$$

Introducing the following notation: $[A_1] = (a-x)$ and $[A_2] = [A_3] = x/2$, and if subscript e indicates the concentrations at equilibrium, the solution is

$$\frac{x_e}{2a(a-x_e)} \ln \frac{x(a-2x_e)+ax_e}{a(x_e-x)} = k_1 t \quad (3)$$

The gas phase homogeneous decomposition of hydrogen iodide and the formation of the latter from hydrogen and iodine can be treated according to this analysis. In his classical work on this system, Bodenstein[34] used the above integral expression.

2.1.2 $\quad A_1 \underset{k_2}{\overset{k_1}{\rightleftarrows}} A_2 + A_3$

The integration of the differential rate equation for a mixed, reversible first–second order reaction is not a complicated task (see ref. 35). The expression for the variation in the concentration of A_1 is

$$d[A_1]/dt = -k_1[A_1] + k_2[A_2][A_3] \quad (1)$$

Assuming the initial concentration $[A_1]_0 \neq 0$ and $[A_2]_0 = [A_3]_0 = 0$, then according to the stoichiometric equation

$$[A_1]_0 - [A_1] = [A_2] = [A_3] \quad (2)$$

The equilibrium constant is

$$K = \frac{k_1}{k_2} = \frac{([A_1]_0 - [A_1]_e)^2}{[A_1]_e} = \frac{[A_2]_e[A_3]_e}{[A_1]_e} \quad (3)$$

Then from equation (1), eliminating $[A_2]$ and $[A_3]$

$$d[A_1]/dt = -k_1[A_1] + k_1 \frac{[A_1]_e}{([A_1]_0 - [A_1]_e)^2} ([A_1]_0 - [A_1])^2 \quad (4)$$

Rearranging this equation, we get a form which agrees with the rate expression for a second order reaction of two different reactants with different initial concentrations. Thus, the solution is

$$\frac{x_e}{(2a-x_e)} \ln \frac{ax_e+x(a-x_e)}{a(x_e-x)} = k_1 t \tag{5}$$

where $x = [A_1]_0-[A_1]$, $x_e = [A_1]_0-[A_1]_e$ and $a = [A_1]_0$. If the initial conditions are $[A_1]_0 = 0$ and $[A_2]_0 = [A_3]_0 = b$, introducing $y = [A_2]_0-[A_2] = [A_3]_0-[A_3]$ and $y_e = [A_2]_0-[A_2]_e = [A_3]_0-[A_3]_e$, the result of the integration will be

$$\frac{y_e}{(b^2-y_e^2)} \ln \frac{y_e(b^2-yy_e)}{b^2(y_e-y)} = k_2 t \tag{6}$$

Equations (5) and (6) are suitable for the determination of k_1 and k_2, respectively. Knowing the equilibrium concentrations, the rate coefficients can be calculated by plotting the logarithmic term against time.

Reactions the kinetics of which may be treated according to the reversible mixed first–second order scheme are the dissociation of hexaphenyl ethane into radicals, radical recombination in the gas phase at high pressure, the interaction of alkyl iodides and trimethylamine as well as the simultaneous formation and hydrolysis of esters.

The kinetics are much more complicated if the reaction takes place on a solid surface, for example on the surface of a catalyst. In this case the reaction scheme includes (besides the chemical process) the adsorption and desorption processes. Depending on the conditions, the mechanism can take different forms, and limiting cases are possible according to whether the desorption of substance A_1, the surface reaction itself or the desorption of a product (or products) determine the overall rate.

The deduction of the kinetic equations is undertaken by supposing that the assumption of quasi-stationary concentrations can be applied to the adsorbed substances. The considerations are essentially the same as described in detail when treating the reactions $A_1 \rightleftarrows A_2$ and $A_1 \rightarrow A_2$, so here we restrict ourselves only to the final results. The notations used are as before, or suitably modified. [S] and $[S]_0$ are the concentrations of free and total adsorption sites, Q is the surface area of the catalyst, $[\ ]_a$ represents the concentration of the adsorbed substance and $[\ ]$ the concentration or the partial pressure of a substance in the gas phase. Further parameters are: $K_{A_1} = (k_{A_1})_a/(k_{A_1})_d$; $K = k_1/k_2$; $K_{A_2} = (k_{A_2})_a/(k_{A_2})_d$; $K_{A_3} = (k_{A_3})_a/(k_{A_3})_d$.

The first type of heterogeneous process is shown by the following reaction scheme

$$A_1 + S \underset{(k_{A1})_d}{\overset{(k_{A1})_a}{\rightleftarrows}} (A_1)_a \tag{7}$$

$$(A_1)_a \underset{k_2}{\overset{k_1}{\rightleftarrows}} (A_2)_a + A_3 \tag{8}$$

$$(A_2)_a \underset{(k_{A2})_a}{\overset{(k_{A2})_d}{\rightleftarrows}} A_2 + S \tag{9}$$

(*i*) When the overall rate of the reaction is determined by the adsorption of component A_1, the kinetic equation takes the form

$$W = (k_{A_1})_a Q[S]_0 \frac{[A_1] - \dfrac{1}{K} \dfrac{K_{A_2}}{K_{A_1}} [A_2][A_3]}{1 + K_{A_2}[A_2] \left(1 + \dfrac{1}{K}[A_3]\right)} \tag{10}$$

(*ii*) When the rate determining process is the surface reaction

$$W = k_1 K_{A_1} Q[S]_0 \cdot \frac{[A_1] - \dfrac{1}{K} \cdot \dfrac{K_{A_2}}{K_{A_1}} [A_2][A_3]}{1 + K_{A_1}[A_1] + K_{A_2}[A_2]} \tag{11}$$

(*iii*) Lastly, when the rate determining process is the desorption of product A_2, the rate expression is

$$W = (k_{A_2})_d K K_{A_1} Q[S]_0 \cdot \frac{[A_1] - \dfrac{1}{K} \cdot \dfrac{K_{A_2}}{K_{A_1}} [A_2][A_3]}{[A_3] \left\{ 1 + K_{A_1}[A_1] \cdot \dfrac{K}{1 + [A_3]} \right\}} \tag{12}$$

The second type is given by the following scheme

$$A_1 + S \underset{(k_{A1})_d}{\overset{(k_{A1})_a}{\rightleftarrows}} (A_1)_a \tag{13}$$

$$(A_1)_a + S \underset{k_2}{\overset{k_1}{\rightleftarrows}} (A_2)_a + (A_3)_a \tag{14}$$

$$(A_2)_a \underset{(k_{A2})_a}{\overset{(k_{A2})_d}{\rightleftarrows}} A_2 + S \tag{15}$$

$$(A_3)_a \underset{(k_{A3})_a}{\overset{(k_{A3})_d}{\rightleftarrows}} A_3 + S \tag{16}$$

Thus, in the surface reaction both product molecules are formed in the adsorbed state. Depending on the rate determining process, the following four limitating cases are possible.

(*i*) When the adsorption of component A_1 is the rate determining process

$$W = (k_{A_1})_a Q[S]_0 \cdot \frac{[A_1] - \dfrac{1}{K} \cdot \dfrac{K_{A_2} K_{A_3}}{K_{A_1}} [A_2][A_3]}{1 + K_{A_2}[A_2]\left(1 + \dfrac{1}{K} \cdot K_{A_3}[A_3]\right) + K_{A_3}[A_3]} \tag{17}$$

(*ii*) When the surface chemical reaction is the rate determining process

$$W = k_1 K_{A_1} Q[S]_0^2 \cdot \frac{[A_1] - \dfrac{1}{K} \cdot \dfrac{K_{A_2} K_{A_3}}{K_{A_1}} [A_2][A_3]}{(1 + K_{A_1}[A_1] + K_{A_2}[A_2] + K_{A_3}[A_3])^2} \tag{18}$$

(*iii*) When the desorption of product A_2 is the rate determining process

$$W = (k_{A_2})_d K \cdot \frac{K_{A_1}}{K_{A_3}} \cdot Q[S]_0 \cdot \frac{[A_1] - \dfrac{1}{K} \cdot \dfrac{K_{A_2} K_{A_3}}{K_{A_1}} [A_2][A_3]}{[A_3]\left\{1 + K_{A_1}[A_1]\left(1 + \dfrac{K}{K_{A_3}[A_3]}\right) + K_{A_3}[A_3]\right\}} \tag{19}$$

(*iv*) When the desorption of product A_3 is the rate determining process

$$W = (k_{A_3})_d K \cdot \frac{K_{A_1}}{K_{A_2}} Q[S]_0 \cdot \frac{[A_1] - \dfrac{1}{K} \cdot \dfrac{K_{A_2} K_{A_3}}{K_{A_1}} [A_2][A_3]}{[A_2]\left\{1 + K_{A_1}[A_1]\left(1 + \dfrac{K}{K_{A_2}[A_2]}\right) + K_{A2}[A_2]\right\}} \tag{20}$$

When the surface reaction is complex, it is necessary to amplify the above schemes. For details the literature[77] should be consulted.

The kinetic equations (11)–(13), (18)–(21) and (25)–(27) can be simplified mathematically, if some terms can be neglected. Thus further limiting cases and rate equations can be obtained. Some of the rate expressions may formally be the same, and for this reason it is obvious that from the experimental rate equation no definite conclusions can be drawn as to the mechanism, the rate determining step and the physical meaning of the empirical constant. In most of the cases these questions can be solved only by other means.

2.1.3 $A_1 \xrightarrow{k_1} A_2 + A_3$

The homogeneous first order conversion of substance A_1 into products A_2 and A_3 does not require any treatment, since the differential equation can be integrated simply. However, if the chemical reaction takes place on a surface, the adsorption and desorption of the starting material and of the products makes the kinetics of the reaction more complicated. The kinetic equations are derived in a way similar to that described in detail in the treatment of the heterogeneous process $A_1 \rightleftarrows A_2$ (p. 10). Thus only the final results are given. The notation used below is identical with that introduced earlier. The reaction $A_1 \xrightarrow{k_1} A_2 + A_3$ is an extreme case of $A_1 \underset{k_2}{\overset{k_1}{\rightleftarrows}} A_2 + A_3$, since the latter changes into the former when $k_1 \gg k_2$.

When only one of the products of the reaction is adsorbed the mechanism can be written as

$$A_1 + S \underset{(k_{A1})_d}{\overset{(k_{A1})_a}{\rightleftarrows}} (A_1)_a \tag{1}$$

$$(A_1)_a \xrightarrow{k_1} (A_2)_a + A_3 \tag{2}$$

$$(A_2)_a \underset{(k_{A2})_a}{\overset{(k_{A2})_d}{\rightleftarrows}} A_2 + S \tag{3}$$

(*i*) The rate of the reaction is determined by the adsorption of component A_1

$$W = (k_{A_1})_a Q[S]_0 \cdot \frac{[A_1]}{1 + K_{A_2}[A_2]} \tag{4}$$

(*ii*) The surface reaction is the rate determining process

$$W = k_1 K_{A_1} Q[S]_0 \cdot \frac{[A_1]}{1 + K_{A_1}[A_1] + K_{A_2}[A_2]} \tag{5}$$

(*iii*) The desorption of component A_2 is rate determining

$$W = (k_{A_2})_d Q[S]_0 \tag{6}$$

When both the product molecules are adsorbed we have

$$A_1 + S \underset{(k_{A1})_d}{\overset{(k_{A1})_a}{\rightleftarrows}} (A_1)_a \tag{7}$$

$$(A_1)_a + S \xrightarrow{k_1} (A_2)_a + (A_3)_a \tag{8}$$

$$(A_2)_a \underset{(k_{A_2})_a}{\overset{(k_{A_2})_d}{\rightleftarrows}} A_2 + S \tag{9}$$

$$(A_3)_a \underset{(k_{A_3})_a}{\overset{(k_{A_3})_d}{\rightleftarrows}} A_3 + S \tag{10}$$

(*i*) The reaction rate is determined by the adsorption of component A_1

$$W = (k_{A_1})_a Q[S]_0 \cdot \frac{[A_1]}{1 + K_{A_2}[A_2] + K_{A_3}[A_3]} \tag{11}$$

(*ii*) The surface chemical reaction determines the overall rate of the process

$$W = k_1 K_{A_1} Q[S]_0^2 \cdot \frac{[A_1]}{(1 + K_{A_1}[A_1] + K_{A_2}[A_2] + K_{A_3}[A_3])^2} \tag{12}$$

(*iii*) The desorption of product A_2 is rate determining

$$W = (k_{A_2})_d Q[S]_0 \tag{13}$$

(*iv*) The desorption of product A_3 is rate determining

$$W = (k_{A_3})_s Q[S]_0 \tag{14}$$

When the surface reaction is complicated, it may be necessary to modify the scheme and derive further rate equations (for details see the literature[77]).

2.1.4 $A_1 + A_2 \overset{k_1}{\rightarrow} A_3$

If this process is homogeneous the rate equation can be integrated directly, of course. The kinetic analysis of the process, however, is more complicated when the chemical reaction takes place on the surface of a catalyst. In this case adsorption and desorption processes have to be considered. On applying the approximation of quasi-stationary concentrations of the adsorbed molecules, we get rate expressions which are easily linearized. The method applied is similar to that illustrated in the treatment of the reaction $A_1 \rightleftarrows A_2$ (p. 10), and the results can be derived directly from the corresponding equations for the scheme $A_1 \underset{k_2}{\overset{k_1}{\rightleftarrows}} A_2 + A_3$ (p. 36) if we take $k_1 = 0$.

Below we give only the final results, the notation being the same as before. When only one of the reactants taking part in the surface reaction is adsorbed the mechanism is

$$A_1 + S \underset{(k_{A_1})_d}{\overset{(k_{A_1})_a}{\rightleftarrows}} (A_1)_a \tag{1}$$

$$(A_1)_a + A_2 \overset{k_1}{\rightarrow} (A_3)_a \tag{2}$$

$$(A_3)_a \underset{(k_{A_3})_a}{\overset{(k_{A_3})_d}{\rightleftarrows}} A_3 + S \tag{3}$$

(*i*) The adsorption of component A_1 is rate determining

$$W = (k_{A_1})_a Q[S]_0 \cdot \frac{[A_1]}{1 + K_{A_3}[A_3]} \tag{4}$$

(*ii*) The overall rate of reaction is determined by that of the surface chemical process

$$W = k_1 K_{A_1} Q[S]_0 \cdot \frac{[A_1][A_2]}{1 + K_{A_1}[A_1] + K_{A_3}[A_3]} \tag{5}$$

(*iii*) The desorption of product A_3 is rate determining

$$W = (k_{A_3})_a Q[S]_0 \tag{6}$$

When both the reactants are adsorbed we have

$$A_1 + S \underset{(k_{A_1})_d}{\overset{(k_{A_1})_a}{\rightleftarrows}} (A_1)_a \tag{7}$$

$$A_2 + S \underset{(k_{A_2})_d}{\overset{(k_{A_2})_a}{\rightleftarrows}} (A_2)_a \tag{8}$$

$$(A_1)_a + (A_2)_a \overset{k_1}{\rightarrow} (A_3)_a + S \tag{9}$$

$$(A_3)_a \underset{(k_{A_3})_a}{\overset{(k_{A_3})_d}{\rightleftarrows}} A_3 + S \tag{10}$$

(*i*) The adsorption of component A_1 is rate determining

$$W = (k_{A_1})_a Q[S]_0 \cdot \frac{[A_1]}{1 + K_{A_2}[A_2] + K_{A_3}[A_3]} \tag{11}$$

(*ii*) The adsorption of component A_2 is the rate determining process

$$W = (k_{A_2})_a Q[S]_0 \cdot \frac{[A_2]}{1 + K_{A_1}[A_1] + K_{A_3}[A_3]} \tag{12}$$

(*iii*) The overall rate of reaction is determined by that of the surface reaction

$$W = k_1 K_{A_1} K_{A_2} Q[S]_0^2 \cdot \frac{[A_1][A_2]}{(1 + K_{A_1}[A_1] + K_{A_2}[A_2] + K_{A_3}[A_3])^2} \tag{13}$$

(*iv*) The desorption of product A_3 is the rate determining process

$$W = (k_{A_3})_d Q[S]_0 \tag{14}$$

There is a further possibility if the surface reaction is

$$2(\tfrac{1}{2}A_1)_a + A_2 \rightarrow (A_3)_a + S \tag{15}$$

The kinetics may also be derived for several extreme cases, depending on the rate determining process.

2.1.5 $A_1 + A_2 \overset{k_1}{\underset{k_2}{\rightleftarrows}} A_3 + A_4$

The differential rate expression is

$$d[A_1]/dt = -k_1[A_1][A_2] + k_2[A_3][A_4] \tag{1}$$

or, introducing the notation $[A_1] = [A_1]_0 - x$

$$dx/dt = k_1([A_1]_0 - x)([A_2]_0 - x) - k_2([A_3]_0 + x)([A_4]_0 + x) \tag{2}$$

This equation, after appropriate mathematical treatment, can be written in the following form

$$\frac{dx}{\alpha + \beta x + \gamma x^2} = dt \tag{3}$$

The integration of (3) after introducing

$$q = \beta^2 - 4\alpha\gamma \tag{4}$$

leads to an expression which can be given in the following form[11]

$$\ln \frac{x+(\beta-q^{\frac{1}{2}})/2\gamma}{x+(\beta+q^{\frac{1}{2}})/2\gamma} = tq^{\frac{1}{2}}+\theta \tag{5}$$

The integration constant can be determined from the initial conditions. When $t = 0$, $x = 0$, and thus

$$\theta = \ln \frac{\beta-q^{\frac{1}{2}}}{\beta+q^{\frac{1}{2}}} \tag{6}$$

The condition for the experimental applicability of equations (5) and (6) is that at least the ratio k_1/k_2 be known.

If at $t = 0$ only substances A_1 and A_2 are present, the concentrations being $[A_1]_0$ and $[A_2]_0$, respectively, equation (1) has the following form

$$d[A_1]/dt = -k_1[A_1]([A_2]_0-[A_1]_0+[A_1])+k_2([A_1]_0-[A_1])^2 \tag{7}$$

After integration we obtain the expression

$$\ln \frac{([A_1]_0-[A_1]_e)([A_1]-[A_1]_e+Q)}{([A_1]-[A_1]_e)([A_1]_0-[A_1]_e+Q)} = (k_1-k_2)Qt \tag{8}$$

where

$$Q = [1/(K-1)]\sqrt{K^2([A_2]_0-[A_1]_0)^2+4[A_1]_0[A_2]_0 K}$$

and $K = k_1/k_2$.

The equilibrium concentration of A_1 is

$$[A_1]_e = \frac{-[K([A_2]_0-[A_1]_0)+2[A_1]_0]+Q(K-1)}{2(K-1)} \tag{9}$$

If A_1 and A_2 are present in equal concentrations, equation (9) becomes

$$\ln \frac{(K-\sqrt{K})\{[A_1](K-1)+[A_1]_0(\sqrt{K}+1)\}}{(K+\sqrt{K})\{[A_1](K-1)-[A_1]_0(\sqrt{K}-1)\}} = (k_1-k_2)\left(\frac{2[A_1]_0\sqrt{K}}{K-1}\right)t \tag{10}$$

The above system is illustrated by the homogeneous gas phase decomposition of HI.

2.II PARALLEL REACTIONS

2.II.1 $A_1 + A_2 \xrightarrow{k_1} A_3$

$A_1 \xrightarrow{k_2} A_4$

Young and Andrews[36] dealt with the above system, but the expressions given are not suitable to check the applicability of the scheme to any appropriate reaction. Widequist[37] suggested a treatment which, according to the author, is suitable for the calculation of k_1 without knowing the initial concentrations of the reactants, and further that it is also suitable for the calculation of k_1 and k_2 from a single experiment. (The application of the method to experimental results is given by these authors.) Widequist started from the basic assumptions that

$$[A_1] = [A_1]_0 - x, [A_2] = [A_2]_0 - x, [A_3] = x \text{ and } [A_4] = x \tag{1}$$

Such assumptions are rather doubtful; at least they cannot be generally true.

Dealing with the analysis of the pressure–time curves of gas phase reactions, Pasfield and Waring[38] also gave in a rather doubtful form the set of differential rate equations for the scheme. They attempted to decide the applicability of the mechanism by means of these expressions.

An exact kinetic treatment of the scheme leads to a graphical determination of the ratio of the rate coefficients, and to the possibility of calculating k_1 by numerical integration.

$$d[A_1]/dt = -k_1[A_1][A_2] - k_2[A_1] \tag{2}$$

$$d[A_2]/dt = -k_1[A_1][A_2] \tag{3}$$

From (2) and (3) one can obtain

$$\frac{d[A_1]}{d[A_2]} = 1 + \frac{k_2}{k_1} \frac{1}{[A_2]} \tag{4}$$

After rearrangement and integration of equation (4) we have

$$[A_1] = [A_1]_0 - [A_2]_0 + [A_2] + \frac{k_2}{k_1} \ln \frac{[A_2]}{[A_2]_0} \tag{5}$$

According to equation (4), k_2/k_1 can be determined graphically from a plot of

$([A_1]_0 - [A_1]) - ([A_2]_0 - [A_2])$ *versus* $\ln \dfrac{[A_2]_0}{[A_2]}$.

(A straight line plot is at the same time the proof of the applicability of the mechanism).

After substitution of the expression for $[A_1]$ into equation (3) and rearrangement

$$\frac{d[A_2]}{[A_2]\left\{[A_1]_0-[A_2]_0+[A_2]+\dfrac{k_2}{k_1}\ln\dfrac{[A_2]}{[A_2]_0}\right\}} = -k_1\,dt \tag{6}$$

which can be integrated numerically. At the same time k_1 can be determined since

$$k_1 = \frac{1}{t}\int_{[A_2]}^{[A_2]_0}\frac{d[A_2]}{[A_2]\left\{([A_1]_0-[A_2]_0)+[A_2]+\dfrac{k_2}{k_1}\ln\dfrac{[A_2]}{[A_2]_0}\right\}} \tag{7}$$

2.II.2 $A_1 \xrightarrow{k_1} A_2$
$$2A_1 \xrightarrow{k_2} A_3$$

The equation describing the consumption of substance A_1 is

$$d[A_1]/dt = -k_1[A_1]-k_2[A_1]^2 \tag{1}$$

After integration we get

$$\ln\left\{\frac{[A_1]_0}{[A_1]}\frac{(k_1+k_2[A_1])}{(k_1+k_2[A_1]_0)}\right\} = k_1 t \tag{2}$$

The rate coefficients cannot be directly determined from equation (2). According to Yerrick and Russell[39], however, values of k_1 and k_2 can separately be calculated in the following way. Putting $[A_1]_0/[A_1] = \alpha$ and expanding in a power series the exponential term in the rearranged form of equation (2)

$$\frac{k_1+k_2[A_1]_0}{\alpha k_1+k_2[A_1]_0} = e^{-k_1 t} \tag{3}$$

we obtain

$$\frac{k_1+k_2[A_1]_0}{\alpha k_1+k_2[A_1]_0} = 1-k_1 t+\frac{(k_1 t)^2}{2!}-\frac{(k_1 t)^3}{3!}+\cdots \tag{4}$$

This equation can be rearranged in the following form

$$[A_1]_0 = \frac{\alpha - 1}{k_2 t\{1 - k_1 t/2! + (k_1 t)^2/3! - \ldots\}} - \alpha k_1/k_2 \tag{5}$$

If $\{1 - k_1 t/2! + (k_1 t)^2/3! - \ldots\}$ can approximately be taken as 1, we finally get

$$[A_1]_0 = \frac{\alpha - 1}{k_2 t} - \alpha \frac{k_1}{k_2} \tag{6}$$

From (6) the value of k_1 and k_2 can be determined by iteration. However, it must be noted that the approximation, by which equation (6) can be derived from (5) is fairly rough.

The "apparent" overall order of the system consisting of mixed first- and second-order reactions is between 1 and 2 and is a function of the relative importance of the first and second order processes. If the reaction order is close to 1, in the initial stage of the reaction (small conversion), approximate values for the coefficients k_1 and k_2 can be determined graphically by measuring the "pseudo-first order" rate coefficient.

Yerrick and Russell[39] treated the kinetics of the decomposition of dimethyl mercury according to this scheme and by the method described above. It followed from the magnitude of the "rate coefficients" that neither k_1 and k_2 (and certainly k_2) can be regarded as referring to elementary reactions, but are only combinations of the elementary rate coefficients. This is a good instance in which it is impossible to derive information on the mechanism directly from the order of reaction.

2.II.3 $A_1 + A_2 \xrightarrow{k_1} A_4 + A_5$

$A_1 + A_3 \xrightarrow{k_2} A_5 + A_6$

This scheme consists of two reactions where A_2 and A_3 compete for A_1. The solution of the kinetics, in spite of the practical importance of the reaction, has no special significance, since the two competitive processes can be dealt with separately. However, it should be pointed out that the change with time of the concentrations of A_2 and A_3 in the system can be found even when only the change of $[A_1]$ with time is known[40,41].

The kinetics of the above scheme can be derived as an extreme case of the solution for the scheme previously described, viz.

$$A_1 + A_2 \xrightarrow{k_1'} A_3 + A_4 \tag{1}$$

$$A_1 + A_3 \xrightarrow{k_2'} A_4 + A_5 \tag{2}$$

Introducing variables B and C, defined by concentrations A_2 and A_3 of this latter scheme and by the rate coefficients, as follows

$$[A_2] = \left(\frac{k'_1 - k'_2}{k'_1 - 2k'_2}\right) B \tag{3}$$

$$[A_3] = C - \left(\frac{k'_1}{k'_1 - 2k'_2}\right) B \tag{4}$$

and substituting these expressions into the set of differential rate equations (1) and (2) given previously (p. 43), we get

$$d[A_1]/dt = -k'_1[A_1]B - k'_2[A_1]C \tag{5}$$

$$dB/dt = -k'_1[A_1]B \tag{6}$$

$$dC/dt = -k'_2[A_1]C \tag{7}$$

It can thus be seen that the kinetics of the competitive reaction treated here is a special case of that of the previous scheme.

The saponification reactions of mixtures of simple esters can be mentioned as examples.

2.II.4 $A_1 \xrightarrow{k_1} A_3$

$\quad\quad\quad A_1 + A_2 \xrightarrow{k_2} A_4$

$\quad\quad\quad 2A_1 \xrightarrow{k_3} A_5$

The above scheme includes competitive reactions in which substance A_1 can react either in a first, or in two second order reactions. The rate of consumption of substance A_1 is

$$d[A_1]/dt = -k_1[A_1] - k_2[A_1][A_2] - 2k_3[A_1]^2 \tag{1}$$

Benson[11] deals with the solution of equation (1). By putting

$$[A_1] = [A_1]_0 - x \text{ and } [A_2] = [A_2]_0 - x \tag{2}$$

and by resolving into partial fractions, the expression can be integrated. The final equation, however, is not directly suitable either to determine the rate coefficients, or even to decide the applicability of the given mechanism to the experimental results.

Benson's treatment is, in fact, not correct, since the introduction of the relationships (2) is not applicable. Instead the following considerations are valid. By dividing the expression for $d[A_1]/dt$ by that for $d[A_2]/dt$ we get

$$\frac{d[A_1]}{d[A_2]} = \frac{k_1 + k_2[A_2] + 2k_3[A_1]}{k_2[A_2]} \tag{3}$$

Introducing

$$\alpha = [A_1] + \frac{k_1}{2k_3}, \beta = [A_2], \kappa = \frac{2k_3}{k_2} \text{ and } z = \frac{\alpha}{\beta} \tag{4}$$

we obtain

$$\frac{d\alpha}{d\beta} = \beta \frac{dz}{d\beta} + z \tag{5}$$

After substitutions, rearrangement and finally integration, we have

$$\ln \frac{\beta}{\beta_0} = \frac{1}{\kappa - 1} \ln \frac{(\kappa - 1)z + 1}{(\kappa - 1)z_0 + 1} \tag{6}$$

or

$$\left(\frac{\beta}{\beta_0}\right)^{\kappa - 1} = \frac{(\kappa - 1)z + 1}{(\kappa - 1)z_0 + 1} \tag{7}$$

Using equation (7) the expressions $z = \alpha/\beta$ and α can be obtained (α_0, β_0 and z_0, respectively, are values of α, β and z at time $t = 0$).

$$z = \frac{\alpha}{\beta} = \frac{1}{\kappa - 1} \left\{ \left[1 + \frac{(\kappa - 1)\alpha_0}{\beta_0} \right] \left(\frac{\beta}{\beta_0}\right)^{\kappa - 1} - 1 \right\} \tag{8}$$

$$\alpha = z\beta \tag{9}$$

Equations (8) and (9) give a relationship between $[A_1]$ and $[A_2]$.

From the scheme consisting of three competitive reactions, other more simple reaction mechanisms may be derived if one of the rate coefficients is taken equal to zero. These cases are treated separately.

2.III PARALLEL REACTIONS WITH REVERSIBLE REACTION STEP

2.III.1 $A_1 + A_2 \overset{k_1}{\underset{k_2}{\rightleftarrows}} A_3 + A_4$

$A_1 + A_5 \overset{k_3}{\underset{k_4}{\rightleftarrows}} A_3 + A_6$

The above scheme describes the isotope exchange reaction between alkyl iodide and I_2 labelled with ^{135}I. Darbee *et al.*[42] carry out the treatment of the set of differential rate equations in the following manner.

Let a, b, and c denote the total molar concentrations of substances A_1, A_2 and A_3, and x, y and z the concentrations of the corresponding labelled substances, at time t. Introducing $w(= x+y+z)$, the set of first order differential equations approximately describing this exchange is the following

$$dx/dt = R_1/ac\{aw-(a+c)x-ay\} \tag{1}$$

$$dy/dt = R_2/bc\{bw-bx-(b+c)y\} \tag{2}$$

where R_1 and R_2 denote the rates of the exchange.

These equations can be integrated to give

$$x = x_\infty + C_1 \exp\left\{-\frac{p-q}{2}\cdot t\right\} + C_2 \exp\left\{-\frac{p+q}{2}\cdot t\right\} \tag{3}$$

$$y = y_\infty + C_3 \exp\left\{-\frac{p-q}{2}\cdot t\right\} + C_4 \exp\left\{-\frac{p+q}{2}\cdot t\right\} \tag{4}$$

where

$$x_\infty = aw/(a+b+c)$$

$$y_\infty = bw/(a+b+c)p+R_1\left(\frac{a+c}{ac}\right)+R_2\left(\frac{b+c}{bc}\right)$$

$$q = \left[\left\{R_1\left(\frac{a+c}{ac}\right)-R_2\left(\frac{b+c}{c}\right)\right\}^2 + \frac{4R_1 R_2}{c^2}\right]^{\frac{1}{2}}$$

$$C_1 = -x_\infty$$

$$-C_2 = 1/q\left[\frac{R_1 y_\infty}{c} + \frac{x_\infty}{2}\left\{R_1\left(\frac{a+c}{ac}\right)-R_2\left(\frac{b+c}{bc}\right)-q\right\}\right] \tag{5}$$

$$C_3 = -y_\infty$$

$$-C_4 = 1/q\left[\frac{R_2 x_\infty}{c} - \frac{y_\infty}{2}\left\{R_1\left(\frac{a+c}{ac}\right)-R_2\left(\frac{b+c}{bc}\right)+q\right\}\right] \tag{6}$$

If $R_1 \gg R_2$, R_1 can be directly determined. Then, at the beginning, x increases rapidly and exceeds the equilibrium value. By defining t_0 as the time when $x = x_\infty$, it can be shown that

$$R_1 = \frac{ac}{t_0(a+c)}\ln\left(\frac{a+b+c}{b}\right) \tag{7}$$

It can be seen that by using an appropriate ratio $a:b:c$, R_1 can be determined for a reaction which proceeds too rapidly on its own. The following expression gives the ratio $t_0/t_{\frac{1}{2}}$

$$t_0/t_{\frac{1}{2}} = \ln \left(\frac{a+b+c}{c}\right) / \ln 2 \tag{8}$$

2.IV CONSECUTIVE REACTIONS

2.IV.1 $A_1 \xrightarrow{k_1} A_2$
$$2A_2 \xrightarrow{k_2} A_3$$

The set of differential equations

$$d[A_1]/dt = -k_1[A_1] \tag{1}$$

$$d[A_2]/dt = k_1[A_1] - k_2[A_2]^2 \tag{2}$$

$$d[A_3]/dt = -\tfrac{1}{2}k_2[A_2]^2 \tag{3}$$

is solved according to Talat-Erben[43] in the following way.

Let $[A_1] = [A_1]_0$ and $[A_2] = [A_3] = 0$ at time $t = 0$.
The concentration of the intermediate product $[A_2]$ is given by

$$[A_2] = [A_1]_0 \left(\frac{\tau}{\kappa}\right)^{\frac{1}{2}} \frac{iJ_1[2i(\kappa\tau)^{\frac{1}{2}}] - \beta H_1^{(1)}[2i(\kappa\tau)^{\frac{1}{2}}]}{J_0[2i(\kappa\tau)^{\frac{1}{2}}] + \beta_i H_0^{(1)}[2i(\kappa\tau)^{\frac{1}{2}}]} \tag{4}$$

where

$$\kappa = [A_1]_0 \frac{k_2}{k_1}, \tau = e^{-\kappa t} \text{ and } \beta = \frac{iJ_1(2i\sqrt{\kappa})}{H_1^{(1)}(2i\sqrt{\kappa})} \tag{5}$$

and J and H refer to zero- or first-order Bessel and Hankel functions.

For $[A_3]$ the following equality is valid

$$[A_3] = \tfrac{1}{2}([A_1]_0 - [A_1] - [A_2]) \tag{6}$$

while the maximum concentration of the intermediate A_2 can be given, after introducing $[A_1] = [A_1]_0$ and $[A_2]_0 = 0$ at time $t = 0$, as

$$\frac{[A_2]_{max}}{[A_1]_0} = \left(\frac{\tau_{max}}{\kappa}\right)^{\frac{1}{2}} \tag{7}$$

Thus for any given uni–bimolecular consecutive reaction pair, at a constant temperature, κ changes only with $[A_1]_0$.

In the same paper Talat-Erben considered the dependence of the expressions $k_1 t_{max}$ and $[A_2]_{max}/[A_1]_0$ on temperature, assuming constant initial concentration of A_1, for hypothetical consecutive reactions with identical pre-exponential factors. He stated that these expressions are independent of temperature in the range where the rates of these consecutive reactions can be measured.

The coefficient k_1 may be obtained from the experimental determination of $[A_1]$ as a function of time, while the value of κ and thus of k_2 can easily be calculated from the variation of $[A_2]$ with time.

In practice, to differentiate among different types of consecutive reactions, according to Chien[44] it is necessary to investigate the dependence of the concentrations of the reactants on $(1+k_1 t)^{-1}$. By comparing the experimental data with general curves corresponding to different κ values, it is possible to establish not only the type of the consecutive reaction investigated, but also to estimate the rate coefficients of the stages.

Stepukhovich and Bakhareva[45] dealt with a similar reaction scheme

$$A_1 \rightarrow 2A_2 \rightarrow A_3$$

They studied in detail the kinetics of this system of consecutive reactions. Putting $[A_1] = a-x$, $[A_2] = x-y$ and $[A_3] = y$, the following differential equations can be written

$$dx/dt = k_1 \left(\frac{a}{2} - x\right) \text{ and } dy/dt = \frac{k_2}{2}\left(x - \frac{y}{2}\right)^2 \tag{8}$$

Thus

$$x = \frac{a}{2}\left(1-e^{-k_1 t}\right) \tag{9}$$

while

$$dy/dt = k_2\{a^2(1-e^{-k_1 t})^2 - 2a(1-e^{-k_1 t})y + y^2\} \tag{10}$$

Introducing $u = x-y/2$, after differentiation we get the following simpler expression

$$\frac{du}{dt} = ak_1 e^{-k_1 t} - k_2 u^2 \tag{11}$$

After differentiating twice and expanding the exponential function as a series, the following expressions are obtained

$$z = 1 + bk_1 t + b\,\mathrm{e}^{-k_1 t} - b \tag{12}$$

and

$$u = x - y = \frac{\dfrac{a}{2}(1 - \mathrm{e}^{-k_1 t})}{1 - b + bk_1 t + b\,\mathrm{e}^{-k_1 t}} \tag{13}$$

where $b = ak_2/k_1$ and $u = 2\left(\dfrac{\mathrm{d}z}{\mathrm{d}t}\right)/k_2 z$.

Differentiating these equations with respect to time and simplifying gives the following expressions, from which t_{max} (the time for the maximum concentration of A_2 to be attained) can be calculated

$$1 + 1/b + k_1 t_{max} = \mathrm{e}^{-k_1 t_{max}} \tag{14}$$

$$y = x - u = a(1 - \mathrm{e}^{-k_1 t}) \frac{bk_1 t - b + b\,\mathrm{e}^{-k_1 t}}{1 + bk_1 t - b + b\,\mathrm{e}^{-k_1 t}} \tag{15}$$

2.IV.2 $2A_1 \overset{k_1}{\rightarrow} A_2$

   $A_2 \overset{k_2}{\rightarrow} A_3$

Putting

$$[A_1] = a - 2x, \quad [A_2] = x - y \quad \text{and} \quad [A_3] = y \tag{1}$$

We obtain the following differential rate equations

$$\mathrm{d}x/\mathrm{d}t = 2k_1(a - 2x)^2 \tag{2}$$

$$\mathrm{d}y/\mathrm{d}t = k_2(x - y) \tag{3}$$

The solution can be given[43] as

$$A_1 = \frac{[A_1]_0}{\tau} \tag{4}$$

$$[A_2] = \tfrac{1}{2}[A_1]_0 \left\{ \mathrm{e}^{-\kappa(\tau - 1)} + \kappa\,\mathrm{e}^{-\kappa\tau}[\mathrm{Ei}(\kappa\tau) - \mathrm{Ei}(\kappa)]\,\frac{1}{\tau} \right\} \tag{5}$$

$$[A_3] = \tfrac{1}{2}([A_1]_0 - [A_1]) - [A_2] \tag{6}$$

where

$$\kappa = (k_2/k_1)[A_1]_0, \tau = 1+[A_1]_0 k_1 t \text{ and } Ei(x) = \int_{-\infty}^{x} \frac{e^y}{y} dy \qquad (7)$$

The maximum concentration of A_2 is given by

$$[A_2]_{max}/[A_1]_0 = \tfrac{1}{2}\kappa\tau_{max}^2 \qquad (8)$$

These results show that $k_1 t_{max}$ and $[A_2]_{max}/[A_1]_0$ are independent of temperature and decrease with increasing $[A_1]_0$. These functions are shown in Figs. 2 and 3, where curve (1) refers to uni–bimolecular, (2) to bi–bimolecular and (3) to bi–unimolecular consecutive reactions.

2.IV.3 $2A_1 \xrightarrow{k_1} A_2$

$$2A_2 \xrightarrow{k_2} A_3$$

The differential rate equations of this second order consecutive system are

$$d[A_1]/dt = -k_1[A_1]^2 \qquad (1)$$

$$d[A_2]/dt = \frac{k_1}{2}[A_1]^2 - k_2[A_2]^2 \qquad (2)$$

$$d[A_3]/dt = \frac{k_2}{2}[A_2]^2 = -\tfrac{1}{4}(d[A_1]/dt + 2d[A_2]/dt) \qquad (3)$$

Dividing equation (2) by (1) we get

$$d[A_2]/d[A_1] = -\tfrac{1}{2} + (k_2/k_1)([A_2]/[A_1])^2 \qquad (4)$$

Putting $[A_2] = [A_1] y$ and $\kappa = k_2/k_1$, the following equation is obtained

$$dy/d \ln [A_1] = \kappa y^2 - y - \tfrac{1}{2} \qquad (5)$$

On separating the variables and integrating

$$\ln \frac{1-(y/y_1)}{1-(y/y_2)} = (1+2\kappa)^{\frac{1}{2}} \ln \frac{[A_1]}{[A_1]_0}, \qquad (6)$$

where y_1 and y_2 are the positive and negative roots, respectively, of the equation $y^2 - (y/\kappa) - (\tfrac{1}{2}\kappa) = 0$. The initial condition is that $y = 0$ at $t = 0$. From this $[A_1]/[A_1]_0$ can be calculated as a function of y. If it is assumed that there is a

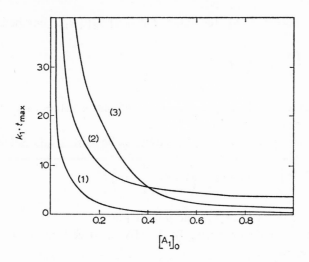

Fig. 2. $k_1 t_{max}$ as a function of initial concentration of A_1 for hypothetical consecutive reactions with equal rate coefficients. Curve (1) refers to uni-, bi-; curve (2) to bi-, bi-; and curve (3) to bi-, uni- molecular consecutive reactions.

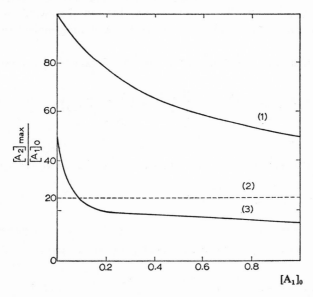

Fig. 3. $[A_2]_{max}/[A_1]_0$ as a function of $[A_1]_0$, assuming equal rate coefficients. Curve (1) refers to uni-, bi-; curve (2) to bi-, bi-; and curve (3) to bi-, uni- molecular consecutive reactions.

stationary concentration of A_2, the value of $y_s = ([A_2]/[A_1])_s$ can be calculated

$$\left(\frac{[A_2]}{[A_1]}\right)_s = \left(\frac{k_1}{2k_2}\right)^{\frac{1}{2}} = (2\kappa)^{-\frac{1}{2}} \tag{7}$$

Benson[7] tabulates and plots graphically y_s for different κ values as a function of the percentage consumption of A_1.

The change of concentration of A_2 with time can be written, after solving (5), as

$$[A_2] = \frac{[A_1]_0}{2\kappa\tau}\left[(\mu+1) - 2\mu \left/ \left(1 + \frac{\mu-1}{\mu+1} \cdot \tau\mu\right)\right.\right] \tag{8}$$

where $\kappa = k_2/k_1$, $\tau = 1 + [A_1]_0 k_1 t$ and $\mu = (1+2\kappa)^{\frac{1}{2}}$.

Putting $d[A_2]/dt = 0$ and introducing $[A_1] = [A_1]_0/\tau$ (see below), the following expression is obtained

$$\tau_{max}\mu = \left(1 \mp \frac{(2\kappa)^{\frac{1}{2}}}{\mu-1}\right) \left/ \left(1 \pm \frac{(2\kappa)^{\frac{1}{2}}}{\mu+1}\right)\right. \tag{9}$$

This can easily be solved for τ_{max}, giving

$$[A_2]_{max}/[A_1]_0 = (1/2\kappa)^{\frac{1}{2}}(\tau_{max}) \tag{10}$$

From the results it appears that $k_1 t_{max}$ and $[A_2]_{max}/[A_1]_0$ are independent of temperature. $k_1 t_{max}$ is inversely proportional to the initial concentration of the reactant and the maximum yield of the intermediate is independent of the concentration of A_1.

Solving the differential equations (1) and (3) we obtain for $[A_1]$ and $[A_3]$

$$[A_1] = [A_1]_0/\tau \text{ and } [A_3] = \frac{1}{2}\left(\frac{[A_1]_0}{2} - \frac{[A_1]}{2} - [A_2]\right) \tag{11}$$

2.V CONSECUTIVE REACTIONS WITH REVERSIBLE REACTION STEP

No example considered.

2.VI COMPETITIVE-CONSECUTIVE REACTIONS

2.VI.1 $A_1 \xrightarrow{k_1} A_2$

$A_1 + A_2 \xrightarrow{k_2} A_3$

The differential rate equations are

$$d[A_1]/dt = -k_1[A_1] - k_2[A_1][A_2] \tag{1}$$

$$d[A_2]/dt = k_1[A_1] - k_2[A_1][A_2] \tag{2}$$

The set has no explicit solution, as is the case generally with consecutive reaction systems of higher order. However, it is possible to establish certain kinetic relationships even in this case[7]. Dividing equation (2) by (1) and substituting α for the ratio k_1/k_2 gives the connection between the concentrations of A_1 and A_2

$$\frac{d[A_2]}{d[A_1]} = \frac{k_1[A_1] - k_2[A_1][A_2]}{-k_1[A_1] - k_2[A_1][A_2]} = \frac{[A_2] - \alpha}{[A_2] + \alpha} \tag{3}$$

Rearranging (3) gives

$$\frac{\alpha + [A_2]}{\alpha - [A_2]} d[A_2] = -d[A_1] \tag{4}$$

which can be directly integrated, the result being

$$[A_2] + 2\alpha \ln (\alpha - [A_2]) = [A_1] + \text{const.} \tag{5}$$

Determining the value of the integration constant from the condition that $[A_2] = 0$ when $[A_1] = [A_1]_0$, we get the following expression

$$[A_2] + 2\alpha \ln \frac{\alpha - [A_2]}{\alpha} = [A_1] - [A_1]_0 \tag{6}$$

This equation can be rearranged into the form below which is easier to use

$$\frac{[A_2]}{\alpha} + 2 \ln \left(1 - \frac{[A_2]}{\alpha}\right) = -\frac{[A_1]_0}{\alpha} \left(1 - \frac{[A_1]}{[A_1]_0}\right) \tag{7}$$

If A_2 attains a stationary state, the value of the stationary concentration is

$$[A_2]_s = \alpha \tag{8}$$

The extent of reaction at which $[A_2]$ reaches a certain fraction of $[A_2]_s$ is easily calculated; it is a function of α. Benson[7] lists for different values of $[A_1]_0/\alpha$ the degrees of consumption of A_1 when $[A_2]$ reaches 0.1, 0.5, 0.9 etc. of its stationary concentration.

From equation (7) it follows that the concentration of $[A_2]$ does not decrease to zero when component A_1 is completely consumed ($[A_1] = 0$). The graphical analysis of equation (7) shows that $[A_2]$ reaches a limiting value ($[A_2] = \alpha$) only when $[A_1]_0/\alpha$ is fairly large compared to 1. Otherwise the value of $[A_2]$ remains much below the possible concentration, $[A_2] = \alpha$, during the course of the reaction.

As has been said, the set of differential equations for the above scheme cannot be solved explicitly. However, the method described here can be applied in any case when, after elimination of time as an independent variable from the original equations, we get differential equations in which the variables can be separated.

$2.VI.2$ $A_1 \xrightarrow{k_1} A_2$

$$A_1 + A_2 \xrightarrow{k_2} 2A_2 + A_3$$

This is a reaction scheme for a simple autocatalytic process. The kinetics have been dealt with by several workers, *e.g.* Tobin[35].

Assuming that the initial concentrations $[A_2]_0 = [A_3]_0 = 0$ and $[A_1]_0 \neq 0$, and introducing $[A_1]_0 - [A_1] = x$, the rate expression is

$$dx/dt = k_1([A_1]_0 - x) + k_2 x([A_1]_0 - x) \tag{1}$$

After rearrangement and integration

$$\ln \frac{(k_1/k_2) + x}{[A_1]_0 - x} = \ln \frac{k_1}{k_2[A_1]_0} + k_2 \left(\frac{k_1}{k_2} + [A_1]_0 \right) t \tag{2}$$

If x_{max}, the value of x at the maximum rate, is determined, k_1 and k_2 can be obtained, since differentiation of equation (1) gives

$$x_{max} = \tfrac{1}{2}([A_1]_0 - \kappa)$$

where

$$\kappa = \frac{k_1}{k_2} \tag{3}$$

Thus, plotting x_{max} against $[A_1]_0$ we get κ from the intercept. (Slope $\tfrac{1}{2}$ is the criterion of correct mechanism). The left hand side of equation (2) is a linear function of t. Thus, from the plot of the right hand side (knowing κ) *versus* t, k_2 and k_1/k_2 can be determined. (A further test of mechanism is the agreement of the values of k_1/k_2 obtained in the two ways).

Further characteristic rate parameters are the maximum rate $(dx/dt)_{max}$, and the time required to reach it, t_{max}

$$(dx/dt)_{max} = k_2([A_1]_0 + \kappa)^2/4 \tag{4}$$

$$t_{max} = -\frac{\ln\dfrac{\kappa}{[A_1]_0}}{k_2([A_1]_0 + \kappa)} \tag{5}$$

Knowing these, the rate coefficients k_1 and k_2 can again be determined.

The oxidation of oxalic acid by potassium permanganate, catalyzed by Mn^{2+} ions, is an example of this type of autocatalysis.

$$2.VI.3 \quad A_1 + A_2 \xrightarrow{k_1} A_3$$

$$A_2 + A_3 \xrightarrow{k_2} A_4$$

The set of differential equations describing the change of the concentrations with time is

$$d[A_1]/dt = -k_1[A_1][A_2] \tag{1}$$

$$d[A_2]/dt = -k_1[A_1][A_2] - k_2[A_2][A_3] \tag{2}$$

$$d[A_3]/dt = k_1[A_1][A_2] - k_2[A_2][A_3] =$$
$$d[A_2]/dt - 2d[A_1]/dt \tag{3}$$

$$d[A_4]/dt = k_2[A_2][A_3] \tag{4}$$

On dividing equation (3) by (1)

$$d[A_3]/d[A_1] = -1 + \frac{k_2}{k_1}[A_3]/[A_1] \tag{5}$$

Putting $[A_3] = [A_1] y$ and $\kappa = k_2/k_1$

$$dy/d \ln [A_1] = -1 + (\kappa - 1)y \tag{6}$$

Considering the initial conditions, $y = 0$ and $[A_1] = [A_1]_0$, the solution of equation (6) is

$$y = \left(\frac{[A_3]}{[A_1]}\right) = \frac{1}{\kappa - 1}\left[1 - \left(\frac{[A_1]}{[A_1]_0}\right)^{\kappa - 1}\right] \tag{7}$$

Assuming that the stationary state approximation is applicable to A_3, the following expression is obtained for the ratio $([A_3]/[A_1])_s$

$$([A_3]/[A_1])_s = \frac{k_1}{k_2} = \frac{1}{\kappa} \tag{8}$$

From equations (7) and (8) it is easy to obtain the ratio of the correct and stationary values of $[A_3]/[A_1]$

$$R = \frac{([A_3]/[A_1])}{([A_3]/[A_1])_s} = \frac{1}{1-(1/\kappa)} \left\{ 1 - \left(\frac{[A_1]}{[A_1]_0}\right)^{\kappa-1} \right\} \tag{9}$$

The problem has been analyzed in detail by Benson[7] who tabulated the values of R (the measure of the correctness of the approximation of stationary concentration) as a function of the percentage conversion for a wide range of values of κ.

Similarly it can be shown that

$$[A_2]_0 - [A_2] = \left(\frac{2\kappa-1}{\kappa-1}\right) \left\{ [A_1]_0 - [A_1] - \frac{[A_1]_0}{\kappa-1} \left[1 - \left(\frac{[A_1]}{[A_1]_0}\right)^{\kappa} \right] \right\} \tag{10}$$

$$[A_4] = [A_1]_0 - [A_1] - [A_3] = [A_1]_0 - \frac{[A_1]}{\kappa-1} \left[\kappa - \left(\frac{[A_1]}{[A_1]_0}\right)^{\kappa-1} \right] \tag{11}$$

Kolesnikov[46] describes a simple method for the calculation of the ratio of the rate coefficients of consecutive bimolecular homogeneous reactions of the above type. He studied the alkylation of isopropylbenzene with n-butylene (in the presence of HF) as a model process.

Riggs[47] considered the possibility of solving the kinetics of this mechanism in a special case when the reaction steps take place at low rates. If a and b are the initial molar concentrations of substances A_1 and A_2, x and y the amounts transformed at time t and $\kappa = k_2/k_1$, then

$$-d[A_1]/dt = dx/dt = k_1(b-y)(a-x) \tag{12}$$

$$-d[A_2]/dt = dy/dt = k_1(b-y)[a-x+\kappa(2x-y)] \tag{13}$$

$$d[A_3]/dt = d(2x-y)/dt = k_1(b-y)[a-x-\kappa(2x-y)] \tag{14}$$

From the rate equations one obtains for the maximum concentration of A_3

$$\begin{aligned} &\text{if } \kappa \neq 1, [A_3]_{max}/a = \kappa^{\kappa/(1-\kappa)} \\ &\text{if } \kappa = 1, [A_3]_{max}/a = e^{-1} \end{aligned} \tag{15}$$

$[A_3]$ is proportional to the initial concentration of A_1, but is independent of that of A_2.

As to the final state, Riggs discussed three cases. If $\kappa = 0$, $y = x$, and then the rate equation has the usual second order form. When κ is extremely large, *i.e.* all the A_3 molecules change almost immediately into A_4, and $[A_3]$ is very small

$$k_1 = \left[\frac{1}{t}(2a-b)\right] \left[\ln\{b(2a-y)/2a(b-y)\}\right] \tag{16}$$

When $\kappa = \frac{1}{2}$

$$k_1 = \left[\frac{2}{t}(2a-b)\right] \left[\ln\{b(2a-y)/2a(b-y)\}\right] \tag{17}$$

and the time for A_3 to reach its maximum concentration is given by the following equation

$$\left[\frac{2}{k_1}(2a-b)\right] \left[\ln\{b/2(b-a)\}\right] \tag{18}$$

The author used his method when evaluating the results of a study of the reaction between phenyliodoso acetate and aceto-*p*-toluidine.

The solution of the differential equations for the above scheme has also been discussed by Sievert *et al.*[13] who used dimensionless variables.

2.VI.4 $A_1 + A_2 \xrightarrow{k_1} A_3 + A_4$

$A_1 + A_3 \xrightarrow{k_2} A_4 + A_5$

This scheme is extremely important from a practical point of view[2, 50, 51, 70, 71]. An example is the saponification of symmetrical diesters, such as ethyl succinate. The set of differential rate equations is

$$d[A_1]/dt = -k_1[A_1][A_2] - k_2[A_1][A_3] \tag{1}$$

$$d[A_2]/dt = -k_1[A_1][A_2] \tag{2}$$

$$d[A_3]/dt = -k_1[A_1][A_2] - k_2[A_1][A_3] \tag{3}$$

$$d[A_4]/dt = k_2[A_1][A_3] \tag{4}$$

The solution is simpler when substance A_1 is in large excess and thus we have in effect two successive first order reactions.

It is useful to examine the following limiting cases.

(*i*) $k_1 \gg k_2$. In such cases the first step is almost complete before the second one starts[72]. Then the steps of the reaction can be separately treated as simple second-order reactions. This is the case with the alkaline saponification of oxalic acid esters[73].

(ii) $k_1 \ll k_2$, i.e. the second step is extremely rapid. This is the case with the hydrolysis of acetals, where the hemiacetal reacts much more rapidly.

(iii) $k_1 = 2k_2$. This may be the case for the saponification of glycerol diacetate[74].

The system has been dealt with in its most general form by Frost and Schwemer[48,49]. Assuming the initial conditions $[A_1]_0 = 2[A_2]_0$ and $[A_3]_0 = [A_4]_0 = 0$, the expression for $d[A_1]/dt$ has the following form

$$d[A_1]/dt = (2k_2 - k_1)[A_1][A_2] - k_2[A_1]^2 \tag{5}$$

Introducing the following dimensionless variables α, β and τ and a parameter κ

$$\alpha_1 = \frac{[A_1]}{[A_1]_0} \qquad \alpha_2 = \frac{[A_2]}{[A_2]_0} \tag{6}$$

and

$$\tau = [A_2]_0 k_1 t \qquad \kappa = k_2/k_1$$

we get

$$\frac{d\alpha_1}{d\tau} = (2\kappa - 1)\alpha_1 \alpha_2 - 2\alpha_1^2 \tag{7}$$

$$\frac{d\alpha_2}{d\tau} = -2\alpha_1 \alpha_2 \tag{8}$$

Solving the above differential equation system for $\alpha_1 = \alpha_2 = 1$ at $t = 0$, the following expression is obtained for α_1

$$\alpha_1 = \frac{1-2\kappa}{2(1-\kappa)}\alpha_2 + \frac{1}{2(1-\kappa)}\alpha_2^\kappa \tag{9}$$

and for τ

$$\tau = \frac{1-\kappa}{1-2\kappa} \int_{\alpha_2}^{\alpha_1} \frac{d\alpha_2}{\alpha_2^2 \left(1 + \dfrac{1}{1-2\kappa}\right) \alpha_2^{\kappa-1}} \tag{10}$$

From equations (9) and (10) curves of $\alpha_1 = f(\tau)$ and $\alpha_2 = f(\tau)$ can be plotted (Figs. 4 and 5) and the dependence of the variations of α_1 and α_2 on κ have been tabulated.

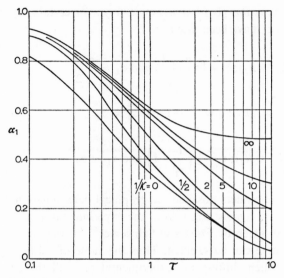

Fig. 4. Plot of the relative concentration of A_1 as a function of the dimensionless time.

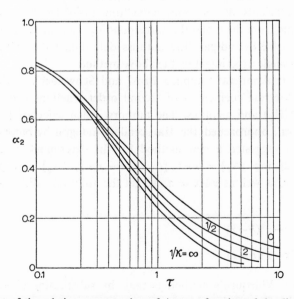

Fig. 5. Plot of the relative concentration of A_2 as a function of the dimensionless time.

The integral of equation (10) has a simple form for certain special values of κ. The following cases have been discussed

$$\kappa = 0 \ (i.e.\ k_1 \gg k_2); \tau = \ln \frac{\alpha_2+1}{2\alpha_2}, \text{and } \alpha_1 = \frac{\alpha_2+1}{2} \tag{11}$$

$$\kappa = \tfrac{1}{2} \ (i.e.\ k_1 = 2k_2); \tau = \frac{1}{\alpha_1} -1 \text{ and } \alpha_1^2 = \alpha_2 \tag{12}$$

$$\kappa = \infty \ (i.e.\ k_1 \ll k_2); \tau = \tfrac{1}{2}\left(\frac{1}{\alpha_2} -1\right) \text{ and } \alpha_1 = \alpha_2 \tag{13}$$

The above method renders possible the graphical determination of k_1, k_2 and κ by plotting α_1 or α_2 against $\log \tau$ from the experimental data. On comparing this plot with curves given by Frost and Schwemer[48,49], the one most similar to the experimental curve can be selected. Thus the value of κ can approximatively be determined. Then shifting the experimental plot till the coordinate axes coincide, the distance between the two curves gives $\log [A_2]_0 k_1$, from the condition that

$$\log \tau = \log [A_2]_0 k_1 + \log t \tag{14}$$

Application of the "time-ratio" method yields more accurate results. This method is based on the fact that the ratio of the times required for the conversion of two different percentages of substance A_1 depends only on κ, that is on the ratio of the rate coefficients. By determining the "time-ratio" from one experiment, it is easy to calculate from the tables the corresponding value of κ, k_1 and k_2. For the saponification of ethyl adipate the authors determined the values of both coefficients from a single experiment by the above method.

Reikhsfeld and Prokhorova[50] applied Frost and Schwemer's method to the determination of the rate coefficients of second order reactions between mono-organosilanes ($RSiH_3$) and alcohols. The authors, when calculating Frost and Schwemer's integral, approached the transcendent integral by binom-integrals.

Widequist[51] has suggested a new method for the determination of the rate coefficients. Introducing $[A_1] = a-x$, $[A_2] = b-x-y$, $[A_3] = x-y$, $[A_4] = x+y$ and $[A_5] = y$, as well as $y+x = z$, we get the following differential equations

$$dx/dt = k_1(a-x)(b-z) \tag{15}$$

$$dy/dt = k_2(x-y)(b-z) \tag{16}$$

The main point of Widequist's method is that, by substituting $(b-z)dt = d\theta$, equations (15) and (16) are reduced to the linear system

$$\mathrm{d}x/\mathrm{d}\theta = k_1(a-x) \tag{17}$$

$$\mathrm{d}y/\mathrm{d}\theta = 2k_2(x-y) \tag{18}$$

which is treated accordingly.

In his paper he also gives a numerical method for the determination of k_1 and k_2 from the expression

$$k_1 = \frac{z}{2a\theta - a - \dfrac{a}{k_2}(1-e^{-k_2\theta})} \tag{19}$$

Widequist applied his method to experimental data for the alkaline saponification of diethyl adipate and found good agreement between his rate coefficients and Frost and Schwemer's.

In a more recent publication Widequist[52] deals with the reaction scheme

$$A_1 + A_2 \xrightarrow{k_1} A_3 + A_5$$

$$A_1 + A_3 \xrightarrow{k_2} A_4 + A_6$$

The mathematical treatment of this scheme is the same as for the previous one (it is therefore not dealt with in detail) although there are differences in the chemistry, since the two reactions have no common product. Widequist applied his method with success to the experimental data of Regna and Caldwell[53] on the conversion of 2-ketopolyhydroxy acid into furfural. The process can be regarded as a consecutive first order process where ascorbic acid is the intermediate.

McMillan[54] has also studied this variant of the reaction scheme and pointed out that it is easy to determine κ, *i.e.* the ratio of the two rate coefficients, when the concentrations of the two components of the reacting system are both known. The variation of $[A_2]$ and $[A_3]$ can be described by the following equations

$$\mathrm{d}[A_2]/\mathrm{d}t = -k_1[A_1][A_2] \tag{20}$$

$$\mathrm{d}[A_3]/\mathrm{d}t = k_1[A_1]([A_2]-\kappa[A_3]) \tag{21}$$

Introducing the following reduced concentrations

$$\beta = [A_2]/[A_2]_0, \gamma = [A_3]/[A_2]_0 \tag{22}$$

and dividing equation (21) by (20), we get

$$\frac{\mathrm{d}\gamma}{\mathrm{d}\beta} = (\kappa\gamma - \beta)/\beta \tag{23}$$

After integration, expression (23) has the following form

$$\kappa \ln \beta = \ln [\beta + (1-\kappa)\gamma] \tag{24}$$

or, after rearrangement

$$\gamma = \beta(1-\beta^{\kappa-1})/(\kappa-1) \tag{25}$$

To make the approximate determination of κ easier, the author gives a graph showing the connection between β and γ for different values of κ. However, if κ is less than 0.05, equation (24) can be written in the following explicit, approximate form

$$\kappa = [\ln (\beta+\gamma)]/[\ln \beta + \gamma/(\beta+\gamma)] \tag{26}$$

A scheme similar to the former, but containing less restrictions as to the nature of the products, was studied by Friedman and White[55]

$$A_1 + A_2 \xrightarrow{k_1} P_1$$

$$P_1 + A_2 \xrightarrow{k_2} P_2$$

$$P_2 + A_3 \xrightarrow{k_3} P_3 \text{ etc.}$$

Introducing dimensionless variables, they followed a method similar to that previously described. They solve the set of equations for $d[A_i]/d[A_1]$. The distribution of $[A_1] \ldots . [A_n]$ is derived from a system of linear equations. They applied their formulae to the determination of the rate coefficients of the synthesis of ethanolamine, where suitable experimental data were available.

A scheme agreeing in principle with the original one in this section, but having one more step

$$A_1 + A_2 \xrightarrow{k_1} A_3 + A_6$$

$$A_1 + A_3 \xrightarrow{k_2} A_4 + A_6$$

$$A_1 + A_4 \xrightarrow{k_3} A_5 + A_6$$

has been dealt with by Svirbely et al.[56-59]. They have used their analysis when calculating the rate coefficients for the hydrolysis of 1, 3, 5–tris(4–carbomethoxy-phenyl)benzene.

By introducing a new variable, λ, defined by equation (27)

$$\lambda = \int_0^t [A_1] dt \tag{27}$$

$$d\lambda = [A_1]dt \tag{28}$$

the set of differential rate equations can be written in the following form

$$\frac{d[A_1]}{d\lambda} = -k_1[A_2] - k_2[A_3] - k_3[A_5] \tag{29}$$

$$\frac{d[A_2]}{d\lambda} = -k_1[A_2] \tag{30}$$

$$\frac{d[A_3]}{d\lambda} = -k_1[A_2] - k_2[A_3] \tag{31}$$

$$\frac{d[A_5]}{d\lambda} = k_2[A_3] - k_3[A_5] \tag{32}$$

$$\frac{d[A_6]}{d\lambda} = k_3[A_5] \tag{33}$$

From equations (29–33) the concentrations being, at $t = 0$: $\lambda = [A_3] = [A_5] = 0$, $[A_1] = [A_1]_0$ and $[A_2] = [A_2]_0$, we obtain after integration

$$\frac{[A_1]}{[A_2]_0} = \left\{ 3 + \frac{2k_1}{k_2 - k_1} + \frac{k_1 k_2}{(k_2 - k_1)(k_3 - k_1)} \right\} e^{-k_1 \lambda}$$

$$- \left\{ \frac{2k_1}{k_2 - k_1} + \frac{k_1 k_2}{(k_2 - k_1)(k_3 - k_2)} \right\} e^{-k_2 \lambda} + \frac{k_1 k_2}{(k_3 - k_1)(k_3 - k_2)} e^{-k_3 \lambda} \tag{34}$$

assuming $[A_1]_0 = 3[A_2]_0$. Two methods are suggested for the determination of the rate coefficients[56-59].

Some reactions of mono-organosilanes take place according to this scheme. The calculation of the rate coefficients of such reactions has been carried out using the method above[60].

2.VI.5 $A_1 + A_2 \xrightarrow{k_1} A_3 + A_6$

$A_1 + A_3 \xrightarrow{k_2} A_4 + A_7$

$A_1 + A_2 \xrightarrow{k_3} A_5 + A_7$

$A_1 + A_5 \xrightarrow{k_4} A_4 + A_6$

The above system consisting of two pairs of competitive and consecutive reactions has been considered by Chao-Tung[61]. The differential rate equations are

$$d[A_1]/dx = -k[A_2]-k_2[A_3]-k_4[A_5] \tag{1}$$

$$d[A_2]/dx = -k[A_2] \tag{2}$$

$$d[A_3]/dx = k_1[A_2]-k_2[A_3] \tag{3}$$

$$d[A_4]/dx = k_2[A_3]+k_4[A_5] \tag{4}$$

$$d[A_5]/dx = k_3[A_2]-k_4[A_5] \tag{5}$$

where $k = k_1+k_3$. Assuming the initial conditions

$$[A_1] = [A_1]_0, [A_2] = [A_2]_0, [A_3]_0 = [A_4]_0 = [A_5]_0 = 0$$

equation (2) is readily integrated (a, b, c and d are constants) to give

$$[A_1]/[A_2]_0 = a+b\,e^{-kx}+c\,e^{-k_2x}+d\,e^{-k_4x} \tag{6}$$

and from equation (5) the following results can be obtained

$$[A_1]/[A_2]_0 = ([A_1]_0/[A_2]_0-2)+[2+k_1/(k_2-k)$$
$$+k_3/(k_4-k)]\,e^{-kx}-[k_1/(k_2-k)]\,e^{-k_2x}-[k_3/(k_4-k)]\,e^{-k_4x} \tag{7}$$

$$[A_2]/[A_2]_0 = e^{-kx} \tag{8}$$

$$[A_3]/[A_2]_0 = [k_1/(k_2-k)][e^{-kx}-e^{-k_2x}] \tag{9}$$

where

$$x = \int_0^t [A_1]\,dt.$$

The ratios of rate coefficients have been determined by a graphical method. From the above equations we find:

$$\ln([A_2]/[A_2]_0) = (k/k_2)\ln\{([A_2]/[A_2]_0)+([A_3]/[A_2]_0)(k-k_2)/k_1\} \tag{10}$$

$$\ln([A_2]/[A_2]_0) = (k/k_4)\ln\{([A_2]/[A_2]_0)+([A_5]/[A_2]_0)(k-k_4)/k_3\} \tag{11}$$

A plot according to equations (9) or (10) with the correct value of the parameter $(k-k_2)/k_1$ yields a straight line passing through the origin. Thus the corresponding complex rate coefficients can be calculated.

The practical application of this reaction scheme is illustrated by the successive and competitive saponification of non-symmetrical diesters.

$2.VI.6$ $A_1 + A_1 \xrightarrow{k_1} A_2$

$A_1 + A_2 \xrightarrow{k_2} A_3$

$A_1 + A_3 \xrightarrow{k_3} A_4$

$\vdots$

$A_1 + A_n \xrightarrow{k_n} A_{n+1}$

The differential rate equation set may be written[7] as

$$d[A_1]/dt = -k_1[A_1]^2 - k_2[A_1][A_2] - k_3[A_1][A_3] - \ldots - k_n[A_1][A_n]$$
$$d[A_2]/dt = k_1[A_1]^2 - k_2[A_1][A_2]$$
$$\vdots$$
$$d[A_n]/dt = k_{n-1}[A_1][A_{n-1}] - k_n[A_1][A_n]$$

(1)

Differential equation sets for schemes consisting of consecutive second order reactions in which each reaction has a common reactant as above can be reduced to a set of linear, first order equations by dividing by the concentration of the common member and introducing a new variable. Let us divide the equations (1) by $[A_1]$ and denote $[A_1]dt$ by dy. Then we obtain the following system of equations

$$d[A_1]/dy = -k_1[A_1] - k_2[A_2] - k_3[A_3] - \ldots - k_n[A_n]$$
$$d[A_2]/dy = k[A_1] - k_2[A_2]$$
$$\vdots$$
$$d[A_n]/dy = k_{n-1}[A_{n-1}] - k_n[A_n]$$

(2)

This system can easily be solved by approximation[11]. Assuming that, in the expression for $d[A_1]/dy$, all terms other than the first can be neglected, the set agrees with that for a system of first order consecutive reactions $A_1 \to A_2 \to A_3 \to \ldots$, treated previously (p. 17 *et seq.*). Such an approximation is justified in the initial stage of the reaction when the concentrations of the higher homologues $(A_2, A_3 \ldots)$ are much smaller than that of A_1, assuming that $[A_1]_0 \neq 0$, $[A_2]_0 = [A_3]_0 = \ldots = 0$ when $t = 0$. As to the solution for the change of $[A_2], [A_3] \ldots$ with time, we refer to what has been said in connection with the scheme $A_1 \to A_2 \to A_3 \to$ (p. 17 *et seq.*).

A solution for the variation of the concentration of A_1 can be derived in the following manner[11]. Let us suppose that $k_2 = k_3 = \ldots = k_n$.

By adding the set of equations (2), except the first one, we obtain

$$\sum_2^\infty \frac{d[A_n]}{dy} = k_1[A_1]$$

(3)

and, after differentiation of equation (1) with respect to y

$$\frac{d^2[A_1]}{dy^2} = -k_1 \frac{d[A_1]}{dy} - \sum_{2}^{\infty} k_n \frac{d[A_n]}{dy} \tag{4}$$

The sum term can be eliminated, and a second order linear differential equation obtained

$$\frac{d^2[A_1]}{dy^2} + k_1 \frac{d[A_1]}{dy} k_1 k_2[A_1] = 0 \tag{5}$$

the solution of which is

$$[A_1] = C_1 e^{-k_1 y/2} \sin\left\{\frac{k_1}{2}\left(\frac{4k_2}{k_1} - 1\right)^{\frac{1}{2}} y + C_2\right\} \tag{6}$$

(C_1 and C_2 are integration constants). C_1 and C_2 can be determined from the initial conditions. The above solution for $[A_1]$ includes the supposition that $4k_2 > k_1$, that is the rate of "propagation" is greater than the rate of "initiation".

From the equation defining y, viz.

$$y = \int_0^t [A_1] dt \tag{7}$$

it follows that $y = 0$ when $t = 0$ and y is essentially the area under the curve of $[A_1]$ versus t. Thus y can be determined graphically when the relation between the concentration of A_1 and time is known.

A kinetic treatment of the scheme has also been given by Natta and Mantica[62]. The elementary steps can be considered as the initiation, propagation and termination reactions of a polymerization reaction, for example; thus the stationary state approximation can be applied.

2.VI.7 $A_1 + A_1 \overset{k_1}{\rightarrow} A_2 + A_4$

$A_1 + A_2 \overset{k_2}{\rightarrow} A_3 + A_4$

The following simple method has been evolved by Hayman and Hayman[63] for the treatment of competitive, consecutive second order reactions. The scheme is a special case of the preceding one, so that a detailed treatment of the solution of the system of differential equations is not given here.

These rate equations can be written in the following form

$$-\frac{d[A_1]}{dt} = 2k_1[A_1]^2 + k_2[A_1][A_2] \tag{1}$$

$$\frac{d[A_2]}{dt} = k_1[A_1]^2 - k_2[A_1][A_2] \tag{2}$$

Introducing the following parameters

$$x = [A_1]/[A_1]_0, y = [A_2]/[A_1]_0, \tau = t[A_1]_0 \tag{3}$$

then

$$x - \frac{dx}{d\tau} = k_1(2x^2 + xy/\kappa) \tag{4}$$

or

$$\frac{dy}{d\tau} = k_1(x^2 - xy/\kappa) \tag{5}$$

where $\kappa = k_1/k_2$. On dividing equation (5) by (4), we obtain the following homogeneous differential equation

$$\frac{dy}{dx} = (y - \kappa x)/(y + 2\kappa x) \tag{6}$$

This may be solved in the usual manner.

Starting from (6) the ratio of the rate coefficients, κ, can be determined, and knowing this the numerical values of the individual coefficients are easily be calculated from equations (4) and (5).

The application of the method has been extended to the case when it is not possible to determine $[A_2]$. The results were compared with those obtained using the "time-ratio" method.

A system, similar to the above, is found in practice in the decomposition of NaOBr, where the two end-products are bromate and bromide, and the intermediate is bromite.

2.VI.8 $A_1 \xrightarrow{k_1} A_2 \xrightarrow[A_2]{k_2} A_4$
$\downarrow k_3$
A_3

This scheme, which has been considered by Weller and Berg[64] consists of consecutive and competitive steps. The following rate equations are derived

$$-d[A_1]/dt = k_1[A_1] \tag{1}$$

$$-d[A_2]/dt = -k_1[A_1] + k_2[A_2]^2 + k_3[A_2] \tag{2}$$

$$d[A_3]/dt = k_3[A_2] \tag{3}$$

$$d[A_4]/dt = \tfrac{1}{2}k_2[A_2]^2 \tag{4}$$

When the initial conditions are $[A_1] = [A_1]_0$ and $[A_2]_0 = [A_3]_0 = [A_4]_0 = 0$, then

$$[A_1] = [A_1]_0 \, e^{-k_1 t} \tag{5}$$

$$d[A_2]/dt = -k_3[A_2] - k_2[A_2]^2 + k_1[A_1]_0 \, e^{-k_1 t} \tag{6}$$

The differential equations are solved after repeated transformation. The transformation

$$[A_2](t) = \frac{1}{k_2} \, \dot{z}(t)/z(t) \tag{7}$$

leads to the following second-order linear differential equation

$$\ddot{z} + k_3 \dot{z} - k_1 k_2 [A_1]_0 \, e^{-k_1 t} \, z = 0 \tag{8}$$

Introducing the variable $-k_1 t = \tau$

$$\ddot{z} - (k_3/k_1)\dot{z} - (k_2/k_1)[A_1]_0 \, e^{\tau} \, z = 0 \tag{9}$$

The solution of this equation for whole number values of $1 - a$ is

$$[A_2](t) = -\frac{k_1\sqrt{b}}{k_2} \cdot \frac{N_{-a}(2\sqrt{b})J_{-a}(2\sqrt{b}\sigma) - J_{-a}(2\sqrt{b})N_{-a}(2\sqrt{b}\sigma)}{N_{-a}(2\sqrt{b})J_{1-a}(2\sqrt{b}\sigma) - J_{-a}(2\sqrt{b})N_{1-a}(2\sqrt{b}\sigma)} \cdot \sigma \tag{10}$$

and, for values of $1 - a$ which are not whole numbers, is

$$[A_2](t) = -\frac{k_1\sqrt{b}}{k_2} \cdot \frac{J_a(2\sqrt{b})J_{-a}(2\sqrt{b}\sigma) - J_{-a}(2\sqrt{b})J_a(2\sqrt{b}\sigma)}{J_a(2\sqrt{b})J_{1-a}(2\sqrt{b}\sigma) + J_{-a}(2\sqrt{b})J_{-(1-a)}(2\sqrt{b}\sigma)} \cdot \sigma \tag{11}$$

where J denotes the Bessel and N the Neumann functions, $a = 1 - (k_3/k_1)$, $b = k_2/k_1 \cdot [A_1]_0$ and $\sigma = \sqrt{e^{\tau}}$.

For calculating the value of k_2, the authors suggest the following approximative method. By selecting an appropriate time t after the concentration of A_2 has attained its maximum value, they suppose that the mechanism reduces to the following

$$A_2 \xrightarrow[\substack{k_2 \\ A_2}]{} A_4$$
$$\downarrow k_3$$
$$A_3$$

The rate equation for this is (putting $t-t_0 = t'$)

$$-d[A_2]/dt' = k_2[A_2]^2 + k_3[A_2]$$ (12)

For the initial conditions $t-t_0 = t' = 0$, $[A_2] = [A_2]_0$, the solution can be expressed as

$$t' = (1/k_3) \left(\ln \frac{k_2[A_2]_0}{k_2[A_2]_0 + k_3} - \ln \frac{k_2[A_2]}{k_2[A_2] + k_3} \right)$$ (13)

The expression for the half-life, in this case $t'_{\frac{1}{2}}$, is

$$t'_{\frac{1}{2}} = \frac{1}{k_3} \left(\ln \frac{k_2[A_2]_0}{k_2[A_2]_0 + k_3} - \ln \frac{k_2[A_2]_0}{k_2[A_2]_0 + 2k_3} \right)$$ (14)

and this is suitable for calculating the rate coefficients.

Weller and Berg[64] have considered the reduction of ketones and quinones by hydrogen catalyzed by Pd in terms of this scheme. Fig. 6 shows the variation of the concentration of the intermediate A_2 with time.

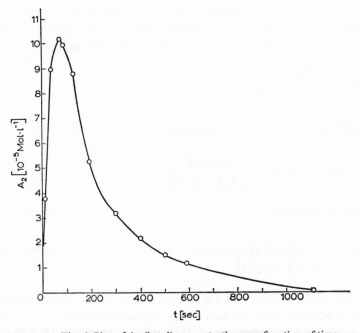

Fig. 6. Plot of A_2 (ketyl) concentration as a function of time.

2.VII COMPETITIVE-CONSECUTIVE REACTIONS WITH REVERSIBLE REACTION STEP

2.VII.1 $A_1 + A_2 \underset{k_2}{\overset{k_1}{\rightleftarrows}} A_3$

$A_1 + A_3 \overset{k_3}{\rightarrow} A_4$

For this system the reaction rates are given by the following kinetic equations[10]

$$\frac{d[A_1]}{dt} = -k_1[A_1][A_2] - k_3[A_1][A_3] + k_2[A_3] \tag{1}$$

$$\frac{d[A_2]}{dt} = -k_1[A_1][A_2] + k_2[A_3] \tag{2}$$

$$\frac{d[A_3]}{dt} = -k_2[A_3] - k_3[A_1][A_3] + k_1[A_1][A_2] \tag{3}$$

$$\frac{d[A_4]}{dt} = k_3[A_1][A_3] \tag{4}$$

Let us suppose that A_3 is, in fact, the initial reactant present, *i.e.* that $[A_1]_0 = [A_2]_0 = [A_4]_0 = 0$ at $t = 0$. Furthermore, it is assumed that after a short induction period the concentration of substance A_1 reaches a stationary value. Considering the stationary state, when $d[A_1]/dt = 0$, equation (1) gives

$$[A_1] = \frac{k_2[A_3]}{k_1[A_2] + k_3[A_3]} \tag{5}$$

Substituting equation (5) into (4) we get equation (6)

$$\frac{d[A_4]}{dt} = k_3[A_1][A_3] = \frac{k_2 k_3[A_3]^2}{k_1[A_2] + k_3[A_3]} \tag{6}$$

In a similar way we find that

$$-\frac{d[A_3]}{dt} = \frac{2k_2 k_3[A_3]^2}{k_1[A_2] + k_3[A_3]} \tag{7}$$

From equation (7) it follows that the rate of change of the concentration of A_3 depends not only on $[A_3]$ but also on the concentration of A_2. Equation (7) becomes simpler in two extreme cases. Thus, when $k_3[A_3] \ll k_1[A_2]$, the rate of consumption of A_3 is proportional to $[A_3]^2$; in this case we are dealing with a reaction of the second order with respect to A_3. On the other hand, when $k_3[A_3] \gg k_1[A_2]$, equation (7) takes the form

$$-\frac{d[A_3]}{dt} = 2k_2[A_3]$$

(8)

In this case the decrease of $[A_3]$ follows first-order kinetics. The order of a reaction of this type with respect to A_3 may vary in the range from 1 to 2.

For examination and application of the scheme see refs. 65 and 66. The exchange reaction of *tert*-butyl bromide (or iodide) with ICl has this mechanism[67].

2.VII.2 $A_1 + A_2 \overset{k_1}{\underset{k_2}{\rightleftarrows}} A_3$

$$A_1 + A_4 \overset{k_3}{\rightarrow} A_5$$

$$A_3 + A_4 \overset{k_4}{\rightarrow} A_6$$

The kinetics of this system has been solved by Overbeek and Konigsberger[68], on the assumption that $[A_3]$ reaches a stationary value. The rate of disappearance of A_1 is given by

$$-\frac{d[A_1]}{dt} = k_1[A_1][A_2] + k_3[A_1][A_4] - k_2[A_3]$$

(1)

For $[A_3]$ we get a similar rate expression

$$\frac{d[A_3]}{dt} = k_1[A_1][A_2] - k_2[A_3] - k_4[A_3][A_4]$$

(2)

Under stationary conditions, $(d[A_3]/dt = 0)$, we obtain from equation (2)

$$[A_3] = \frac{k_1[A_1][A_2]}{k_2 + k_4[A_4]}$$

(3)

Substituting for $[A_3]$ into equation (1) and rearranging we obtain equation (4)

$$V = -\frac{d[A_1]}{dt} = k_1' \cdot \frac{k_4}{k_2} \cdot [A_1][A_4] \frac{1}{1 + (k_4/k_2)[A_4]} + k_3[A_1][A_4]$$

(4)

where $k_1' = k_1[A_2]$.

Taking the concentration of A_4 as constant, then $k' = V/[A_1]$ (which can be determined experimentally) and

$$k' = [A_4] \left\{ k'_1 \cdot \frac{k_4}{k_2} \cdot \frac{1}{1+(k_4/k_2)[A_4]} + k_3 \right\} \tag{5}$$

If the value of $[A_4]$ is sufficiently small, or on the contrary sufficiently large, equation (5) takes much simpler forms

$$\lim_{[A_4] \to 0} k' = [A_4] \left\{ k'_1 \cdot \frac{k_4}{k_2} + k_3 \right\} \tag{6a}$$

$$\lim_{[A_4] \to \infty} k' = k'_1 + k_3[A_4] \tag{6b}$$

Rate coefficients k'_1 and k_3 may be determined graphically, starting from the variation of $k'/[A_4]$ with $1/[A_4]$, equation (6b); at the same time the ratio k_4/k_2 is obtained by plotting $k'/[A_4]$ *versus* $[A_4]$ (equation (6a)).

3. Third order reactions

3.1 REVERSIBLE REACTIONS

3.1.1 $A_1 + A_2 + A_3 \overset{k_1}{\underset{k_2}{\rightleftarrows}} A_4 + A_5 + A_6$

The following special case of reversible third order reactions has been dealt with[75,76]

$$2A_1 + A_2 \overset{k_1}{\underset{k_2}{\rightleftarrows}} 2A_3$$

Assuming that the initial concentrations of reactants are $[A_1] = 2[A_2]_0$, $[A_2] = [A_2]_0$, $[A_3]_0 = 0$ at $t = 0$, and further that the concentrations of A_1, A_2 and A_3 at time t are $2([A_2]_0 - x)$, $[A_2]_0 - x$ and $2x$, respectively, the rate equation can be written as

$$\frac{dx}{dt} = 4\{k_1([A_2]_0 - x)^2([A_2]_0 - x) - k_2 x^2\} \tag{1}$$

It is supposed[75,76] that the equilibrium constant is known, so that k_2 can be eliminated and from the rate equation obtained, k_1 can be determined from the differential quotient.

4. *n*th order reactions

4.I PARALLEL REACTIONS

4.I.1 $n_1 A_1 + n_2 A_2 \xrightarrow{k_1} A_3 + \ldots$

             $n_1 A_1 + n_2 A_2 \xrightarrow{k_2} A_4 + \ldots$

             $n_1 A_1 + n_2 A_2 \xrightarrow{k_3} A_5 + \ldots$

Let us suppose that, for all the three simultaneous reactions, the order with respect to substance A_1 is the same (n_1), and similarly the order with respect to A_2 (n_2). Then the rate of consumption of A_1 (or A_2) can be written in the following simple form[2]

$$\frac{1}{n_2} \cdot \frac{d[A_2]}{dt} = \frac{1}{n_1} \cdot \frac{d[A_1]}{dt} = -(k_1 + k_2 + k_3)[A_1]^{n_1}[A_2]^{n_2} \tag{1}$$

In such a case the concentrations of the products are very simply related to one another, the ratios being independent of the magnitude of n_1 and n_2. The differential rate equation set is

$$d[A_3]/dt = k_1[A_1]^{n_1}[A_2]^{n_2} \tag{2}$$

$$d[A_4]/dt = k_2[A_1]^{n_1}[A_2]^{n_2} \tag{3}$$

$$d[A_5]/dt = k_3[A_1]^{n_1}[A_2]^{n_2} \tag{4}$$

Therefore, if initially $[A_3]_0 = [A_4]_0 = [A_5]_0 = 0$

$$d[A_4]/d[A_3] = k_2/k_1 \text{ or } [A_4]/[A_3] = k_2/k_1 \tag{5}$$

$$d[A_5]/d[A_3] = k_3/k_1 \text{ or } [A_5]/[A_3] = k_3/k_1 \tag{6}$$

Thus the ratio of the amounts of the products is constant during the reaction and independent of its order.

When the above conditions are fulfilled, the ratios of the rate coefficients can be determined by product analysis without knowing the order of the reaction. From the temperature dependences of the ratios of rate coefficients, the difference in the activation energies and the ratio of the pre-exponential factors can be calculated. The method can be applied to the determination of relative rate coefficients of the formation of isomers.

4.II CONSECUTIVE REACTIONS

4.II.1 $a_1 A_1 \xrightarrow{k_1} A_2$ (*m*th order)

$a_2 A_2 \xrightarrow{k_2} A_3$ (*n*th order)

The following differential rate equations can be written

$$\frac{-d[A_1]}{dt} = k_1[A_1]^m \tag{1}$$

$$\frac{d[A_2]}{dt} = \frac{1}{a_1} k_1[A_1]^m - k_2[A_2]^n \tag{2}$$

$$\frac{d[A_3]}{dt} = \frac{1}{a_2} k_2[A_2]^n \tag{3}$$

Introducing the dimensionless variables[69] $[A_1]^*$, $[A_2]^*$ and $[A_3]^*$, respectively $[A_1]/[A_1]_0$, $a_1[A_2]/[A_1]_0$ and $a_1 a_2[A_3]/[A_1]_0$;

$$\tau = k_1[A_1]_0^{m-1} t \qquad \text{(dimensionless time);} \tag{4}$$

$$S = a_1^{n-1}[A_1]_0^{m-n} \cdot \frac{k_1}{k_2} \qquad \text{(selectivity parameter)} \tag{5}$$

we get the following equations

$$\frac{d[A_1]^*}{d\tau} = -[A_1]^{*m} \tag{6}$$

$$\frac{d[A_2]^*}{d\tau} = [A_1]^{*m} - \frac{1}{S}[A_2]^{*n} \tag{7}$$

$$\frac{d[A_3]^*}{d\tau} = \frac{1}{S}[A_2]^{*n} \tag{8}$$

Chermin and Van Krevelen[69] point out that, according to equations (6), (7) and (8), the reaction becomes more selective towards removal of A_2 as S increases. The variation of the selectivity parameter with the activity and the concentration was given. These considerations are important for the operation of large scale equipment.

ACKNOWLEDGEMENT

The author is sincerely grateful to his co-workers, Dr. T. Bérces, for collecting and systematising the literature, and Dr. P. Huhn for supervising the mathematical treatment.

REFERENCES

1 K. MICKA, *Chem. Listy*, 48 (1954) 355.
2 A. A. FROST AND R. G. PEARSON, *Kinetics and Mechanism*, Wiley, New York, 1961.
3 W. ESSON, *Phil. Trans. Roy. Soc. London*, 156 (1866) 220.
4 E. KLEIN AND T. F. FAGLEY, *J. Phys. Chem.*, 58 (1954) 447.
5 F. THIERSCH, *Z. Physik. Chem.*, 111 (1924) 175.
6 A. E. R. WESTMAN AND D. B. DE LURI, *Can. J. Chem.*, 34 (1956) 1134.
7 S. W. BENSON, *J. Chem. Phys.*, 20 (1952) 1605.
8 A. RAKOWSKI, *Z. Physik. Chem.*, 57 (1906) 321.
9 C. G. SWAIN, *J. Am. Chem. Soc.*, 66 (1944) 1696.
10 R. ZAHRADNIK, *Uspekhi Khim.*, 10 (1961) 1272.
11 S. W. BENSON, *The Foundations of Chemical Kinetics*, McGraw-Hill, New York, 1960.
12 B. V. EROFEEV, *Zhur. Fiz. Khim.*, 24 (1950) 721.
13 H. R. SIEVERT, P. N. TENNEY AND T. XERMEULEN, *U.S. Atomic Energy Comm. UCRL* 10575, 1962.
14 D. H. MCDANIEL AND C. R. SMOOT, *J. Phys. Chem.*, 60 (1956) 966.
15 T. M. LOWRY AND W. T. JOHN, *J. Chem. Soc.*, 97 (1910) 2634.
16 B. J. ZWOLINSKI AND H. EYRING, *J. Am. Chem. Soc.*, 69 (1947) 2702.
17 F. HALLA, *Monatsh.*, 33 (1956) 1448.
18 J. M. LOS, L. B. SIMPSON AND K. WIESNER, *J. Am. Chem. Soc.*, 78 (1956) 1564.
19 A. SKRABAL, *Z. Physik. Chem.*, B3 (1929) 247.
20 A. SKRABAL, *Z. Elektrochem.*, 42 (1936) 228.
21 A. SKRABAL, *Z. Elektrochem.*, 43 (1937) 309.
22 G. N. VRIENS, *Ind. Eng. Chem.*, 46 (1954) 669.
23 R. WEGSCHEIDER, *Z. Physik. Chem.*, 39 (1902) 257.
24 R. A. ALBERTY AND W. G. MILLER, *J. Chem. Phys.*, 26 (1957) 1231.
25 R. A. ALBERTY, W. G. MILLER AND H. F. FISCHER, *J. Am. Chem. Soc.*, 79 (1957) 3973.
26 A. A. FROST AND M. TAMRES, *J. Chem. Phys.*, 15 (1947) 383.
27 G. E. HAY AND H. J. MCDONALD, *J. Phys. Chem.*, 45 (1941) 1177.
28 A. SKRABAL, *Monatsh.*, 81 (1950) 239.
29 D. F. ABELL, N. A. BONNER AND W. GOISHI, *J. Chem. Phys.*, 27 (1957) 658.
30 D. KALLÓ AND G. SCHAY, *Acta Chim. Acad. Sci. Hung.*, 39 (1963) 183.
31 J. A. BERSON AND R. D. REYNOLDS, *J. Am. Chem. Soc.*, 77 (1955) 4434.
32 F. A. MATSEN AND J. L. FRANKLIN, *J. Am. Chem. Soc.*, 72 (1950) 3337.
33 R. V. LEMIEUX, W. P. SHYLUK AND G. HUBER, *Can. J. Chem.*, 33 (1955) 148.
34 M. BODENSTEIN, *Z. Physik. Chem.*, 13 (1894) 56; 22 (1897) 1.
35 M. C. TOBIN, *J. Phys. Chem.*, 59 (1955) 799.
36 W. G. YOUNG AND L. J. ANDREWS, *J. Am. Chem. Soc.*, 66 (1944) 421.
37 S. WIDEQUIST, *Arkiv Kemi*, 2 (1950) 303.
38 W. H. PASFIELD AND R. WARING, *J. Am. Chem. Soc.*, 78 (1956) 2698.
39 K. B. YERRICK AND M. E. RUSSELL, *J. Phys. Chem.*, 68 (1964) 3752.
40 T. S. LEE, *Anal. Chem.*, 21 (1949) 537.
41 T. S. LEE AND I. M. KOLTHOFF, *Ann. N.Y. Acad. Sci.*, 53 (1951) 1093.
42 L. R. DARBEE, F. E. JENKINS AND G. M. HARRIS, *J. Chem. Phys.*, 25 (1956) 605.
43 M. TALAT-ERBEN, *J. Chem. Phys.*, 26 (1957) 75.

44 JEN-YUAN CHIEN, *J. Am. Chem. Soc.*, 70 (1948) 2256.
45 A. D. STEPUKHOVICH AND I. F. BAKHAREVA, *Zhur. Fiz. Khim.*, 28 (1954) 970.
46 I. M. KOLESNIKOV, *Zhur. Fiz. Khim.*, 34 (1960) 1069.
47 N. V. RIGGS, *Australian J. Chem.*, 11 (1958) 86.
48 A. A. FROST AND W. C. SCHWEMER, *J. Am. Chem. Soc.*, 73 (1951) 4541.
49 A. A. FROST AND W. C. SCHWEMER, *J. Am. Chem. Soc.*, 74 (1952) 1268.
50 V. O. REIKHSFELD AND V. A. PROKHOROVA, *Kinetika i Kataliz*, 4 (1963) 483.
51 S. WIDEQUIST, *Arkiv Kemi*, 8 (1956) 545.
52 S. WIDEQUIST, *Acta Chem. Scand.*, 16 (1962) 1119.
53 P. P. REGNA AND B. P. CALDWELL, *J. Am. Chem. Soc.*, 66 (1944) 246.
54 W. G. MCMILLAN, *J. Am. Chem. Soc.*, 79 (1957) 4848.
55 H. FRIEDMAN AND R. R. WHITE, *Am. Inst. Chem. Engrs. Journal*, 8 (1962) 581.
56 W. J. SVIRBELY, *J. Phys. Chem.*, 62 (1958) 380.
57 W. J. SVIRBELY, *J. Am. Chem. Soc.*, 81 (1959) 255.
58 W. J. SVIRBELY AND H. E. WEISBERG, *J. Am. Chem. Soc.*, 81 (1959) 257.
59 W. J. SVIRBELY AND J. A. BLAUER, *J. Am. Chem. Soc.*, 83 (1961) 4115.
60 V. O. REIKHSFELD, V. A. PROKHOROVA AND V. A. PUNINA, *Kinetika i Kataliz*, 6 (1965) 171.
61 CHAO-TUNG, *J. Phys. Chem.*, 62 (1958) 639.
62 G. NATTA AND E. MANTICA, *J. Am. Chem. Soc.*, 74 (1952) 3152.
63 H. J. G. HAYMAN AND D. P. HAYMAN, *Bull. Res. Council Israel, Sec. A*, 8 (1959) 95.
64 K. WELLER AND H. BERG, *Ber. Bunsenges.*, 68 (1964) 33.
65 E. I. ADIROVICH, *Doklady Akad. Nauk SSSR*, 61 (1948) 467.
66 S. C. WALEY AND J. WATSON, *Proc. Roy. Soc. (London), Ser. A*, 199 (1949) 499.
67 R. M. KEEFER AND L. J. ANDREWS, *J. Am. Chem. Soc.*, 76 (1954) 253.
68 J. T. G. OVERBEEK AND V. V. KONINGSBERGER, *Proc. Koninkl. Ned. Akad. Wetenschap.*, B57 (1954) 311.
69 H. A. G. CHERMIN AND D. V. VAN KREVELEN, *Chem. Eng. Sci.*, XIV (1961) 58.
70 J. C. MORROW, *J. Chem. Phys.*, 23 (1955) 2452.
71 S. V. DOBROVOLSKII AND V. YA. POLOTNYUK, *Zhur. Fiz. Khim.*, 32 (1958) 2792.
72 O. Z. KNOBLAUCH, *Phys. Chem.*, 26 (1898) 96.
73 A. SKRABAL, *Monatsh.*, 38 (1917) 29, 159.
74 J. MEYER, *Z. Physik. Chem.*, 67 (1909) 272.
75 E. A. MOELWYN-HUGHES, *Physical Chemistry*, Cambridge University Press, Cambridge, 1940.
76 F. E. E. GERMANN, *J. Phys. Chem.*, 32 (1928) 1748.
77 F. NAGY, *The Kinetics of Contact Catalytic Reactions* in Z. G. SZABÓ (Ed.), *Contact Catalysis*, Publishing House of the Hungarian Academy of Sciences, Budapest, 1966, p. 387. (*In Hungarian*).

Chapter 2

Chain Reactions

V. N. KONDRATIEV

1. Introduction

Chain reactions were discovered in 1913 by Bodenstein[27] who found that the quantum yields of some photochemical reactions were very high. Since according to the Stark–Einstein photochemical law[52,192] no more than one or two molecules would react per every quantum absorbed, Bodenstein came to the conclusion that the primary photochemical step was followed by a long *chain* of chemical conversions involving many molecules of primary compounds, and this would give rise to the high quantum yield.

Of decisive importance in the development of the chain theory was the mechanism proposed by Nernst[149] for the reaction between chlorine and hydrogen that was discovered and studied later in detail by Bodenstein. According to the Nernst mechanism which was generally accepted, the *active centres* in this process are free chlorine and hydrogen atoms, *i.e. valence-unsaturated* particles formed in the course of the reaction.

Later on Christiansen[42], Herzfeld[79], and Polanyi[157] applied the essential features of the Nernst mechanism for the reaction of chlorine with hydrogen to the thermal reaction of bromine with hydrogen that was investigated earlier by Bodenstein and Lind[29].

The next important stage in the development of chain theory was the discovery of *branched chain reactions* by Semenov and Hinshelwood [†]. The theory of branched chain reactions was clearly formulated in a paper by Semenov[173], and in his monograph "Chain Reactions"[174]. The theory implies that, on the average, one step of a branched chain reaction involves the formation of *more than one* new active centre per every disappearing centre. In other words, the rate of regeneration of active centres in the course of a branched chain reaction can *exceed* that of their consumption.

The scheme given in Fig. 1 may be considered as the most simple formal one for a chain reaction. In accordance with this scheme an active centre (A) is formed as a result of thermal or external (for example, photochemical) activation (reaction route 0), *initiation*. Entering into reaction 1, the active centre yields a reaction product (C) and a certain number ε of active centres (A), *propagation*. Thus, we

---

[†] For references see Semenov[174].

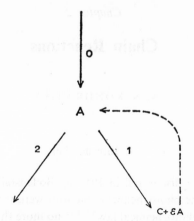

Fig. 1. Simplified formal scheme of a chain reaction.

have *two* sources of active centres here: primary, thermal or external activation (route 0), and the reaction as such (route 1) resulting in regeneration of active centres. Disappearance of the active centres to give inert species (*termination*) occurs by route 2.

Denoting by v_1 the *specific rate* or *frequency* of reaction (1) and by v_2 that of reaction (2), we may express the ratio of the rate of regeneration of active centres to that of their consumption as

$$\frac{\varepsilon v_1}{v_1 + v_2} = \varepsilon w_1,$$

where $w_1 = v_1/(v_1 + v_2)$ is the probability for reaction (1). Then the probability for the disappearance of an active centre by route (2) will be $w_2 = v_2/(v_1 + v_2)$, and $w_1 + w_2 = 1$. In accordance with the above, for a branched chain reaction we have $\varepsilon v_1 > (v_1 + v_2)$, *i.e.* $\varepsilon w_1 > 1$, and for an unbranched chain reaction $\varepsilon v_1 \leqslant (v_1 + v_2)$, *i.e.* $\varepsilon w_1 \leqslant 1$.

Whereas an unbranched chain reaction may be conditionally plotted as a saw-like line (Fig. 2a), every straight section denoting one *chain cycle*, and the number of such sections (chain cycles) representing the *chain length v*, a branched chain reaction, $2 > \varepsilon > 1$, would look like a branched saw-like line (Fig. 2b), as branching occurs from time to time at individual chain propagation steps, *i.e.* two chains start propagating instead of one. In a limiting case of $\varepsilon \geqslant 2$ branching will occur at *every* chain step. Such chains are called *continuously branched* (Fig. 2c).

According to Semenov there are two types of branched chain reactions: branched reactions as such and degenerate branched reactions. The first involve branching as a result of interaction between active centres and molecules, the branching step being a definite *component* of the chain propagation. In reactions of the

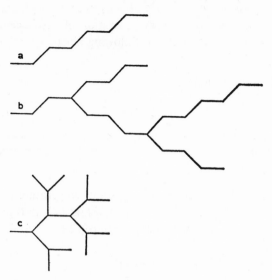

Fig. 2. Scheme of (a) an unbranched, (b) a branched, and (c) a continuously branched chain reaction.

second type the chain is propagated in the ordinary way and is not accompanied by branching in the usually accepted sense[174]. However, the reaction yields relatively stable *intermediate* products that accumulate in the gas and then enter into independent slow reactions to form end products. From time to time active centres are formed at the expense of the energy of this reaction. They are capable of starting additional reaction chains. Semenov defines these secondary chains as those due to *degenerate* branching, sometimes called *delayed branching*.

It has become evident lately that the importance of the *energy factor* in branched chain reactions pointed out by Semenov some thirty years ago[174] did not receive due attention in the development of the chain theory. However, the number of branched chain reactions of the kind in which branching is accounted for by *hot* (energy rich) particles is constantly increasing. The energy factor seems to be of especial importance in fast reactions occurring at high temperatures.

2. Formal kinetics of chain reactions

The following elementary steps representing the mechanism of a reaction may be written in accordance with the scheme in Fig. 1

$$
\begin{aligned}
\text{initiation} &\rightarrow A, & 0 \\
A \quad &\rightarrow C + \varepsilon A, & 1 \\
A \quad &\rightarrow \text{chain termination,} & 2
\end{aligned}
$$

Denoting by a and c the concentrations of active centres and reaction products, respectively, by W_0 the rate of primary generation of active centres and by W the rate of the reaction, we may write the following kinetic equations

$$\frac{da}{dt} = W_0 - [v_1(1-\varepsilon) + v_2]a \tag{2.1}$$

and

$$W = \frac{dc}{dt} = v_1 a \tag{2.2}$$

Assuming that the reaction occurs at a constant temperature (isothermal reaction) and that the values of W_0, v_1 and v_2 are constant (which is justifiable for the initial reaction period), we obtain by integration of (2.1)

$$a = a_0 \exp(-t/\tau) + W_0 \tau[1 - \exp(-t/\tau)] \tag{2.3}$$

where

$$\tau = \frac{1}{v_1(1-\varepsilon) + v_2} \tag{2.4}$$

and a_0 is the concentration of active centres at $t = 0$. For simplification we shall henceforth suppose that $a_0 = 0$, and eqn. (2.3) will be rewritten as

$$a = W_0 \tau[1 - \exp(-t/\tau)] \tag{2.5}$$

Substituting (2.5) into (2.2) we obtain the following expression for the reaction rate

$$W = \frac{dc}{dt} = v_1 W_0 \tau[1 - \exp(-t/\tau)] \tag{2.6}$$

It follows from (2.6) and (2.5) that close to $t = 0$ the reaction rate and the concentration of active molecules will increase with time following the law of *proportionality*

$$W = v_1 W_0 t \quad \text{and} \quad a = W_0 t \tag{2.7}$$

However, further variations in W and a will depend on whether the reaction is unbranched or branched. Let us consider first an unbranched chain reaction.

2.1 UNBRANCHED CHAIN REACTIONS

2.1.1 Development of an unbranched chain reaction in time

According to Semenov[174] an unbranched chain reaction is that for which the probability of branching defined as $\delta = \varepsilon - 1$ is zero, *i.e.* $\varepsilon = 1$. Consequently, in this case (2.4) will be rewritten as

$$\tau = \frac{1}{v_2} \tag{2.8}$$

We have seen above (eqns. (2.7)) that close to $t = 0$ the reaction rate (W) and the concentration of active particles (a) increase proportionally with time; it follows from (2.1) and (2.2) that, for unbranched chain reactions, the maximum acceleration corresponds to $t = 0$. After a certain time the rate of increase of W and a becomes lower and at a sufficiently great time ($t \gg \tau$) W and a attain maximum values

$$W_{max} = W_0 \frac{v_1}{v_2} \tag{2.9}$$

and

$$a_{max} = \frac{W_0}{v_2} \tag{2.10}$$

However, W_{max} and a_{max} do not remain constant, as W_0 and v_1 continuously decrease during the course of the reaction due to consumption of the initial substances. Thus, after having attained the maximum, the reaction rate will decrease as will the concentration of active centres, tending to zero as $t \to \infty$. The variation in the rate of an unbranched chain reaction with time is shown in Fig. 3 [†].

Thus, the rate of an unbranched chain reaction increases obeying law (2.6) until it reaches a maximum and is represented by expression (2.9), after which the consumption of initial products causes its slow decrease with time.

It will be noted that, as the chain length v is by definition equal to the ratio $W : W_0$, we have for an unbranched chain reaction

$$v = \frac{v_1}{v_2} \tag{2.11}$$

[†] The curve for Fig. 3 is plotted for a case where the rate of chain initiation is constant, and the factor $k_1 W_0 / v_a^2 = 0.1$ (k_1 is the rate coefficient of reaction 1).

References pp. 183–188

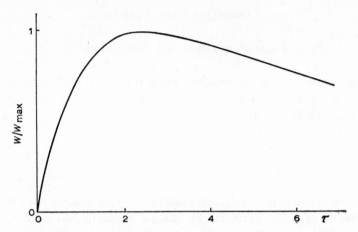

Fig. 3. Development of a chain reaction with time. τ, dimensionless time $= v_2 t$.

The constant, or more explicitly the slowly decreasing concentration of active centres, expressed by (2.10), may be obtained directly from eqn. (2.1) on the assumption that $da/dt = 0$ (provided $\varepsilon = 1$). This is called a *steady-state concentration*, and the equality

$$\frac{da}{dt} = 0 \qquad\qquad (2.12)$$

the *steady-state* condition.

2.1.2 Method of steady-state concentrations

The method of steady-state concentrations plays an important part in chemical kinetics and, in particular, in the kinetics of chain reactions. This method, proposed for the first time by Bodenstein[27], was developed upon by Semenov[176] and is known at present as the Bodenstein–Semenov method. In order to make it clear, let us consider a reaction involving the conversion of a certain substance X to an *intermediate* Y, and of the latter to Z.

The kinetic equations for X, Y, and Z will be

$$-\frac{dx}{dt} = v_1 x$$

$$\frac{dy}{dt} = v_1 x - v_2 y \qquad\qquad (2.13)$$

$$\frac{dz}{dt} = v_2 y$$

where x, y, and z are the concentrations of the species X, Y, and Z at time t, and v_1 and v_2 are the frequencies of the respective elementary reactions[†].

From these equations we obtain

$$x = \exp(-v_1 t)$$

$$y = \frac{v_1}{v_2-v_1}[\exp(-v_1 t) - \exp(v_2 t)] \tag{2.14}$$

$$z = 1 - \frac{v_2}{v_2-v_1}\exp(-v_1 t) + \frac{v_1}{v_2-v_1}\exp(-v_2 t)$$

It is assumed that when $t = 0$, $x = 1$, $y = 0$, and $z = 0$.

The method of steady-state concentrations, as applied to the two-step reaction considered, implies the assumption that the rate of change of y is zero, *i.e.*

$$\frac{dy}{dt} = v_1 x - v_2 y = 0 \tag{2.15}$$

This assumption is based on the fact that the second term in the right hand side of eqn. (2.15) which is zero at the start of the reaction, increases with time, while the first term decreases. As a result of this, after a certain time t' we have $dy/dt = 0$. From this moment the concentration of Y decreases in parallel with that of the initial substance X, the value of dy/dt being kept *automatically* close to zero. Thus, at $t > t'$, instead of two independent differential equations (2.13)[††] we will have one. The second equation, that for concentration of the intermediate, *viz.* $dy/dt = v_1 x - v_2 y$, will be replaced by the algebraic equation (2.15). This simplifies calculation of x, y and z as functions of time, and is particularly important when the frequencies $v_1, v_2 \ldots$ are not constant but depend upon concentrations of certain substances, which vary during the course of the reaction.

In the case considered, use of condition (2.15) yields the following approximate expressions for the concentrations of X, Y and Z

$$x = \exp(-v_1 t)$$

$$y = \frac{v_1}{v_2}\exp(-v_1 t) \tag{2.16}$$

$$z = 1 - \exp(-v_1 t)$$

[†] When these reactions are first order, v_1 and v_2 will represent the *rate coefficients*.
[††] Since $dx/dt + dy/dt + dz/dt = 0$, one of three equations (2.13) is not independent.

Comparison with the precise expressions (2.14) shows that the approximate Y and Z concentrations become closer to those given by (2.14) as v_2 becomes increasingly greater than v_1.

It was mentioned above (p. 86) that the concentration of Y satisfying condition (2.15) is called a *steady-state concentration*, and (2.15) the *steady-state condition*. As the latter may be fulfilled starting from time t' only (*i.e.* for $t \geqslant t'$), two stages of the reaction can be distinguished, that before the attainment of steady-state concentrations, and that during which the concentrations of intermediates remain steady. During the first, the "pre-steady-state" stage, the concentration of an intermediate grows from zero to a certain maximum value governed by the steady-state condition, after which, during the second stage (stationary reaction), the intermediate concentration will gradually drop in parallel with that of the initial substance.

The reaction discussed above belongs to the class of less complicated non-chain reactions but all that was said about it may pertain to chain reactions as well. However, it will be noted that two or more active intermediates are formed in the course of all chemical chain reactions. For example, with two intermediates Y_1 and Y_2 one may consider the steady-state concentration of either substance, attained in times t_1' and t_2', respectively. In this case there will be a "pre-steady-state" and a "steady-state" stage of the reaction, just as for one intermediate. The duration of the first stage will evidently be governed by the highest t'.

On the other hand the reaction may be such that the concentration of only one intermediate will be steady. This may occur in branched chain reactions owing to their high rates. Such reactions were discussed for the first time by Semenov. The method used for theoretical treatment of their kinetics was called the *method of partial steady-state concentrations*. Some examples will be given in Section 3. Finally, at high reaction rates there might appear to be conditions such that the concentrations of *all* intermediates would be far from steady. This was found, for instance, for a reaction between bromine and hydrogen and for ethane thermal decomposition[69a].

2.2 BRANCHED CHAIN REACTIONS

2.2.1 Autocatalytic acceleration of a reaction

In accordance with the definition given above, the condition $\varepsilon w_1 > 1$ will be characteristic of a branched chain reaction within the scope of the scheme given in Fig. 1. Then, the kinetic equation for the concentration (a) of active centres in a branched chain reaction will be written as

$$\frac{da}{dt} = W_0 + (\varepsilon w_1 - 1)(v_1 + v_2)a = W_0 + \varphi a \tag{2.17}$$

where

$$\varphi = (v_1 + v_2)(\varepsilon w_1 - 1) \tag{2.18}$$

Integrating eqn. (2.17) we obtain (taking $a_0 = 0$)

$$a = \frac{W_0}{\varphi} [\exp(\varphi t) - 1] \tag{2.19}$$

and

$$W = \frac{W_0 v_1}{\varphi} [\exp(\varphi t) - 1] \tag{2.20}$$

It follows from (2.20) that the rate of a branched chain reaction close to $t = 0$ is proportional to t

$$W = W_0 v_1 t$$

(as with an unbranched chain reaction). However, at high t, when $\exp(\varphi t)$ becomes considerably greater than unity, the reaction rate will obey the exponential law

$$W = \frac{W_0 v_1}{\varphi} \exp(\varphi t) \tag{2.21}$$

This law (*Semenov law*) established theoretically by Semenov is obeyed experimentally by diverse chemical reactions. Exponential *autocatalytic acceleration of the reaction*, following from this law, is a characteristic feature of branched chain reactions.

Variations in rates of unbranched and branched chain reactions with time are shown in Fig. 4. It was mentioned above that close to $t = 0$ the rates of both types of reactions increase with time obeying the law of direct proportionality. However, acceleration of an unbranched chain reaction will gradually *slow down* and when the reaction rate becomes a maximum, it will be zero, whereas a branched chain reaction will *accelerate continuously*.

After a certain (generally speaking short) time the rate of a branched chain reaction increasing exponentially (2.21) becomes so high that the reaction terminates in explosion. However, this will be the case only if the value of φ is positive ($\varepsilon w_1 > 1$) *throughout* the reaction.

Change of the sign of φ in the course of the reaction may be accounted for by a change in the ratio of branching to termination rates. Substituting $w_1 = v_1/(v_1 + v_2)$ into (2.18) we obtain

$$\varphi = (\varepsilon - 1)v_1 - v_2 \tag{2.22}$$

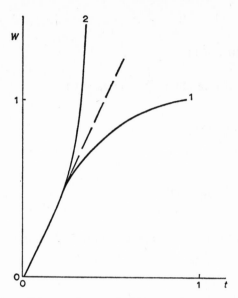

Fig. 4. Time dependence of the reaction rate for (1) an unbranched, and (2) a branched chain
reaction.

The frequency of the step resulting in branching, *i.e.* $(\varepsilon-1)v_1 = \delta v_1$, is proportional
to the concentration of one of the initial compounds. Thus, in a reaction at
constant temperature this value will decrease with increasing time, *i.e.* with consumption of the reactants. The frequency of steps resulting in chain termination,
v_2, either remains practically constant, or increases with accumulation of some
products formed in the course of the reaction. In hydrogen combustion, for
instance, the probability of chain termination becomes greater with formation
of water, as water molecules are more effective partners than H_2 and O_2 in the
three-body collision $H + O_2 + H_2O \rightarrow HO_2 + H_2O$ that results in chain termination
in the gas phase (see ref. 146, p. 145). As a result of decrease in δv_1, or of its
decrease accompanied by simultaneous increase in v_2, the value of φ will become
lower in the course of the reaction. This leads to reduction in the acceleration of
the reaction, and even to a decrease in its rate corresponding to a change to the
opposite sign of φ. In the latter case there will be no explosion. Semenov has called
this phenomenon *degenerate explosion*.

It will be noted that transition to explosion in the course of the reaction is usually
favoured by the temperature increase which occurs even at relatively low reaction
rates and results in an acceleration of the reaction; this frequently terminates in
thermal explosion or ignition.

2.2.2 Thermal ignition

Van't Hoff suggested at the end of the last century[209] that the increase in temperature due to a higher rate of heat release than that of heat transfer to the surroundings might be the reason for an initial acceleration of the reaction ending in explosion. In particular, Van 't Hoff defined the ignition temperature as one at which the loss of heat by conduction at the start of the reaction is equal to the heat released by the reaction. He suggested that this could be easily expressed by a mathematical formula. A formula of this kind was first derived by Nernst[148] in 1906. Van't Hoff's ideas were developed by Taffanel and Le Floch[197,198] and more explicitly by Semenov[172]. Semenov has worked out a detailed quantitative theory of thermal explosion that was further developed in the research carried out by Todes[203,204] and Frank-Kamenetskii[62,64].

The rate of heat release in a homogeneous gas reaction, *i.e.* the amount of heat liberated in one second per cm^3 of the reaction zone, may be taken as

$$\Phi_+ = QW \tag{2.23}$$

where Q is the heat, and W the rate, of a reaction. On the other hand, the rate of heat transfer may be taken as proportional to the difference in temperatures of the reaction zone T and of the reactor wall T_0, *i.e.*

$$\Phi_- = \alpha \frac{S}{V}(T - T_0) \tag{2.24}$$

where α is the heat transfer coefficient, and S and V are the surface area and volume of the reactor.

Since in the initial stages of many chain reactions the consumption of initial compounds is not great, the reaction rate as a function of temperature and pressure may be expressed by

$$W = CN^n \exp\left(-E/RT\right) \tag{2.25}$$

where C is a constant dependent upon the nature and composition of the gas, n is a number defining a certain effective reaction order, N is the concentration, and E the effective activation energy. It follows from (2.25) that the heat input Φ_+ as a function of temperature will be represented by a curve convex to the temperature axis (x-axis). When this curve is intersected by the straight line Φ_- *versus* T (2.24), heat loss is higher than or equal to heat gain. In this case we have a stationary reaction occurring at a certain constant temperature. When the curve Φ_+ *versus* T lies *higher* than the straight line Φ_- *versus* T, *i.e.* when heat input is greater than heat loss at any temperature, there will be a continuous increase in temperature resulting in continuous acceleration and subsequent explosion.

Evidently, the heat loss line (Φ_- *versus* T) tangent to the heat input curve (Φ_+ *versus* T) will correspond to transition from a stationary reaction to explosion. Thus, the tangency conditions

$$(\Phi_+)_{T_i} = (\Phi_-)_{T_i} \text{ or } QW_{T_i} = \alpha \frac{S}{V}(T_i - T_0) \tag{2.26}$$

and

$$\left(\frac{d\Phi_+}{dT}\right)_{T_i} = \left(\frac{d\Phi_-}{dT}\right)_{T_i} \text{ or } Q\left(\frac{dW}{dT}\right)_{T_i} = \alpha \frac{S}{V} \tag{2.27}$$

are those required for this transition, the temperature (T_i) that corresponds to the point of tangency being *that of ignition*. The relevant pressure will be the *minimum ignition pressure* p_i (at a temperature T_i or initial temperature T_0).

In accordance with (2.25) we obtain from (2.26) and (2.27) the ignition temperature

$$T_i = \frac{E}{2R}\left\{1 - \sqrt{1 - \frac{4RT_0}{E}}\right\} \tag{2.28}$$

Since usually $E > 20000$ cal. mole^{-1}, and $T_0 < 1000\,°K$, this expression may be reduced with fair accuracy to

$$T_i = T_0 + \frac{RT_0^2}{E} \tag{2.29}$$

$T_i - T_0$ is the *pre-ignition temperature rise* of the mixture. It will be noted that with the values of E and T_0 taken above, the pre-ignition rise will not exceed 100°.

It will be seen that, to within the accuracy of the above calculation, the increase in temperature from T_0 to T_i results in a rate increase by a factor of e. Thus, substituting $T_i = T_0 + RT_0^2/E$ in (2.25) and making the following transformations that are permissible due to the low value of RT_0^2/E as compared to T_0, *viz.*

$$\exp\left(-E/RT_i\right) = \exp\left[(-E/RT_0)\Big/\left(1 + \frac{RT_0}{E}\right)\right]$$

$$\simeq \exp\left[(-E/RT_0)\left(1 - \frac{RT_0}{E}\right)\right] = e\exp\left(-E/RT_0\right)$$

we obtain

$$W_{T_i}/W_{T_0} = e$$

The conditions for transition to ignition (2.26) and (2.27) may also be used for calculation of the ignition pressure p_i, connected with concentration N by the relation $p_i = NRT_0$. Substituting $W_{T_i} = eW_{T_0}$ and $T_i - T_0 = RT_0^2/E$ into (2.26) and (2.27), after some manipulation, we have

$$\ln \frac{p_i}{T_0^{1+2/n}} = \frac{A'}{T_0} + B' \tag{2.30}$$

where

$$A' = \frac{E}{nR} \quad \text{and} \quad B' = \frac{1}{n} \ln \left(\frac{\alpha S}{V} \cdot \frac{R^{n+1}}{E} \cdot \frac{1}{eQC} \right)$$

However, it will readily be seen that (2.30) may also be written as

$$\ln \frac{p_i}{T_0} = \frac{A}{T_0} + B \tag{2.31}$$

Thus, as ignition of a gas mixture is studied over a rather narrow temperature range, a mean value of T_0 for this range, $\overline{T}_0$, may be introduced, and as the difference $(T_0 - \overline{T}_0)$ is low compared to T_0, it may be written with fair accuracy as

$$T_0 = \overline{T}_0 \exp[(T_0 - \overline{T}_0)/T_0]$$

Making use of this equation, we obtain (2.31) from (2.30), where

$$A = \frac{E - 2RT_0}{nR} \quad \text{and} \quad B = \frac{1}{n} \ln \left(\frac{\alpha S}{V} \cdot \frac{R^{n+1}}{E} \cdot \frac{e\overline{T}_0^2}{QC} \right).$$

Relations of type (2.31) between p_i and T_0 were established empirically for many reactions. Let us mention the unbranched chain reaction $H_2 + Br_2 = 2HBr$. The variation of the minimum ignition pressure, p_i, with temperature T_0 for the mixtures $H_2 + Br_2$ and $H_2 + 2Br_2$, established by Zagulin[220], is shown in Fig. 5. It will be seen that the experimental results are consistent with (2.31) for both hydrogen/bromine mixtures studied.

Frank-Kamenetskii has made an essential contribution to the theory of thermal ignition[62, 64]. A disadvantage of Semenov's theory discussed above is the assumption of constant gas temperature throughout the reaction vessel. This is the case only when there is intensive convection in the pre-ignition period resulting in uniform temperature of the mixture. However, temperature distribution in the gas should be allowed for in most real cases, and it is taken into account in the Frank-Kamenetskii theory, which implies that heat transfer to the wall occurs by conduc-

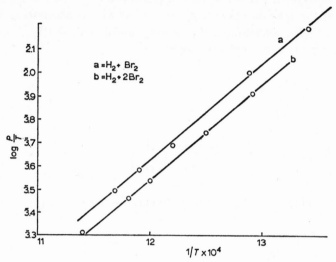

Fig. 5. The minimum ignition pressure as a function of temperature for mixtures of H_2 with Br_2 [220].

tion. Here the initial equation, similar to (2.26)

$$\kappa \Delta T + QW = 0 \tag{2.32}$$

represents transition from a stationary to an autoaccelerated reaction terminating in explosion (κ is the heat conduction coefficient and ΔT the Laplace operator).

Frank-Kamenetskii made use of eqn. (2.32) for reactors of various shapes, assuming the reaction rate W to be that expressed by (2.25). For instance, for a cylindrical vessel eqn. (2.32) yields the following expression for transition to thermal ignition

$$QW_{T_i} = \frac{2\kappa}{r^2} \frac{RT_0^2}{E} \ ^\dagger \tag{2.33}$$

The only difference between this expression and (2.26) is that the term $\alpha S/V$ in (2.26) is replaced by $2\kappa/r^2$. This is a great advantage, as in Semenov's theory the heat transfer coefficient remained indefinite. Moreover the theory of Frank-Kamenetskii implies a definite dependence of critical conditions on the *size* and *shape* of the reactor.

The dependence of the ignition temperature on the vessel diameter was confirmed

---

† For a flat and a spherical reactor the value 2 in expression (2.33) will be replaced by 0.88 and 3.32, respectively.

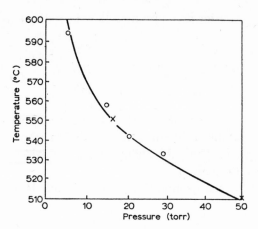

Fig. 6. Ignition temperature at the third limit for an oxyhydrogen mixture at various reactor diameters[64].

in studies of the thermal ignition of a hydrogen–oxygen mixture. The full curve in Fig. 6 represents the theoretical dependence of T_i on the reactor diameter, as calculated by Frank-Kamenetskii[63]; circles and crosses denote experimental values obtained by various workers[156, 223].

Due to the close resemblance of (2.23) and (2.26) we obtain from (2.33) and (2.25) an expression that differs from (2.31) only in the B value, which is now

$$B = \frac{1}{n} \ln \left(\frac{2\kappa}{r^2} \cdot \frac{R^{n+1}}{E} \cdot \frac{e\overline{T}_0^2}{QC} \right)$$

Correlation of the theory with the experimental results obtained for various systems mostly shows good agreement. For instance, the p_i values for various temperatures calculated for a stoichiometric mixture of hydrogen with oxygen in a spherical vessel practically coincide with those obtained experimentally, as may be seen from the following results[213]

$T°K$	851	864	870
p_{exp} (torr)	700	660	590
p_{calc} (torr)	685	655	588

Again, for many reactions such as azomethane decomposition, $(CH_3)_2N_2 = C_2H_6 + N_2$, methyl nitrate decomposition, $2CH_3ONO_2 = CH_3OH + CH_2O + 2NO_2$, hydrogen sulphide oxidation, $2H_2S + 3O_2 = 2H_2O + 2SO_2$, etc., the ignition temperatures calculated for various pressures appeared to agree with those measured. The accuracy of theoretical calculation may be seen from the following data on nitrous oxide decomposition $(2N_2O = 2N_2 + O_2)$

p (torr)	170	330	590
$T_{i\ exp}$ [222]	1285	1195	1100
$T_{i\ calc}$ [64]	1255	1175	1110

It is of interest to note that the theoretical T_i values were obtained by Frank-Kamenetskii from the kinetic results of Volmer[214] before relevant experimental determinations were made. Thus, the ability of nitrous oxide to ignite spontaneously was predicted theoretically.

A full discussion of thermal ignition theories has recently been given by Gray and Lee[69a].

2.2.3 Chain ignition limits

The *ignition limits* are another feature of branched chain reactions. Since according to the previous definition if $\varepsilon w_1 < 1$ the reaction is stationary and relatively slow, and if $\varepsilon w_1 > 1$ it is fast, autoaccelerated, and ends in explosion, the condition for transition from a stationary to a non-stationary regime, *i.e.* the *condition for an ignition limit* may be written as

$$\varepsilon w_1 = 1 \tag{2.34}$$

or

$$\delta v_1 = v_2 \tag{2.35}$$

Assuming that

$$v_1 = ap \text{ and } v_2 = bp^2 + c$$

where a, b, and c are constants depending on temperature and p is the pressure, we transform (2.35) into a quadratic equation

$$p^2 - \frac{\delta a}{b} p + \frac{c}{b} = 0 \tag{2.36}$$

(In the expression for v_2 the first term corresponds to chain termination in the gas phase by a three-body collision, and the second term to chain termination at the wall in the kinetic region.)

Eqn. (2.36) yields the following pressure values for the *lower* (p_1) and the *upper* (p_2) ignition limits

$$p_1 = \frac{\delta a}{2b} - \sqrt{\frac{\delta^2 a^2}{4b^2} - \frac{c}{b}} \tag{2.37}$$

and

$$p_2 = \frac{\delta a}{2b} + \sqrt{\frac{\delta^2 a^2}{4b^2} - \frac{c}{b}} \qquad (2.38)$$

It will be noted that

$$p_1 + p_2 = \frac{\delta a}{b} \qquad (2.39)$$

The approximate values for the lower and upper limits may be obtained on the assumption that chain termination at the wall will be predominant close to the lower limit (the bp^2 term is small compared to c), whereas close to the upper limit the chains will terminate mainly in the gas phase ($bp^2 \gg c$). In this way we obtain

$$p_1 \simeq \frac{c}{\delta a} \qquad (2.40)$$

and

$$p_2 \simeq \frac{\delta a}{b} \qquad (2.41)$$

It will be seen from comparison of (2.39) and (2.41) that the approximate values of p_1 (2.40) and p_2 (2.41) may be considered as close to the correct ones only as long as $p_1 \ll p_2$.

Fig. 7 shows the rate of the reaction between hydrogen and oxygen at a constant

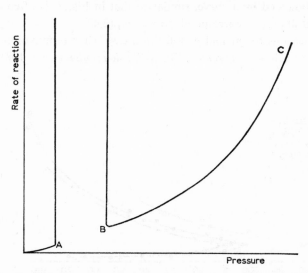

Fig. 7. Ignition limits (with respect to pressure) for a mixture of hydrogen with oxygen at a constant temperature[200].

temperature as a function of total pressure of the mixture (according to Thompson and Hinshelwood[200]). The abrupt change in the reaction rate at the ignition limits will be noted.

p_1 and p_2 depend on temperature in a definite way. Since the value of δa, which is in fact the branching rate coefficient, increases with temperature, and the value of b (proportional to the rate coefficient of a termolecular reaction) depends only slightly on temperature, as does the value of c (proportional to the rate coefficient for chain termination at the wall), $p_1 (\simeq c/\delta a)$ will drop, and $p_2 (\simeq \delta a/b)$ will rise with increasing temperature. Consequently, as the temperature decreases, p_1 and p_2 become closer, and at a certain temperature determined by the condition

$$\frac{\delta^2 a^2}{4b^2} = \frac{c}{b}$$

their values will coincide at p_M given by

$$p_M = p_1 = p_2 = \frac{1}{2}\frac{\delta a}{b} = \sqrt{\frac{c}{b}} \tag{2.42}$$

This temperature dependence of p_1 and p_2 is shown in Fig. 8 for carbon monoxide combustion. The figure was taken from a paper by Semenov et al.[117] reporting pioneer research in the field of branched chain reactions, carried out at the same time as work by Zagulin[220] and by Thompson and Hinshelwood[200]. The ignition region bounded by a curve, similar to that in Fig. 8, is often called the *ignition peninsula*. Its vertex corresponds to $p = p_M$ (2.42).

The above equations for p_1 and p_2 refer to a case when termination of chains at the wall occurs in the *kinetic region*. For a *diffusion region* (see p. 108), c will be

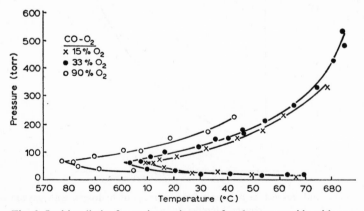

Fig. 8. Ignition limits for various mixtures of carbon monoxide with oxygen[117].

replaced by c'/p, and the equation for pressures at the upper and lower limits will be cubic and not quadratic, *viz.*

$$p^3 - \frac{\delta a}{b} p^2 + \frac{c'}{b} = 0 \tag{2.43}$$

This equation yields the following approximate values for p_1 and p_2

$$p_1 \simeq \sqrt{\frac{c'}{\delta a}} \tag{2.44}$$

and

$$p_2 \simeq \frac{\delta a}{b} \tag{2.45}$$

It will be readily seen that the third root of the cubic equation is negative and, thus, has no physical significance.

An additional *third ignition limit* (p_3) is observed in many cases besides the first (p_1) and the second (p_2) limits, the magnitudes being $p_3 > p_2 > p_1$. When passing through the second ignition limit towards higher pressures, the reaction rate will show an abrupt decrease due to termination being predominant over branching of chains. However, with higher pressures the reaction rate will pass through a *minimum* and then start increasing again. One of the reasons for this increase in the reaction rate at sufficiently high pressures is the occurrence of additional propagation reactions involving relatively unreactive radicals (for example, HO_2 in the reaction between H_2 and O_2); at low pressures the diffusion of these radicals to the wall with their subsequent destruction is predominant. Usually this acceleration of the reaction results in an increase in the mixture temperature, leading to thermal explosion. This is the origin of the third ignition limit for hydrogen shown in Fig. 9 together with the first and second limits. All experimental data available show that, in practically all the cases studied, the third ignition limit is of a thermal nature.

The thermal ignition limit is common both to unbranched and branched chain reactions. It is the only limit for unbranched reactions, and represents one of the three limits (p_3) for branched chain reactions; the other two $(p_1$ and $p_2)$ are those for chain ignition, *i.e.* for ignition under isothermal conditions. However, Voevodskii has shown theoretically that, in specific cases, the third limit may be of a chain nature as well. This was confirmed experimentally by himself and Poltorak[213] for hydrogen combustion. It appeared possible to observe such a limit for the ignition of oxyhydrogen gas in reactors pretreated with potassium chloride, which favours HO_2 destruction at the wall and thus slows the reaction

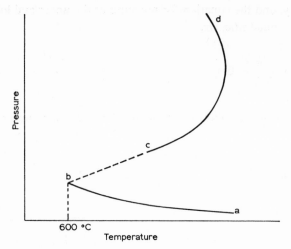

Fig. 9. Three limits for the ignition of a hydrogen–oxygen mixture given by Semenov *et al.*[117] on the basis of ref. 220 and ref. 66 (region a–b) and ref. 50 (region c–d).

down. Voevodskii[210] has shown also that, in this case, the third hydrogen ignition limit lies considerably lower than that for thermal ignition.

The rate of a branched chain reaction is initially low and becomes measurable after a certain period of time, $t = \tau_i$, called the *induction period*. It follows from the above that the value of the induction period, which is of course dependent on the sensitivity of the method used to follow the course of the reaction, is of a conditional nature only. This may be seen from Fig. 10 (ref. 109, p. 533) showing

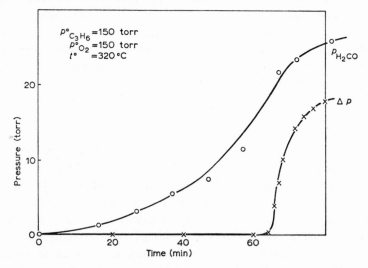

Fig. 10. Course of reaction as shown by the pressure variation and by the formaldehyde formed in propene oxidation (according to Karmilova and Kondratiev).

the progress of propene oxidation determined both from variations in pressure of a propene–oxygen mixture, and from the yield of formaldehyde. Using the first method $\tau_i = 65$ min, and using the second it is practically zero.

Despite the formal sense of the induction period, it may be used for determination of individual rate coefficients. For example, the rate coefficient of the branching step

$$H+O_2 \rightarrow OH+O$$

in the H_2+O_2 reaction[169] and that of

$$CO+O_2 \rightarrow CO_2+O$$

in the $CO+O_2$ reaction[196] were determined from the measured induction periods.

Let us consider in some detail the induction period for the simplest formal scheme of a chain reaction that was described in Fig. 1. Integrating the equation for the reaction rate

$$W = \frac{dc}{dt} = \frac{W_0 v_1}{\varphi} \left[\exp(\varphi t) - 1\right] \tag{2.20}$$

at constant values of W_0, v_1, and φ, which is permissible for the initial stage of the reaction, during the time from 0 to τ_i, we obtain

$$\Delta c = \frac{W_0 v_1}{\varphi^2} \left[\exp(\varphi t) - (1+\varphi t)\right] \tag{2.46}$$

where Δc is the amount of the reaction product accumulated by time t. When t is sufficiently high, the second term in the square brackets may be neglected and

$$\Delta c = \frac{W_0 v_1}{\varphi^2} \exp(\varphi t) \tag{2.46'}$$

Denoting by Δc_{min} the minimum measurable amount of the reaction product, which is present when $t = \tau_i$, we obtain from (2.46')

$$\tau_i = \frac{1}{\varphi} \ln \frac{\Delta c_{min}}{A} \tag{2.47}$$

where

$$A = \frac{W_0 v_1}{\varphi^2}$$

References pp. 183–188

In principle, expression (2.47) may be used for determination of relevant rate coefficients from the induction period. An example illustrating this method will be given in the section dealing with hydrogen combustion (p. 132).

It will be noted that an induction period may be observed with both branched and unbranched chain reactions, as the rates of the latter also increase in the initial stage of the reaction (from zero at $t = 0$). However, very often the reaction is accompanied by a *rise in temperature* of the mixture, the rate of which increases with time and results in autoacceleration of the process. This may lead to the appearance of an induction period formally indistinguishable from that observed for branched chain reactions under isothermal conditions. Heating of the mixture in the course of the reaction, which would mask the true (isothermal) induction period, is certainly also possible for branched chain reactions; this may be seen from temperature measurements during the induction period.

The theory of reactions involving non-isothermal chain branching has been considered but is difficult to develop mathematically (see, for example, Ben Aim et al.[72]).

Moreover, the true induction period is often masked by the action of impurities, often present in uncontrollable amounts. For instance, it was found as far back as 1801[45] that the photochemical reaction of chlorine with hydrogen as well as that of chlorine with hydrocarbons, does not start immediately, but becomes measurable only after a certain period of time. Further research has shown that the duration of this period, which was defined as the induction period[35], depended upon preparation of the mixture and upon other factors. Discovery of the inhibiting action of oxygen was of great importance in the interpretation of this effect[35]. It was shown that the presence of oxygen or of other impurities in the reacting mixture is the main factor responsible for the initial delay, that is, for the induction period. In the presence of inhibitors the active centres (for example Cl atoms) react mainly with the impurity molecules; as a result the chains cannot propagate, and the reaction rate is vanishingly small. Only at the end of the induction period, when most of the inhibitor has been consumed, does chain propagation begin playing an important part and thus accounting for the measurable reaction rate.

2.3 EFFECT OF THE WALL ON CHAIN REACTIONS

2.3.1 Homogeneous and heterogeneous steps of chain reactions

Certain individual steps of a chain reaction may occur at the reaction vessel surface, the other steps being homogeneous. Thus, in order to establish the mechanism of a reaction, it is of great importance to find out the part played by the wall.

One of the widely used methods for estimation of the wall effect in a chemical reaction consists in studying the latter in vessels with different ratios of the surface to the reactor volume, S/V, or in a reactor filled with solid fragments that would markedly increase the S/V ratio. If it appears that the reaction rate or the composition of products depend on the S/V value, this is considered as evidence for the occurrence of heterogeneous reaction steps. When the process appears to be independent of S/V, this is taken as an indication of a completely homogeneous reaction. However, Rice and Herzfeld[163] have shown that, in certain cases, the reaction rate will be independent of S/V despite the occurrence of heterogeneous individual steps. For instance, this is so for heterogeneous generation and termination of chains (see p. 104). It follows that the method would be unambiguous proof of the contribution of the wall to the reaction mechanism only when the dependence of the reaction rate on the S/V value is established.

Another method for evaluation of the effect of the heterogeneous factor in the mechanism of a chemical reaction is that of *differential calorimetry*, worked out by Bogoyavlenskaya and Kovalskii[32]. This involves simultaneous temperature measurements at the centre of the reactor (T_c) and at its wall (T_w). The difference $\Delta T = T_c - T_w$ gives the extent of heating of the gas mixture at the reactor centre relative to the reactor wall. Calculation of ΔT from the equation for heat conduction shows that for a reaction occurring in the gas phase only (in a cylindrical reactor)

$$\Delta T = \frac{q}{4\pi\kappa} \tag{2.48}$$

where q is the amount of heat released per second per cm^3 of the reaction zone, and κ is the heat conduction coefficient. When the reaction occurs in part at the reactor wall and at the surface of the capillaries enclosing the thermocouples, the value of ΔT appears to be dependent on the proportion of heat released both in the homogeneous and heterogeneous reaction steps, and also on the reactor and capillary radii.

Let us consider the following example. Making use of the differential calorimetry technique, Markevich[129] has shown that under his experimental conditions (a 1 : 1 mixture of $H_2 + Cl_2$, $p = 8$ to 100 torr, $T = 270$ to 370 °C, in the absence of impurities) the thermal reaction of chlorine with hydrogen will be homogeneous and its rate will be uniform throughout the reactor.

However, it does not follow that the *generation* and *termination* of chains necessarily occur in the gas phase. Investigation of the oxygen effect seems to show the reverse. Thus, Markevich[130] has established that, in the presence of oxygen giving rise to strong inhibition, the reaction (that is accompanied by heat release) occurs within a narrow zone close to the reactor wall. A direct conclusion would be that chlorine atoms are formed at the wall and, being in general scavenged by oxygen, in its presence these atoms would succeed in entering into reaction with hydrogen only close to the place of their origin. This was confirmed by Chaikin[40], who investigated the reaction between chlorine and hydrogen in a $1:1$ mixture at $p = 115$ torr and $T = 286\,°C$, and found that in the absence of oxygen the change in the reaction rate induced by an eight-fold increase in S/V did not exceed 10 per cent, whereas for a mixture containing 15 per cent of O_2 an increase in S/V by a factor of 8.5 increased the rate by a factor of six.

Both the independence of the rate on S/V in the absence of inhibiting impurities, and the apparent homogeneity of the reaction as shown by the differential calorimetry technique in such cases when heterogeneous generation of active centres is an established fact, provide direct evidence for chain termination at the wall in the given case as well. Indeed, in accordance with the reaction mechanism (p. 113 *et seq.*), the steady-state conditions for a reaction involving formation and destruction of chlorine atoms at the reactor wall will be

$$\frac{d[Cl]}{dt} = k_0[Cl_2] - k_1[H_2][Cl] + k_2[Cl_2][H] - k[Cl_{ads}][Cl] = 0$$

and

$$\frac{d[H]}{dt} = k_1[H_2][Cl] - k_2[Cl_2][H] = 0$$

Here, k_0, k_1, etc. are rate coefficients for relevant elementary reactions, the generation and destruction of chlorine atoms being assumed to occur by the steps

$$Cl_2 + wall \rightarrow Cl_{ads} + Cl \qquad k_0$$
$$Cl + Cl_{ads} \rightarrow Cl_2 + wall \qquad k$$

(Cl_{ads} denotes an adsorbed chlorine atom). Adding the two equations we find

$$k_0(Cl_2) = k(Cl_{ads})(Cl)$$

i.e. when the reaction is in a stationary state, the rate of formation of chlorine atoms equals that of their destruction. Since the formation of a chlorine atom requires the absorption of the same amount of energy as that released by its destruction in heterogeneous recombination, the whole process involving direct and back reactions will be thermoneutral and, thus, will have no effect on the heat balance.

The heterogeneous generation and destruction of active centres will be considered below in more detail.

2.3.2 Heterogeneous generation of active centres

The first suggestion of the possibility of heterogeneous generation of radicals seems to have been made by Polyakov[160]. First experimental proof of this phenomenon was given by Kovalskii et al.[32,129], by means of the differential calorimetry technique. Of later work let us cite that of Patrick and Robb[155], who have found that pretreatment of the reactor wall with potassium chloride inhibits the thermal, but does not affect the photochemical reaction of H_2 with O_2. It follows that in this case the inhibiting effect of KCl consists in suppression of chain initiation, in other words of the generation of active centres at the surface.

Voevodskii[211] provided theoretical backing for the hypothesis of the heterogeneous generation of active centres, having shown that in addition to the known recombination of atoms at the surface there must occur, in accordance with the detailed balancing principle, a reverse reaction of dissociation with partial transfer of atoms to the gas phase. As an example let us take the generation of halogen atoms at a solid surface. Voevodskii has shown, making use of the results obtained by Markevich[129], that at $T = 310\,°C$ and $p = 60$ torr the rate of heterogeneous generation of chlorine atoms will be approximately 1.2×10^{11} $cm^{-2} \cdot sec^{-1}$.

For comparison let us calculate the rate of homogeneous formation of chlorine atoms from the dissociation of Cl_2 molecules, i.e. the rate of the reaction

$$Cl_2 + M \rightarrow 2Cl + M$$

The rate coefficient for this reaction was determined by many workers at temperatures higher than $1500\,°K$[34,81,86]. Assuming the possibility of extrapolation to a temperature of $583\,°K$ ($310\,°C$), we obtain from the expressions given by these workers a mean value of $\sim 7 \times 10^{-29}$ $cm^3 \cdot molecule^{-1} \cdot sec^{-1}$ at $583\,°K$. Thus the rate of homogeneous dissociation of Cl_2 under the above conditions will be $\sim 1.7 \times 10^7$ $cm^{-3} \cdot sec^{-1}$, i.e. $\sim 1.7 \times 10^7$ $cm^{-2} \cdot sec^{-1}$ for 1 cm^2 of the wall surface, a value four powers of ten lower than that for the rate of heterogeneous generation of chlorine atoms.

Heterogeneous generation of chlorine atoms (of X atoms in the general case) may be conceived as involving two steps, that of chemisorption of X_2 molecules

$$X_2 + surface \rightarrow X_{ads} + X_{ads} \tag{I}$$

and that of release of the adsorbed X atoms

$$X_{ads} \rightarrow X + surface \tag{II}$$

Denoting by S the heat of adsorption of X and by D the heat of dissociation of

X_2, we obtain for the heat change in step (I)

$$Q_I = -D + 2S$$

The heat effect of step (II) will evidently be

$$Q_{II} = -S$$

Let us take the activation energies for (I) and (II) as equal in absolute magnitude to their heat changes, $E_I = D - 2S$ and $E_{II} = S$, and assume that the surfaces are not metallic, in other words are such that the heat of X-atom adsorption would be such that $E_I > E_{II}$, i.e. $S < D/3$. It would appear then that of the two consecutive steps (I) and (II) of the initiation process, step (I) is limiting and determines the overall rate of heterogeneous generation of atoms.

An attempt may be made to estimate the activation energy for reaction (I) from the rate of heterogeneous generation of chlorine atoms, determined from the results of Markevich. The rate coefficient of this reaction in the kinetic region (p. 108) may be taken as

$$k = \frac{\gamma c}{4}$$

where γ is the probability of chemisorption and c the velocity of chlorine molecules. Expressing the reaction rate as

$$W_0 = k[Cl_2] = \frac{\varepsilon c}{4} \exp(-E/RT)[Cl_2] \tag{2.50}$$

(where ε is a certain steric factor that we shall take as 0.1) and substituting into the above expression the velocity of Cl_2 molecules and their concentration at 583 °K and $W_0 = 2 \times 10^{11}$ cm^{-2}. sec^{-1}, we have $E = 25$ kcal. mole^{-1}. This value seems to be quite reasonable. We obtain from it $S = 16$ kcal. mole^{-1} for the heat of chlorine adsorption on glass. Only one estimate of this value seems to have been reported, namely $\geqslant 20$ kcal. mole^{-1} for a pyrex vessel[5].

The value of S may be determined in another way. The expression

$$\ln \frac{p_i}{T_0} = \frac{A}{T_0} + B \tag{2.31}$$

(p. 93) represents the relation between the temperature (T_0) and pressure (p_i) corresponding to thermal ignition, the value of A being $(E - 2RT_0)/nR$.

A linear relationship between $\ln p_i/T_0$ and $1/T_0$, satisfying expression (2.31) and similar to that for the reaction $H_2 + Br_2 = 2HBr$ (Fig. 5), was found[220] for the reaction $H_2 + Cl_2 = 2HCl$. For the ignition limit the theoretical rate expressions (3.3) and (3.5), pp. 114 and 116, may be rewritten as

$$W = 2k_1 \sqrt{K_{Cl}} [H_2][Cl_2]^{\frac{1}{2}}$$

and

$$W = \frac{2k_0 k_1}{k} [H_2][Cl_2]$$

The first expression refers to the case of initiation in the gas phase giving Cl atoms, and the second to heterogeneous generation. In the first case the effective activation energy E and the reaction order n in the expression for A will be given by $E_I + D/2$ (D is the heat of dissociation of Cl_2) and $\frac{3}{2}$, respectively, and for the second by $E_I + D - 2S$ (S is the heat of Cl-atom adsorption) and 2. Assuming that $E_I = 5.5$ kcal. mole^{-1} (Ashmore and Chanmugam[4]), $D = 57.1$ kcal. mole^{-1}, and the mean temperature $\overline{T}_0 = 630 °K$[220], we obtain $A = 10.5 °K$ (generation in the gas phase) and $A = 15.0 - S/2 °K$ (generation at the wall). The experimental value is 5.8 °K. As $5.8 < 10.5 \simeq 2 \times 5.8$, it follows that the formation of Cl atoms in the initiation step in this system is *heterogeneous*. From the relation $15.0 - S/2 = 5.8$ we have $S = 18.4$ kcal. mole^{-1}, as compared with 16 kcal. mole^{-1} obtained above†.

2.3.3 Heterogeneous chain termination

It has been found for a great number of chemical reactions, including photochemical processes in which the initial production of active centres is known to be homogeneous, that chain termination reactions often involve a step dependent on the nature, condition and pretreatment of the reactor wall. Trifonov[206] was the first to study the inhibiting effect of the wall on the photochemical reaction between chlorine and hydrogen. Making use of reactors of various diameters, Trifonov found that, at total pressure $\leqslant 30$ torr, the rate of a reaction initiated by a narrow beam of light directed along the axis of a cylindrical reactor was proportional to the square of the reactor diameter. This could be explained only on the assumption that chains were terminated at the wall to which the active particles penetrated by diffusion. The square dependence may be readily understood, as the time of diffusion from the axis of the vessel to the surface is also proportional to d^2.

It appears from these results that allowance should be made for termination of chains both in the gas phase and at the wall. Heterogeneous termination

---

† The heat of adsorption of Br atoms determined in a similar way is 31.5 kcal. mole^{-1}.

becomes predominant at low pressures; its mechanism presumably involves surface adsorption of active centres. Let us denote by γ the probability of an active centre being destroyed by hitting the wall; the rate of this process, w_2 (referred to a unit volume of a cylindrical reactor of diameter d), will be

$$w_2 = \frac{\gamma c}{d} a = v_2 a \qquad (2.51)$$

where c is the velocity and a the concentration of active particles. It will be noted that (2.51) will be valid only for low γ ($\ll 1$), when the active centre usually rebounds from the wall rather than being adsorbed, as a result of which the concentrations of centres in the gas phase and at the wall are practically indistinguishable. In this case a heterogeneous reaction (here a heterogeneous chain termination) is said to occur in the *kinetic region*.

When the value of γ is high (only just less than unity), the concentration of active particles at the surface will be considerably *lower* than that in the gas phase. Thus, there will be a *gradient* in the concentration of chain centres, directed normal to the wall surface, and their rate of adsorption will be essentially that of their diffusion to the wall. For this reason it is said that the reaction occurs in the *diffusion region*. The relationship between kinetic and diffusion factors responsible for the rate of heterogeneous chain termination was discussed by Frank-Kamenetskii[64]. When the reaction is in a stationary state, the rate of destruction of active particles at the surface, ka_s (a_s is the active centre concentration at the surface), will be that of the diffusion flow, *i.e.*

$$ka_s = -D_A \frac{da}{dx} \qquad (2.52)$$

(D_A is the active particle diffusion coefficient.) According to Frank-Kamenetskii, the concentration gradient, $-da/dx$ can be represented as $(a-a_s)/\Delta$, where a is the concentration of active particles in the gas phase, and Δ is the thickness of a certain "reduced film". Then (2.52) will be rewritten as

$$ka_s = \beta(a-a_s) \qquad (2.53)$$

where $\beta = D/\Delta$ is a parameter that may be called the *diffusion rate coefficient* by analogy with the reaction rate coefficient. From (2.53) we find

$$a_s = \frac{\beta}{k+\beta} \cdot a \qquad (2.54)$$

and, thus, for the rate of destruction of active centres, *i.e.* for that of chain termination, we obtain

$$w_2 = \frac{k\beta}{k+\beta} \cdot a = k^* a \tag{2.55}$$

Here the effective rate coefficient for chain termination is

$$k^* = \frac{k\beta}{k+\beta} \quad \text{or} \quad \frac{1}{k^*} = \frac{1}{k} + \frac{1}{\beta} \tag{2.56}$$

It follows from the above that there are two extreme cases: $k \ll \beta$ and $k \gg \beta$. For the first we have $k^* = k, a_s = a$, and

$$w_2 = ka \tag{2.57}$$

i.e. an expression the same as (2.51). The low k value (equal to v_2 for low γ) and a constant concentration of active particles over the reactor cross-section are characteristic of the kinetic region. For the second extreme case $k^* = \beta, a_s \ll a$, and

$$w_2 = \beta a \tag{2.58}$$

i.e. chains are terminated in the diffusion region.

Formal treatment of heterogeneous chain termination naturally gives only a qualitative picture of the "regions" in which it occurs and of the conditions for transition from one region to another. No definite value can be obtained for the "coefficient" β. However, rigorous quantitative solution of the problem is possible. This was first done by Bursian and Sorokin[36] who have shown that the kinetic equations for the variation of the active centre concentrations

$$\frac{\partial a}{\partial t} = W_0 - v_1(1-\varepsilon)a + D_A \frac{\partial^2 a}{\partial x^2} \tag{2.59}$$

and

$$\frac{da}{dt} = W_0 - v_1(1-\varepsilon)a - v_2 a \tag{2.60}$$

are equivalent, just as the relevant equations for a stationary state

$$W_0 - v_1(1-\varepsilon)a + D_A \frac{d^2 a}{dx^2} = 0 \tag{2.61}$$

and

$$W_0 - v_1(1-\varepsilon)a - v_2 a = 0 \tag{2.62}$$

This permits representation of the rate of chain termination in the diffusion region as

$$w_2 = \beta a$$

as before.

Detailed solution of the problem of heterogeneous chain termination for the most usual case of a cylindrical reactor was made by Nalbandyan and Voevodskii[146]. The stationary-state equation will be

$$W_0 - (1-\varepsilon)v_1 a + D_A \left(\frac{d^2 a}{dr^2} + \frac{1}{r}\frac{da}{dr} \right) = 0 \tag{2.63}$$

where r is the distance from the axis of the vessel. More detailed information may be found in their book[146]; here we shall give only the expression for the mean concentration of active particles $\bar{a}$. By integrating (2.63) and neglecting the small terms, Nalbandyan and Voevodskii[146] have obtained[†]

$$\bar{a} = \frac{W_0}{\dfrac{4u_1^2}{d^2} D_A + v_1(1-\varepsilon)} \tag{2.64}$$

where u_1 is the first positive root of the equation

$$J(u) = \frac{2S}{d} u J_1(u)$$

J is the Bessel function, J_1 is a first-order function of the first kind, d the reactor diameter, and

$$S = \frac{4}{3}\frac{\lambda}{\gamma} \tag{2.65}$$

(the diffusion coefficient D_A is taken as equal to $\lambda c/3$, where λ is the mean free path of an active centre A).

Two extreme cases of high and low values of the $2S/d$ term are considered[146].

---

† The difference $g-f$ instead of $v_1(1-\varepsilon)$ was used, g and f denoting kinetic coefficients (frequencies) for termination (in the gas phase) and branching of chains.

The first (high $2S/d$) refers to the kinetic region (low γ, d, and p), and the second to the diffusion region. For $2S/d \gg 1$ the value of u_1 may be taken as equal to $\sqrt{d/S}$. Substituting this value together with (2.65) into (2.64) we obtain

$$\bar{a} = \frac{W_0}{\dfrac{c\gamma}{d} + v_1(1-\varepsilon)} \tag{2.66}$$

or, since $c\gamma/d = v_2$

$$W_0 - v_1(1-\varepsilon)\bar{a} - v_2 a = 0 \tag{2.67}$$

As the concentration of active particles in the kinetic region is the same at any point of the reactor cross-section, i.e. $\bar{a} = a$, expression (2.67) is similar to that obtained from (2.60) for the stationary state. Nalbandyan and Voevodskii[146] show that (2.67) may be considered as sufficiently accurate provided that

$$\frac{\lambda}{\gamma d} \geqq 2$$

If this condition is not fulfilled, the reaction will occur in a region intermediate between the kinetic and diffusion regions.

The second extreme case ($2S/d \ll 1$) corresponds to the diffusion region. In this case the u_1 value may be taken as $2.405/(1+2S/d)$ and the rate coefficient $v_2 = 4u_1^2 D_A/d^2$ will be given by

$$v_2 = \frac{23.2 D_A}{d^2 \left(1 + \dfrac{2S}{d}\right)^2} \tag{2.68}$$

This may be reduced to

$$v_2 = \frac{23.2 D_A}{d^2} \tag{2.69}$$

in view of the low value of $2S/d$. Nalbandyan and Voevodskii[146] have shown that eqn. (2.68) will be valid on condition that $\gamma p d \geqslant 5\lambda_1$, where λ_1 is the mean free path of the particle at a pressure of 1 torr, and eqn. (2.69) on condition that $\gamma p d \geqslant 50\lambda_1$.

Similar calculation for a one-dimensional problem (flat reactor) and for a spherical reactor gives, for the kinetic region

$$v_2 = \frac{\gamma c}{2d} \qquad \text{(flat reactor}^\dagger) \tag{2.70}$$

---

† d is the distance between parallel walls of a flat reactor.

and

$$v_2 = \frac{3\gamma c}{2d} \qquad \text{(spherical reactor}^\dagger) \qquad\qquad (2.71)$$

and for the diffusion region

$$v_2 = \frac{9.9\, D_A}{d^2 \left(1 + \dfrac{2S}{d}\right)^2} \qquad \text{(flat reactor)} \qquad\qquad (2.72)$$

and

$$v_2 = \frac{39.5\, D_A}{d^2} \left(1 - \frac{2S}{d}\right)^2 \qquad \text{(spherical reactor)} \qquad\qquad (2.73)$$

It follows from the above that the kinetic equations for concentrations of active centres in the kinetic and diffusion regions may be reduced to an equation such as (2.55). However, while in the kinetic region the concentration of active particles (A) is the same for any point of the reactor cross-section, the expression for the diffusion region will involve the mean concentration $\bar{a}$ instead of a. Naturally, the rate coefficients k_2 (frequencies v_2) will also be different for the two cases and will be represented by equations (2.51) and (2.68) for a cylindrical reactor, and by relevant expressions for a flat and a spherical reactor. Heterogeneous chain termination is treated in detail by Semenov[175,179] with respect to the lower ignition limit, for a reaction occurring in the diffusion region.

It will be noted that the simplified expression (2.69) for the diffusion region permits the calculation of rate coefficients for destruction of active centres without the necessity for a knowledge of the diffusion coefficient D_A. This is of essential importance in the determination of rate coefficients of elementary reaction steps, since, unless the stationary concentration of active particles is known, kinetic studies yield only the ratio of rate coefficients for two steps involving the same active centre, say k_1/k_2. When k_2 can be determined independently (for a reaction in the diffusion region), k_1 may be obtained directly from the ratio. This method was widely used by Nalbandyan et al. (see, for example, ref. 12).

3. Radical chain reactions

The great amount of work carried out since the discovery of chain reactions has shown convincingly that both unbranched and branched chain reactions occur

---

† d is the sphere diameter.

via *valence-unsaturated* particles, *radicals*, as suggested by Nernst for the reaction between chlorine and hydrogen. For this reason the mechanism of a chain reaction is usually called a *radical* or a *chemical mechanism*. It implies that chains are initiated by formation of free radicals or atoms in a thermal, catalytic or photochemical process, or in some other way. In the course of chain propagation the reacting monoradical is replaced by another monoradical, owing to valence conservation in such reactions, whereas branching of a radical chain is always connected with increase in the free valence: for instance, one monoradical is replaced by three others. Finally, chains are terminated when radicals recombine, add to a molecule to form an inactive species, or else are adsorbed at the wall.

Further consideration of the characteristics of chain reactions, in particular derivation of mechanism and determination of the kinetics of individual elementary processes, is best undertaken by discussion of certain specific examples, although these will be dealt with again in later volumes.

3.1 REACTIONS OF HALOGENS WITH HYDROGEN

The reaction $H_2 + Cl_2 = 2HCl$ may be taken as an example of an unbranched chain reaction. According to Nernst[149], the mechanism of this reaction involves propagation by alternation of the steps

$$Cl + H_2 \rightarrow HCl + H - 1.1 \text{ kcal. mole}^{-1}$$

and

$$H + Cl_2 \rightarrow HCl + Cl + 45.1 \text{ kcal. mole}^{-1}$$

Thus, *two different active centres* are involved. It will be noted that the formation of a second active centre here, as in many other cases, is a simple consequence of the fact that the Cl atom is a *monovalent* species. By virtue of *valence conservation*, the reaction of a Cl atom with a saturated H_2 molecule is bound to give another *monovalent* species, the H atom. For this reason chain reactions occur, with rare exceptions, through the agency of *two* active centres.

As the dissociation energy for a chlorine molecule (57.1 kcal. mole^{-1}) is almost half of that for hydrogen (103.3 kcal. mole^{-1}), initiation of chains in thermal and photochemical reactions of chlorine with hydrogen is connected with generation of *chlorine atoms*. The mechanism of chain generation in this reaction has already been considered to some extent (p. 105) and will be further discussed later.

The disappearance of Cl and H atoms without regeneration may occur by *recombination, e.g.*

$$Cl + Cl + M \rightarrow Cl_2 + M,$$

M being any particle, or by *adsorption* of active centres at the reactor wall or by their reaction with molecules of impurities.

Thus, *termination of chains* will be *homogeneous* for recombination or interaction with impurity molecules in the gas phase, and *heterogeneous* for surface reactions (such as adsorption).

The mechanism of a chain reaction of a halogen X_2 with hydrogen may be conceived to be

$$
\begin{array}{ll}
X_2 \rightarrow 2X & k_0 \\
X+H_2 \rightarrow HX+H & k_1 \\
H+X_2 \rightarrow HX+X & k_2 \\
H+HX \rightarrow H_2+X & k_{-1} \\
X+HX \rightarrow X_2+H & k_{-2} \\
X+X \rightarrow X_2 & k
\end{array}
$$

where $k_0, k_1 \ldots, k$ are rate coefficients of relevant elementary steps. The recombination of H atoms is neglected here in view of their high reactivity; this may be considered as justified under conditions when there are no impurities that would react with H.

Taking H and X atom concentrations as stationary, we find from the steady-state conditions

$$
\frac{d[H]}{dt} = 0 \text{ and } \frac{d[X]}{dt} = 0
$$

that

$$
[H] = \frac{k_1[H_2]+k_{-2}[HX]}{k_2[X_2]+k_{-1}[HX]}[X] \tag{3.1}
$$

and

$$
[X] = \sqrt{\frac{k_0}{k}[X_2]} = \sqrt{K_x[X_2]} \tag{3.2}
$$

where K_x is the equilibrium constant for $X_2 \rightleftarrows 2X$. From (3.1) and (3.2) we obtain the following expression for the reaction rate

$$
W = \frac{d[HX]}{dt} = 2\sqrt{K_x}\frac{k_1 k_2[H_2][X_2]^{\frac{3}{2}}-k_{-1}k_{-2}[HX]^2[X_2]^{\frac{1}{2}}}{k_2[X_2]+k_{-1}[HX]} \tag{3.3}
$$

This expression differs from the empirical formula obtained experimentally for $X = Cl, Br, I$ in that there is a second term in the numerator: the difference is accounted for by the very low value of this term. Indeed, in the range of temperatures lower than 1000 °K the value $k_{-1}k_{-2}/k_1 k_2 = K$, the equilibrium constant for $H_2 + X_2 \rightleftarrows 2HX$, will be lower than $\frac{1}{30}$ for all three halogens. Thus, even at comparable concentrations of HX, H_2 and X_2 the term in question may be neglected, and eqn. (3.3) will be reduced to

$$W = \frac{2k_1\sqrt{K_x}[H_2][X_2]^{\frac{1}{2}}}{1 + \dfrac{k_{-1}}{k_2}\dfrac{[HX]}{[X_2]}} \tag{3.3'}$$

The values of k_1 and k_{-1}/k_2 may be obtained from the reaction rates W for various temperatures and mixture compositions, provided K is known. The expression

$$k_1 = 3.4 \times 10^{12}\sqrt{T}\exp\left(-17640/RT\right)\ \text{cm}^3.\ \text{mole}^{-1}.\ \text{sec}^{-1}$$

for the reaction between hydrogen and bromine was obtained in this way by Bodenstein and Lütkemeyer[30], making use of the results of kinetic investigations on the dark reaction[29]. The value of k_{-1}/k_2 was found to be 0.12 and independent of temperature over the range of 433–575 °K.

The above reaction mechanism for $H_2 + X_2 = 2HX$ is valid for the generation and disappearance of X atoms in the gas phase, *i.e.* for a completely homogeneous process. However, under certain conditions the two reactions could occur at the reactor wall (*i.e.* by heterogeneous steps). This is the case for a reaction between chlorine and hydrogen in a glass reactor at a pressure of 60 torr and a temperature of 310 °C (p. 105). Here the initial formation of Cl atoms occurs by the reaction

$$Cl_2 + \text{wall} = Cl + Cl_{ads}$$

(Cl_{ads} denotes a chlorine atom adsorbed at the wall), and the disappearance of Cl atoms by adsorption, considering it to be a first order reaction. Consequently, the first step of the reaction mechanism (p. 114) should be replaced by

$$X_2 \rightarrow X \qquad k_0$$

and the last by

$$X \rightarrow \text{disappearance} \qquad k$$

Nevertheless, because of the steady-state H and X atom concentrations the relation connecting [H] with [X] appears to be identical to (3.1). However, instead of (3.2) we obtain, in this case, the expression

$$[X] = \frac{k_0}{k}[X_2]$$ (3.4)

and from (3.1) and (3.4) the reaction rate will be

$$W = 2\frac{k_0}{k} \cdot \frac{k_1 k_2 [H_2][X_2]^2 - k_{-1} k_{-2}[HX]^2[X_2]}{k_2[X_2] + k_{-1}[HX]}$$ (3.5)

For the initial stage of the reaction, when [HX] may be neglected, the latter equation reduces to

$$W = \frac{2k_0 k_1}{k}[H_2][X_2]$$ (3.5')

or

$$W = k_{eff}[H_2][X_2]$$

This equation, with a characteristic rate constant $k = k_{eff}$, is formally the same as that for the rate of a non-chain bimolecular reaction $H_2 + X_2 \rightarrow 2HX$[†]. This is not the only example where the macrokinetic chain reaction law could be that of a non-chain reaction (see p. 136). Thus, agreement between the theoretical and experimental kinetic expressions is far from being an undeniable criterion for the true reaction mechanism.

We can show that eqn. (3.5') is identical to the general equation (2.9) obtained for the simplest chain reaction scheme. Thus, we have here

$$W_0 = k_0[X_2], \quad v_1 = 2k_1[H_2] \quad \text{and} \quad v_2 = k$$

i.e.

$$W = W_0 \frac{v_1}{v_2}$$ (2.9)

Coincidence of the two equations, one obtained for a reaction involving two active centres (H and X) and the other a general equation for the simplest chain process occurring with participation of one active centre (A), is a direct consequence of treating the reaction $H_2 + X_2 \rightarrow 2HX$ as a one-centre reaction,

---

[†] An exception is the reaction $H_2 + I_2 \rightarrow 2HI$, which is not a chain reaction at temperatures lower than 600 °K, and only becomes a chain reaction with the above mechanism at higher temperatures[23].

which is possible because of the low steady-state concentration of H atoms.

The mechanism of the reaction between a halogen and hydrogen, either entirely homogeneous (p. 113) or involving the generation and disappearance of X atoms at the wall (p. 115), will be valid as long as there are no impurities inhibiting the reaction[†]. It was mentioned above (p.102) that oxygen is such an inhibitor; other examples are ClO_2, NCl_3, NH_3 (see ref.166, p.307–310). If an inhibitor is present, the mechanism involves reactions of H and Cl atoms with the additive molecules, resulting in chain termination. When the reaction is inhibited by molecular oxygen, these will be [31, 119, 188]

$$H + O_2 \rightarrow HO_2$$

and

$$Cl + O_2 \rightarrow ClO_2$$

followed by

$$H, Cl + HO_2 \rightarrow H_2, HCl + O_2$$

and

$$H, Cl + ClO_2 \rightarrow HCl, Cl_2 + O_2$$

Similar reactions probably account for the effect of NCl_3, one of the most active inhibitors. According to Griffiths and Norrish[71], small amounts of NCl_3 decrease the quantum yield of HCl from tens of thousands (*viz.* of HCl molecules formed per quantum of light absorbed) to a value of 2, *i.e.* result in the transition of a chain to a non-chain reaction.

3.2 REACTION OF HYDROGEN WITH OXYGEN

3.2.1 The reaction mechanism

The reaction of hydrogen with oxygen is an example of a branched chain reaction. Owing to the great amount of research carried out on hydrogen combustion, as well as to the relative chemical simplicity of the H_2–O_2 system, the mechanism of this reaction has been elucidated in greater detail than those of other branched chain reactions[146] (see also ref. 109, pp. 513–529).

[†] This mechanism does not seem to be valid if there is a great excess of chlorine (see ref. 166, p. 307).

The mechanism of the hydrogen–oxygen reaction involves the following elementary steps[†]

$H_2 + O_2 \rightarrow 2OH - 18.6$ kcal. mole^{-1} k_0 chain initiation

$OH + H_2 \rightarrow H_2O + H + 14.7$ kcal. mole^{-1} k_1 chain propagation

$H + O_2 \rightarrow OH + O - 16.7$ kcal. mole^{-1} k_2 ⎫
$O + H_2 \rightarrow OH + H - 1.9$ kcal. mole^{-1} k_3 ⎬ chain branching

$O + H_2O \rightarrow 2OH - 16.6$ kcal. mole^{-1} k_4

$H \rightarrow$ k_5 ⎫
$O \rightarrow$ k_6 ⎬ chain termination at the wall
$OH \rightarrow$ k_7 ⎭

$H + O_2 + M \rightarrow HO_2 + M + 47$ kcal. mole^{-1} k_8 chain termination in the gas phase

$HO_2 \rightarrow$ k_9 HO$_2$ radical destruction at the wall

$HO_2 + H \rightarrow 2OH$ k_{10} ⎫
$HO_2 + H_2 \rightarrow H_2O_2 + H - 16$ kcal. mole^{-1} k_{11} ⎬ chain propagation via
$HO_2 + H_2O \rightarrow H_2O_2 + OH - 30.7$ kcal. mole^{-1} k_{12} ⎭ the HO$_2$ radical

According to this mechanism the active centres are H and O atoms, and OH and HO$_2$ radicals. However, the last, due to its low reactivity, is of secondary importance for propagating the chain.

The four active centres, H, O, OH and HO$_2$, have been detected during hydrogen combustion using various techniques, the H, O, and OH concentrations being determined quantitatively. The concentration of H was measured by the catalytic probe[116] and the ESR[154] techniques, that of O by the ESR[14], and that of hydroxyl by the spectroscopic[107, 113] and ESR[15] techniques. The HO$_2$ radical was detected along with H, O, and OH using mass spectrometry[60, 84].

Fig. 11 shows H, O, and OH concentrations determined by the ESR technique, in a rarefied hydrogen flame ($p = 2.86$ torr, $T = 993\,°K$) as functions of the H$_2$–O$_2$ mixture composition[13]. It will be seen, in particular, that while the concentration of atomic hydrogen in rich mixtures is considerably higher than that of O, the reverse may be observed with very lean mixtures; the hydroxyl concentration is always much lower than that of H and O. It should be noted also that the concentration of H atoms in rich mixtures is comparable to that of H$_2$O. Com-

---

[†] This mechanism does not account for any contribution from hydrogen peroxide. It follows from experiment that H$_2$O$_2$ may play a dual role. On the one hand, dissociation of H$_2$O$_2$ by the reaction H$_2$O$_2$ + M → 2OH + M results in additional chain branching[16a] (that may apparently be called degenerate branching). On the other hand, it can act as an inhibitor, for instance by the reaction H + H$_2$O$_2$ → H$_2$ + HO$_2$, where the reactive H atom is replaced by an HO$_2$ radical of low reactivity. The inhibiting effect of hydrogen peroxide was observed by Forst and Giguère[60a].

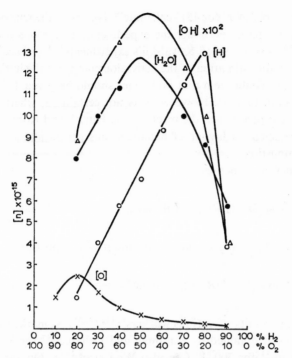

Fig. 11. The concentrations of H, O, and OH (as well as of H$_2$O) in a rarefied hydrogen flame as functions of the hydrogen–oxygen mixture composition[13]. Temp., 993 °K; pressure, 2.86 torr.

parison with the H, O and OH concentrations at thermodynamic equilibrium shows that those in Fig. 11 are higher by factors of tens, hundreds, and even thousands.

3.2.2 Rate coefficients for the elementary reactions

The elementary reaction

$$H_2 + O_2 \rightarrow 2OH - 18.6 \text{ kcal. mole}^{-1} \qquad (0)$$

is considered to be responsible for the primary generation of active centres. However, the possibility is not excluded that this will be due to

$$H_2 + O_2 \rightarrow HO_2 + H - 56.3 \text{ kcal. mole}^{-1} \qquad (0')$$

Reaction (0) seems to have been suggested for the first time by Semenov[178] as the most likely on energetic grounds; he suggested that its activation energy was about 45 kcal. mole^{-1}. According to Ripley and Gardiner[164] the rate coefficient for an initiation process, such as the tentatively chosen reaction (0), would be

$$k_0 = 10^{9.4} \exp\left(-39000/RT\right) \text{ cm}^3. \text{ mole}^{-1}. \text{ sec}^{-1}$$

If the activation energy of 39 (or 45) kcal. mole^{-1} is close to the true one, and if the generation of active particles, under the experimental conditions used by Ripley and Gardiner[164], and those of Kovalskii's experiments[118] (whose results were taken by Semenov for estimation of the activation energy for the initiation) occurs in the gas phase, it would seem that preference should be given to reaction (0), since $E < 56.3$ kcal. mole^{-1}. However, it is not certain yet whether chains are generated in the gas phase or at the reactor wall. Patrick and Robb[155] have given a convincing argument in favour of heterogeneous generation of active centres in hydrogen combustion.

The rate coefficient of the reaction

$$OH + H_2 \rightarrow H_2O + H + 14.7 \text{ kcal. mole}^{-1} \tag{1}$$

was considered until recently as closely consistent with expression

$$k_I = 10^{14.5} \exp(11000/RT) \text{ cm}^3 \text{. mole}^{-1} \text{. sec}^{-1}$$

obtained from the results of various workers[6, 11, 13, 58]. At marked variance with this is the coefficient obtained by Kaufman and DelGreco[94] for 310 °K, and by Dixon-Lewis et al.[51] for 300 °K (see also Wise et al.[218]). On the basis of the value they obtained and of that found by Fenimore and Jones[58] for a temperature close to 1400 °K, Kaufman and DelGreco proposed that

$$k_I = 10^{10.8 \pm 0.7} \exp(-5900 \pm 1000/RT) \text{ cm}^3 \text{. mole}^{-1} \text{. sec}^{-1}$$

At present the rate coefficient

$$k_{II} = 10^{14.34} \exp(-16500/RT) \text{ cm}^3 \text{. mole}^{-1} \text{. sec}^{-1}$$

may be considered as reliably established[65] for

$$H + O_2 \rightarrow OH + O - 16.7 \text{ kcal. mole}^{-1} \tag{II}$$

and[65]

$$k_{III} = 10^{13.04} \exp(-9400/RT) \text{ cm}^3 \text{. mole}^{-1} \text{. sec}^{-1}$$

for

$$O + H_2 \rightarrow OH + H - 1.9 \text{ kcal. mole}^{-1} \tag{III}$$

The rate coefficient[92]

$$k_{IV} = 10^{13.84} \exp\left(-17750/RT\right) \text{ cm}^3 . \text{ mole}^{-1} . \text{ sec}^{-1}$$

seems to be fairly reliable for

$$O + H_2O \rightarrow 2OH - 16.6 \text{ kcal. mole}^{-1} \tag{IV}$$

Termination of chains at the wall by adsorption of H atoms, O atoms, OH radicals or HO_2 radicals (reactions V, VI, VII and IX respectively) occurs in the kinetic or diffusion region, depending upon the conditions under which the overall reaction occurs. In the first case with the condition $8\lambda_A/3\gamma d \gg 1$, where λ_A is the mean free path of an active particle, the rate coefficient of the termination process will be

$$k = \frac{\gamma c}{d} \tag{3.6}$$

where γ is the adsorption probability, expressed as

$$\gamma = \gamma_0 \exp(-E/RT) \tag{3.7}$$

c is the velocity of the particles undergoing adsorption, and d is the reactor diameter. The γ value differs for different surfaces and this is one of the reasons for the marked dependence of reaction rate on the nature and pretreatment of the reactor wall often observed. For example, using the technique developed by Smith[189], the γ value for adsorption of H atoms on pure quartz over the range from 20° to 600°C has been obtained[69c] and may be represented by the following approximate expression

$$\gamma_H = 0.5 \exp\left(-5600/RT\right)$$

Greaves and Linnett[69b] have obtained the value $\gamma_0 = 1.6 \times 10^{-4}$ for O atom adsorption at 20 °C, and 1.4×10^{-2} for 600 °C. They emphasize that, according to their data, log γ is not a linear function of $1/T$. On making allowance for a rather great (up to five-fold) deviation of measured values from those calculated making use of (3.7) the value may be represented as

$$\gamma_0 = 0.2 \exp(-4600/RT)$$

In the diffusion region, where the reaction occurs under the condition $8\lambda_A/3\gamma d \ll 1$, the rate coefficient for the chain termination step in a cylindrical reactor will be

$$k = \frac{23.2D_A}{d^2} \tag{3.8}$$

where D_A is the diffusion coefficient for the active centres in the given mixture. For example, according to Azatyan[7] the diffusion coefficient for H atoms in a $2H_2 + O_2$ mixture at a pressure of 760 torr can be expressed as

$$D_H = 3.1 \times 10^{-4} T^{1.5} \text{ cm}^2. \text{ sec}^{-1}$$

When the reaction occurs in the diffusion region, the condition of the vessel wall has no effect on the reaction rate, provided the generation of active centres is independent of the surface. This was found, for instance, from the shift of the upper (second) ignition limit for mixtures of H_2 and O_2, induced by aging of the reactor[16a].

When the conditions $8\lambda_A/3\gamma d \gg 1$ or $8\lambda_A/3\gamma d \ll 1$ are not fulfilled, the reaction occurs in an intermediate region. The rate coefficients of chain termination for this region are given by Nalbandyan and Voevodskii[146].

The value of the rate coefficient for chain termination in the gas phase:

$$H + O_2 + M \rightarrow HO_2 + M + 47 \text{ kcal. mole}^{-1} \tag{VIII}$$

depends on the nature of the third particle M. For $M = H_2$, Kondratiev[111] suggested the approximate expression

$$k_{VIII, H_2} = 2.2 \times 10^{20} T^{-1.5} \text{ cm}^6. \text{ mole}^{-2}. \text{ sec}^{-1}$$

The relative efficiencies of various gases determined for a temperature range close to 800 °K are[146] (see also ref. 121)

M	H_2	O_2	N_2	H_2O	CO_2
$k_{VIII, M}/k_{VIII, H_2}$	1	0.35	0.47	5.8	1.46

The rate coefficient for reaction

$$H + HO_2 \rightarrow 2OH + 38 \text{ kcal. mole}^{-1} \tag{X}$$

has not been determined. Semenov (ref. 179, p. 557) assumes that for a temperature of about 700 °K it is $\approx 10^{12.8} \text{ cm}^3. \text{ mole}^{-1}. \text{ sec}^{-1}$.

According to Voevodskii[212] the rate coefficient for

$$HO_2 + H_2 \rightarrow H_2O_2 + H - 16 \text{ kcal. mole}^{-1} \tag{XI}$$

can be expressed approximately as [†]

$$k_{XI} = 10^{11.8} \exp(-24000/RT) \text{ cm}^3. \text{ mole}^{-1}. \text{ sec}^{-1}$$

and according to Kondratiev[110] the rate coefficient for

$$HO_2 + H_2O \rightarrow H_2O_2 + OH - 30 \text{ kcal. mole}^{-1} \tag{XII}$$

will be approximately

$$k_{XII} = 10^{13.25} \exp(-30000/RT) \text{ cm}^3. \text{ mole}^{-1}. \text{ sec}^{-1}.$$

3.2.3 Reaction kinetics at low pressures and temperatures

With relatively low pressures and temperatures the HO_2 reactions (XI and XII), as well as reactions between radicals and atoms (X and the reverse of II, III, IV) and also reaction IV, may be omitted from the mechanism (p. 118). Moreover, chain termination due to O atoms, and OH and HO_2 radicals (reactions VI, VII, and IX) may also be neglected; this is permissible for stoichiometric and rich mixtures. On the basis of this simplified reaction mechanism we obtain the following system of differential equations for H, O, and OH concentrations

$$\frac{d[H]}{dt} = k_I[H_2][OH] - \{k_{II}[O_2] + k_V + k_{VIII}[M][O_2]\}[H] + k_{III}[H_2][O]$$

$$\frac{d[O]}{dt} = k_{II}[O_2][H] - k_{III}[H_2][O]$$

$$\frac{d[OH]}{dt} = 2W_0 - k_I[H_2][OH] + k_{II}[O_2][H] + k_{III}[H_2][O]$$

First of all let us consider the case, when the concentrations of all three particles are stationary, *i.e.* when

$$\frac{d[H]}{dt} = 0, \frac{d[O]}{dt} = 0, \frac{d[OH]}{dt} = 0$$

[†] Voevodskii does not rule out the possibility that the reaction between HO_2 and H_2 is
$$HO_2 + H_2 \rightarrow H_2O + OH + 53 \text{ kcal. mole}^{-1}.$$

Instead of the three differential equations we shall have three algebraic equations, *viz.*

$$[H]_{st} = -\frac{1}{\varphi} 2W_0$$

$$[O]_{st} = -\frac{k_{II}[O_2]}{k_{III}[H_2]} \frac{1}{\varphi} 2W_0 \tag{3.9}$$

$$[OH]_{st} = -\frac{k_V + k_{VIII}[M][O_2]}{k_I[H_2]} \frac{1}{\varphi} 2W_0$$

where

$$\varphi = 2k_{II}[O_2] - k_V - k_{VIII}[M][O_2] \tag{3.10}$$

It follows that a stationary state would be possible only when $\varphi < 0$, *i.e.* when

$$2k_{II}[O_2] < k_V + k_{VIII}[M][O_2]$$

Since the probability of chain propagation via H atoms (which is at the same time the probability of branching) evidently is

$$\frac{2k_{II}[O_2]}{2k_{II}[O_2] + k_V + k_{VIII}[M][O_2]}$$

whereas that of chain termination is

$$\frac{k_V + k_{VIII}[M][O_2]}{2k_{II}[O_2] + k_V + k_{VIII}[M][O_2]}$$

the condition $\varphi < 0$ means that a stationary reaction will be an *attenuating* process.

Let us now consider the case of $\varphi > 0$. In the initial stage of the reaction the H_2 and O_2 concentrations may be considered as approximately constant. With this approximation the solution of three differential equations and the use of the appropriate rate coefficients for the elementary reactions (discussed above) shows (ref. 109, pp. 517–520) that the O and OH concentrations must be close to the steady-state values, at least for temperatures lower than 800 °K [†]. The zero values of certain concentration derivatives (with respect to time) are an illustration

---

[†] The attainment of steady-state concentrations by various active particles in a thermal reaction between hydrogen and oxygen is discussed in detail by Blackmore[24a]. It is shown, in particular, that the approximate times of attaining steady-state concentrations in a system containing $0.28\ H_2 + 0.14\ O_2 + 0.58\ N_2$ at 540 °C are: 10^{-6} (OH), 10^{-4} (O), 10^{-3} (H), and 10^{-1} (HO$_2$) sec.

of Semenov's concept of partial steady-state concentrations (p. 88). The fact that the steady-state (or close to steady-state) concentrations are those of O and OH, rather than of H is accounted for by the relation between constants k_I, k_{II}, and k_{III} (at $T \leqslant 800\,°K$) and by the mixture composition $[O_2] \leqslant \frac{1}{2}[H_2]$. Under these conditions the reactions involving O or OH have higher rates than those of H atoms reactions, and this is the reason for fast attainment of the O and OH steady-state concentrations. Under different conditions, for example at higher temperatures and with leaner mixtures, it might be the H atom concentration that would become stationary.

Assuming $d[O]/dt = 0$ and $d[OH]/dt = 0$, we obtain

$$\frac{d[H]}{dt} = 2W_0 + \varphi[H]$$

i.e. one instead of the three differential equations. In other words, the kinetic problem for three active centres becomes that for one centre. As stated above, the possibility of such an approximation is of great help in the treatment of problems connected with chemical kinetics.

The detailed mechanism of hydrogen combustion given above provides a quantitative account of all macroscopic features of this reaction. Let us consider first of all the *ignition limits*; as already stated, ignition limits (rather than a single explosion limit at any temperature) are a feature peculiar to branched chain reactions.

We have seen that the sign of φ in (3.10) determines the type of reaction occurring: if $\varphi < 0$ it is attenuating, and if $\varphi > 0$ it is an autoaccelerated reaction terminating in ignition (explosion). Thus, the equality

$$\varphi = 2k_{II}[O_2] - k_V - k_{VIII}[M][O_2] = 0 \tag{3.11}$$

is a condition for transition from an attenuating reaction (in fact, from a virtual absence of reaction) to a non-stationary reaction (ignition). Taking $[O_2] = f[M]$, where f is a quantity dependent upon the mixture composition, and $p = [M]RT$, we obtain from (3.11) the equation for the kinetic region

$$p^2 - 2\,ap + b^2 = 0 \tag{3.12}$$

and for the diffusion region

$$p^3 - 2\,ap^2 + b_0^2 = 0 \tag{3.13}$$

where

$$a = \frac{k_{II}}{k_{VIII}}\,RT,\ b^2 = \frac{k_V}{f\,k_{VIII}}\,(RT)^2 \text{ and } b_0^2 = \frac{k_V^0}{f\,k_{VIII}}\,(RT)^3$$

Let us take the first of these cases (the kinetic region). Having solved the quadratic equation (3.12), we find the pressures at the lower (p_1) and the upper (p_2) ignition limits

$$p_1 = a - \sqrt{a^2 - b^2} \quad \text{and} \quad p_2 = a + \sqrt{a^2 - b^2} \tag{3.14}$$

the sum of these being

$$p_1 + p_2 = 2a \tag{3.15}$$

and the pressure corresponding to the *vertex* of the ignition peninsula

$$p_M = p_1 = p_2 = a \tag{3.16}$$

As the temperature is raised from that at the vertex, p_2 will become considerably greater than p_1. This may be explained on the assumption that b^2 in (3.14) is small compared to a^2. In this case, p_1 and p_2 are given by

$$p_1 \simeq \frac{b^2}{2a} \quad \text{and} \quad p_2 \simeq 2a - \frac{b^2}{2a} \simeq 2a \tag{3.14'}$$

On the basis of the above expressions, the relations connecting k_{II}, k_V, and k_{VIII}, and p_1 and p_2 will be

$$\frac{k_{II}}{k_{VIII}} = \frac{p_1 + p_2}{2RT} \quad \text{and} \quad \frac{k_V}{k_{VIII}} = \frac{f}{(RT)^2} p_1 p_2 \tag{3.17}$$

Thus, k_{II}, k_V, and k_{VIII} may be obtained from these relations and from the experimental values of p_1 and p_2, if one of the three coefficients is known.

Let us consider now eqn. (3.13) for the diffusion region. It also yields two limiting pressures, p_1 and p_2, as its third root p_3 is negative and so has no physical significance. Thus, it follows from the cubic equation that

$$p_1 + p_2 + p_3 = 2a, \quad p_1 p_2 + p_3(p_1 + p_2) = 0, \quad \text{and} \quad p_1 p_2 p_3 = b_0^2$$

The last relation means that either one or all three roots, p_1, p_2, p_3, will be negative, but since from the first relation we have $p_1 + p_2 + p_3 > 0$, there will be only one negative root.

Eliminating the third root we find

$$\frac{(p_1 + p_2)^2 - p_1 p_2}{p_1 + p_2} = 2a \quad \text{and} \quad \frac{p_1^2 p_2^2}{p_1 + p_2} = b_0^2$$

and, consequently

$$\frac{k_{II}}{k_{VIII}} = \frac{(p_1 + p_2)^2 - p_1 p_2}{2RT(p_1 + p_2)} \quad \text{and} \quad \frac{k_V^0}{k_{VIII}} = \frac{f}{(RT)^3} \cdot \frac{p_1^2 p_2^2}{p_1 + p_2} \tag{3.18}$$

Thus, as with eqns. (3.17), k_{II}, k_V^0, and k_{VIII} may be obtained from (3.18), provided p_1 and p_2 are determined, and one of the coefficients is known. As k_V^0 (for a cylindrical reactor) is given by

$$k_V^0 = \frac{23.2 D_H}{d^2}$$

(*cf.* 3.8), where D_H is the diffusion coefficient for H atoms in the mixture of hydrogen with oxygen and d is the diameter of the reactor, it may be calculated to the accuracy of the diffusion coefficient. Knowing k_V^0, we may obtain k_{II} and k_{VIII} from (3.18).

As already stated (p. 99), besides the first (lower) and the second (upper) ignition limit, there is a *third ignition limit* (p_3) for the combustion of hydrogen with oxygen at pressures higher than p_2. Usually it is of a thermal nature. However, when the reaction is artificially slowed down (by appropriate pre-treatment of the reactor wall) the third limit may be of a chain nature (as observed for isothermal conditions).

Nalbandyan and Voevodskii[146] have shown that the above reaction mechanism gives a correct description of the experimental features of the reaction between hydrogen and oxygen not only inside the ignition peninsula, but outside it as well. In particular, this mechanism gives a correct quantitative account of slow hydrogen oxidation between the second and the third ignition limits and shows the possibility of chain ignition at the third limit. This appears to be possible if steps XI and XII are included in the reaction mechanism.

The rate of a slow reaction occurring outside the ignition limits may be obtained from the steady-state conditions

$$\frac{d[H]}{dt} = 0, \frac{d[O]}{dt} = 0, \frac{d[OH]}{dt} = 0, \frac{d[HO_2]}{dt} = 0$$

Assuming that the heterogeneous disappearance of H and O atoms and OH radicals may be neglected at high pressures, we obtain from the above relations the following expression for the rate of a stationary reaction

$$W = \frac{d[H_2O] + d[H_2O_2]}{dt} = 2W_0 \frac{\frac{3}{2} + \frac{k_{XI}}{k_{IX}}[H_2]}{1 - \frac{2k_{II}}{k_{VIII}[M]}\left\{1 + \frac{k_{XI}}{k_{IX}}[H_2]\right\}} \tag{3.19}$$

It will be seen that the reaction remains stationary until the denominator is higher than zero, *i.e.* until inequality

$$\frac{2k_{II}}{k_{VIII}[M]} \left\{ 1 + \frac{k_{XI}}{k_{IX}} [H_2] \right\} < 1$$

is satisfied. With increasing pressure the value of the left hand side of this inequality increases and at a certain pressure it may become unity[†]. At this moment the mixture will ignite. Thus, under these circumstances the condition for chain ignition may be written as

$$\frac{2k_{II}}{k_{VIII}\,p} RT \left\{ 1 + \frac{k_{XI}}{k_{IX}^0} \cdot \frac{1-f}{(RT)^2} p^2 \right\} = 1 \tag{3.20}$$

Introducing

$$a = \frac{k_{II}}{k_{VIII}} RT \text{ and } c = \frac{k_{XI}}{k_{IX}^0} \cdot \frac{1-f}{(RT)^2}$$

we rewrite (3.20) as

$$p^2 - \frac{1}{2ac} p + \frac{1}{c} = 0 \tag{3.21}$$

The roots of this equation evidently represent pressures at the second (p_2) and third (p_3) ignition limits, and it is found that $p_2 + p_3 = 1/2ac$. Thus, if $p_3 \gg p_2$ the following approximate expression will be obtained for the third limit

$$p_3 = \frac{1}{2ac} = \frac{k_{VIII} k_X^0}{2k_{II} k_{XI}} \cdot \frac{RT}{1-f} \tag{3.22}$$

Since $p_2 p_3 = 1/c$

$$p_2 = 2a = \frac{2k_{II}}{k_{VIII}} RT \tag{3.23}$$

which is the same as the approximate expression (3.14′) obtained by neglecting the value of p_1.

Let us consider also the problem of *initial acceleration* characteristic of this

[†] It will be noted that due to the high pressures, slow oxidation occurs in the diffusion region, whence $k_{IX}^0 \sim 1/p$.

branched chain reaction under isothermal conditions. The problem was treated in a general way earlier (p. 101).

Integrating equation

$$\frac{d[H]}{dt} = 2W_0 + \varphi[H]$$

we obtain

$$[H] = \frac{2W_0}{\varphi}\{\exp(\varphi t) - 1\} \tag{3.24}$$

an expression identical to (2.18). For the concentrations of the other two active centres, O and OH, we find on the basis of relevant stationary-state relations and (3.24)

$$[O] = \frac{k_{II}[O_2]}{k_{III}[H_2]}[H] = \frac{k_{II}[O_2]}{k_{III}[H_2]} \cdot \frac{2W_0}{\varphi}\{\exp(\varphi t) - 1\} \tag{3.25}$$

and

$$[OH] = \frac{2W_0 + 2k_{II}[O_2][H]}{k_I[H_2]} = 2W_0 \cdot \frac{1 + 2k_{II}[O_2] \cdot \dfrac{\exp(\varphi t) - 1}{\varphi}}{k_I[H_2]} \tag{3.26}$$

Thus, we may see that in the initial stage of the reaction the concentrations of all three active centres will increase following practically the same exponential law. This is particularly evident at sufficiently high t, when the above equations reduce to

$$[H] = \frac{2W_0}{\varphi}\exp(\varphi t) \tag{3.24'}$$

$$[O] = \frac{k_{II}[O_2]}{k_{III}[H_2]} \cdot \frac{2W_0}{\varphi}\exp(\varphi t) \tag{3.25'}$$

and

$$[OH] = \frac{2k_{II}[O_2]}{k_I[H_2]} \cdot \frac{2W_0}{\varphi}\exp(\varphi t) \tag{3.26'}$$

From the simplified reaction mechanism (p. 123) by summation

$$\frac{d[M]}{dt} = \frac{d[H_2]}{dt} + \frac{d[O_2]}{dt} + \frac{d[H_2O]}{dt}$$

$$+ \frac{d[H]}{dt} + \frac{d[O]}{dt} + \frac{d[OH]}{dt} + \frac{d[HO_2]}{dt}$$

we obtain

$$\frac{d[M]}{dt} = -\{\tfrac{1}{2}k_V + k_{VIII}[M][O_2]\}[H] \tag{3.27}$$

from which, making use of (3.24) and changing to pressure p, we have

$$\frac{dp}{dt} = A\varphi[\exp(\varphi t) - 1] \tag{3.28}$$

where

$$A = \frac{2W_0}{\varphi^2}\{\tfrac{1}{2}k_V + k_{VIII}[M][O_2]\}RT$$

Integration of (3.28) at constant A and φ, which is permissible for the initial stage of the reaction, gives [†]

$$\Delta p = A[\exp(\varphi t) - (1 + \varphi t)] \tag{3.29}$$

Assuming that t is sufficiently high, (3.29) reduces to

$$\Delta p = A \exp(\varphi t) \tag{3.29'}$$

When the reaction rates are determined from the pressure variations, as is often the case, denoting by Δp_{min} the minimum change in pressure measurable (at $t = \tau_i$), we obtain from (3.29')

$$\tau_i = \frac{1}{\varphi} \ln \frac{\Delta p_{min}}{A} \tag{3.30}$$

As at $p = p_1$ and $p = p_2$ the induction period τ_i will be infinite, and at $p_1 < p < p_2$ it has finite values, the curve of τ_i *versus* p will have a minimum at a certain pres-

---

[†] Strictly speaking, integration from $t = 0$ is not precise, as at $t = 0$ the O and OH concentrations are not stationary, and this was allowed for in (3.24) and thus in (3.28). However, it may be shown that the time for attainment of steady-state concentrations is sufficiently short so that the lower integration limit may be taken as zero.

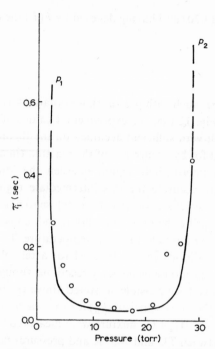

Fig. 12. The induction period as a function of the $2H_2+O_2$ pressure at 731 °K[143].

sure somewhere between p_1 and p_2, and two branches tending to infinity close to $p = p_1$ and $p = p_2$. This was the dependence of τ_i on p established by Nalbandyan[143] for hydrogen–oxygen mixtures of various compositions, in different reactors and at various temperatures. Fig. 12 shows induction periods (circles) as functions of pressure for a stoichiometric mixture of hydrogen with oxygen at 731 °K. The full curve was plotted by Nalbandyan making use of the expression

$$\tau_i = \frac{F(T)}{(p-p_1)(p_2-p)} \tag{3.31}$$

with his measured values of $p_1 = 1.6$ and $p_2 = 28.4$ torr, and with an arbitrary value of $F(T)$.

The above expression was obtained on the assumption that, within the range from p_1 to p_2, the reaction occurs in the kinetic region. The expression

$$\varphi = 2k_{II}[O_2]-k_V-k_{VIII}[M][O_2]$$

for this region may be written as

$$\varphi = \frac{k_{VIII}f}{(RT)^2}(p-p_1)(p_2-p) \tag{3.32}$$

Substituting (3.32) into (3.30) and having denoted by $F(T)$ the quantity

$$\ln \frac{\varDelta p}{A} \bigg/ \frac{k_{\text{VIII}}}{(RT)^2}$$

(ln $\varDelta p/A$ does not change much with pressure), we obtain (3.31).

It will be seen from Fig. 12 that the experimental values of τ_i (here and in all similar cases) do not fall with sufficient accuracy on the theoretical curve (3.31). This might be accounted for by occurrence of the reaction (in all the cases studied by Nalbandyan) in an intermediate region between the kinetic and diffusion regions, (3.32) not being accurate for this intermediate region[†]. Yet the main reason seems to be that at pressures of tens of torr (close to p_2) there is appreciable heating of the mixture during the induction period and, consequently, p_1 and p_2 cannot be considered as referring to the same temperature. Thus the $\tau_i : p$ relation obtained under these conditions cannot be used for a reliable determination of the rate coefficients of appropriate elementary reactions, though in principle such determinations might have been possible, as stated above (p. 102).

In order to determine the rate coefficient (k_{II}) for the reaction $H + O_2 = OH + O$ the induction periods for a $2H_2 + O_2$ mixture were measured[91] close to the lower limit at temperatures between 733 and 783 °K and pressures between 0.33 and 1.1 torr; under these conditions heating of the mixture will not occur. In this case we obtain instead of (3.32)

$$\varphi = \frac{2k_{\text{II}}f}{RT} (p - p_1) \tag{3.34}$$

whence

$$\varphi \tau_i = \frac{2k_{\text{II}}f}{RT} (p - p_1)\tau_i = \ln \frac{\varDelta p_{\min}}{A} \tag{3.35}$$

A is given by

$$A = \frac{W_0 k_{\text{V}}}{\varphi^2} RT \tag{3.36}$$

[†] For the diffusion region we have

$$\tau_i = \frac{F(T)}{\left(p_1 + p_2 - \dfrac{p_1 p_2}{p_1 + p_2}\right) p - p_2^2 - \dfrac{p_1^2 p_2^2}{p_1 + p_2} \dfrac{1}{p}} \tag{3.33}$$

For the data in Fig. 12, calculation making use of this expression gives τ_i values in fact coinciding with those obtained from (3.31).

(see p. 130) or, since $k_V = 2k_{II} f p_1/RT$, by

$$A = \frac{2k_{II} f W_0 p_1}{\varphi^2} \tag{3.36'}$$

Then, taking $W_0 = k_0[H_2]_0[O_2]_0 = k_0 f(1-f) p^2/(RT)^2$ and making use of the value of φ (3.34), we obtain

$$A = \tfrac{1}{3} p_1 \frac{p^2}{(p-p_1)^2} \frac{k_0}{k_{II}} \tag{3.36''}$$

(for a stoichiometric mixture, $f = \tfrac{1}{3}$), and

$$\ln \frac{\Delta p_{min}}{A} = \ln \left\{ \frac{3\Delta p_{min}}{p_1} \left(\frac{p-p_1}{p} \right)^2 \frac{k_{II}}{k_0} \right\} \tag{3.37}$$

It may be shown that $\ln \Delta p_{min}/A$ and consequently $\varphi \tau_i$ do not change much under the experimental conditions[91]: deviations of individual logarithms from their mean value do not exceed 5%. Thus, assuming that $\varphi \tau_i$ is constant, the temperature dependence of the product $(p-p_1)\tau_i$ should be identical to that of $1/(k_{II}/T)$. In fact, it was shown[91] that $\log(p-p_1)\tau_i$ is a linear function of $1/T$ (see Fig. 13), *i.e.*

$$(p-p_1)\tau_i = B \exp (E/RT)$$

a value of 15.4 kcal. mole^{-1}, close to the activation energy for reaction II, being obtained for E.

A value practically the same as 16.5 kcal. mole^{-1} for E_{II} may be obtained in the following way. The quantity $\ln \Delta p_{min}/A$ (3.37) may be derived from the expressions for k_0 and k_{II} (pp. 119, 120) and from the results reported by Semenov *et al.*[91], taking that $\Delta p_{min} = 10^{-4}$ (see ref. 179, p. 508). Such a calculation yields

$$\ln \frac{\Delta p_{min}}{A} = 12.3 \exp (710/RT)$$

Then, from the results given in Fig. 13, we find by the least squares procedure

$$(p-p_1)\tau_i = 6.06 \times 10^{-6} \exp (15675/RT) \text{ torr. sec}$$

Knowing $\ln \Delta p_{min}/A$ and $(p-p_1)\tau_i$ we obtain from (3.35)

$$\frac{k_{II}}{T} = 10^{11.28} \exp (-14965/RT) \text{ cm}^3. \text{ mole}^{-1}. \text{ sec}^{-1}. \text{ deg}^{-1}$$

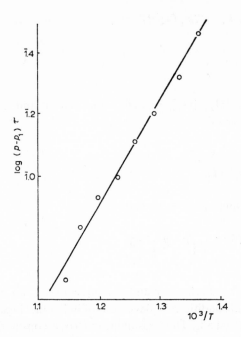

Fig. 13. Dependence of log $(p-p_1)\tau$ on temperature[91].

or, representing T as $10^{3.34} \exp(-1595/RT)$, we have

$$k_{II} = 10^{14.62} \exp(-16560/RT) \text{ cm}^3. \text{ mole}^{-1}. \text{ sec}^{-1}$$

Comparing this expression with that on p. 120 we see that the difference in the values of k_{II} calculated from the two expressions is less than a factor of 2.

Let us consider also the *macrokinetic law* of a reaction making allowance for consumption of the reactants. In all the above expressions the H_2 and O_2 concentrations were taken as constant, since it was the initial reaction period that was discussed. Consumption of these gases should be allowed for at higher degrees of conversion. Thus, the problem of differential equations with varying coefficients arises; they cannot be solved analytically for a general case. However, by making use of the steady-state approximation it is often possible to reduce the problem to solution of a system of two equations, which would permit a relatively simple analytical solution for certain individual cases. As a result of the solution, definite relations between the concentrations of active centres and those of initial substances would be obtained, and these would make possible the expression of reaction rates in terms of concentrations of initial substances only. This is how the *macrokinetic law* of a reaction is obtained.

Let us illustrate this procedure for hydrogen combustion, taking for simplicity the case of high pressures, when termination of chains at the wall may be neglected.

From the simplified reaction mechanism adopted and considering the O, OH, and HO_2 concentrations as stationary, we obtain the following differential equations for H and O_2

$$\frac{d[H]}{dt} = 2W_0 + \{2k_{II}[O_2] - k_{VIII}[M][O_2]\}[H]$$

and

$$\frac{d[O_2]}{dt} = W_0 - \{k_{II}[O_2] + \tfrac{1}{4}k_{VIII}[M][O_2]\}[H]$$

After a certain very short time, t_x, the rate of initiation W_0, may be considered as low compared with the overall rate, and so

$$\frac{d[H]}{dt} = \{2k_{II} - k_{VIII}[M]\}[O_2][H]$$

and

$$-\frac{d[O_2]}{dt} = \{k_{II} + \tfrac{1}{4}k_{VIII}[M]\}[O_2][H]$$

Dividing one of these equations by the other and integrating, we find

$$[O_2]_x - [O_2] = C\{[H] - [H]_x\}$$

where

$$C = \frac{k_{II} + \tfrac{1}{4}k_{VIII}[M]}{2k_{II} - k_{VIII}[M]}$$

and $[O_2]_x$ and $[H]_x$ are the O_2 and H concentrations at t_x. On the assumption that $[O_2]_x = [O_2]_0$ and $[H]_x = 0$, this equation reduces to

$$[O_2]_0 - [O_2] = C[H]$$

Substituting [H] into the equation for the variation in oxygen concentration and taking this as a measure of the reaction rate, we obtain

$$W = -\frac{d[O_2]}{dt} = \{2k_{II} - k_{VIII}[M]\}\{[O_2]_0 - [O_2]\}[O_2]$$

or, denoting $[O_2]/[O_2]_0$ by $1-\eta$ and

$$\{2k_{II}-k_{VIII}[M]\}[O_2]_0$$

by k, we have

$$\frac{d\eta}{dt} = k\eta(1-\eta) \tag{3.38}$$

Expression (3.38) is formally identical to the usual expression for the rate of an *autocatalytic* reaction. The similarity of the macrokinetic law for a branched chain reaction and that for an autocatalytic reaction was shown by Semenov[177] for the general case, and he has established the limits of this similarity.

3.2.4 Concentration limits for ignition of mixtures of hydrogen with oxygen

Experiment shows that the ignition limits, p_1 and p_2, for a mixture of hydrogen with oxygen, show a definite dependence on the mixture composition. The dependence may also be established theoretically on the basis of a reaction mechanism. However, in view of the wide range of variation in H_2 and O_2 concentrations, or in the value of $f = [O_2]/[H_2]+[O_2]$, the simplified reaction mechanism cannot be the starting point, as it allows for the heterogeneous destruction of one active centre only (H atoms). Let us consider the overall reaction mechanism (p. 118), neglecting for simplicity the reactions of HO_2 with H, H_2, and H_2O (X, XI, and XII), which are unimportant in the pressure range of p_2 to p_1, as well as reaction IV, *i.e.* considering as negligible the water content before ignition of a mixture. In addition no account will be taken of the initial generation of active centres, *i.e.* of W_0. Then from the conditions for a stationary state, *viz.*

$$\frac{d[H]}{dt} = 0, \quad \frac{d[O]}{dt} = 0, \quad \frac{d[OH]}{dt} = 0, \quad \text{and} \quad \frac{d[HO_2]}{dt} = 0,$$

we obtain the following equation

$$k_I k_{III} k_{VIII}[M][H_2]^2[O_2]+(k_I k_{VI}+k_{III} k_{VII})k_{VIII}[M][H_2][O_2]$$

$$-2k_I k_{II} k_{III}[H_2]^2[O_2]+k_I k_{III} k_V[H_2]^2+k_{VI} k_{VII} k_{VIII}[M][O_2]$$

$$+(k_I k_{VI}+k_{III} k_{VII})k_V[H_2]+k_{II} k_{VI} k_{VIII}[O_2]+k_V k_{VI} k_{VII} = 0.$$

Let us consider first the region close to the lower ignition limit. This problem was solved by Nalbandyan and Voevodskii[146], allowance being made for termination involving two active centres. Termination by O atoms was not taken into account on the assumption (which proved to be wrong later on) that their concentration was low compared to that of hydroxyl[†]. However, the problem may be solved without making this restriction.

Neglecting chain termination in the gas phase, we obtain from the above equation

$$p^3 - ap^2 - bp - c = 0 \qquad (3.39)$$

where

$$a = \frac{k_V}{2k_{II}f} RT$$

$$b = \frac{(k_I k_{VI} + k_{III} k_{VII})k_V(1-f) + k_{II} k_{VI} k_{VII} f}{2k_I k_{II} k_{III} f(1-f)^2} (RT)^2$$

$$c = \frac{k_V k_{VI} k_{VII}}{2k_I k_{II} k_{III} f(1-f)^2} (RT)^3$$

and

$$f = \frac{[O_2]}{[H_2] + [O_2]}$$

It may readily be shown that equation (3.39) has one positive and two negative roots, the first being higher in absolute magnitude than the sum of the two others. Thus, as an approximation the constant term in this equation may be neglected, whence it follows that

$$p^2 - ap - b = 0 \qquad (3.40)$$

Equation (3.40) yields

$$p_1 = \frac{\beta}{f} \left\{ 1 + \sqrt{1 + \omega \frac{f}{1-f} + \omega' \left(\frac{f}{1-f}\right)^2} \right\} \qquad (3.41)$$

[†] The heterogeneous destruction of two active centres, H and O atoms, was taken into account for the first time by Kassel[93]. However, he limited himself to qualitative investigation of two extreme cases only: the destruction of H atoms neglecting that of O, and the destruction of O atoms neglecting that of H.

where

$$\beta = \frac{k_V}{4k_{II}} RT$$

$$\omega = \frac{8k_{II}(k_I k_{VI} + k_{III} k_{VII})}{k_I k_{III} k_V}$$

$$\omega' = \frac{8k_{II}^2 k_{VI} k_{VII}}{k_I k_{III} k_V^2}$$

The root in expression (3.41) has a third term and this is the difference between (3.41) and the equation obtained by Nalbandyan and Voevodskii[146] assuming heterogeneous destruction of only two active centres, H atoms and OH radicals, *i.e.* $k_{VI} = 0$.

The theoretical expression (with $\omega' = 0$) was verified by Nalbandyan and Voevodskii, making use of experimental results reported by various authors. We shall mention here only those of Ivanov and Nalbandyan[85] since they were obtained for pressures at the lower limit of less than 1 torr, when the reaction may be considered to occur in the kinetic region. Fig. 14 shows the results obtained for three temperatures over a wide range of f-values (0.05 to 0.95). These results will be used here for the determination of the coefficient (k_V) for heterogeneous destruction of H atoms.

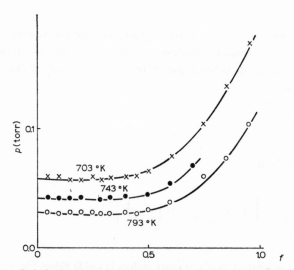

Fig. 14. The lower limit for ignition of a hydrogen–oxygen mixture as a function of the oxygen content (f) at different temperatures. The drawn curves are calculated making use of eqn. (3.41) (see Nalbandyan and Voevodskii[146], p. 65).

It follows from (3.41) that as $f \to 0$ the product $p_1 f$ tends to a value of

$$2\beta = \frac{k_V}{2k_{II}} RT.$$

We find from Fig. 14 that $2\beta = 0.0594$ (703 °K), 0.0419 (743 °K), and 0.0295 (793 °K). Substituting the values of k_{II} (p. 120) and RT into the expression for β, we obtain $k_V = 4.41$ (703 °K), 5.48 (743 °K) and 7.37 (793 °K). Then, making use of eqn. (3.6), we find the adsorption probability γ_H [†] to be

$$\gamma_H = 3.6 \times 10^{-3} \exp\left(-5700/RT\right)$$

Assuming that chain termination is of secondary importance at pressures close to the upper ignition limit, we obtain from the general equation (p. 136)

$$\frac{p_2}{RT} = \frac{2k_{II}}{k_{VIII}} - \frac{k_V}{2k_{II}} \cdot \frac{1}{f} - \frac{k_{VI}}{k_{III}} \cdot \frac{1}{1-f} - \frac{k_{VII}}{k_I} \cdot \frac{1}{1-f} \tag{3.42}$$

However, close to the upper ignition limit the reaction will most probably occur in the diffusion region. Consequently, it may be assumed that:

$$k_V = \frac{k_V^0}{p_2}, \qquad k_{VI} = \frac{k_{VI}^0}{p_2} \quad \text{and} \quad k_{VII} = \frac{k_{VII}^0}{p_2}$$

Then, since heterogeneous chain termination is of secondary importance, (3.42) may be transformed to

$$p_2 = \frac{2k_{II}}{k_{III}} RT - \left[\frac{k_V^0}{2k_{II}} \cdot \frac{1}{f} + \frac{k_{VI}^0}{k_{III}} \cdot \frac{1}{1-f} + \frac{k_{VII}^0}{k_I} \cdot \frac{1}{1-f} \right] \frac{k_{VIII}}{2k_{II}} \tag{3.42'}$$

The dependence of k_{VIII}, k_V^0, k_{VI}^0, and k_{VII}^0 on f must be allowed for in this expression. k_{VIII} is taken as

$$k_{VIII} = k_{VIII, H_2}(1 - 0.65f)$$

where k_{VIII, H_2} is the rate coefficient for the reaction

$$H + O_2 + H_2 = HO_2 + H_2 \text{ (p. 122)}$$

[†] A quartz reactor (diam. 5.8 cm) pretreated with hydrofluoric acid and potassium tetraborate was used in the experiments[85].

As for k_V^0, etc., these coefficients are proportional to the relevant diffusion coefficients (3.8) and consequently use may be made of the expressions[217] (see also ref. 65, p. 266)

$$1/(D_{760}^{273})_H = \frac{1-f}{1.70} + \frac{f}{1.01} \, cm^{-2}. \, sec$$

$$1/(D_{760}^{273})_O = \frac{1-f}{0.98} + \frac{f}{0.30} \, cm^{-2}. \, sec$$

and

$$1/(D_{760}^{273})_{OH} = \frac{1-f}{1.03} + \frac{f}{0.33} \, cm^{-2}. \, sec$$

Then we obtain

$$k_V^0 = \frac{6.65}{d^2} \frac{T^{1.5}}{1+0.68f}, \qquad k_{VI}^0 = \frac{3.84}{d^2} \frac{T^{1.5}}{1+2.27f}$$

and

$$k_{VII}^0 = \frac{4.03}{d^2} \frac{T^{1.5}}{1+2.12f}$$

The condition for the maximum value of p_2 at a particular temperature is:

$$\frac{dp_2}{df} = F_1(k_I, k_{II}, k_{III}, k_{VIII}, f_M) = 0$$

and from equation (3.42) for p_2:

$$(p_2)_{max} = F_2(k_I, k_{II}, k_{III}, k_{VIII}, f_M)$$

Substituting into these expressions k_I, k_{II}, and k_{III} at 704 °K calculated from the formulae given on p. 120, we find $f_M = 0.86$ and $k_{VIII, H_2} = 0.93 \times 10^{16} \, cm^6.$ $mole^{-2}. \, sec^{-1}$. It will be noted that the latter value practically coincides with that for k_{VIII, H_2} ($1.18 \times 10^{16} \, cm^6. \, mole^{-2}. \, sec^{-1}$) calculated from the expression given on p. 122.

Fig. 15 shows the results obtained by Chirkov[41] at 704 °K. The curve is plotted making use of (3.42') and the values of k_I, k_{II}, k_{III}, and k_{VIII}. Thus, theory and experiment are in pretty good agreement.

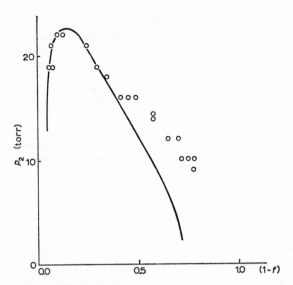

Fig. 15. The upper ignition limit as a function of the hydrogen content $(1-f)$ at 431 °C[41]. Drawn curve calculated making use of eqn. (3.42′).

3.2.5 Shifts of ignition limits by impurities (or additives)

Two classes of impurities affecting the development of a branched chain reaction and, thus, the ignition limits will be distinguished: those entering into no chemical reactions with the components of the mixture, and those entering into such reactions. The first will be called *inert*, and the second *active impurities*.

The inert substances have an effect on the lower and upper ignition limits. That on the lower limit is due to hindrance of the diffusion of active centres to the reactor wall; this leads to a *decrease* in the lower limit value, *i.e.* to an extension of the ignition zone. This effect will be expected when the reaction occurs in the diffusion or intermediate regions. Very instructive in this connection is the result obtained by Biron and Nalbandyan[24] who studied the effect of argon on the lower limit for ignition of a $2H_2 + O_2$ mixture in a quartz and in a pyrex reactor. In the first case, at a temperature of 791 °K, the lower limit lies at pressures of the order of 10 torr, in the second at those of about 0.1 torr. In agreement with the above, the lower limit pressure with a pyrex reactor, where the reaction occurs in the kinetic range, appears to be independent of addition of argon, whereas an appreciable drop of p_1 on dilution of the mixture by argon is observed for a quartz reactor. This will be seen from Table 1.

The inert impurities might also have an effect on the adsorption capacity of the reactor wall, and this may show in a change of the γ-value and, thus, of the coefficient for heterogeneous termination. Such an effect of inert impurities would

TABLE 1

EFFECT OF ADDITION OF ARGON IN PYREX AND QUARTZ REACTORS

Pyrex		Quartz	
p_{Ar} (torr)	p_1 (torr)	p_{Ar} (torr)	p_1 (torr)
0.00	0.15	0.0	11.2
0.04	0.15	8.5	8.5
0.07	0.14	14.5	7.3
0.10	0.14	20.5	6.9
0.15	0.15		

be expected for the kinetic region. However, this possibility does not seem to have been investigated.

Inert impurities have a direct effect on the rate coefficient for chain termination in the gas phase and, thus, on the upper ignition limit (p_2). For example, the rate coefficient k_{VIII} for the reaction of hydrogen with oxygen in the presence of an impurity may be written as

$$k_{VIII} = \frac{k_{VIII, H_2} p_{H_2} + k_{VIII, O_2} p_{O_2} + k_{VIII, x} p_x}{p} \tag{3.44}$$

where k_{VIII, H_2}, k_{VIII, O_2}, $k_{VIII, x}$ are individual coefficients for the reaction

$$H + O_2 + M = HO_2 + M \tag{VIII}$$

when M is H_2, O_2, or X, and p_{H_2}, p_{O_2}, p_x, are the partial pressures ($p_{H_2} + p_{O_2} + p_x = p$).

Denoting by p_2 and p_2^0 the pressures at the upper limit in the presence and absence of an inert impurity, and by k_{VIII} and k_{VIII}^0 the rate coefficients for homogeneous chain termination, we obtain

$$\frac{p_2}{p_2^0} = \frac{k_{VIII}^0}{k_{VIII}} = \frac{f_{H_2}^0 + 0.35 f_{O_2}^0}{f_{H_2} + 0.35 f_{O_2} + \kappa f_x} \tag{3.45}$$

where

$$f_{H_2} = \frac{p_{H_2}}{p_{H_2} + p_{O_2} + p_x},$$

etc., and $\kappa = k_{VIII, x}/k_{VIII, H_2}$. For a mixture of a composition

$$f_{H_2}^0 : f_{O_2}^0 = f_{H_2} : f_{O_2} = \frac{1-f}{f}$$

(3.45) reduces to

$$\frac{p_2}{p_2^0} = \frac{1+\alpha_X}{1 + \dfrac{\kappa}{1-0.65f}\alpha_X} \tag{3.46}$$

where $a_X = f_X/f_{H_2}+f_{O_2}$. It will be seen from (3.46) that for a stoichiometric mixture ($f = \frac{1}{3}$) the upper limit will become lower or higher if α_X decreases or increases depending on whether the κ value is higher or lower than $1-0.65f$. This has been confirmed experimentally. For instance, comparing the effect on the upper limit of nitrogen and CO_2, inert impurities for which the κ values are 0.43 and 1.47, respectively, it will be expected that with a $2H_2 + O_2$ mixture the presence of nitrogen will make p_2 higher, *i.e.* the ignition region will become wider, whereas carbon dioxide will make p_2 lower. In order to illustrate this effect on the upper limit Nalbandyan and Voevodskii[146] cite the following results.

f_{N_2}	f_{CO_2}	p_2 (torr)
0	—	85.6
0.25	—	94.5
1.05	—	108.8
2.35	—	124.5
—	0.26	72.8
—	1.05	60.07
—	2.30	52.1

In considering the effect of *chemically active impurities* let us distinguish that leading to acceleration of the reaction and, thus, to extension of the ignition zone, and that resulting in inhibition of the reaction (narrowing of the ignition zone), as well as that which might result either in acceleration or in inhibition depending upon the impurity concentration.

The acceleration effect is characteristic of active impurities, in particular of H and O atoms introduced into the reaction zone either from an electric discharge or by UV illumination of the mixture. Figs. 16 and 17 show the results obtained by Nalbandyan[141] on the ignition of a $2H_2+O_2$ mixture upon admission of H and of O atoms from an electric discharge, and Fig. 18 those[142] for illumination by UV light of various intensities. In the latter case O is the active additive.

Theoretical interpretation of the ignition zone extension by active particles (shift of the upper limit), based on *positive chain interaction* as suggested by Semenov, was given by Nalbandyan[144] and by Semenov (ref. 179, pp. 554–560) for the hydrogen/oxygen system. Nalbandyan believes that the effect is due to the

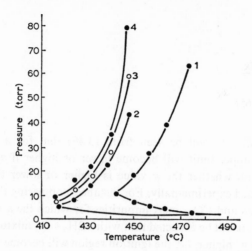

Fig. 16. Extension of the ignition region for a $2H_2+O_2$ mixture under the action of hydrogen atoms: (1) H atoms absent, (2) in the presence of H atoms obtained at a current of 0.25 amp in the discharge tube, (3) same at a current of 0.35 amp, (4) same at 0.5 amp[141].

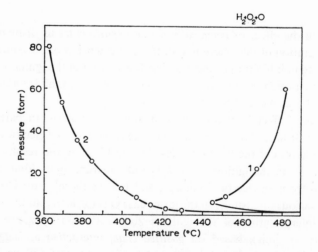

Fig. 17. Extension of the ignition region for a $2H_2+O_2$ mixture. (1) without additive, (2) in the presence of oxygen atoms obtained from an electric discharge[141].

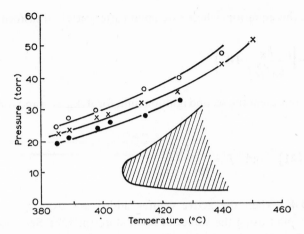

Fig. 18. Extension of the ignition region for a $2H_2+O_2$ mixture under the action of UV light of various intensities[142]. The ignition region for absence of illumination is shaded.

basic reaction

$$HO_2 + HO_2 \rightarrow H_2O + O_2 + O + 7 \text{ kcal. mole}^{-1}$$

whereas Semenov considers that it is due to

$$HO_2 + H \rightarrow 2OH + 38 \text{ kcal. mole}^{-1}$$

Semenov's view will be adopted here. When there is an additional source of active centres entering the reaction zone, the interaction of two active particles is assumed to become of importance. Then, introducing into the reaction mechanism step (X) from the hydrogen combustion scheme (p. 118), as well as step (IX), from the conditions

$$\frac{d[H]}{dt} = 0, \qquad \frac{d[O]}{dt} = 0, \qquad \frac{d[OH]}{dt} = 0, \qquad \frac{d[HO_2]}{dt} = 0$$

and neglecting heterogeneous destruction of H, O, and OH, we obtain

$$[H]^2 - \frac{\{k_{VIII}[M][O_2] - 2k_{II}[O_2]\}k_{IX} - 2\Delta W_0 k_X}{2k_{II}k_X[O_2]}[H] + \frac{\Delta W_0 k_{IX}}{k_{II}k_X[O_2]} = 0$$

where ΔW_0 is the number of additional active centres admitted into the mixture.

According to Semenov, an additional condition for ignition is the equality of the two roots of this equation, whence we obtain after certain transformations

$$\frac{p_2}{p_2^0} = \left\{ 1 + \sqrt{\frac{k_X}{k_{II} k_{IX}^0 f} \Delta W_0} \right\}^2 \tag{3.47}$$

where p_2 and p_2^0 represent the second ignition limit for $\Delta W_0 \neq 0$ and for $\Delta W_0 = 0$, respectively,

$$k_{IX}^0 = k_{IX}[M] \quad \text{and} \quad f = \frac{[O_2]}{[H_2] + [O_2]}$$

Estimating ΔW_0 when the active centres added are oxygen atoms formed photochemically, and using the values of k_{IX} and k_X (p. 122), Semenov obtained for the limit shift a value close to that observed by Nalbandyan.

Obviously, eqn. (3.47) provides a reasonable explanation of the effect of relatively low concentrations of the active impurity (low ΔW_0). At high concentrations, particularly such as those when O atoms are obtained from an electric discharge (Fig. 17), the heating of the gas in the pre-ignition period, and the reactions accounting for the third (chain) ignition limit, particularly reaction XI, as well as that of heterogeneous chain termination, must be taken into account. At high concentrations, Fig. 17 shows that the three limits seem to merge into one.

Substances yielding *less active radicals* by reactions with the atoms and radicals which are the normal chain carriers belong to chemically active impurities that narrow the ignition zone. An example of such impurities in the combustion of hydrogen are the hydrocarbons RH that react with hydrogen atoms

$$H + RH \rightarrow H_2 + R \tag{XIII}$$

to form radicals R. The reaction of these with O_2 molecules resulting in chain branching

$$R + O_2 \rightarrow RO + O$$

is rather slow compared to a similar reaction (II) involving H atoms.

When the mixture of hydrogen with oxygen contains a small amount of a hydrocarbon, the condition $\varphi = 0$ for chain ignition limits could be written as

$$\varphi = 2k_{II}[O_2] - k_V - k_{VIII}[M][O_2] - k_{XIII}[RH] = 0 \tag{3.48}$$

Let us discuss the effect of impurities on the lower (p_1) and upper (p_2) ignition limits.

For the first case (the lower limit), neglecting chain termination in the gas phase, we obtain from (3.48) for the kinetic region

$$\frac{p_1 - p_1^0}{p_1^0} = \frac{k_{XIII}}{k_V}[RH] \qquad (3.49)$$

where p_1^0 is the pressure at the lower limit in the absence of an impurity. It follows from (3.49) that the relative shift of the lower ignition limit is proportional to the impurity concentration.

A somewhat more complicated expression is obtained for the diffusion region, viz.

$$\Delta p_1 \frac{p_1 + p_1^0}{p_1} = \frac{k_{XIII}}{2k_{II}f} RT[RH] \qquad (3.50)$$

that may also be written

$$\frac{\Delta p_1}{p_1^0} \cdot \frac{p_1 + p_1^0}{p_1 p_1^0} = \frac{k_{XIII}}{k_V^0 RT}[RH] \qquad (3.50')$$

or, taking a molar fraction of the mixture $f_X = p_{RH}/p_1$

$$\frac{p_1^2 - p_1^{02}}{p_1^2} = \frac{k_{XIII}}{2k_{II}} \cdot \frac{f_X}{f} \qquad (3.50'')$$

Fig. 19 shows the result of determining p_1 for a mixture of $2H_2 + O_2$ in the presence of methane ($T = 570\,°C$). In agreement with (3.50''), $(p_1^2 - p_1^{02})/p_1^2$ is a linear function of the molar methane content f_X. However, the experimental line does not pass through the origin, but this is accounted for by the approximate nature of (3.50''). Full agreement between theory and experiment is obtained by making allowance for chain termination in the gas phase. In this case we obtain instead of (3.50'') a more accurate expression

$$\frac{p_1^2 - p_1^{02}}{p_1^2} = \frac{k_{XIII}}{2k_{II}} \frac{f_X}{f} + \alpha \qquad (3.50''')$$

where

$$\alpha = \frac{p_1^0}{p_2^0}.$$

For the upper limit, heterogeneous chain termination will be neglected. Then

$$p_2(p_2^0 - p_2) = \frac{k_{XIII}}{k_{VIII}f}(RT)^2[RH] \qquad (3.51)$$

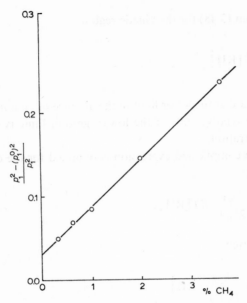

Fig. 19. Shift of the lower ignition limit for a mixture of $2H_2+O_2$ in the presence of methane at 570 °C[7].

For $f_X = p_{RH}/p_2$ this expression reduces to

$$p_2^0 - p_2 = \frac{k_{XIII}}{k_{VIII}} RT \frac{f_X}{f} \qquad (3.51')$$

It follows from (3.51) and (3.51') that addition of RH will *lower* the upper limit.

Many inhibitors, in particular those widely used in studying the oxidation of liquid hydrocarbons, belong to the class of chemically active species. Their effect will be discussed in the next section. Here it will only be noted that, as in the preceding case, the part played by the inhibitor (InH) usually consists in replacement of one of the radicals that is a chain carrier in the reaction by a less active inhibitor radical (In)

$$R + InH \rightarrow RH + In$$

It will also be noted that one and the same compound may appear to be active for one system and inactive for another. For instance, sulphur dioxide markedly narrows the ignition region of a $CO-O_2$ mixture, but has almost no effect on the reaction between hydrogen and oxygen[202a]. The reason for this is that in the first case O atoms are efficiently removed by sulphur dioxide by the reaction $O+SO_2 \rightarrow SO_3$, whereas in the second this step cannot compete with $O+H_2 = OH+H$ which is fast under the given conditions.

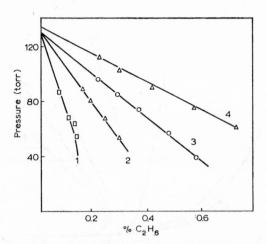

Fig. 20. Shift of the second ignition limit for an H_2–O_2–N_2 mixture in the presence of ethane at $T = 540\,°C$ and at various hydrogen contents[16].

$\dfrac{[H_2]}{[O_2]} = \tfrac{1}{2}\ (1),\ 1\ (2),\ 2\ (3),\ \text{and}\ 4\ (4);$ $[O_2] = 0.14$ mole fraction.

The above expressions for shifts in ignition limits caused by impurities are used for the determination of rate coefficients. The method was first used by Tikhomirova and Voevodskii[202] to obtain rate coefficients for reactions of the type

$$H + RH = H_2 + R$$

from the upper limit shift[†]. Later, this method was extensively used by Voevodskii, Nalbandyan and their coworkers. They studied the effect of impurities on the lower ignition limit for mixtures of hydrogen with oxygen, and of carbon monoxide with oxygen and hydrogen. By pretreatment of the reactor wall Azatyan et al.[12] succeeded in making determinations in the diffusion region close to the low ignition limit, so that one of the constants in the expression for the p_1 shift could be calculated by making use of the diffusion coefficient (p. 140).

Baldwin et al.[16] studied the effect of impurities on the upper ignition limit. Fig. 20 shows the p_2 value for various mixtures of hydrogen with oxygen and nitrogen as a function of the concentration of the ethane added. In agreement with (3.51') p_2 decreased linearly with increasing concentration of C_2H_6. Later, in order to obtain quantitative agreement between theory and experiment, Baldwin and Simmons[17] introduced additional steps into the reaction mechanism (for example, $OH + C_2H_6 \rightarrow H_2O + C_2H_5$).

[†] A similar method was used earlier[108] for qualitative investigation of the mechanism of carbon monoxide combustion in the presence of minor amounts of hydrogen.

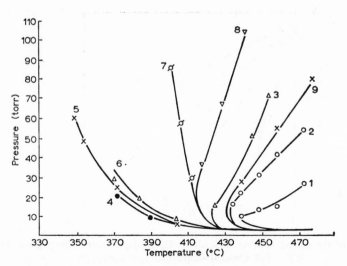

Fig. 21. Shifts of the ignition limits for a mixture of $2H_2 + O_2$ in the presence of NO_2 at following contents (%) of the latter[145]. (1) 0; (2) 0.008; (3) 0.026; (4) 0.24; (5) 0.5; (6) 0.7; (7) 1.0; (8) 3.4 and (9) 7.1.

At the end of this section let us consider active impurities displaying a dual effect depending on their concentrations. Nitrogen dioxide may serve as a well known example. Addition of small amounts of NO_2 to a combustible mixture, for instance of hydrogen with oxygen, will *extend* the ignition zone. However, after a certain NO_2 concentration, further increase in the amount results in *narrowing* of the ignition zone. The pattern, as established by Nalbandyan[145] for a $2H_2 + O_2$ mixture, is shown in Fig. 21. It will be seen that addition of NO_2 in amounts of 0.008, 0.026, and 0.24 per cent results in a great extension of the ignition zone (with disappearance of the upper limit, as in the presence of large amounts of atomic oxygen), whereas further increase in NO_2 (from 0.5 to 7.1 per cent) narrows the ignition zone. Though the effect of NO_2 on the reaction between hydrogen and oxygen was studied quite extensively[†], the chemical mechanism of its action is far from being completely elucidated. The only thing evident is that NO_2 is capable both of accelerating and of inhibiting the oxidation. Acceleration seems to be accounted for by the formation of active species in a reaction of NO_2 itself, for example

$$NO_2 + HO_2 \rightarrow NO_3 + OH - 14 \text{ kcal. mole}^{-1}$$

or by a reaction involving nitric oxide, formed from NO_2 and assumed to be in

---

[†] A review of all data available is given in the well known book by Lewis and von Elbe[120] and in that by Nalbandyan and Voevodskii[146].

equilibrium with the latter, *e.g.*

$$NO + HO_2 \rightarrow NO_2 + OH + 8 \text{ kcal. mole}^{-1}$$

(Semenov[179], p. 560). Inhibition by NO_2 might be due to the fast reaction

$$H + NO_2 \rightarrow OH + NO + 29.5 \text{ kcal. mole}^{-1}$$

as a result of which the H atom is replaced by an OH radical incapable of causing chain branching.

A similar dual effect of hydrocarbons (alkanes and olefins) depending upon their concentration, was established recently for the reaction between carbon monoxide and oxygen[8-10, 83]. Fig. 22 shows the lower ignition limit for a mixture of $2CO + O_2$ as a function of the ethane content at three temperatures.

The reason for the dual effect of hydrocarbon on this oxidation seems to be the following. Whereas the interaction between oxygen atoms and RH

$$O + RH \rightarrow OH + R$$

accelerates the reaction and lowers the ignition limit, that involving a hydrogen atom

$$H + RH \rightarrow H_2 + R$$

results in replacement of the active H atom by a less active radical and this leads to a higher ignition limit.

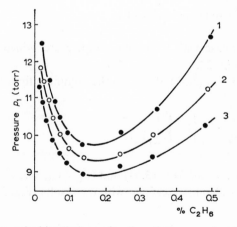

Fig. 22. Shift of the lower ignition limit as a function of ethane content in a mixture of $2\,CO + O_2$ at (*1*) 610°, (*2*) 630° and (*3*) 650 °C [10].

On the basis of the reaction mechanism close to the lower ignition limit, *viz.*

$$O + RH \rightarrow OH + R \tag{1}$$
$$OH + CO \rightarrow CO_2 + H \tag{2}$$
$$H + O_2 \rightarrow OH + O \tag{3}$$
$$H + RH \rightarrow H_2 + R \tag{4}$$
$$H \rightarrow \text{termination} \tag{5}$$
$$O \rightarrow \text{termination} \tag{6}$$

and of the conditions

$$\frac{d[H]}{dt} = 0, \qquad \frac{d[O]}{dt} = 0 \quad \text{and} \quad \frac{d[OH]}{dt} = 0$$

the pressure of oxygen at the limit p_{1,O_2} will be expressed as

$$p_{1,O_2} = \frac{1}{RT}\frac{k_5}{k_3}\left\{1 + \frac{k_4}{k_5}[RH]\right\}\left\{1 + \frac{k_6}{k_1[RH]}\right\} \tag{3.52}$$

From the condition for the minimum value of p_{1,O_2}, $dp_{1,O_2}/d[RH] = 0$, the minimum RH concentration will be

$$[RH]_{min} = \sqrt{\frac{k_5 k_6}{k_1 k_4}} \tag{3.53}$$

whence

$$[RH]_{min} \sim T^{\frac{1}{2}} \exp\left[(E_1 + E_4)/2RT\right]$$

i.e. $[RH]_{min}$ will decrease with increasing temperature, which is consistent with the results given in Fig. 22.

Substituting (3.53) into (3.52) we obtain the minimum limit pressure

$$(p_{1,O_2})_{min} = \frac{1}{RT} \cdot \frac{k_5}{k_3}\left\{1 + \sqrt{\frac{k_4 k_6}{k_1 k_5}}\right\}^2 \tag{3.54}$$

Since $E_3 > E_1, E_4$, it follows from (3.54) that $(p_{1,O_2})_{min}$ decreases with increasing temperature, which is also in agreement with the results (Fig. 22). The expression for k_1 $(R = C_2H_5)$

$$k_1 = (8.4 \pm 3.6)10^{13} \exp\left(-7500 \pm 600/RT\right) \text{cm}^3 . \text{mole}^{-1} . \text{sec}^{-1}$$

was obtained[10] from the data given in Fig. 22, making use of the known values of k_3 and k_4 and values of k_5 and k_6 determined from diffusion coefficients.

3.3 OXIDATION OF HYDROCARBONS

Let us consider the oxidation of hydrocarbons in the gas and liquid phases, as processes illustrative of degenerate branching. Two regimes of the gaseous oxidation of hydrocarbons will be distinguished: *slow oxidation* usually occurring over the temperature range of 20 to 500 °C, and combustion at *high temperatures* accompanied by ordinary *hot flames*. Besides, so-called *cool flames* are observed with all hydrocarbons (except methane) at certain temperatures and pressures. Elevated temperatures exceeding by 100 to 200° the initial temperature of the mixture and luminescence exhibiting a band spectrum from electronically-excited formaldehyde[55, 105] are characteristic of cool flames.

The distribution of three regions of hydrocarbon oxidation may be illustrated by Fig. 23, pertaining to a reaction between propene and oxygen[158, 184]: the regions of slow oxidation and of hot flames may be seen to the left and to the right, respectively; that for cool flames is dashed. There is a distortion in the smooth shape of the curve separating the hot flame region from that of slow oxidation; it is accounted for by the cool flame that favours ignition. As a result of this, the ignition boundary becomes shifted towards lower pressures. It will be noted that in the given case ignition undoubtedly is of a thermal and not of a chain nature.

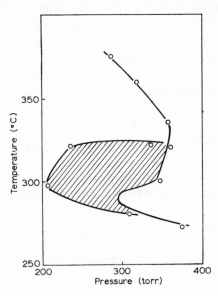

Fig. 23. The regions of hot and cool (shaded) flames in an equimolar mixture of propene with oxygen[158, 184].

Moreover, since slow oxidation and cool-flame combustion occur at close, relatively low temperatures, these reactions must follow practically the *same* mechanism, and a *different* one from that for hot flames.

The mechanism for hot flames involves ordinary chain branching, many elementary steps being similar to those of hydrogen combustion. Indeed, both processes may involve the following steps

$$OH + H_2 \rightarrow H_2O + H \qquad \text{and} \quad OH + RH \rightarrow H_2O + R$$
$$O + H_2 \rightarrow OH + H \qquad \text{and} \quad O + RH \rightarrow OH + R$$
$$H + O_2 \rightarrow OH + O \qquad \text{and} \quad R + O_2 \rightarrow RO + O$$
$$H + O_2 \rightarrow HO_2 \qquad \text{and} \quad R + O_2 \rightarrow RO_2$$
$$HO_2 + H_2 \rightarrow H_2O_2 + H \qquad \text{and} \quad RO_2 + RH \rightarrow ROOH + R$$

However, the mechanism of the high-temperature oxidation of hydrocarbons has its specific features accounted for by the diversity of elementary steps and fuels, which makes this mechanism considerably more complicated than that of hydrogen oxidation. In particular, steps involving thermal decomposition, greatly complicating the reaction mechanism, occur along with those involving oxidation. Yet, with a sufficient oxygen content and when the reaction is not inhibited, the diversity of compounds taking part in the overall process (peroxides, aldehydes, alcohols, acids, in addition to the initial compounds) does not affect the composition of the *end* products, water and carbon dioxide.

In contrast to a high-temperature reaction, the slow (low-temperature) oxidation of hydrocarbons in the gas as well as in the liquid phase will involve *degenerate branching*. The concept of degenerate branching was introduced into chain theory by Semenov[174] in order to explain the kinetics of many reactions (such as the slow oxidation of organic compounds) displaying characteristic features of branched chain reactions but an unusually slow development of chains. Certain features of the gas- and liquid-phase oxidation of hydrocarbons will be discussed below.

3.3.1 Oxidation of hydrocarbons in the gas phase

The gas phase oxidation of hydrocarbons is slow owing to the fact that at relatively low temperatures (20 to 400 °C) the branching step

$$R + O_2 \rightarrow RO + O \tag{I}$$

with an activation energy[†] no lower than 30 kcal. mole^{-1}, is incapable of competing

---

[†] Soloukhin[191] has obtained the value of $E = 34.4$ kcal. mole^{-1} for the reaction $CH_3 + O_2 \rightarrow CH_3O + O -26 \pm 2$ kcal. mole^{-1}.

with the other possible reactions between R and O_2 which have lower activation energies. Such are primarily the exothermic reactions, for example the formation of peroxy radicals

$$R + O_2 \rightarrow RO_2 \qquad\qquad\qquad (II)$$

a reaction accompanied by the release[21,101] of about 29 kcal. mole^{-1}, or

$$R + O_2 \rightarrow R'CHO + R''O \qquad\qquad\qquad (III)$$

the exothermicity of which exceeds 50 kcal. mole^{-1}.

At pressures of $p \leqslant 200$ to 300 torr[†], reaction (II) for R = CH_3 will be third order, *viz.*

$$CH_3 + O_2 + M \rightarrow CH_3O_2 + M \qquad\qquad\qquad (II')$$

and over the temperature range of 298 to 434 °K its rate coefficient may be expressed as $k = 4 \times 10^{17} \exp(-1400/RT)$ cm^6. mole^{-2}. sec^{-1} [M = $(CH_3N)_2$ and $(CH_3)_4C$][188]. For R = C_2H_5, (II) will be second order even at a $(C_2H_5N)_2$ pressure of 4.8 torr[48]

$$C_2H_5 + O_2 \rightarrow C_2H_5O_2 \qquad\qquad\qquad (II'')$$

According to Dinglady and Calvert[48] at a temperature of 295 °K the rate coefficient for this reaction will be 4×10^{12} cm^3. mole^{-1} sec^{-1}.

The peroxy radical formed by reaction (II) undergoes further conversions. One of the primary products obtained in many oxidations at moderate temperatures is the hydroperoxide. Its primary nature may be seen from Fig. 24[128] showing that ethyl peroxide, C_2H_5OOH, is the sole product of ethane photo-oxidation at temperatures below 150 °C.

Various workers propose different mechanisms of hydroperoxide formation. For instance, Nalbandyan *et al.*[59] believe that at 100 to 150 °C the main and only reaction will be[††]

$$RO_2 + RH \rightarrow ROOH + R \qquad\qquad\qquad (IV)$$

Watson and Darwent[216] suggest that, over a temperature range of 40 to 200 °C,

---

[†] A value of 200 was obtained for M = CH_3COCH_3 [82] and of 300 for M = $(CH_3N)_2$ and $(CH_3)_4C$ [188].
[††] This reaction together with II was first suggested by Ubbelohde[208], and later on by Walsh[215] and Hinshelwood[80].

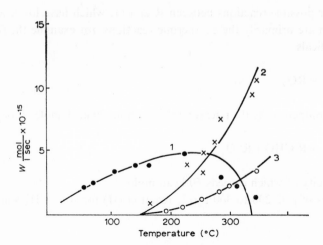

Fig. 24. The yields of (*1*) C₂H₅OOH, (*2*) HCHO and (*3*) CH₃CHO in the photochemical oxidation of ethane at various temperatures[128].

in the oxidation of ethane photosensitized by mercury (as in Nalbandyan's experiments) the hydroperoxide is formed by ($R = C_2H_5$)

$$RO_2 + HO_2 \rightarrow ROOH + O_2 \qquad\qquad (V)$$

Subbaratnam and Calvert[194,195] think that the very small amount of hydroperoxide they detected in the photochemical oxidation of azomethane at 25 °C is formed by reaction

$$CH_3O_2 + CH_3O \rightarrow CH_3OOH + HCHO \qquad\qquad (VI)$$

the methoxy radical CH_3O being produced by

$$2CH_3O_2 \rightarrow 2CH_3O + O_2 \qquad\qquad (VII)$$

The photochemical oxidation of azoethane[39] and the oxidation of alkyl iodides[75,134] yield either very small or no measurable amounts of hydroperoxides. This seems to be an argument in favour of reaction (V) rather than of (IV). However, this would mean the absence of chains or a quantum yield lower than or equal to unity in the temperature range where hydroperoxide is the only product of mercury-sensitized oxidation. Yet, according to Mantashyan and Nalbandyan[128] at 100 °C the hydroperoxide quantum yield in the oxidation of ethane is 3 and according to Fok and Nalbandyan[59] the quantum yield of propane oxidation will be 2 even at room temperature. If these values are correct the absence of, or the very low concentrations of hydroperoxides in experiments involving $(RN)_2$ and RI should be ascribed to specific experimental conditions (relatively high degrees of conversion, the formation of hot radicals[48], etc.) different from those in Nalbandyan's experiments.

Another objection to reaction (IV) might be that it is a slow reaction at temperatures close to room temperature, as its rate coefficient seems to be such that, below 150 °C, it would not account for (with methane and ethane, for example) the observed rate of hydroperoxide formation[†]. However, the k_{IV} value given in the footnote refers to a reaction involving *stabilized* RO_2. It may be conceived that, in fact, hydroperoxide is formed by a termolecular exothermic reaction

$$R + O_2 + RH \rightarrow ROOH + R \qquad \text{(VIII)}$$

the enthalpy change of which is higher than that of (IV) by the value of the heat of formation of RO_2 from R and O_2. Consequently the activation energy for reaction (VIII) must be lower than that for (IV). Since stabilization of the peroxy radical [reaction (II′)] must be more rapid than the formation of hydroperoxide in reaction (VIII), this would lead to accumulation of peroxy radicals in the reacting mixture (see, for example Thomas *et al.*[37]). As a result, reaction (VII) may become of essential importance in the reaction mechanism and this might account for the well known fact that the yield of alcohol formed by

$$RO + RH \rightarrow ROH + R \qquad \text{(IX)}$$

increases with increasing pressure, which favours the production of RO_2 radicals.

Decomposition of peroxy radicals becomes more important as the temperature increases. It may be suggested again that along with the generally accepted decomposition of stabilized RO_2 radicals, decomposition may occur according to

$$R + O_2 \rightarrow R'CHO + R''O \qquad \text{(III′)}$$

involving no RO_2 stabilization step. Reaction (III′) was first suggested by Norrish[150], who assumed that R″O = OH.

According to Mantashyan and Nalbandyan[128] the ratio of rate coefficients for the formation of aldehydes and hydroperoxides during the photochemical oxidation of ethane in the temperature range 170 to 390 °C will be expressed as

$$\frac{k_{III}}{k_{IV}} = 8.2 \times 10^{21} \exp\left(-8900/RT\right) \mathrm{cm^3. \; molecule}$$

[†] In accordance with the approximate estimation made by Benson[22], for ethane
$$k_{IV} = 10^{11} \exp\left(-14000/RT\right) \mathrm{cm^3. \; mole^{-1}. \; sec^{-1}}$$

A similar ratio for the photochemical oxidation of methane would be expressed as[127]

$$\frac{k_{III}}{k_{IV}} = 2.5 \times 10^{22} \exp\left(-8500/RT\right) \text{ cm}^3. \text{ molecule}$$

Naturally, in the latter case formaldehyde is the only aldehyde present[†]. It follows from these expressions that, at atmospheric pressure in an equimolecular mixture of $C_2H_6 + O_2$ or of $CH_4 + O_2$ at 350 °C, hydroperoxides and aldehydes will form at approximately the same rate. At lower temperatures the formation of peroxide will predominate over that of aldehydes. Indeed, this may be seen in Fig. 24.

According to Mantashyan and Nalbandyan[128] the ratio of the rates of formaldehyde and acetaldehyde formation, *i.e.* of the rates of the reactions

$$C_2H_5OO \rightarrow HCHO + CH_3O$$

and

$$C_2H_5OO \rightarrow CH_3CHO + OH$$

may be expressed as

$$\frac{W_{HCHO}}{W_{CH_3CHO}} = 0.43 \exp\left(2340/RT\right)$$

Consequently, the first of these reactions has a somewhat higher rate than the second.

It must be borne in mind that the mechanism of oxidation of hydrocarbons cannot be considered as remaining the same for all experimental conditions and for the whole temperature range from 20 °C to that corresponding to transition from a slow reaction to combustion. It was stated before (p. 157) that reaction (VII) might very probably become considerably more important if peroxy radicals accumulate in the system. As a result, the nature of the radicals propagating the reaction would also become different. It will be noted, in particular, that decomposition of peroxy radicals may yield hydroxyl besides alkoxy radicals.

According to the alkane oxidation mechanism proposed by Knox[100, 101] the reaction

$$R + O_2 \rightarrow \text{olefin} + HO_2 \tag{X}$$

[†] Nalbandyan *et al.*[127, 128] believe that the formation of hydroperoxide and aldehydes is preceded by stabilization of peroxy radicals. However, the same expressions would be obtained on the assumption that these compounds are formed by reactions (VIII) and (III′).

becomes of importance at temperatures higher than 300 °C. Since, at these temperatures, olefins are the *main* products of the early stages of alkane oxidation[†], Knox suggests that (X) is the main reaction between R and O_2, and olefins are the primary reaction products.

Olefins undergo further conversions, as their reactivity is higher than that of alkanes. At 300 to 350 °C carbonyl compounds are the primary products of olefin oxidation[99]. Knox and Turner[103] suggest that these compounds are formed by the overall reaction

$$>C = C< + O_2 = >CO + >CO \qquad\qquad (XI)$$

involving the following sequence of elementary radical reactions

The appearance of hydroxyl formed at the expense of HO_2 radicals at a certain stage of the reaction is, according to Knox and Wells[104], an important feature of the reaction mechanism. At this stage hydroxyl becomes the main radical attacking the alkane. In particular, Knox suggests[98, 100, 101] that the *negative temperature coefficient*, *i.e.* the decrease in the reaction rate with increasing temperature (usually over a certain range of some 150° around 400 °C), with its subsequent increase at still higher ˌtemperatures (see Shtern[183], pp. 81–84) is connected with interchange of the HO_2 and OH radicals. At a certain temperature (about 370 °K according to Knox) the rate of the reaction

determining the rate of replacement of HO_2 radicals by hydroxyl, becomes equal to that of reaction

$$HO_2 + HO_2 \rightarrow H_2O_2 + O_2$$

At $T > 370$ °C the latter reaction becomes predominant and this accounts for the decrease in the reaction rate with increasing temperature (negative temperature

---

[†] Olefins represent about 80 per cent of the primary reaction products in the oxidation of ethane[104], propane[97, 98] and isobutane[74, 221].

coefficient). The normal (positive) temperature coefficient observed at higher temperatures is ascribed by Knox to the reactions

$$HO_2 + RH \rightarrow H_2O_2 + R \quad \text{or}$$
$$H_2O_2 + M \rightarrow 2OH + M$$

that are possible at these temperatures (about 500 °C). We shall not mention here the other explanations for the negative temperature coefficient proposed by various workers: these may be found in refs. 183, pp. 81–84; 139, p. 130 and refs. 56, 102.

It follows from the above discussion that, in spite of the great amount of research dealing with oxidation of hydrocarbons, the mechanism of this important class of reactions cannot be considered as definitely established. Correlation of the rate coefficients of individual elementary steps (as far as these are known) with the mechanism for the overall reaction is not easy in many cases. However, the main feature of the oxidation mechanism, its degenerately branched chain nature, is not affected.

The formation of primary intermediates of considerable reactivity is responsible for the possibility of *degenerate* (*delayed*) branching in the slow oxidation of hydrocarbons. Hydroperoxide is often considered to be an intermediate of this kind. Its decomposition with cleavage of the O–O bond yields two monoradicals

$$ROOH \rightarrow RO + OH \tag{XII}$$

capable of initiating new chains. However, in view of the relatively high dissociation energy of the O–O bond (some 40 kcal. mole^{-1})[95], reactions involving other intermediates, such as aldehydes, would seem to be the most probable source of branching. For instance, according to Norrish[150b] and Nalbandyan et al.[90], formaldehyde is the intermediate responsible for degenerate branching in the slow oxidation of methane. Nalbandyan et al. suggested that the branching step was

$$HCHO + O_2 = HO_2 + HCO - 28 \pm 5 \text{ kcal. mole}^{-1}$$

One of the arguments in favour of branching being due to aldehydes stems from a result obtained by Shtern and Polyak[184], namely that addition of acetaldehyde to the initial mixture completely eliminates the induction period in the slow oxidation of propene. This was found for other hydrocarbons as well[183].

A suggestion has also been made[170] that the reaction

$$ROO + O_2 \rightarrow RO + O_3 \tag{XIII}$$

is basically responsible for branching in the oxidation of hydrocarbons (see also ref. 73a). Other possible modes of branching will not be discussed here. Undoubtedly many branching reactions might be conceived and the predominance of one of these over the others would depend on the experimental conditions.

As to hydroperoxides, it may be suggested that these would react at moderate temperatures with peroxy and other radicals according to

$$ROOH + RO_2 \rightarrow R' + \text{products} \tag{XIV}$$

as well as decompose at the wall[†], as may hydrogen peroxide be formed, for instance by

$$HO_2 + RH \rightarrow H_2O_2 + R$$

Heterogeneous decomposition of peroxy radicals also plays an important part in the reaction mechanism. We shall mention here only the results obtained for isobutane and isobutene oxidation at 250 to 350 °C[74]. These showed that the yield of the *main* oxidation products (80 % of isobutene and 75 % of acetone plus formaldehyde) is almost unaffected by pretreatment of the reaction vessel wall, whereas the composition of products of secondary importance, such as propene, ethylene, methyl alcohol, aldehydes, acetone, olefin oxides, etc. is very sensitive to the condition of the wall (and also to pressure and temperature). It is suggested[74] that these products of secondary importance are produced mainly by heterogeneous decomposition of peroxy radicals $C_4H_9O_2$ or C_4H_8OOH.

Several cool flames succeeding each other are often observed under static conditions over a certain range of temperatures, pressures and mixture compositions. Thus, for instance, the oxidation of propene can give rise to three cool flames in the course of the reaction (Fig. 25)[158, 159, 184]. The sharp pressure peaks are due to an abrupt rise of temperature, and this leads to the conclusion that heat input becomes predominant over heat loss as a result of autocatalytic acceleration of the reaction. The part played by the thermal factor in the appearance of cool flames is considered in more detail by Minkoff and Tipper[139].

However, the high temperature would cause the appearance of an ordinary hot flame succeeding the cool flame only when the initial pressure and temperature

---

[†] According to the results of Kirk and Knox[95], over the temperature range 280 to 380 °C, the ratio of rate coefficients for heterogeneous and homogeneous unimolecular decomposition of hydroperoxides will be

$$\frac{k_{\text{het}}}{k_{\text{hom}}} = 10^{\bar{5}.6} \exp (19700/RT) \text{ for } C_2H_5OOH, \ 10^{\overline{10}.8} \exp (23000/RT)$$

for iso-C_3H_7OOH and $10^{\bar{6}.3} \exp(12800/RT)$ for tert-C_4H_9OOH. Thus, it follows that the rates of the heterogeneous and homogeneous reactions become equal at 310, 275, and 220 °C, respectively.

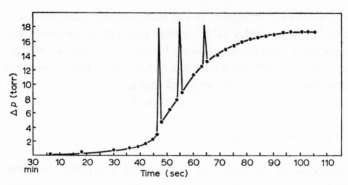

Fig. 25. Three consecutive cool flames in the slow oxidation of propene at 300 °C [158, 184].

of the combustible mixture correspond to the ignition region, *e.g.* on the right-hand side of Fig. 23. When this condition is not fulfilled, the cool flame is extinguished as abruptly as it appears, unless, of course, measures are taken for flame stabilization, which is possible using a flow system. The reason for cool flame extinction is not completely clear. It might be conceived that a certain part is played by radiation from the flames as a result of which the excess energy of radicals is lost as radiation energy (or is dissipated as thermal energy owing to chemiluminescence quenching).

It will be noted that in the majority of formaldehyde excitation reactions proposed (activation energy about 80 kcal. mole^{-1}) the excited HCHO molecule is assumed to be formed by interaction of two radicals (Kondratiev[109], p. 581)[†], which would lead to chain termination[††]. Chain termination by destruction of two active centres is called by Semenov *negative chain interaction* to distinguish it from *positive interaction*; an example of the latter was given by discussing the shifts of ignition limits induced by the addition of active impurities (p. 143).

Pease advanced the hypothesis that extinction of cool flames is a result of transition of the system into the region of the negative temperature coefficient[156a].

[†] Postnikov and Shlyapintokh[161] have recently investigated the weak chemiluminescence observed in the slow oxidation of hydrocarbons and found that excited formaldehyde was formed by the reaction

$$RO + OH \rightarrow HCHO + R'OH$$

suggested before by Knox and Norrish[102]. The exothermicity of this reaction is 90 kcal. mole^{-1} for R = CH_3, and 72 kcal. mole^{-1} for R = C_2H_5.

[††] However, an opposite suggestion was made as well (Kondratiev[109], page 531) implying that the electronically-excited formaldehyde molecule, which may be in fact a biradical, is capable of yielding two radicals

$$HCHO' + RH \rightarrow CH_3O + R \text{ or } CH_2OH + R,$$

and this would mean the possibility of branching.

Due to the low probability of branching in this region, the reaction rate would fall abruptly and the system would return to the *state preceding* the cool flame appearance. This hypothesis was developed further by Norrish[102, 150c].

The sequence of cool flames is one of the manifestations of the *periodicity* of reaction steps accounted for by the strictly kinetic peculiarities of the overall chemical reaction. The possibility of periodical occurrence of a chemical reaction was suggested for the first time by Lotka[123] as long ago as 1920, and later it was investigated in detail by Frank-Kamenetskii[61, 64] (for further references and discussion of the nature of the cool flame periodicity see Minkoff and Tipper[139], pp. 211–214). Assuming that the oxidation of hydrocarbons is autocatalytic and involves two consecutively formed intermediates, X and Y, Frank-Kamenetskii derives from the formal reaction scheme

$$A \xrightarrow{k_1} X \xrightarrow{k_2} Y \xrightarrow{k_3} C$$

(A is the initial compound, C the reaction product) a system of kinetic equations

$$\frac{dx}{dt} = k_1 ax - k_2 xy$$

$$\frac{dy}{dt} = k_2 xy - k_3 ay$$

(a, x, y are the concentrations of A, X, Y). Then, introducing the deviations of x and y from their steady-state values, $\varepsilon_x = x - x_{st}$, $\varepsilon_y = y - y_{st}$, where x_{st} and y_{st} are the values of x and y under steady-state conditions, $dx/dt = 0$ and $dy/dt = 0$, and assuming that these deviations are not very significant, we obtain

$$\frac{d^2\varepsilon}{dt^2} = -a^2 k_1 k_3 \varepsilon \tag{3.55}$$

where $\varepsilon = \varepsilon_x = \varepsilon_y$. The latter equation represents an ordinary equation for harmonic frequencies

$$v = \frac{a}{2\pi} \sqrt{k_1 k_3} \tag{3.56}$$

In deriving his formal reaction scheme Frank-Kamenetskii proceeded from Neiman's concepts[147] that a cool flame appears as a result of the accumulation of peroxides to certain critical concentrations, and that the oxidation of hydrocarbons involves two consecutive intermediates: peroxides (X) and aldehydes (Y). Grey[70] made use of the Frank-Kamenetskii theory in explaining the cool flames of ethane he observed in a flow apparatus, and suggested that X and Y were the C_2H_5 and HO_2 radicals. Without dwelling on other attempts to give a thorough chemical interpretation of the periodicity of cool flames (see, for example, ref. 139, pp. 211–214 and also refs. 124, 57), let us mention only that all these attempts were of a purely formal nature, as the mechanism of hydrocarbon oxidation, and in particular the part played by various intermediates, cannot be considered as definitely established.

It was mentioned above that a hot flame may follow a cool flame only provided that the initial temperature and pressure lie in the ignition region. However, even in this case the transition from the cool flame to the hot is often not immediate, the hot flame appearing only after extinction of the cool flame. This seems to be an indication of *different* reaction mechanisms for production of the two flames.

When a cool flame is followed by a hot flame, this may be treated as *two-stage ignition*. In a certain time, τ_1, after the mixture is admitted to a heated reactor, the initial acceleration induces the appearance of a cool flame; the ordinary hot flame appears after the lapse of another period τ_2.

It was repeatedly reported by Neiman *et al.* (see, for example, ref. 147), as well as by other workers, that the cool flame induction period changes with temperature and pressure according to:

$$\tau_1 = Ap^{-n} \exp(B/T) \tag{3.57}$$

(A, B, and n are constants, $B > 0$). This is identical in form with the theoretical expression for the induction period of branched chain reactions (see p. 130). Thus the concept of the direct relation between the induction period and the rate of branching, basic for branched chain reactions, may be considered as valid for expression (3.57) as well. In particular, an attempt may be made to treat the constant B as E/R, where E is the activation energy for the branching step.

However, in view of the non-isothermal nature of the oxidation during τ_1 (*cf.* p. 153) there might be a possibility that the resemblance between (3.57) and a similar expression for the induction period of isothermal branched chain reactions is only apparent. This resemblance would be expected taking into account that modification of the Arrhenius law to allow for the continuous increase in temperature in the course of the reaction formally gives the same exponential increase in the reaction rate as that following from Semenov's law (for a constant temperature)[72].

While the induction period, τ_1, for a cool flame decreases monotonically with increasing temperature, the temperature dependence of the hot flame induction period, τ_2, is of a more complicated nature. According to Aivazov *et al.*[3] the value of τ_2 for the mixture $C_4H_{10} + O_2$ (0.1 sec at 280°C) increases with temperature and reaches a maximum at 320 °C ($\tau_{2\,max} \simeq 0.47$ sec), after which it starts decreasing. According to Neiman[147] the increase in τ_2 with temperature is related to the effect of intermediates (peroxides) formed in the cool flame: the importance of these intermediates accelerating the reaction becomes greater with decreasing temperature, and this seems to result in the lowering of τ_2.

3.3.2 Oxidation of hydrocarbons in the liquid phase

As a result of a vast amount of research on the liquid-phase oxidation of hydrocarbons, many trustworthy facts have been accumulated. These make it possible to draw a fairly reliable picture of the mechanism of oxidation, to identify the basic elementary steps it involves, and often to determine quantitatively the rate coefficients of individual reactions.

Firstly, it was established that the liquid-phase oxidation of hydrocarbons is a chain process involving *degenerate branching*. The quantum yields of the photochemical oxidation of hydrocarbons, often tens, hundreds and even thousands, as well as the specific action of initiators and inhibitors, are evidence for the chain mechanism of this reaction. The occurrence of degenerate branching follows, in particular, from the *autocatalytic nature* of the oxidation which is apparent from the initial acceleration, *i.e.* from the S-shaped curves for the accumulation of reaction products (see Fig. 27, p. 172) *Critical phenomena* characteristic of branched and degenerate branched chain reactions were also observed for liquid phase oxidations. One such is the *critical concentration* of an *inhibitor*: at a concentration higher than the critical, the reaction is *stationary* and proceeds at a constant rate, whereas a non-stationary autoaccelerated reaction occurs with a concentration below the critical value. The rate of the latter reaction obeys the Semenov law, $W \sim \exp(\varphi t)$ (see ref. 54, p. 283; 179, p. 632–633).

It was established that in the liquid-phase oxidation of hydrocarbons, chains are propagated by *radicals*, mostly by hydrocarbon and peroxy radicals, R and RO_2. The basic reactions involving these radicals are

$$R + O_2 \rightarrow RO_2 \tag{I}$$

and

$$RO_2 + RH \rightarrow ROOH + R \tag{II}$$

Other radicals appear in the system in the course of the reaction. These are formed by reactions of R and RO_2 with saturated molecules (for example, OH and RO radicals), or by their interaction to give RO, or else as a result of decomposition of the RO_2 and RO radicals, and partake in chain propagation along with R and RO_2.

In the absence of peroxides and other initiators, the liquid-phase generation of radicals initiating chains may obey the quadratic law[132,190], *viz.* the step

$$RH + O_2 \rightarrow R + HO_2 \tag{III}$$

or the third order law[46], *viz.* the step

$$2RH + O_2 \rightarrow 2R + H_2O_2 \tag{IV}$$

One of the two reactions will predominate depending upon the structure of the molecule, in particular upon the R–H bond dissociation energy. For instance, the kinetics of ethyl linoleate oxidation in its initial stage seems to show that in this case chains are initiated by reaction (III)[33] (see also Semenov[179], p. 634). On the other hand, it was suggested (Emanuel *et al.*[54], p. 83) that for many hydrocarbons the rate of (IV) might be higher than that of (III). The validity of this suggestion was supported in many cases by determination of the chain generation rates. For example, it was shown[47] that the rate of chain initiation in the oxidation of tetralin (at 140 °C) can be expressed as

$$W_0 = k_0[RH]^2[O_2]$$

where

$$k_0 = 3.5 \times 10^9 \exp(-20700/RT) \text{ cm}^6. \text{ mole}^{-2}. \text{ sec}^{-1}$$

With accumulation of hydroperoxide the initiation reactions (III) and (IV) become less important giving way to branching involving the hydroperoxide molecule. Moreover, it was established that hydroperoxide is the main *branching agent* responsible for the autocatalytic nature of the reaction, and Semenov[179] suggests that the most probable branching step would be

$$ROOH + R'H \rightarrow RO + H_2O + R' \tag{V}$$

where R'H is a molecule of the solvent. Addition of this step to the reaction mechanism would explain, in particular, the observed dependence of the rate of radical generation in the presence of hydroperoxide on the nature of the solvent[126, 199]. Bateman *et al.*[18] believe that at high hydroperoxide concentrations radicals are formed by the bimolecular reaction

$$2ROOH \rightarrow RO_2 + H_2O + RO \tag{VI}$$

It was established as well that, whatever the nature of the initial hydrocarbon, *hydroperoxides* represent the *primary product of oxidation*. With compounds which undergo ready oxidation at relatively low temperatures, such as alkyl aromatics, this follows directly from the fact that in the early stages of the reaction the amount of hydroperoxide formed practically equals that of the oxygen consumed (see, for example, Emanuel *et al.*[54], Table 1). Saturated hydrocarbons oxidize at higher temperatures, due to their relatively greater stability. Thus, other products will be observed, along with hydroperoxides, even at an early stage of their oxidation. However, since these products are formed by further conversions of the hydroperoxide, the latter should again be considered as the sole primary product of the interaction between peroxide radicals and the initial hydrocarbon[†].

The rate coefficients for the reactions

$$R + O_2 \rightarrow RO_2 \tag{I}$$

and

$$RO_2 + RH \rightarrow ROOH + R \tag{II}$$

have been determined experimentally. The first is of the order of 10^{10} to 10^{11} cm^3. mole^{-1}. sec^{-1} and is taken as temperature independent. For various hydrocarbons the activation energy for reaction (II) is 4 to 12 kcal. mole^{-1}, and the pre-exponential factor is[53] of the order of 10^7 to 10^{10} cm^3. mole^{-1}. sec^{-1}. If the activation energy for (I) is zero, the steric factor for this reaction will be 10^{-3} to 10^{-4}. The steric factor of (II) will be of the order of 10^{-4} to 10^{-7}. The reason for the low values of the steric factor is far from clear. They might be accounted for by the same steric hindrance as that induced by formation of solvation shells at the expense of the hydrogen bonding. However, there is a possibility that the low values are merely a function of the complex structures of the reacting entities. This may be shown by calculation of pre-exponential factors on the basis of the theory of absolute reaction rates[47a] (see also ref. 54, p. 104).

After a certain period of time (dependent on the reaction conditions, such as temperature, and on the nature of the hydroperoxide) during which hydroperoxide is practically the sole reaction product[††], the curve for accumulation of hydroperoxide begins to lag behind that for oxygen consumption. The two curves obtained[135] for tetralin are shown in Fig. 26.

It was said earlier (p. 157) that in the gas phase peroxy radical decomposition

$$RCH_2R'CHOO \rightarrow RCH_2O + R'CHO \tag{VIII}$$

will compete with the formation of hydroperoxide (II), the competition becoming

---

[†] In the case of olefins it must be assumed that, in competition with reaction (II), the peroxy radical may react with the initial hydrocarbon as follows[131, 207]

$$RO_2 + \overset{}{\underset{}{>}}C=C\overset{}{\underset{}{<}} \rightarrow RO + \overset{\overset{O}{\triangle}}{>C-C<} \tag{VII}$$

This is apparent from the formation of olefin oxides often observed, and also from the fact that even at a low extent of conversion the amount of oxygen consumed is considerably higher than that of the hydroperoxides formed (see ref. 54, p. 11).

[††] For olefins see above.

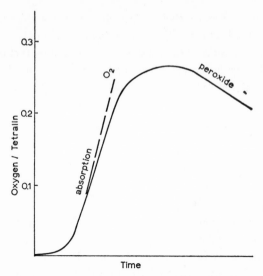

Fig. 26. Oxygen consumption and hydroperoxide accumulation in the oxidation of tetralin at 78 °C [135].

more effective as the temperature increases. In the liquid phase, due to the relatively lower temperatures, this reaction would seem to be possible only at the vessel wall. Thus, for example, in the oxidation of *n*-butane the products of n-C_4H_9OO radical decomposition were present only when the reaction was carried out in a steel reactor (at 145 °C), while with a glass reactor under the same conditions they were absent[25]. Thus, it would seem that homogeneous decomposition of peroxy radicals does not occur to any appreciable extent in the liquid-phase oxidation of hydrocarbons. It may be concluded therefore, that *chemical conversions of the hydroperoxide* play the main part in the formation of products of hydrocarbon oxidation in the liquid phase. It follows that the composition and nature of these products will be determined by the structure of the initial hydroperoxide.

Decomposition may be considered as one of the main routes of hydroperoxide conversion. For instance, the oxidation products include compounds observed also in the products of hydroperoxide decomposition in nitrogen. As to the mechanism of hydroperoxide decomposition, the assumption is made that bimolecular and other second order reactions between hydroperoxide molecules and other molecules and radicals occur along with the unimolecular reaction[†]

$$ROOH \rightarrow RO + OH \qquad\qquad (IX)$$

[†] According to Bell *et al.*[20] the rate coefficient for t-C_4H_9OOH decomposition in the liquid phase is $k_9 = 10^{15.1} \exp(-39000/RT) \sec^{-1}$.

Thus, the formation of alcohol and oxygen in the thermal decomposition of tertiary butyl hydroperoxide is assumed to occur by [140]

$$2\ t\text{-}C_4H_9OOH \rightarrow 2\ t\text{-}C_4H_9OH + O_2$$

Interaction between hydroperoxide and aldehyde molecules (present as labile intermediates)[219] according to

$$RCH_2OOH + RCHO \longrightarrow RCH \overset{O\cdots O}{\underset{H\cdots H}{<\quad>}} CR \longrightarrow RCOOH + H_2 + RCHO$$

has been suggested as being responsible for the composition of the products of the thermal decomposition of primary hydroperoxides.

Under the conditions of oxidation an important part is played by interaction between the hydroperoxide and R and RO_2 radicals. For example, it was shown[186] that the formation of alcohol in the oxidation of cymene occurs by the reaction

$$R + ROOH \rightarrow ROH + RO$$

The reaction[187]

$$ROOH + RO_2 \rightarrow ROH + RO + O_2$$

also seems to be possible. The formation of ketones in the decomposition of secondary hydroperoxides is assumed to occur[165,205] (see also ref. 54, p. 149) by the step

$$R_1R_2HCOOH + RO_2 \rightarrow R_1COR_2 + ROOH + OH$$

As for (IX), its high endothermicity (about 40 kcal. mole^{-1}) would make this unimolecular reaction important only at very low hydroperoxide concentrations and relatively high temperatures.

Finally, it was established that *termination of chains* in the liquid-phase oxidation of hydrocarbons is *second order* and occurs by one of the reactions

$$R + R \rightarrow R_2 \tag{X}$$

$$R + RO_2 \rightarrow ROOR \tag{XI}$$

$$RO_2 + RO_2 \rightarrow ROH + R_1COR_2 + O_2 \tag{XII}$$

This follows, in particular, from the observed proportionality between the rate of the photochemical oxidation of hydrocarbons and the square root of the light intensity, and between the rate of an initiated reaction and the square root of the initiator concentration. Thus, writing the mechanism for the early stages (hydroperoxide formation) as

$$R + O_2 \rightarrow RO_2 \qquad\qquad\qquad k_I$$

$$RO_2 + RH \rightarrow ROOH + R \qquad\qquad\qquad k_{II}$$

$$RO_2 + RO_2 \rightarrow \text{molecular products} \qquad\qquad\qquad k_{XII}$$

and assuming stationary concentrations of R and RO_2, we obtain the following expression for the reaction rate

$$W = \frac{d(ROOH)}{dt} = k_{II}(RH) \sqrt{\frac{W_0}{k_{XII}}} \tag{3.58}$$

where W_0 is the rate of initiation (giving R). As the value of W_0 for a photochemical reaction will be proportional to the light intensity I, and for an initiated reaction to the initiator concentration [J], we obtain $W \propto \sqrt{I}$, and correspondingly $W \propto \sqrt{[J]}$, in agreement with experiment. It will be noted that the absence of linear (heterogeneous) chain termination characteristic of gas reactions is quite natural for the liquid phase, since diffusion of radicals to the wall is slow.

The rate coefficient of reaction (XII) was determined for many hydrocarbons, and those for reactions (X) and (XI) in a limited number of cases only. The k_{XII} values for many hydrocarbons determined over a temperature range of 0° to 65 °C are of the order 10^7 to 10^{10} cm^3. mole^{-1}. sec^{-1} (see ref. 54, Table 21). Thus, here, as for the reaction $R + O_2 = RO_2$ (I) (see p. 167) the steric factor is very low, about 10^{-2} to 10^{-6}. The activation energy for radical recombination is 1 to 2 kcal. mole^{-1}.

It will be noted in connection with chain termination by interaction of two RO_2 radicals, that besides reaction (XII) yielding molecular products, the reaction

$$2RO_2 \rightarrow 2RO + O_2 \tag{XIII}$$

giving RO radicals and oxygen may also occur. The possibility of (XIII) follows, in particular, from the fact that $^{16}O^{18}O$ was detected in the oxidation of cumene by oxygen containing the $^{18}O^{18}O$ isotope[165]. It might have been formed by

$$R^{16}O_2 + R^{18}O_2 \rightarrow R^{16}O + {}^{16}O^{18}O + R^{18}O$$

Leaving aside many other reactions involved in the mechanism of liquid-phase oxidation of hydrocarbons, such as those between hydroperoxides and various reaction products, or those of R and RO_2 with various compounds[†], we shall note only that the main reaction products – peroxides, alcohols, carbonyl compounds, acids, as well as H_2O and CO_2 – are the same both in the gas- and in the liquid-phase oxidation of hydrocarbons, though the latter occurs over the range of 0° to 150 °C, *i.e.* under milder conditions. Some of these reactions, and also the consecutive appearance of the products of liquid-phase oxidation of hydrocarbons, are considered in the book by Emanuel *et al.*[54].

It will be noted also that, despite the unusually low values of steric factors of the elementary steps involved, the oxidation of liquid hydrocarbons occurs at much higher rates than those observed for the same hydrocarbons in gaseous oxidation. Semenov suggests that this might be accounted for by the higher concentration of the substance undergoing oxidation, by a considerable decrease in the rate of termination due to its becoming homogeneous and second order (slow diffusion of free radicals to the wall), and by more favourable conditions for accumulation of hydroperoxides owing to less heterogeneous decomposition (see ref. 179, p. 615).

To end this section let us consider in brief the oxidation of *n*-butane, which has been studied in great detail[26] and represents an example of the liquid-phase oxidation of paraffins (see also ref. 54, pp. 344–355). Fig. 27 shows curves for butane consumption, and for the variations in concentrations of products during a reaction carried out at 145 °C. It will be seen that while certain products attain a steady concentration that remains, in fact, unchanged (acids, CO_2, etc.), the concentrations of other products pass through a maximum and then drop (hydroperoxides, alcohols, ketones, etc.). The former may, evidently, be called end products, and the latter intermediates.

The observed composition of reaction products may be obtained by assuming that interaction between peroxy radicals and butane gives hydroperoxide, *viz.*

$$RO_2 + RH \rightarrow ROOH + R \tag{II}$$

and that decomposition of the peroxy radical

$$RO_2 \rightarrow R'CHO + R''O \tag{VIII}$$

can occur. Hydroperoxide reacts with RO_2 radicals yielding secondary butyl alcohol, RO radicals and O_2. Methyl ethyl ketone ($+ H_2O$) is formed by decomposition of ROOH, or by interaction between ROOH and RO_2. Further oxidation of methyl ethyl ketone gives acetaldehyde and acetic acid.

Depending upon the bonds broken, decomposition of peroxy radicals would

---

[†] A certain part in the liquid-phase oxidation of hydrocarbons is played by reactions involving hydrogen ions from the dissociation of acids formed (see ref. 54, pp. 148 and 164).

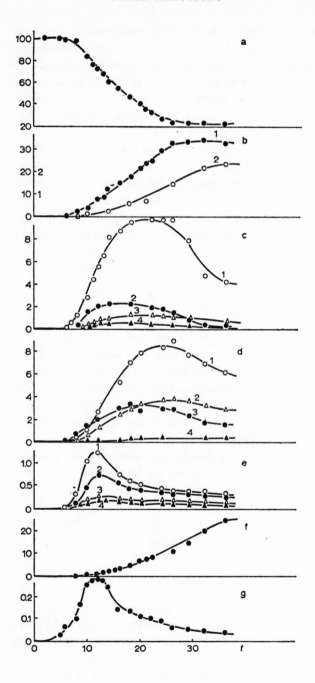

result in the appearance of acetaldehyde, acetone, methyl and ethyl alcohols, and products of their subsequent conversions (for example, methyl- and ethyl acetate formed as a result of esterification of the appropriate alcohols by acetic acid). It was mentioned above (p. 168) that decomposition of peroxy radicals is catalyzed by the surface of a steel reactor.

The liquid-phase oxidation of butane or some other paraffin may be conceived as involving the following sequence:

$$\text{hydrocarbon} \rightarrow \text{hydroperoxide} \begin{array}{c} \nearrow \ \text{alcohols} \\ \downarrow \\ \searrow \ \text{ketones} \rightarrow \text{acids.} \end{array}$$

4. The energy factor in chain reactions

4.1 ENERGY EXCHANGE IN THE CHEMICAL INTERACTION OF MOLECULES

The theory of radical chain reactions tacitly assumes that all reacting particles have an energy distributed *in equilibrium* over all degrees of freedom. Yet any chemical reaction is inevitably accompanied by *disturbance* of the equilibrium energy distribution, and this may be accounted for by two factors. The first is that there is a decrease in concentration of the most energy-rich molecules which are consumed by reaction. This factor is evident in *endothermic* reactions, in particular in the thermal dissociation of molecules. The second is the liberation of energy in *exothermic steps* and its distribution among molecules of reaction products. Since all chain reactions are, in fact, exothermic, we shall consider here only the second factor and its effect on the kinetics and mechanism of a chain reaction.

There are direct experimental results permitting consideration of the nature of the primary distribution of energy among products of exothermic reactions. For

---

Fig. 27. Consumption of n-butane and the reaction product yields in the liquid-phase oxidation of n-butane at 145 °C.

(*a*) Butane.

(*b*) Accumulation of acids; acetic (*1*), formic (*2*).

(*c*) Accumulation of carbonyl compounds; methyl ethyl ketone (*1*), diacetyl (*2*), acetone (*3*), acetaldehyde (*4*).

(*d*) Accumulation of esters; ethyl acetate (*1*), methyl acetate (*2*), butyl acetate (*3*), 2-butanone-3-ol ester and acetic acid ester (*4*).

(*e*) Accumulation of alcohols; *sec* butyl alcohol (*1*), ethyl alcohol (*2*), 2-butanone-3-ol (*3*), methyl alcohol (*4*).

(*f*) CO_2.

(*g*) Hydroperoxides.

Pressure 50 atm; rate of air admission 20 l.hr^{-1}; temp. 145 °C.

References pp. 183–188

instance, determination[68] of the "translational temperature" from the Doppler broadening of the emission lines of the CH radical, in rarefied oxygen flames of acetylene, formed by the reaction

$$C_2 + OH \rightarrow CO + CH + 91.2 \text{ kcal. mole}^{-1}$$

gave a value of 4000 °K at a maximum theoretical temperature of the flame of 2500 to 2800 °K. Hence, the translational energy of electronically-excited CH[†] radicals appearing in the acetylene flame will be ~ 1.5 times higher than the average thermal energy.

The distribution of rotational line intensities in emission or absorption spectra of flames shows that often the rotational energies of particles formed in exothermic reactions also exceed the average thermal energy. Since the distribution of intensities usually follows the Boltzmann law, the "rotational temperature" of molecules may be taken as a measure of their "rotational energy". As an example, let us take the "rotational temperature" of hydroxyl in an emission spectrum of acetylene — oxygen flames. The "rotational temperature" of electronically-excited hydroxyl[††], probably formed by the reaction

$$CH + O_2 \rightarrow CO + OH + 159.2 \text{ kcal. mole}^{-1},$$

as measured by Gaydon and Wolfhard[69] for a pressure range of 1 atm to 1.5 torr, appeared to increase as the pressure decreased, from 5400 °K at 1 atm to 8750 °K at 1.5 torr (the maximum "flame temperature" at 1 atm is 3400 °K; at lower pressures it drops correspondingly). Taking into account the rotational energy exchange by collisions of hydroxyl with other molecules, and also the chemiluminescence quenching, it may be shown that the energy corresponding to the maximum "rotational temperature" (8750 °K) should be close to the rotational energy of excited hydroxyl at the moment of its formation. This energy (proportional to "rotational temperature") thus appears to exceed the mean rotational energy of OH at the flame temperature by a factor of three and more.

Spectroscopic investigation of flames shows that the vibrational energy of molecules in the reaction zone also often exceeds the equilibrium energy at the flame temperature. For instance, measurement of the relative intensity of hydroxyl bands corresponding to $v' = 0, 1$ and 2 (v' is the vibrational quantum number in the upper electronic level) in the emission spectrum of a rarefied hydrogen flame[114] gave values similar to those corresponding to 9000 °K (at a flame temperature of 1000 °K).

The excitation of high vibrational levels in the course of a chemical reaction is

---

[†] Excitation energy 66.2 and 74.2 kcal. mole⁻¹.
[††] Excitation energy 93.4 kcal. mole⁻¹.

evident from investigation of absorption, as well as of rotational–vibrational emission, spectra of the molecules formed. For example, Norrish *et al.*[122] observed intense bands due to transitions from vibrational levels $v'' = 4, 5, 6, 7$ and 8 in the absorption spectrum of molecular oxygen formed by the reactions

$$O + NO_2 \rightarrow NO + O_2 + 46.6 \text{ kcal. mole}^{-1}$$

and

$$O + ClO_2 \rightarrow ClO + O_2 + 50.9 \text{ kcal. mole}^{-1}$$

These cannot be observed under ordinary conditions due to the negligible concentration of O_2 molecules in these levels. It follows from the above that an O_2 molecule formed in the above reactions may remove 74 and 67 % of the heat of reaction, respectively, as vibrational energy.

A still greater amount of vibrational energy is imparted to a hydroxyl radical formed in the reaction

$$H + O_3 \rightarrow O_2 + OH + 76.8 \text{ kcal. mole}^{-1}$$

This reaction was first used[19,78] to explain the origin of rotational–vibrational hydroxyl emission bands observed in the night sky spectrum[136]. Later on these bands were detected during the reaction of H atoms with ozone under laboratory conditions[133]. Both night sky and laboratory spectra exhibit bands corresponding to transitions from the $v'' = 9$ and lower levels. It follows that OH radicals formed by the $H + O_3$ reaction may remove 75.0 kcal. mole^{-1} ($v'' = 9$), *i.e.* 98 % of the heat of reaction as vibrational energy (see also ref. 152).

Bands and lines due to electronic excitation of reaction products and intermediates are often observed in chemiluminescence spectra, along with rotational–vibrational bands. These are evidence for the appearance of electronically-excited species in the course of the reaction. We have mentioned above the electronically-excited CH and OH radicals observed in acetylene flames. Let us consider some other examples.

Besides the rotational–vibrational OH bands mentioned above, the spectra obtained during the reactions of H atoms with ozone showed[67] O_2 bands connected with the electronic transition $^1\Sigma_g^+ - ^3\Sigma_g^-$. The electronically-excited O_2 molecules (excitation energy 37.7 kcal. mole^{-1}) are also formed by the reaction

$$H + O_3 = O_2 + OH$$

The carbon monoxide flame spectrum exhibits bands from electronically-excited

CO_2 molecules formed by

$$O + CO \rightarrow CO_2 + 125.8 \text{ kcal. mole}^{-1}$$

Measurements of the absolute intensity of the spectrum at a pressure of 100 torr and a temperature of 1000 °K[115] have shown that under these conditions one excited molecule will appear for every 125 CO_2 molecules formed. The measured intensity is immensely higher than that of the equilibrium emission at the flame temperature, and this seems to be directly related to the non-equilibrium O atom concentrations in the flame, exceeding by a factor of several thousands the thermodynamic equilibrium concentration. The effective rate coefficient for the formation of electronically-excited CO_2 molecules by collisions between CO and O has been estimated[44, 112, 125]. Thrush has recently reported[201] rate coefficients for the reactions

$$NO + O_3 \rightarrow NO_2' + O_2 \quad \text{and} \quad SO + O_3 \rightarrow SO_2' + O_2$$

giving electronically-excited NO_2 and SO_2 molecules. (For further examples and references see Rabinovitch and Flowers[162]).

Molecules with excess energy will, naturally, be more reactive than those with energy equal to the mean thermal energy. This is shown by the frequent appearance of "hot" species in photochemical reactions. Let us mention here one example only. Hot CH_3 radicals are formed in the photodissociation of CH_3I using the 2537 A mercury line. This may be deduced from the fact that the rate of methane formation in this reaction is temperature independent over the temperature range 40 to 100 °C[171]. Since the activation energy of the reaction

$$CH_3 + CH_3I \rightarrow CH_4 + CH_2I + \sim 5 \text{ kcal. mole}^{-1}$$

responsible for methane formation under the experimental conditions is about 10 kcal. mole^{-1} for "cool" CH_3[193], the rise of temperature from 40 to 100 °C would be expected to result in a more than ten-fold increase in the rate of CH_4 formation. The temperature independence of the latter thus is evidence that "hot" CH_3 radicals, considerably more reactive than normal ones, are formed in the photochemical process.

It follows from the above that, since the assumption of equilibrium energy distribution excludes the participation of energy-rich "hot" species, this assumption would be valid only when the *time between formation and reaction* (for a particular species) *exceeds that of thermal relaxation*. When the latter is greater than the former, energy-rich particles accumulate in the reaction system and may have a marked effect on the kinetics and mechanism of the overall process. The importance of this energy factor is most evident in so-called *energy chains*.

4.2 RADICAL-ENERGY CHAINS

Bodenstein[28] was the first to make use of the energy factor in 1916, when putting forward the chain mechanism of the photochemical chlorine–hydrogen reaction. He suggested that electronically-excited chlorine molecules acted as active centres in this reaction. By reacting with hydrogen molecules they would yield excited HCl molecules transmitting their energy to chlorine molecules by collisions, and this would result in regeneration of active (excited) chlorine molecules. However, the energy chain suggested by Bodenstein for the chlorine–hydrogen reaction proved to be invalid since it was at variance with spectroscopic data, and gave place to the radical mechanism proposed by Nernst.

The concept of active centres as energy-rich valence-saturated particles may be encountered also in a paper by Christiansen and Kramers[43] who attempt to make use of chain concepts in explaining the high rate of unimolecular reactions. Christiansen and Kramers suggest that reaction products with excess energy are capable of transmitting this energy to molecules of the initial substance[†]. New active molecules are thus formed and the reaction proceeds by a chain mechanism (*energy chain*). However, this energy chain concept obtained no confirmation in the further development of chemical kinetics.

Only recently, with increase of interest in chain reactions occurring at *high temperatures*, involving large amounts of energy in the reacting system and consequently high concentrations of energy-rich particles, the energy concept was again resuscitated[185]. Since high temperatures favour the dissociation of molecules into free atoms and radicals, it has to be expected that both the energy and chemical (or radical) factors would be involved in chain propagation under high-temperature conditions. In other words, at temperatures corresponding to thermal relaxation times higher than that for reaction (considered as one elementary step), mixed *radical-energy chains* would be the most probable.

Until recently the only example of such reactions was the decomposition of ozone induced by UV radiation ($\lambda < 2500$ A). In this case an electronically-excited (metastable) O atom in the ^1D state (excitation energy 45.4 kcal. gr. atom^{-1}) will be produced in the primary photochemical step

$$O_3 + h\nu \rightarrow O_2 + O(^1D)$$

Reaction and regeneration of the $O(^1D)$ atoms seems to be responsible for the quantum yield of between 8[76, 77] and 17[153], since the photochemical decomposition of ozone using visible (orange) light, with formation of O atoms in the ground state (^3P), has a quantum yield of 2[38, 96]. Interaction between an $O(^1D)$ atom and

---

[†] The idea that the energy released by a reaction as translational energy was utilized for acceleration of the reaction was also put forward by Dixon[49].

an ozone molecule releases $94.1 + 45.4 = 139.5$ kcal. mole^{-1}, since

$$O(^3P) + O_3 \rightarrow 2O_2 + 94.1 \text{ kcal. mole}^{-1}$$

i.e. an average of 70 kcal. mole^{-1} per O_2. This energy is sufficient for chain propagation by

$$O_2 + O_3 \rightarrow O_2 + O_2 + O(^1D) - 70.9 \text{ kcal. mole}^{-1}$$

Thus, in this case the chain centres are O atoms in an electronically-excited state, and O_2 molecules with excess energy. According to Norrish[151], these are vibrationally-excited molecules, but according to Semenov[180] they are electronically-excited molecules in the $^3\Sigma_u^-$ state (excitation energy 142.4 kcal. mole^{-1}) produced by interaction between $O(^1D)$ atoms and O_3 molecules.

It has been mentioned above that, as far back as 1916, Bodenstein attempted to explain the mechanism of the chlorine–hydrogen reaction from an energy chain viewpoint. In 1934 Semenov[174] making use of a mechanism similar to that suggested by Bodenstein, showed, in principle, the possibility of branching in this reaction, which usually involves a simple (unbranched) chain. Semenov's idea was that energy-rich HCl molecules formed by the reaction

$$H + Cl_2 \rightarrow HCl + Cl + 45.1 \text{ kcal. mole}^{-1}$$

may collide with chlorine molecules with a sufficient energy and dissociate these into atoms (the Cl_2 dissociation energy is 57.1 kcal. mole^{-1}). This would result in the initiation of two additional chains (branching).

Recently Semenov[180] attempted to substantiate quantitatively the occurrence of "energy branching" in a chlorine–hydrogen reaction, but came to the conclusion that in the present state of knowledge on the dynamics of an elementary chemical reaction and of energy exchange by molecular collision this was impossible. Determination[1] of relative rate coefficients for the process

$$H + Cl_2 \rightarrow HCl_v^* + Cl + 45.1 - E_v \text{ kcal. mole}^{-1}$$

(HCl_v^* is a vibrationally-excited HCl molecule in the v state, and E_v its vibrational energy) has shown that most HCl molecules appear in the zero vibrational state ($E_0 = 0$), and only 0.015 % of the HCl molecules formed occupy the $v = 6$ level corresponding to an energy of 45.5 kcal. mole^{-1}. The small fraction of HCl molecules carrying away a large amount of vibrational energy at the moment of their formation, as well as the high probability of translational and rotational energy dissipation by molecular collisions, make the suggestion of Semenov that energy branching is important in the chlorine–hydrogen reaction rather improbable.[†]

Things are different for a similar reaction between fluorine and hydrogen

$$H + F_2 \rightarrow HF + F + 96.6 \text{ kcal. mole}^{-1}$$

where the heat of reaction is 2.5 times higher than the endothermicity of fluorine molecule dissociation. Thus it may be assumed that the number of excited HF molecules with an energy sufficient to dissociate F_2 on collision will be considerably higher than in the case of HCl. Consequently energy branching in the fluorine–hydrogen reaction would seem to be probable[180]. The ready ignition of fluorine–hydrogen mixtures, occurring even at very low temperatures[73], may be considered as supporting this view. The ready occurrence of a chemical reaction between fluorine and hydrogen (and also between fluorine and hydrocarbons) is obviously related to the low activation energy for generation of chains by the step

$$H_2 + F_2 \rightarrow H + HF + F - 6.7 \text{ kcal. mole}^{-1}$$

(see Semenov[179], Chap. III, Sections 8 and 9).[††]

Since a reaction of a radical with a molecule occurs, as a rule, more readily than that between two saturated molecules, that of a saturated molecule with *two* radicals

$$R + AB + R' \rightarrow AR + BR'$$

[†] The same negative result was obtained in the attempt[2] to find out what was the contribution from hot atoms formed in the reaction

$$H + X_2 \rightarrow HX + X \ (X = Cl, Br)$$

to the overall H_2–halogen reaction. It appeared that the X atoms succeed in thermalizing before entering into the reaction

$$X + H_2 \rightarrow HX + H$$

However, this does not seem to occur in the photochemical decomposition of hydrogen in the presence of HI, yielding hot hydrogen atoms by reaction

$$HI + h\nu \rightarrow H^* + I$$

In this case the reaction rate increases with increasing energy of the H atoms[106].

[††] A similar initiation step for the reaction of F_2 with hydrocarbons of the paraffin series, *viz.*

$$RH + F_2 \rightarrow R + HF + F$$

seems to have been postulated for the first time by Miller *et al.*[137, 138] who obtained convincing evidence for it by analysis of the reaction products.

It will be noted that in comparison with the reaction between fluorine and hydrogen that of chlorine with hydrogen

$$H_2 + Cl_2 \rightarrow H + HCl + Cl - 58.2 \text{ kcal. mole}^{-1}$$

is strongly endothermic.

References pp. 183–188

would be even easier (from the point of view of activation energy). Semenov's postulate was[179] that the reverse reaction to the above would be one of the main steps leading to radical formation in the interaction between saturated molecules. A high AB bond dissociation energy is a necessary condition for the ready occurrence of this reaction, the activation energy of which Semenov considers to be close to the value of ΔH for endothermic reactions, and tending to zero for exothermic reactions. Diverse examples supporting this idea are given in the book by Semenov[179] and in papers that have appeared later†.

Returning to the H_2–F_2 reaction, the H and F atoms formed in the initiation step interact, in turn, with fluorine and hydrogen molecules, thus propagating an unbranched chain (the Nernst chain). However, the energy-rich HF molecules formed by interaction between H and F_2 appear to be capable of dissociating fluorine molecules (see also ref. 34a)

$$F_2 + HF^* \rightarrow F + FH + F$$

and this provides an additional source of free radicals and leads to the initiation of two new chains. The chains are branched and due to this explosion may occur††. The branched chain nature of a fluorine–hydrogen reaction is quite evident from the recently discovered first and second ignition limits[89]. The first limit was determined by observing the ignition of a fluorine–hydrogen mixture flowing into an evacuated reactor. Ignition took place only after the mixture pressure attained a limiting minimum value p_1. The second limit was established in two ways. One was by observation of ignition in a fluorine–hydrogen mixture occurring when the pressure was decreased to a certain limiting value p_2. The other consisted in observation of the ignition of a mixture prepared at 77° K and then rapidly heated to a certain temperature. Ignition occurred only when the mixture pressure was lower than p_2. The second ignition limit, $[F_2] : [H_2] = 0.09$, for various temperatures is shown in Fig. 28.

In order to explain the ignition limits, chain termination steps should be introduced into the fluorine–hydrogen reaction scheme. It would be natural to connect the lower ignition limit with termination of chains at the reactor wall. As to the upper limit, all the features observed may be explained[89] by considering chain termination as due to oxygen (present in minor amounts) which removes

---

† Shilov et al.[87] obtained a rate coefficient:

$k = 3 \times 10^{10} \exp(-3500/RT)$ cm³. mole⁻¹. sec⁻¹

for the reaction

$HI + F_2 \rightarrow F + HF + I + 2.2$ kcal. mole⁻¹

†† It would be expected that due to the high heat of the reaction

$H_2 + F_2 = 2HF + 128.4$ kcal. mole⁻¹

the chain ignition will readily change into thermal ignition.

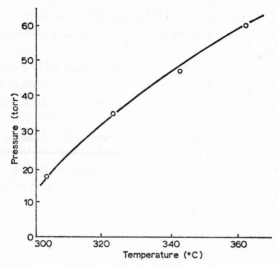

Fig. 28. The second ignition limit for a mixture of hydrogen with fluorine[89].

H atoms by the reaction

$$H + O_2 + M \rightarrow HO_2 + M$$

While branching in the fluorine–hydrogen reaction is accounted for by *intermolecular energy exchange* ($HF^* + F_2 \rightarrow HF + 2F$), that in the fluorine–$CH_3I$ reaction occurs by *intramolecular energy exchange*[88, 168]. The mechanism of this reaction may be conceived as

$$F + CH_3I \rightarrow HF + CH_2I \tag{I}$$

$$CH_2I + F_2 \rightarrow CH_2IF^* + F \tag{II}$$

$$CH_2IF^* \rightarrow CH_2F + I \tag{III}$$

$$CH_2F + F_2 \rightarrow CH_2F_2 + F \tag{IV}$$

Here branching occurs as a result of the energy liberated in reaction (II) (some 80 kcal. mole^{-1}) being 25 kcal. mole^{-1} higher than the C–I bond dissociation energy in the CH_2IF molecule (some 55 kcal. mole^{-1}). When the energy removed by the excited (CH_2IF^*) molecule is higher than 55 kcal. mole^{-1}, and the molecule has no time to become stabilized, it will dissociate by (III) owing to *internal redistribution of energy*. Reaction (IV), which increases the number of fluorine atoms in the system, is responsible for branching.

In agreement with the reaction mechanism given above, molecular iodine and methylene fluoride, CH_2F_2, were observed in the reaction products, iodine atoms were detected by the ESR spectrum[88, 168] (Fig. 29), and IF bands in the flame emission spectrum; this is evidence for the presence of iodine and fluorine atoms in the reaction zone[182].

The reactions of fluorine with hydrogen and methyl iodide certainly do not represent the sole examples of radical-energy branched chain reactions involving fluorine. The reactions of fluorine with hydrogen halides: HCl, HBr, and HI, and with a number of organic halogen derivatives, in particular with methyl chloride and halogenated olefins, have also been studied[167]. High concentrations of Cl and F, and of Br and F atoms were detected by means of the ESR technique in reactions of fluorine with C_2Cl_4 and $C_2H_2Br_2$, respectively. Fig. 30 represents an ESR spectrum due to fluorine atoms obtained in a reaction of fluorine with C_2Cl_4[168].

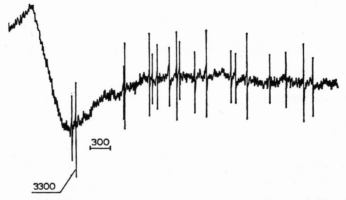

Fig. 29. ESR spectrum exhibited by iodine atoms in the reaction of fluorine with methyl iodide[88, 168].

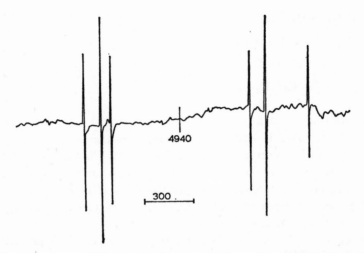

Fig. 30. ESR spectrum exhibited by fluorine atoms in the reaction of C_2Cl_4 with fluorine[168].

Semenov[181] gives the following table

RH	$\Delta E = Q - D$ kcal. mole^{-1}	RH	$\Delta E = Q - D$ kcal. mole^{-1}
H_2	-37	CH_3Br	$+10$
CH_4	-21	CH_3NO_2	$+19$
CH_3OH	-13	CH_3I	$+24$
CH_3Cl	-3	CH_3ONO_2	$+39$
CH_3SH	$+7$	CH_3ONO	$+41$

The first column represents RH molecules reacting with fluorine atoms by $F + RH = R + HF$, and the second the value $\Delta E = Q - D$, where Q is the exothermicity of the reaction $R + F_2 = RF + F$, and D the energy for dissociation of an RF molecule to two new radicals ($RF = R_1 + R_2$). Semenov suggests that in radical-energy chain reactions of fluorine with compounds for which $\Delta E < 0$, branching is possible owing to intermolecular energy exchange, and for reactions with $\Delta E > 0$ to intramolecular energy exchange.

Thus the number of possible branched chain reactions appears to be considerably higher than was thought not long ago, due to underestimation of the energy factor in branched chain reactions. The part played by this factor seems to be of extreme importance in chemical reactions occurring at high temperatures.

REFERENCES

1 J. R. AIREY, R. R. GETTY, J. C. POLANYI AND D. R. SNELLING, *J. Chem. Phys.*, 41 (1964) 3255.
2 J. R. AIREY, J. C. POLANYI AND D. R. SNELLING, *Symp. Combustion, 10th*, (1965) 403.
3 B. V. AIVAZOV, M. B. NEIMAN AND I. I. KHANOVA, *Izv. Akad. Nauk SSSR, Otd. Khim. Nauk*, (1938) 307; *Acta Physicochimica URSS*, 10 (1938) 765.
4 P. G. ASHMORE AND J. CHANMUGAM, *Trans. Faraday Soc.*, 49 (1953) 254.
5 P. G. ASHMORE AND M. S. SPENCER, *Trans. Faraday Soc.*, 55 (1959) 1868.
6 L. I. AVRAMENKO AND R. V. LORENTSO, *Zhur. Fiz. Khim.*, 24 (1950) 207.
7 V. V. AZATYAN, *D.Ph. Thesis*, Inst. of Chem. Physics, Acad. Sci. USSR, Moscow, 1963.
8 V. V. AZATYAN AND A. B. NALBANDYAN, *Zhur. Vsesoyuznogo Khim. Obshchestva im. Mendeleeva*, 11 (1966) 162.
9 V. V. AZATYAN, A. B. NALBANDYAN AND N. T. SILAKHTARYAN, *Izv. Akad. Nauk Arm. SSR*, 17 (1964) 117.
10 V. V. AZATYAN, A. B. NALBANDYAN AND MEI-YUAN' TSUI, *Dokl. Akad. Nauk SSSR*, 147 (1962) 361; *Dokl. Akad. Nauk Arm. SSR*, 36 (1963) 23; *Kinetika i Kataliz*, 5 (1964) 120.
11 V. V. AZATYAN, L. B. ROMANOVICH AND S. G. SYSOEVA, *Fizika Goreniya i Vzryva*, (1967) 77.
12 V. V. AZATYAN, V. V. VOEVODSKII AND A. B. NALBANDYAN, *Kinetika i Kataliz*, 2 (1961) 340.
13 V. P. BALAKHNIN, *D.Ph. Thesis*, Inst. of Chem. Physics, Acad. Sci. USSR, Moscow, 1966.
14 V. P. BALAKHNIN, YU. M. GERSHENSON, V. N. KONDRATIEV AND A. B. NALBANDYAN, *Dokl. Akad. Nauk SSSR*, 154 (1964) 883.
15 V. P. BALAKHNIN, YU. M. GERSHENSON, V. N. KONDRATIEV AND A. B. NALBANDYAN, *Dokl. Akad. Nauk SSSR*, 154 (1964) 1142.
16 R. R. BALDWIN, N. S. CORNEY AND R. F. SIMMONS, *Symp. Combustion, 5th*, (1955) 502.
16a R. R. BALDWIN AND P. DORAN, *Trans. Faraday Soc.*, 57 (1961) 1578.
17 R. R. BALDWIN AND R. F. SIMMONS, *Trans. Faraday Soc.*, 53 (1957) 955, 964.

18 L. BATEMAN, H. HUGHES AND A. L. MORRIS, *Discussions Faraday Soc.*, 14 (1953) 190.
19 D. R. BATES AND M. NICOLET, *Publ. Astron. Soc. Pacific*, 62 (1950) 106; *Compt. Rend. Acad. Sci.*, *Paris*, 230 (1950) 1943; *J. Geogr. Research*, 55 (1950) 301.
20 E. R. BELL, J. H. RALEY, F. F. RUST, F. H. SEUBOLD AND W. E. VAUGHAN, *Discussions Faraday Soc.*, 10 (1951) 242.
21 S. W. BENSON, *J. Chem. Phys.*, 40 (1964) 1007.
22 S. W. BENSON, *J. Am. Chem. Soc.*, 87 (1965) 972.
23 S. W. BENSON AND R. SRINIVASAN, *J. Chem. Phys.*, 23 (1955) 200.
24 A. E. BIRON AND A. B. NALBANDYAN, *Zhur. Fiz. Khim.*, 9 (1937) 132.
24a D. R. BLACKMORE, *Trans. Faraday Soc.*, 62 (1966) 415.
25 E. A. BLYUMBERG, YU. D. NORIKOV AND N. M. EMANUEL, *Dokl. Akad. Nauk SSSR*, 151 (1963) 1127.
26 E. A. BLYUMBERG, G. E. ZAIKOV AND N. M. EMANUEL, *Neftekhimiya*, 1 (1961) 235.
27 M. BODENSTEIN, *Z. Phys. Chem.*, 85 (1913) 329.
28 M. BODENSTEIN, *Z. Elektrochem.*, 22 (1916) 53.
29 M. BODENSTEIN AND C. LIND, *Z. Phys. Chem.*, 57 (1906) 168.
30 M. BODENSTEIN AND H. LÜTKEMEYER, *Z. Phys. Chem.*, 114 (1925) 208.
31 M. BODENSTEIN AND P. W. SCHENK, *Z. Phys. Chem.*, B20 (1933) 420.
32 M. L. BOGOYAVLENSKAYA AND A. A. KOVALSKII, *Zhur. Fiz. Khim.*, 20 (1946) 1325.
33 J. L. BOLLAND AND G. GEE, *Trans. Faraday Soc.*, 42 (1946) 236.
34 R. K. BRINTON, *Can. J. Chem.*, 38 (1960) 1339.
34a R. S. BROKAW, *J. Phys. Chem.*, 69 (1965) 2488.
35 R. W. BUNSEN AND H. ROSCOE, *Phil. Trans.*, (1857) 381.
36 V. R. BURSIAN AND V. S. SOROKIN, *Z. Phys. Chem.*, B12 (1931) 247.
37 R. BURT, F. SKUSE AND A. THOMAS, *Combust. Flame*, 9 (1965) 159.
38 E. CASTELLANO AND H. J. SCHUMACHER, *J. Chem. Phys.*, 36 (1962) 2238.
39 H. CERFONTAIN AND K. O. KUTSCHKE, *J. Am. Chem. Soc.*, 84 (1962) 4017.
40 A. M. CHAIKIN, *D. Ph. Thesis*, Moscow State University, 1955.
41 N. M. CHIRKOV, *D. Ph. Thesis*, Inst. of Chem. Physics, Leningrad, 1935; *Acta Physicochim. URSS*, 6 (1937) 915.
42 J. A. CHRISTIANSEN, *Dansk. Vid. Math. Phys. Med.*, 1 (1919) 14.
43 J. A. CHRISTIANSEN AND H. A. KRAMERS, *Z. Phys. Chem.*, 104 (1923) 451.
44 M. A. A. CLYNE AND B. A. THRUSH, *Proc. Roy. Soc. (London), Ser. A*, 269 (1962) 404.
45 W. CRUICKSHENK, *Nicholson's Journal*, 5 (1801) 201.
46 E. T. DENISOV, *Dokl. Akad. Nauk SSSR*, 130 (1960) 1055.
47 E. T. DENISOV, *Kinetika i Kataliz*, 4 (1963) 53.
47a E. T. DENISOV AND V. P. KOSAREV, *Zhur. Fiz. Khim.*, 38 (1964) 2875.
48 D. P. DINGLADY AND J. G. CALVERT, *J. Am. Chem. Soc.*, 85 (1963) 856.
49 H. B. DIXON, *Ber.*, 38 (1905) 2419.
50 H. B. DIXON, *J. Chem. Soc.*, 97 (1910) 661.
51 G. DIXON-LEWIS, W. E. WILSON AND A. A. WESTENBERG, *J. Chem. Phys.*, 44 (1966) 2877.
52 A. EINSTEIN, *Ann. Phys.*, 37 (1912) 832; 38 (1912) 881; *J. Phys.*, 3 (1913) 277.
53 N. M. EMANUEL, *Zhur. Vsesoyuznogo Khim. Obshchestva im. Mendeleeva*, 11 (1966) 186.
54 N. M. EMANUEL, E. T. DENISOV AND Z. K. MAIZUS, *Tsepnye Reaktsii Okisleniya Uglevodorodov v Zhidkoi Faze (Chain Reactions in the Liquid Phase Oxidation of Hydrocarbons)*, Nauka, Moscow, 1965.
55 H. EMELEUS, *J. Chem. Soc.*, 128 (1926) 2948.
56 N. S. ENIKOLOPYAN, *Dokl. Akad. Nauk SSSR*, 119 (1958) 520.
57 N. S. ENIKOLOPYAN, V. YA. SHTERN AND S. S. POLYAK, *Zhur. Fiz. Khim.*, 32 (1958) 2224.
58 C. P. FENIMORE AND G. W. JONES, *J. Phys. Chem.*, 62 (1958) 693.
59 N. V. FOK AND A. B. NALBANDYAN, *Dokl. Akad. Nauk SSSR*, 86 (1952) 589; 89 (1953) 125; *Dokl. Akad. Nauk SSSR*, 85 (1952) 1093; *Voprosy Khimicheskoi Kinetiki, Kataliza i Reaktsionnoi Sposobnosti (Problems of Chemical Kinetics, Catalysis, and Reactivity)*, Acad. Sci. USSR, Moscow, 1955, p. 219.
60 S. N. FONER AND R. L. HUDSON, *J. Chem. Phys.*, 21 (1953) 1374, 1608.
60a W. FORST AND P. A. GIGUÈRE, *J. Phys. Chem.*, 62 (1958) 340.

61 D. A. FRANK-KAMENETSKII, *Dokl. Akad. Nauk SSSR*, 25 (1939) 672; *Zhur. Fiz. Khim.*, 14 (1940) 30.
62 D. A. FRANK-KAMENETSKII, *Zhur. Fiz. Khim.*, 13 (1939) 738; *Acta Physicochim. URSS*, 10 (1939) 365; 16 (1942) 357; 20 (1945) 729.
63 D. A. FRANK-KAMENETSKII, *Dr.Sci. Thesis*, Inst. of Chem. Physics, Acad. Sci. USSR, Kazan', 1943.
64 D. A. FRANK-KAMENETSKII, *Diffuziya i Teploperedacha v Khimicheskoi Kinetike*, Acad. Sci. USSR, Moscow, 1947.
65 R. M. FRISTROM AND A. A. WESTENBERG, *Flame Structure*, McGraw-Hill, New York, 1965.
66 W. E. GARNER AND A. S. GOMM, *Trans. Faraday Soc.*, 24 (1928) 470.
67 D. GARVIN, H. P. BROIDA AND H. J. KOSTKOWSKI, *J. Chem. Phys.*, 32 (1960) 880.
68 A. G. GAYDON AND H. G. WOLFHARD, *Proc. Roy. Soc. (London), Ser. A*, 199 (1949) 89.
69 A. G. GAYDON AND H. G. WOLFHARD, *Flames*, Chapman and Hall, London, 1953, pp. 247–249.
69a J. C. GIDDINGS AND HYUNG KYU SHIN, *J. Chem. Phys.*, 36 (1962) 640;
P. GRAY AND P. R. LEE, *Combust. Flame*, 9 (1965) 201;
P. GRAY AND P. R. LEE, *Oxidation and Combustion Reviews*, Vol. 2, Elsevier, Amsterdam, 1967, p. 1.
69b J. C. GREAVES AND J. W. LINNETT, *Trans. Faraday Soc.*, 55 (1959) 1355.
69c M. GREEN, K. R. JENNINGS, J. W. LINNETT AND D. SCHOFIELD, *Trans. Faraday Soc.*, 55 (1959) 2152.
70 J. A. GREY, *J. Chem. Soc.*, (1953) 741.
71 J. G. A. GRIFFITHS AND R. G. W. NORRISH, *Proc. Roy. Soc. (London), Ser. A*, 147 (1934) 140.
72 M. GROS, A. M. DIAMY AND R. BEN-AIM, *Symp. Combustion, 11th*, (1967).
73 A. V. GROSSE AND A. D. KIRSCHENBAUM, *J. Am. Chem. Soc.*, 77 (1955) 512.
73a F. L. HANST AND J. G. CALVERT, *J. Phys. Chem.*, 63 (1959) 71.
74 J. HAY, J. H. KNOX AND J. M. C. TURNER, *Symp. Combustion, 10th*, (1965) 331.
75 J. R. HEICKLEN AND H. S. JOHNSTON, *J. Am. Chem. Soc.*, 84 (1962) 4030, 4394.
76 L. J. HEIDT, *J. Am. Chem. Soc.*, 57 (1935) 1710.
77 L. J. HEIDT AND G. S. FORBES, *J. Am. Chem. Soc.*, 56 (1934) 2365.
78 G. HERZBERG, *J. Roy. Astron. Soc. Can.*, 45 (1951) 100.
79 K. F. HERZFELD, *Z. Elektrochem.*, 25 (1919) 301.
80 C. N. HINSHELWOOD, *Discussions Faraday Soc.*, 2 (1947) 117.
81 H. HIRAOKA AND R. HARDWICK, *J. Chem. Phys.*, 36 (1962) 1715.
82 D. E. HOARE AND A. D. WALSH, *Trans. Faraday Soc.*, 53 (1957) 1102.
83 D. E. HOARE AND A. D. WALSH, *Trans. Faraday Soc.*, 50 (1954) 37.
84 K. U. INGOLD AND W. A. BRYCE, *J. Chem. Phys.*, 24 (1954) 360.
85 O. A. IVANOV AND A. B. NALBANDYAN, *Kinetika i Kataliz*, 1 (1960) 3.
86 T. A. JACOBS AND R. R. GIEDT, *J. Chem. Phys.*, 39 (1963) 749.
87 G. A. KAPRALOVA, L. YU. RUSIN, A. M. CHAIKIN AND A. E. SHILOV, *Dokl. Akad. Nauk SSSR*, 150 (1963) 1282.
88 G. A. KAPRALOVA, E. A. TROFIMOVA, L. YU. RUSIN, A. M. CHAIKIN AND A. E. SHILOV, *Kinetika i Kataliz*, 4 (1963) 653.
89 G. A. KAPRALOVA, E. A. TROFIMOVA AND A. E. SHILOV, *Kinetika i Kataliz*, 6 (1965) 977.
90 L. V. KARMILOVA, N. S. ENIKOLOPYAN AND A. B. NALBANDYAN, *Zhur. Fiz. Khim.*, 31 (1957) 851.
91 L. V. KARMILOVA, A. B. NALBANDYAN AND N. N. SEMENOV, *Zhur. Fiz. Khim.*, 32 (1958) 1193.
92 W. E. KASCAN AND W. G. BROWNE, cited by B. MYERS, K. P. G. SULZMANN AND E. R. BARTLE, *J. Chem. Phys.*, 43 (1965) 1220.
93 L. S. KASSEL, *Chem. Rev.*, 21 (1937) 333.
94 F. KAUFMAN AND F. P. DELGRECO, *Symp. Combustion, 9th*, (1963) 659.
95 A. D. KIRK AND J. H. KNOX, *Trans. Faraday Soc.*, 56 (1960) 1296.
96 G. B. KISTIAKOWSKY, *Z. Phys. Chem.*, 117 (1925) 337.
97 J. H. KNOX, *Trans. Faraday Soc.*, 55 (1959) 1362.
98 J. H. KNOX, *Trans. Faraday Soc.*, 56 (1960) 1225.
99 J. H. KNOX, *Ann. Rep. Progr. Chem.*, 59 (1962) 18.

100 J. H. KNOX, *Chemical Communications*, 6 (1965) 108.
101 J. H. KNOX, *Combust. Flame*, 9 (1965) 297.
102 J. H. KNOX AND R. G. W. NORRISH, *Proc. Roy. Soc. (London)*, *Ser. A*, 221 (1954) 151.
103 J. H. KNOX AND J. M. C. TURNER, *J. Chem. Soc.*, (1965) 3491.
104 J. H. KNOX AND C. H. J. WELLS, *Trans. Faraday Soc.*, 59 (1963) 2786, 2801.
105 V. N. KONDRATIEV, *Z. Physik*, 63 (1930) 322; *Acta Physicochim. URSS*, 4 (1936) 556.
106 V. N. KONDRATIEV, *Problemy Kinetiki i Kataliza (Problems of Kinetics and Catalysis)*, *Gos. Nauch. Tekh. Izd. Khim. Lit. M.–L.*, Vol. 4, 1940, p. 63.
107 V. N. KONDRATIEV, *Spektroskopicheskoye Izuchenie Khimicheskikh Gazovykh Reaktsii (Spectroscopic Investigation of Chemical Gas Reactions)*, Acad. Sci. USSR, Moscow–Leningrad, 1944.
108 V. N. KONDRATIEV, *Dokl. Akad. Nauk SSSR*, 49 (1945) 38.
109 V. N. KONDRATIEV, *Kinetika Khimicheskikh Gazovykh Reaktsii*, Acad. Sci. USSR, Moscow, 1958; English translation: *Chemical Kinetics of Gas Reactions*, Pergamon Press, Oxford, 1964.
110 V. N. KONDRATIEV, *Dokl. Akad. Nauk SSSR*, 137 (1961) 120.
111 V. N. KONDRATIEV, *Uspekhi Khim.*, 34 (1965) 2081.
112 V. N. KONDRATIEV AND I. I. PTICHKIN, *Kinetika i Kataliz*, 2 (1961) 492.
113 V. N. KONDRATIEV AND M. S. ZISKIN, *Zhur. Fiz. Khim.*, 9 (1937) 542; *Acta Physicochim. URSS*, 6 (1937) 307.
114 V. N. KONDRATIEV AND M. ZISKIN, *Zhur. Fiz. Khim.*, 10 (1937) 360.
115 E. I. KONDRATIEVA AND V. N. KONDRATIEV, *Zhur. Fiz. Khim.*, 9 (1937) 746.
116 E. I. KONDRATIEVA AND V. N. KONDRATIEV, *Dokl. Akad. Nauk SSSR*, 51 (1946) 607; *Acta Physicochim. URSS*, 21 (1946) 1, 629; *Zhur. Fiz. Khim.*, 20 (1946) 1239.
117 D. I. KOPP, A. A. KOVALSKII, A. V. ZAGULIN AND N. N. SEMENOV, *Zhur. Fiz. Khim.*, 1 (1930) 263; *Z. Phys. Chem.*, B6 (1930) 307.
118 A. A. KOVALSKII, *Physik. Z. Sowjetunion*, 4 (1933) 723.
119 K. B. KRAUSKOPF AND G. K. ROLLEFSON, *J. Am. Chem. Soc.*, 56 (1934) 327.
120 B. LEWIS AND G. VON ELBE, *Combustion, Flames, and Explosions of Gases*, Academic Press, New York, 1961.
121 J. W. LINNETT AND T. D. TRIBBECK, *Khimicheskaya Kinetika i Tsepnye Reaktsii (Chemical Kinetics and Chain Reactions)*, Nauka, Moscow, 1966, p. 128.
122 F. J. LIPSCOMB, R. G. W. NORRISH AND B. A. THRUSH, *Proc. Roy. Soc. (London)*, *Ser. A*, 233 (1955) 455.
123 A. J. LOTKA, *J. Am. Chem. Soc.*, 42 (1920) 1595.
124 M. LUCQUIN AND P. LAFFITTE, *Symp. Combustion, 6th*, (1957) 130.
125 B. H. MAHAN AND R. B. SOLO, *J. Chem. Phys.*, 37 (1962) 2669.
126 Z. K. MAIZUS, I. P. SKIBIDA AND N. M. EMANUEL, *Dokl. Akad. Nauk SSSR*, 131 (1960) 880.
127 A. A. MANTASHYAN, R. I. MOSHKINA AND A. B. NALBANDYAN, *Izvest. Akad. Nauk Arm. SSR, Khim. Nauki*, 14 (1961) 185.
128 A. A. MANTASHYAN AND A. B. NALBANDYAN, *Izvest. Akad. Nauk Arm. SSR, Khim. Nauki*, 14 (1961) 517, 527; 15 (1962) 3, 15.
129 A. M. MARKEVICH, *Zhur. Fiz. Khim.*, 22 (1948) 941.
130 A. M. MARKEVICH, *Zhur. Fiz. Khim.*, 30 (1956) 735.
131 F. R. MAYO, *J. Am. Chem. Soc.*, 80 (1958) 2465.
132 C. A. MCDOWELL AND J. R. THOMAS, *J. Chem. Phys.*, 17 (1949) 558.
133 J. D. MCKINLEY, D. GARVIN AND M. J. BOUDART, *J. Chem. Phys.*, 23 (1955) 784.
134 G. R. MCMILLAN, *J. Phys. Chem.*, 63 (1959) 1526.
135 S. S. MEDVEDEV, *Problemy Kinetiki i Kataliza (Problems of Kinetics and Catalysis)*, Gos. Nauch. Tekh. Izd. Khim. Lit., M.–L., Vol. 4, 1940, p. 28.
136 A. B. MEINEL, *Astron. J.*, 111 (1950) 555; 112 (1950) 120.
137 W. T. MILLER AND S. D. KOCH, *J. Am. Chem. Soc.*, 79 (1957) 3084.
138 W. T. MILLER, S. D. KOCH AND F. W. MCLAFFERTY, *J. Am. Chem. Soc.*, 78 (1956) 4992.
139 G. J. MINKOFF AND C. F. H. TIPPER, *Chemistry of Combustion Reactions*, Butterworths, London, 1962, pp. 210–211.
140 B. K. MORSE, *J. Am. Chem. Soc.*, 79 (1957) 3375.

141 A. B. NALBANDYAN, *Physik. Z. Sowjetunion*, 4 (1933) 747; *Acta Physicochim. URSS*, 1 (1934) 305.

142 A. B. NALBANDYAN, *Dr. Sci. Thesis*, Inst. of Chem. Physics, Acad. Sci. USSR, Kazan', 1942; *Zhur. Fiz. Khim.*, 20 (1946) 1259.

143 A. B. NALBANDYAN, *Zhur. Fiz. Khim.*, 19 (1945) 201, 210, 218.

144 A. B. NALBANDYAN, *Zhur. Fiz. Khim.*, 20 (1946) 1273.

145 A. B. NALBANDYAN, *Zhur. Fiz. Khim.*, 20 (1946) 1283.

146 A. B. NALBANDYAN AND V. V. VOEVODSKII, *Mekhanizm Okisleniya i Goreniya vodoroda* (*Mechanism of the Oxidation and Combustion of Hydrogen*), Acad. Sci. USSR, M.–L., 1949.

147 M. B. NEIMAN, *Acta Physicochim. URSS*, 9 (1938) 527.

148 W. NERNST, cited by K. FALK, *J. Am. Chem. Soc.*, 28 (1906) 1517.

149 W. NERNST, *Z. Elektrochem.*, 24 (1918) 335.

150 R. G. W. NORRISH, (a) *Proc. Roy. Soc. (London), Ser. A*, 150 (1935) 36.
(b) *Rev. Instr. Fr. Pétr. et Ann. Comb. Liq.*, 4 (1949) 288.
(c) *Discussions Faraday Soc.*, 10 (1951) 269.

151 R. G. W. NORRISH, *Transfert d'énergie dans les gaz*, 12ème Conseil de Chimie (Solvay), 1962, p. 99.

152 R. G. W. NORRISH, *Chemistry in Britain*, (1965) 289.

153 R. G. W. NORRISH AND R. P. WAYNE, *Proc. Roy. Soc. (London), Ser. A*, 288 (1965) 200.

154 V. N. PANFILOV, YU. D. TSVETKOV AND V. V. VOEVODSKII, *Kinetika i Kataliz*, 1 (1960) 333.

155 C. R. PATRICK AND J. C. ROBB, *Discussions Faraday Soc.*, 17 (1955) 98.

156 R. N. PEASE, *J. Am. Chem. Soc.*, 52 (1930) 5106.

156a R. N. PEASE, *Equilibrium and Kinetics of Gas Reactions*, Princeton University Press, 1942.

157 M. POLANYI, *Z. Elektrochem.*, 26 (1920) 49.

158 S. S. POLYAK AND V. YA. SHTERN, *Zhur. Fiz. Khim.*, 27 (1953) 341.

159 S. S. POLYAK AND V. YA. SHTERN, in *Tsepnye Reaktsii Okisleniya Uglevodorodov* (*Chain Oxidation of Hydrocarbons*), Acad. Sci. USSR, Moscow, p. 5.

160 M. D. POLYAKOV, *Naturwiss.*, 15 (1927) 539.

161 L. M. POSTNIKOV AND V. YA. SHLYAPINTOKH, *Dokl. Akad. Nauk SSSR*, 150 (1963) 340.

162 B. S. RABINOVITCH AND M. C. FLOWERS, *Quart. Rev.*, 18 (1964) 122; *Khimicheskaya Kinetika i Tsepnye Reaktsii* (*Chemical Kinetics and Chain Reactions*), Nauka, Moscow, 1966, p. 61.

163 F. O. RICE AND K. F. HERZFELD, *J. Phys. and Colloid Chem.*, 55 (1951) 975.

164 D. L. RIPLEY AND W. C. GARDINER, JR., *J. Chem. Phys.*, 44 (1966) 2285.

165 A. ROBERTSON AND W. A. WATERS, *J. Chem. Soc.*, (1948) 1578.

166 G. K. ROLLEFSON AND M. BURTON, *Photochemistry and the Mechanism of Chemical Reactions*, Prentice-Hall, New York, 1942.

167 L. YU. RUSIN, *D.Ph. Thesis*, Inst. of Chem Physics, Acad. Sci. USSR, Moscow, 1966.

168 L. YU. RUSIN, A. M. CHAIKIN AND A. E. SHILOV, *Kinetika i Kataliz*, 5 (1964) 1121.

169 G. L. SCHOTT AND J. L. KINSEY, *J. Chem. Phys.*, 29 (1958) 1177.

170 C. C. SCHUBERT AND R. N. PEASE, *J. Am. Chem. Soc.*, 78 (1956) 5553.

171 R. D. SCHULTZ AND H. A. TAYLOR, *J. Chem. Phys.*, 18 (1950) 194.

172 N. N. SEMENOV, *Z. Physik*, 48 (1928) 571; *Uspekhi Fiz. Nauk*, 23 (1940) 251.

173 N. N. SEMENOV, *Z. Phys. Chem.*, B2 (1929) 161.

174 N. N. SEMENOV, *Tsepnye Reaktsii*, ONTI, Leningrad, 1934; Eng. transl. *Chemical Kinetics and Chain Reactions*, Oxford, 1935.

175 N. N. SEMENOV, *Acta Physicochim. URSS*, 18 (1943) 93.

176 N. N. SEMENOV, *Zhur. Fiz. Khim.*, 17 (1943) 187.

177 N. N. SEMENOV, *Dokl. Akad. Nauk SSSR*, 43 (1944) 360.

178 N. N. SEMENOV, *Acta Physicochim. URSS*, 20 (1945) 291.

179 N. N. SEMENOV, *O Nekotorykh Problemakh Khimicheskoi Kinetiki i Reaktsionnoi Sposobnosti*, Acad. Sci. USSR, Moscow, 1958; Engl. transl. *Some Problems in Chemical Kinetics and Reactivity*, Princeton University Press, 1959, Pergamon Press, 1959.

180 N. N. SEMENOV, *Transfert d'énergie dans les gaz*, 12ème Conseil de Chimie (Solvay), 1962, p. 183.

181 N. N. SEMENOV, *On Branched Chain Reactions*, lecture delivered to the Royal Society of London, February, 1965.

182 A. E. SHILOV, *Dr.Sci. Thesis*, Inst. of Chem. Phys., Acad. Sci. USSR, Moscow, 1966.
183 V. YA. SHTERN, *Mekhanizm Okisleniya Uglevodorodov v Gazovoi Faze*, Acad. Sci. URSS, Moscow, 1960; Engl. transl. *The Gas Phase Oxidation of Hydrocarbons*, Pergamon Press, 1964.
184 V. YA. SHTERN AND S. S. POLYAK, *Dokl. Akad. Nauk SSSR*, 65 (1949) 311.
185 K. E. SHULER, *Symp. Combustion, 5th*, (1955) 56.
186 I. P. SKIBIDA AND E. M. GONIKBERG, *Izvest. Akad. Nauk SSSR, Odt. Khim. Nauk*, (1964) 286.
187 I. P. SKIBIDA, Z. K. MAIZUS AND N. M. EMANUEL, *Dokl. Akad. Nauk SSSR*, 149 (1963) 1111.
188 W. C. SLEPPY AND J. G. CALVERT, *J. Am. Chem. Soc.*, 81 (1959) 769.
189 W. V. SMITH, *J. Chem. Phys.*, 11 (1943) 110.
190 N. A. SOKOLOVA, M. A. MARKEVICH AND A. B. NALBANDYAN, *Zhur. Fiz. Khim.*, 35 (1961) 850.
191 R. I. SOLOUKHIN, *Symp. Combustion, 10th*, (1965) 521.
192 J. STARK, *Physik. Z.*, 9 (1908) 889, 894; *Prinzipien der Atomdynamik*, Band II, Hirzel Verlag, Leipzig, 1911, S. 207; *Ann. Phys.*, 38 (1912) 467.
193 E. W. R. STEACIE, *Atomic and Free Radical Reactions*, Reinhold, New York, 1954, pp. 743–744.
194 N. R. SUBBARATNAM AND J. G. CALVERT, in *Chemical Reactions in the Lower and Upper Atmosphere*, Interscience, New York, 1961, p. 109.
195 N. R. SUBBARATNAM AND J. G. CALVERT, *J. Am. Chem. Soc.*, 84 (1962) 1113.
196 K. P. G. SULTZMANN, B. F. MYERS AND E. R. BARTLE, *J. Chem. Phys.*, 42 (1965) 3969.
197 J. TAFFANEL, *Compt. Rend. Acad. Sci. Paris*, 157 (1913) 714.
198 J. TAFFANEL AND LE FLOCH, *Compt. Rend. Acad. Sci. Paris*, 156 (1913) 1544; 157 (1913) 469, 595.
199 J. R. THOMAS AND O. L. HARLE, *J. Phys. Chem.*, 63 (1959) 1027.
200 H. W. THOMPSON AND C. N. HINSHELWOOD, *Proc. Roy. Soc. (London), Ser. A*, 122 (1929) 610.
201 B. A. THRUSH, *Chemistry in Britain*, (1966) 287.
202 N. N. TIKHOMIROVA AND V. V. VOEVODSKII, *Dokl. Akad. Nauk SSSR*, 79 (1951) 993.
202a C. F. H. TIPPER AND R. K. WILLIAMS, *Trans. Faraday Soc.*, 56 (1960) 1805.
203 O. M. TODES, *Zhur. Fiz. Khim.*, 4 (1933) 78; *Acta Physicochim. URSS*, 5 (1936) 785.
204 O. M. TODES AND B. M. MELENTIEV, *Acta Physicochim. URSS*, 11 (1939) 153; *Zhur. Fiz. Khim.*, 13 (1939) 868, 1594.
205 T. G. TRAYLOR AND P. D. BARLETT, *Tetrahedron Letters*, 24 (1960) 30.
206 A. TRIFONOV, *Z. Phys. Chem.*, B3 (1929) 195.
207 G. H. TWIGG, *Chem. Eng. Sci. Spec. Suppl.*, 3 (1954) 5.
208 A. R. UBBELOHDE, *Proc. Roy. Soc. (London), Ser. A*, 152 (1935) 354.
209 J. H. VAN 'tHOFF, *Etudes de Dynamique Chimique*, Frederik Muller and Co., Amsterdam, 1884.
210 V. V. VOEVODSKII, *Zhur. Fiz. Khim.*, 20 (1946) 179.
211 V. V. VOEVODSKII, *Dokl. Akad. Nauk SSSR*, 90 (1953) 815.
212 V. V. VOEVODSKII, *Symp. Combustion, 7th*, (1959) 34.
213 V. V. VOEVODSKII AND V. A. POLTORAK, *Zhur. Fiz. Khim.*, 24 (1950) 299.
214 M. VOLMER, *Z. Phys. Chem.*, 9 (1930) 141.
215 A. D. WALSH, *Trans. Faraday Soc.*, 42 (1946) 269.
216 J. S. WATSON AND B. DE B. DARWENT, *J. Phys. Chem.*, 61 (1957) 577.
217 C. R. WILKE, *Chem. Eng. Progr.*, 46 (1950) 95.
218 H. WISE, C. M. ABLOW AND K. M. SANCIER, *J. Chem. Phys.*, 41 (1964) 3569.
219 C. F. WURSTER, L. J. DURHAM AND H. S. MOSHER, *J. Am. Chem. Soc.*, 80 (1958) 327, 332.
220 A. V. ZAGULIN, *Z. Physik. Chem.*, B1 (1928) 275.
221 A. P. ZEELENBERG AND A. F. BICKEL, *J. Chem. Soc.*, (1961) 4014.
222 YA. B. ZELDOVICH AND V. I. YAKOVLEV, *Dokl. Akad. Nauk SSSR*, 19 (1938) 699.
223 M. S. ZISKIN, *Dokl. Akad. Nauk SSSR*, 34 (1942) 279.

Chapter 3

The Theory of the Kinetics of Elementary Gas Phase Reactions

R. P. WAYNE

1. Introduction

By no means all experimental kineticists measure rates and rate coefficients with the intention that the data should be used to test the validity, or otherwise, of a theory of reaction rates. Indeed, to many the main value of theoretical reaction kinetics is predictive, although the predictions are often themselves of limited application. For this reason, it was felt that *Comprehensive Chemical Kinetics* should include a humble introduction to theoretical kinetics, written from the point of view of an experimentalist who from time to time needs to examine his findings in the light of theoretical predictions.

It is clear that this article cannot hope to give a description of recent advances in theoretical kinetics at the same time as setting out the basic tenets of the theories, nor is it necessarily profitable to do so. A great wealth of material is published in the field of theoretical kinetics which provides increasingly sophisticated approximations to a number of fundamental problems. Other chapters of *Comprehensive Chemical Kinetics* include appropriate discussions of the theoretical interpretations of specific experimental results, and it is the intention of the present author merely to summarise the *concepts* of theoretical gas-phase kinetics within the confines of a single chapter. Other articles will discuss the special considerations applicable to condensed phase systems. A Bibliography is given at the end of the chapter for the reader who wishes to delve more deeply into particular aspects of the theories, and references are given where it is felt that the reader may wish to acquaint himself with the details of a particular piece of research.

A theory of reaction kinetics should ideally predict the rates at which reactants possessing a given energy distributed about specific excitation modes pass over to products of equally well defined energy, and would presumably have as its basis the many-dimensioned potential energy hypersurface linking reactants and products. Such a theory is, however, unrealistic at the present time. In the first place a complete description of the potential surface is not yet possible, although the advent of high speed computers has enabled considerable advances to be made in this respect. Secondly, the experimentally determined rates of reaction are frequently not sufficiently precise, nor the energy distributions of reactants and products well enough known, for unequivocal tests of such theories to be possible.

Thus the theories currently in use are of necessity in the nature of a compromise, and it is hoped that their limitations will become apparent in the following discussion.

The simplest of these theories is the *collision* theory, which considers the collision rate between reactant species assumed to be rigid spheres in thermal equilibrium with their surroundings. It will be shown later that this collision theory can be regarded as a special case of the *transition state* theory, although in accordance with normal practice the theories are treated separately here. This approach is justified in so far as the collision formulation may be quite realistic for certain reaction systems, and the predictions in suitable cases are as good as those obtained from the more complex transition state theory. The transition state theory starts from a consideration of the system at the highest point of the potential surface—the "transition state" or "activated complex"— and its relation to reactants and products. The passage to products could be regarded as a truly unimolecular reaction (as distinct from a kinetically first order reaction) if the activated complex had a real *chemical* identity. Unimolecular reaction of activated chemical species is, of course, well established, and such processes are considered in some detail in a further section of this chapter. The actual activation itself may be derived in some way from a bimolecular process, and these activation steps are discussed, somewhat arbitrarily, together with the subsequent unimolecular reaction. Reactions of molecularity higher than two are relatively rare, and of some interest: they are considered separately.

2. The Arrhenius equation

The main features which a theory of kinetics must interpret are the effect on reaction velocity of variations of reactant concentration and temperature. For an elementary reaction step the molecularity and kinetic order will be numerically identical, and it remains to be seen whether the various theories can provide rate expressions containing the appropriate concentration terms. The experimentally determined rate of an elementary bimolecular reaction

$$A + B \rightarrow \text{products} \tag{1}$$

can be represented at any temperature by the expression

$$\text{Rate} = k[A][B] \tag{2}$$

Temperature usually has a marked effect on the rate of a chemical reaction [and thus on the rate coefficient, k, in equation (2)], although the analytical expression for the relation between rate and temperature is less immediately apparent.

Arrhenius[1] suggested the relation which bears his name as long ago as 1889, and it is still universally regarded as the most satisfactory simple way of expressing the rate coefficient for a reaction as a function of temperature. In the form most frequently used it is

$$k = Ae^{-E/RT} \tag{3}$$

where A is a "pre-exponential" constant, and E is a characteristic energy, for the particular reaction. Arrhenius arrived at the relation by consideration of an equilibrium process

$$A+B \underset{k_r}{\overset{k_f}{\rightleftarrows}} C+D \tag{4}$$

The equilibrium is established by equal rates of forward and reverse reaction, and

$$K = \frac{[C][D]}{[A][B]} = \frac{k_f}{k_r} \tag{5}$$

Since

$$\frac{d \ln K}{dT} = \frac{\Delta H}{RT^2} \tag{6}$$

where ΔH is the heat of reaction,

$$\frac{d \ln k_f}{dT} - \frac{d \ln k_r}{dT} = \frac{\Delta H}{RT^2} \tag{7}$$

Arrhenius suggested that ΔH could be split into two components, E_f and E_r, such that $\Delta H = E_f - E_r$, and argued that (7) could then be separated into two equations of the form

$$\frac{d \ln k}{dT} = \frac{E}{RT^2} \tag{8}$$

Integration of (8) then gives $\ln k = -E/RT + \text{constant}$, or

$$k = Ae^{-E/RT} \tag{3}$$

That (7) can indeed be separated into the two equations (8) is adequately justified by the vast amount of kinetic data which is roughly consistent with equation (3).

It is now realised, on both theoretical and experimental grounds, that neither A nor E is necessarily temperature independent, although the largest contribution to the temperature dependence of k almost always arises from the exponential term. Some discussion of these matters will be found in the following sections.

3. Collision theory

3.1 PRINCIPLES

It seems reasonable to suppose that before two species A and B can participate in a chemical reaction they must approach each other sufficiently closely for there to exist some interaction between them. The *collision theory*, in its simplest form, goes further than this, and suggests that the reactant molecules must have their centres separated by the sum of the gas-kinetic radii before reaction can occur: that is to say, a "collision" must take place. The rate of reaction will therefore be related to the gas-kinetic collision rate, Z_{AB}, between the two species A and B. The kinetic theory of gases can be used to calculate a value for the collision rate

$$Z_{AB} = n_A n_B \left\{ \frac{\sigma_A + \sigma_B}{2} \right\}^2 \left(\frac{8\pi kT}{\mu} \right)^{\frac{1}{2}} \tag{9}$$

where Z_{AB} is the number of collisions per cm^3 per second

n_A, n_B are the numbers of A and B molecules per cm^3

σ_A, σ_B are the molecular diameters of A and B

$\mu = \dfrac{m_A m_B}{m_A + m_B}$ is the "reduced mass" of A and B.

This expression predicts correctly the concentration dependence of the rate of reaction, although the magnitude of Z_{AB} is much greater than the experimentally determined rate for most reactions. Furthermore, the observed variation of reaction rate with temperature bears little relation to the $\sqrt{T}$ dependence suggested by (9). It is clear, therefore, that the rate coefficient for reaction is not to be identified directly with $Z_{AB}/n_A n_B$, and that some modification of the theory is needed to introduce the dependence of rate on temperature suggested by the Arrhenius equation. The required modification is to assume that not all collisions are reactive, but rather that reaction occurs only after those collisions which possess relative kinetic energy greater than a critical value, E_c, along the line of collision centres. It will be shown in the next section that the rate at which such energetic collisions occur is equal to

$$Z_{AB}e^{-\varepsilon_c/kT} = Z_{AB}e^{-E_c/RT} \tag{10}$$

The equation for the rate of "energetic" collisions is now of a form directly equivalent to the Arrhenius expression (ε_c is the excess energy per molecular collision while E_c is a *molar* quantity $E_c = N_0\varepsilon_c$ where N_0 is Avogadro's number). That collisions need to possess energy in excess of ε_c suggests that there is some energy barrier to reaction which must be overcome. This barrier is related to the energy required to form the "activated complex" of the transition state theory (see section 4), and its height is associated with the potential energy of the highest point on the potential energy surface connecting reactants and products. For this reason, the energy is universally known as the *Activation Energy*, E_a; E of the Arrhenius expression (3) and E_c of the rate of collision with energy greater than E_c (10) are usually referred to as activation energies, E_a, although according to the definitions of (3) and (10) they are not necessarily identical. We shall adopt the practice of referring to $E = E_a$ as the activation energy (per mole) since this is normally the experimentally determined quantity.

The simple collision theory clearly has a number of inadequacies. First, the theory treats the reactants as hard spheres; secondly it assumes that collision in all directions is equally favourable; and thirdly it makes no allowance for interactions at a longer range than so-called collision. It is, perhaps, surprising for how many simple reactions the observed pre-exponential factor A is of the same order of magnitude as Z_{AB}. The inadequacies of the theory do, however, lead to an even larger number of incorrect predictions, and it has become common practice to multiply the predicted rate $Z_{AB}e^{-E_c/RT}$ by a probability factor P which is chosen, more or less empirically, to fit the experimentally determined rates. It may then be convenient to define a reaction cross section, σ_r^2, for the reaction such that

$$\sigma_r^2 = P\left\{\frac{\sigma_A + \sigma_B}{2}\right\}^2 \tag{11}$$

This cross section then allows for the geometrical factors and for the unrealistic assumption of a hard collision. Various attempts have been made to modify the collision theory to allow for the factors described above. For example, Hinshelwood derived (see, for example, ref. 2) the fraction of effective collisions if the critical energy, E_c, could be accommodated in degrees of freedom other than the kinetic energy of collision. The derivation, which is discussed in section 3.6.3, gives the chance of a total energy exceeding E_c without restriction as to how the energy is shared among the separate modes. It does not, however, take any account of the possibility that species may react preferentially if the energy is in one mode rather than another. More recent suggestions concerning modifications to the collision theory are discussed in section 3.7.

3.2 RATE OF BIMOLECULAR COLLISIONS

The gas-collision rate Z_{AB} between two hard spheres of diameter σ_A, σ_B and masses m_A, m_B may be derived from the simple kinetic theory of gases, and the derivation is given in most textbooks. In this section, an expression is obtained for the rate of collisions possessing a relative kinetic energy along the line of centres greater than ε_c. The most straightforward way of obtaining the expression is to use the Maxwell distribution law applied to the relative velocity of two colliding species A and B to give the collision rate distribution as a function of velocity. The condition that the energy along the line of centres shall be greater than ε_c is then applied in the form of a velocity restriction.

The distribution law for the relative velocity, r, between any two species A and B may be given in the form

$$dn_{AB} = 4\pi n_A n_B \left(\frac{\mu}{2\pi kT}\right)^{\frac{3}{2}} e^{-\mu r^2/2kT} r^2 dr \tag{12}$$

where dn_{AB} is the number of pairs of A and B whose relative velocity lies between r and $r+dr$. [The other symbols have the meaning given in connection with equation (9)].

Now suppose that this relative velocity lies at an angle between θ and $\theta+d\theta$ to the line of centres. If σ is taken to be $\frac{1}{2}(\sigma_A+\sigma_B)$, then the possible orientations of B during collision are those in which the centre of B lies on a ring-shaped surface on a sphere of radius σ centred on A. The surface has an area $2\pi\sigma^2 \sin\theta \, d\theta$, so that as A moves with respect to B it sweeps out a volume

$$V = 2\pi\sigma^2 \sin\theta \, d\theta \cdot r \cos\theta \tag{13}$$

in unit time. The number of collisions occurring in unit volume and unit time, dZ, whose relative velocity lies between r and $r+dr$ at an angle between θ and $\theta+d\theta$ to the line of centres may now be calculated from equations (12) and (13)

$$dZ = 8\pi^2 n_A n_B \left(\frac{\mu}{2\pi kT}\right)^{\frac{3}{2}} e^{-\mu r^2/2kT} r^3 dr \sigma^2 \sin\theta \cos\theta \, d\theta \tag{14}$$

Now for the kinetic energy to be greater than ε_c along the line of centres of A and B it follows that $\frac{1}{2}\mu r^2 \cos^2\theta > \varepsilon_c$. Thus the total number of collisions of required energy is obtained by integrating (14) between the limits $\theta = 0$, $\theta = \cos^{-1}(2\varepsilon_c/\mu r^2)^{\frac{1}{2}}$; $r = (2\varepsilon_c/\mu)^{\frac{1}{2}}$, $r = \infty$

$$Z(\varepsilon > \varepsilon_c) = n_A n_B \sigma^2 \left(\frac{8\pi kT}{\mu}\right)^{\frac{1}{2}} e^{-\varepsilon_c/kT} \tag{15}$$

The expression is identical with that given by equations (9) and (10) with $\sigma = \frac{1}{2}(\sigma_A + \sigma_B)$. Note that ε_c is here the critical energy per *molecule*; in practice it is more convenient to work in terms of the energy per *mole* so that k is replaced by R in the exponential term.

3.3 COLLISIONS BETWEEN LIKE SPECIES

So far only collisions between unlike species, A and B, have been considered. Bimolecular reaction between like species

$$A + A \rightarrow \text{Products} \tag{16}$$

is of chemical importance, and it is of interest to consider how the rate coefficient, in terms of the collision theory, is to be adjusted for this type of reaction.

The rate of collisions with energy greater than ε_c can be obtained by setting $m_A = \mu/2$ in equation (15) and dividing by two. The factor of two is required since Z_{AB} effectively counts the rate of collision between a given A and all B and then multiplies the result by the number of A. To do this for Z_{AA} would count every collision twice. Thus

$$Z_{AA}(\varepsilon > \varepsilon_c) = 2n_A^2 \left(\frac{\pi k T}{m_A}\right)^{\frac{1}{2}} \sigma_A^2 e^{-\varepsilon_c/kT} \tag{17}$$

There is, however, a problem in setting the rate coefficient for reaction (16) equal to $PZ_{AA} e^{-\varepsilon_c/kT}/n_A^2$. The rate equation for the reaction of unlike species in (1) is universally given in terms of the reactants, *i.e.*

$$\frac{-dn_A}{dt} = \frac{-dn_B}{dt} = k[A][B] \tag{18}$$

It is desirable, therefore, that the rate equation for (16) should be given in the same form

$$\frac{-dn_A}{dt} = k'[A]^2 \tag{19}$$

Every effective collision counted by Z_{AA} removes *two*, rather than one, A, so that the rate of removal of reactant is twice the rate of effective collision. Thus we should write

$$k' = 2PZ_{AA} e^{-\varepsilon_c/kT}/n_A^2 \tag{20}$$

The argument used above is by no means artificial. Consider a system consisting of equal concentrations of an alkyl radical C_nH_{2n+1} and a monodeuterated radical $C_nH_{2n}D$. The following combination reactions may occur

$$C_nH_{2n+1}+C_nH_{2n+1} \rightarrow C_{2n}H_{4n+2} \tag{21}$$

$$C_nH_{2n+1}+C_nH_{2n}D \rightarrow C_{2n}H_{4n+1}D \tag{22}$$

$$C_nH_{2n}D+C_nH_{2n}D \rightarrow C_{2n}H_{4n}D_2 \tag{23}$$

According to our definition for rate coefficients, the rate coefficients for (21), (22) and (23) will be approximately *identical* if the radicals are sufficiently massive. This result accords with the normal idea of a rate coefficient. On the other hand, the rate of formation of the product $C_{2n}H_{4n+1}D$ in (22) will be twice as great as the rate of formation of the other two products. Experimental determinations of rates of product formation in reaction systems involving even two alkyl radicals of *different* masses show that the cross-combination product appears approximately twice as fast as the other combination products.

3.4 COMPARISON WITH EXPERIMENT: THE "P" FACTOR

A complete comparison of the predicted rate coefficients with experimentally determined values would require a knowledge not only of the mass and collision diameter of the reactants, but also of the critical energy for reactive collisions. While m is in principle a known quantity and some estimate of σ may be obtained from, say, viscosity data, the calculation of E_c from data not involving rate determinations presents some difficulty. For the purposes of this section, E_c is set equal to the experimental activation energy, E_a, so that the comparisons to be made are essentially of the Arrhenius pre-exponential "A" factor and the calculated collision rate at unit concentration. Some remarks are offered in section 3.4 about calculations of E_c without reference to rate data.

The calculation of A from the rate of a reaction is sensitive to the value of E_a, and any attempt to calculate A from the activation energy and the rate constant at a given temperature can give a grossly misleading value of A if E_a is in error. It is probably better to obtain A from the logarithmic form of the Arrhenius equation which is used to calculate E_a

$$\ln k = \ln A - E_a/RT \tag{24}$$

The value of A is then the extrapolated value of k at $1/T = 0$. This practice should give the most reliable value of A, although it still depends critically upon the extrapolation and thus on the value chosen for E_a. Since the rate determinations may be over a relatively narrow temperature range, the extrapolation to $1/T = 0$ can introduce severe errors into the calculated value of A. Thus an agreement of an order of magnitude between experimental and theoretical pre-exponential factors is the best that can be expected.

For a typical elementary gas-phase reaction, the molecular or atomic weights will be about 40, so that $\mu = 20$, and the molecular diameter, σ, about 3×10^{-8} A. At 300° K this gives a value of $A_{calculated} = Z_{AB}/n_A n_B \approx 2 \times 10^{-10}$ cm^3. molec^{-1}. sec^{-1}. Changes in either temperature or reduced mass have little effect on this value since they both appear under the square root sign. Even for the reaction

$$H_2 + I_2 \rightarrow 2HI \tag{25}$$

performed at 600° K, for which μ is but slightly greater than two, the value of $A_{calculated}$ is only $\sqrt{20}$ times (i.e. less than five times) faster than that calculated for $\mu = 20$, $T = 300°$ K. Thus under the ordinary conditions of chemical reactions in the gas phase, the pre-exponential factor lies between about 10^{-10} and 10^{-9} cm^3. molec^{-1}. sec^{-1} or about 10^{11} and 10^{12} in l. mole^{-1}. sec^{-1} units.

Table 1 gives values of log A (for A in l. mole^{-1}. sec^{-1} units) and E_a (in kcal. mole^{-1}) for a number of elementary reactions, and it is arranged in order of

TABLE 1

RATE DATA FOR SOME BIMOLECULAR REACTIONS ARRANGED IN ORDER OF DECREASING
PRE-EXPONENTIAL FACTOR (A) [l. mole^{-1}. sec^{-1} units]

Reaction	log A	E_a (kcal. mole^{-1})	Reference
$H + O_2 \rightarrow OH + O$	11.1	18.8	6
† $H_2 + I_2 \rightarrow 2HI$	11.1	39.0	3
$2HI \rightarrow H_2 + I_2$	10.9	44.0	3
$OH + H_2 \rightarrow H_2O + H$	10.9	10.0	7
$Cl + H_2 \rightarrow HCl + H$	10.9	5.5	8
$H + H_2 \rightarrow H_2 + H$	10.7	7.5	9
$CH_3 + CH_3 \rightarrow C_2H_6$	10.3	0	10
$NO_2 + O_3 \rightarrow NO_3 + O_2$	9.8	7.0	11
$NO + Cl_2 \rightarrow NOCl + Cl$	9.6	20.3	12
$H + C_2H_6 \rightarrow H_2 + C_2H_5$	9.5	6.8	13
$NO + O_3 \rightarrow NO_2 + O_2$	8.8	2.5	14
$CH_3 + C_2H_4 \rightarrow C_3H_7$	8.4	~7	15
$CH_3 + C_2H_6 \rightarrow CH_4 + C_2H_5$	8.3	10.4	16
$C_4H_6 + C_4H_6 \rightarrow cyclo\text{-}C_8H_{12}$	8.1	26.8	17
$CH_3 + C_3H_6 \rightarrow C_4H_9$	8.0	~6	15
$C_4H_6 + C_2H_4 \rightarrow cyclo\text{-}C_6H_{10}$	7.5	27.5	17
$O_3 + C_3H_8 \rightarrow C_3H_7O + HO_2$	6.5	12.1	18

† The H_2/I_2 reaction has been included because of its importance in the historical development of the theory of gas-phase kinetics, although recent work[5] suggests that it is not in fact a simple bimolecular process.

References pp. 298–301

decreasing A. It is seen that the pre-exponential factor observed for the first few of these reactions is quite close to that calculated above. The hydrogen–iodine reaction (25) rate data of Bodenstein[3] was in fact used by McC. Lewis[4] in his earliest comparisons (1918) of the collision theory with experiment. The empirical probability factor, P, is defined by

$$P = A_{observed}/A_{calculated} \qquad (26)$$

and for the hydrogen–iodine reaction is about 0.33. P lies in the range 0.1 to 1 for the simplest reactions given in Table 1, but becomes progressively smaller with an increase in complexity of the reaction. Two factors contribute to this result. First is the so-called "steric factor". The collision theory does not consider the necessity for reaction of any specific geometric orientation of the reactants. It is, however, quite apparent that the reactive portions of reactants must be brought geometrically close to each other before chemical change can take place. To some extent, therefore, the probability factor P may be taken as representing the fraction of collisions with the correct orientation. Of the reactions in Table 1, the Diels–Alder cyclizations of butadiene on its own, or with ethylene are good examples of reactions with low steric P factors. The reaction of butadiene with ethylene to form cyclohexene

$$
\begin{array}{ccc}
\mathrm{CH_2} & & \mathrm{CH_2} \\
\| & & / \quad \backslash \\
\mathrm{CH} & \mathrm{CH_2} & \mathrm{HC} \quad \mathrm{CH_2} \\
| \quad + \quad \| \quad \rightarrow & \| \qquad | \\
\mathrm{CH} & \mathrm{CH_2} & \mathrm{HC} \quad \mathrm{CH_2} \\
\backslash & & \backslash \quad / \\
\mathrm{CH_2} & & \mathrm{CH_2}
\end{array}
\qquad (27)
$$

clearly requires rather specifically oriented collisions, and, in agreement with this expectation, the P factor for this reaction (27) is of the order of 10^{-4}. The second factor which contributes towards the overall magnitude of P concerns the distribution of the kinetic energy of collision. There is no reason to suppose that all the kinetic energy of a collision will appear where or when it is needed to overcome the barrier to reaction, and the probability of enough energy being at the right place will presumably decrease as the total number of bonds increases (see, however, section 3.6.3). Once again, this general conclusion is borne out for the reactions shown in Table 1: for example, the P factor of the reaction

$$CH_3 + C_2H_6 \rightarrow CH_4 + C_2H_5 \qquad (28)$$

is about one tenth that of

$$H + C_2H_6 \rightarrow H_2 + C_2H_5 \qquad (29)$$

Certain further, rather special, cases of reactions with low P factors must be discussed. The first of these concerns processes in which there is a change of electronic state during reaction: perhaps the most important members of this class are those reactions in which there is a change of multiplicity. Such reactions are usually known as "non-adiabatic" processes, although this description is rather misleading. An adiabatic process is certainly one in which there is no change in electronic state, and it can be shown that the probability of a non-adiabatic transition between two potential energy surfaces decreases rapidly as the smallest energy separation increases. The perturbation energy of spin–spin or spin–orbital interactions is small, so that for processes involving a change in resultant spin of a system the probability of the process occurring adiabatically but with the change in multiplicity is low. Thus the probability of the adiabatic reaction

$$O(^3P) + CO(^1\Sigma) \rightarrow CO_2(^1\Sigma) \tag{30}$$

is low. Emphasis is placed on this point since reactions such as (30) are frequently said to be non-adiabatic and therefore of low probability. In practice confusion does not arise, since it is relatively easy to decide for which reactions *as written* the probability is low. In particular, processes for which there is a change of multiplicity may be expected to be slow: this postulate is known as the Wigner Spin Conservation rule[19]. For example, the reactions

$$O(^1D) + O_2(^3\Sigma_g^-) \rightarrow O(^3P) + O_2(^1\Delta_g) \tag{31}$$

$$O(^1D) + CO(^1\Sigma) \rightarrow CO_2(^1\Sigma) \tag{32}$$

$$Hg(^3P) + Tl(^2P) \rightarrow Hg(^1S) + Tl(^2X) \tag{33}$$

are "spin-allowed" since it is possible to find a total spin quantum number which is common to both reactants and product systems. On the other hand, processes such as

$$Hg(^3P) + H_2(^1\Sigma_g^+) \rightarrow Hg(^1S) + H_2(^1\Sigma_g^+) \tag{34}$$

$$O_2(^1\Delta_g) + O_2(^1\Delta_g) \rightarrow O_2(^3\Sigma_g^-) + O_2(^1\Sigma_g^+) \tag{35}$$

are "spin-forbidden", and, other things being equal, will be slow. Adiabaticity is discussed further in section 4.5.7.

Reactions involving the exchange of energy in modes other than electronic may also proceed with a reduced probability as a result of restrictions on the exchange of energy. The transfer of vibrational energy to translation is inefficient if the vibrational energy levels are widely spaced (*i.e.* if $h\nu \gg kT$) while vibration–vibration transfer proceeds slowly if the energy discrepancy between vibrational levels of the exchanging species, which must be taken up by translation, is large (see, for example, the review by Callear[20]). Thus in quenching or de-activation reactions the P factor is likely to be small unless certain closely defined conditions are fulfilled.

A reaction in which the product molecules can spontaneously and rapidly revert to reactants will have an overall rate considerably less than that predicted on the basis of the collision rate of reactants. The most important members of this class are the bimolecular association reactions. For example, the hydrogen molecule formed in the process

$$H + H \rightarrow H_2^* \tag{36}$$

possesses as vibration in the newly-formed H–H bond all the heat of reaction. Since (36) proceeds with little or no activation energy, it follows that H_2^* will dissociate in the course of a single vibration unless it can somehow be stabilised in that time. Collision stabilization means that the reaction is effectively termolecular

$$H + H + M \rightarrow H_2 + M \tag{37}$$

and will be treated further in a later section. The true bimolecular rate is therefore very much smaller than the collision frequency. With increasing complexity of the reacting species a greater number of additional bonds will exist in the product which can accommodate the heat of reaction, and bimolecular association may then proceed at a normal rate. For the association of atoms, however, no such vibrational degrees of freedom exist, and the reactions proceed only *via* a termolecular mechanism. It might be noted at this point that the possibility of alternative reaction paths will alter the probability of any one path being followed. For example, in the association of two radicals (doublet species) there is one singlet and a (triply degenerate) triplet product which may not be an attractive state. The chance of a reaction following the path leading to the stable singlet is thus one in four, and a P factor of $\frac{1}{4}$ might be expected for the reaction. The question of statistical weights for reactions is explored further in section 4.3.3.

We have seen that, except for the very simplest of reactions, the empirical P factor tends to be considerably less than unity. A number of reasons for the low probability of reactions has been discussed, although no attempt has been made to make numerical estimates of the value of P. Such estimates are, in fact, not possible on the basis of the simple collision theory, although, of course, it may be possible to guess from experience with similar reactions whether P for a given reaction will be large or small. The transition state theory (section 4) can make allowance for certain of the factors treated empirically by the collision theory, and, in particular, it identifies the "geometric" parts of P with an "entropy of activation".

3.5 CALCULATION OF ACTIVATION ENERGIES

The activation energy of a chemical reaction may be looked upon as the energy of the highest point of the reaction path over a potential surface connecting reactants and products. Prediction of an activation energy requires in principle a method for the determination of the form of the potential energy surface. The calculation of potential energy surfaces is discussed in some detail in section 4.5, and consideration of this topic is deferred. Such calculations are naturally of the greatest interest, since a complete knowledge of the potential surface would enable equally complete description of the reaction. Unfortunately, the methods at present available for the prediction even of just the maximum energy on the reaction path, let alone of the entire potential surface, give results which in many cases differ considerably from experimental data. In view of the great influence that the value chosen for the activation energy has on the calculated rate coefficient, it must be concluded that absolute calculations of activation energies are not sufficiently accurate for rate predictions in most instances. For the purposes of this section, therefore, some *empirical* methods for the calculation of activation energies are presented. The methods have proved successful for a number of reactions, and some examples of predicted and measured activation energies will be discussed.

Hirschfelder[21] has stated rules for the estimation of activation energies from bond energy data. For metathetical reactions of the type

$$AB + CD \rightarrow AC + BD \tag{38}$$

a rule was formulated to give agreement with activation energy and bond energy data for the hydrogen–iodine reaction. The relation used is

$$E_a \sim 0.28(Q_1 + Q_2) \tag{39}$$

where Q_1, Q_2 are the dissociation energies of the bonds being broken. On the basis of semi-empirical calculations Hirschfelder suggested that for an exothermic displacement reaction involving atoms or simple radicals

$$A + BC \rightarrow AB + C \tag{40}$$

the activation energy could be expressed by the relation

$$E_a \sim 0.055 \, Q \tag{41}$$

where, in this case, Q is the B–C bond dissociation energy. An endothermic reaction may be treated by the same methods.

Fig. 1 illustrates diagrammatically the relationship between heat of reaction and

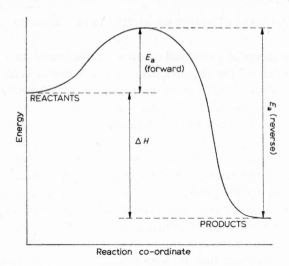

Fig. 1. Relationship between heat of reaction and activation energy for forward and reverse reactions.

activation energy for forward and reverse reactions. ΔH is the enthalpy change of reaction (and is taken to be negative for an exothermic reaction). It is apparent that

$$E_{a(reverse)} = E_{a(forward)} - \Delta H_{(forward)} \tag{42}$$

The Hirschfelder rules may now be applied to endothermic reactions by calculating the activation energy for the reverse (exothermic) reaction and applying (42) to obtain the activation energy for the forward reaction.

Polanyi et al. [22, 23] investigated the series of reactions

$$M + RX \rightarrow MX + R \tag{43}$$

where M = alkali metal and RX = alkyl halide. They suggested that the activation barrier was a result of the energy required to expand the X–R bond so that the postulated transition states MX^-R^+ and M^+X^-R became degenerate. They then obtained the height of the activation barrier from the intersection point of the X–R attractive and repulsive curves, and proposed that

$$E_a = constant - \alpha \Delta H \tag{44}$$

where α is a constant lying in the range 0 to 1. For the series of reactions

$$Na + RCl \rightarrow NaCl + R \tag{45}$$

the experimental results were consistent with (44) with a value of $\alpha = 0.27$. Semenov[24] subsequently empirically extended Polanyi's equation (44) for a number of exothermic atom and radical reactions by using the general relation

$$E_a \sim 0.25\Delta H + 11.5 \tag{46}$$

The appropriate value for endothermic reactions (ΔH positive) can once again be obtained by considering first the reverse exothermic reaction: the result is

$$E_a \sim 0.75\Delta H + 11.5 \tag{47}$$

Both Hirschfelder's and Semenov's empirical estimates of activation energies can only give a rough idea of the activation energy to be expected. In particular, the rules tend to give the same activation energies for a series of related reactions in which similar bonds are broken. Table 2 shows the experimental activation energies for a number of abstraction reactions together with the estimates made using the Hirschfelder and Semenov rules. The calculated activation energies do reflect the trends of the experimental values, although their absolute magnitudes frequently differ by a factor of two or more. For example, for the reaction

$$K + HBr \rightarrow KBr + H \tag{48}$$

the activation energy calculated by the Semenov method is 10 kcal. mole^{-1}, while the measured value is about 3 kcal. mole^{-1}. This difference corresponds to a difference of about 10^5 in the rate of reaction at room temperature, and it is clear

TABLE 2

EXPERIMENTAL AND CALCULATED ACTIVATION ENERGIES FOR SOME ABSTRACTION
REACTIONS

Reaction	E_a(experimental) (kcal. mole^{-1})	ΔH (kcal. mole^{-1})	Calculated activation energies (kcal. mole^{-1})		Reference
			Hirschfelder	Semenov	
$H + H_2 \rightarrow H_2 + H$	7.5	0	6	12	9
$Cl + H_2 \rightarrow HCl + H$	5.5	+1	7	12	8
$Br + H_2 \rightarrow HBr + H$	17.6	+17	22	24	25
$H + CH_4 \rightarrow H_2 + CH_3$	6.6	−2	6	11	26
$Cl + CH_4 \rightarrow HCl + CH_3$	3.9	−1	6	11	27
$Br + CH_4 \rightarrow HBr + CH_3$	18.3	+15	20	23	28
$K + HBr \rightarrow KBr + H$	3.4	−5	5	10	29
$OH + H_2 \rightarrow H_2O + H$	10.0	−15	6	7	7
$CH_3 + C_2H_6 \rightarrow CH_4 + C_2H_5$	13.0	−3	5	11	16
$NO + O_3 \rightarrow NO_2 + O_2$	2.5	−47	1	$\sim$0	14
$NO_2 + O_3 \rightarrow NO_3 + O_2$	7.0	−25	1	5	11

that no great reliance can be placed on rates calculated from estimated activation energies.

3.6 TEMPERATURE DEPENDENCE OF A AND E

The simple form of the Arrhenius equation

$$k = Ae^{-E_a/RT} \tag{3}$$

implies that neither A nor E_a depends upon the temperature. That the implication is unfounded is demonstrated by experimental results which demand that either or both of the "constants" should, in fact, exhibit a temperature dependence. This dependence can arise from a number of causes, operating singly or in combination, and some of these causes are considered here.

3.6.1 Variation of collision frequency with temperature

The basic assumption of the collision theory is that the rate coefficient, k, for a bimolecular reaction is given by the equation

$$k = \frac{Z_{AB}}{n_A n_B} e^{-\varepsilon_c/kT} \tag{49}$$

Since

$$Z_{AB} = n_A n_B \left(\frac{\sigma_A + \sigma_B}{2}\right)^2 \left(\frac{8\pi kT}{\mu}\right)^{\frac{1}{2}} \tag{9}$$

it follows that

$$\frac{d \ln k}{dT} = \frac{1}{2T} + \frac{\varepsilon_c}{kT^2} = \frac{1}{2T} + \frac{E_c}{RT^2} \tag{50}$$

However, according to the Arrhenius expression (3)

$$\frac{d \ln k}{dT} = \frac{E_a}{RT^2} \tag{51}$$

equations (49) and (50) can be reconciled only if the experimental activation energy is temperature dependent in such a way as to accommodate the variation of Z_{AB}

with temperature. That is

$$E_a = E_c + \tfrac{1}{2}RT \tag{52}$$

Most determinations of activation energies employ results obtained over relatively narrow temperature ranges, and deviations from linearity of a $\ln k$–$1/T$ curve may well be masked by scatter of the experimental points. It should, however, be noted that the absolute value of activation energy measured from the curve will be in error by the amount $\tfrac{1}{2}RT$, which in many cases may not be negligible in comparison with E_c.

The temperature dependence of the collision rate might more properly be accommodated by allowing A of the Arrhenius equation to be temperature dependent. That is to say, the Arrhenius equation might be modified to read

$$k = A'\sqrt{T}\,e^{-E_a'/RT} \tag{53}$$

where A' is a temperature invariant pre-exponential factor. The value of E_a' obtained from (53) is *not* the same as E_a of the ordinary equation, and, in fact, quoted activation energies normally apply to E_a of equation (3).

3.6.2 Variation of barrier height with temperature

The relationship between enthalpy change of reaction, ΔH, and the height of the activation barrier, E_c, for forward and reverse reactions has been discussed already in terms of Fig. 1

$$\Delta H_{(\text{forward})} = E_{c(\text{forward})} - E_{c(\text{reverse})} \tag{42a}$$

Kirchoff's well-known law relates the change of ΔH to the specific heats of reactants and products for a chemical reaction

$$\frac{\partial(\Delta H)}{\partial T} = \sum_i C_{pi} \tag{54}$$

and makes it clear that in general ΔH will vary with temperature. In principle, therefore, activation energies for both forward and reverse reactions will exhibit a temperature dependence, although in practice the variation from this cause of E_c with temperature is small. [Equation (54) is itself a simplification, since C_{pi} may be a function of temperature].

3.6.3 Activation in many degrees of freedom

According to the simple collision theory, an activation barrier is overcome by the kinetic energy of collision between two "hard sphere" reaction partners, and takes no account of the energy of excitation which may be possessed by them in internal modes of excitation (in particular, by vibrations). The methods of classical statistical mechanics allow calculation of the number of species possessing a total energy greater than any given amount *in many degrees of freedom, and without reference to the way in which it is shared between them.* The result is

$$\frac{N(E > E_c)}{N} = \frac{1}{\Gamma(\tfrac{1}{2}S)(RT)^{\frac{1}{2}S}} \int_{E_c}^{\infty} e^{-E/RT} E^{(\frac{1}{2}S-1)} dE \tag{55}$$

where S is the number of *squared terms* which describe the energy; that is, each translation or rotation corresponds to one squared term, while each vibration corresponds to two squared terms. If $\tfrac{1}{2}S$ is an integer, we may write:

$$\Gamma(\tfrac{1}{2}S) = (\tfrac{1}{2}S-1)! \tag{56}$$

and, therefore

$$\frac{N(E > E_c)}{N} = e^{-E_c/RT} \left[\frac{1}{(\tfrac{1}{2}S-1)!} \left(\frac{E_c}{RT}\right)^{\frac{1}{2}S-1} + \frac{1}{(\tfrac{1}{2}S-2)!} \left(\frac{E_c}{RT}\right)^{\frac{1}{2}S-2} \cdots +1 \right] \tag{57}$$

If E_c/RT is large, then (57) simplifies to

$$\frac{N(E > E_c)}{N} = \frac{e^{-E_c/RT} \left(\dfrac{E_c}{RT}\right)^{\frac{1}{2}S-1}}{(\tfrac{1}{2}S-1)!} \tag{58}$$

Thus, if activation in many degrees of freedom can, in fact, overcome an energy barrier to reaction, then the reaction rate is equal to

$$Z_{AB} e^{-E_c/RT} \cdot \frac{\left(\dfrac{E_c}{RT}\right)^{\frac{1}{2}S-1}}{(\tfrac{1}{2}S-1)!} \tag{59}$$

It follows that the experimental activation energy differs from E_c, since

$$E_a = RT^2 \frac{d \ln k}{dT} = E_c - (\tfrac{1}{2}S-1)RT \tag{60}$$

and E_a will exhibit a linear temperature dependence.

It is of interest to derive a value for the experimental activation energy from the expression (55) for $N(E > E_c)/N$ which does not employ the simplifications introduced. The rate coefficient, k, may be written

$$k = B \cdot \frac{N(E > E_c)}{N} = B \cdot \frac{1}{\Gamma(\frac{1}{2}S)(RT)^{\frac{1}{2}S}} \int_{E_c}^{\infty} e^{-E/RT} \cdot E^{(\frac{1}{2}S-1)} dE \tag{61}$$

where B is a constant.

Thus

$$\frac{dk}{dT} = \left\{ \frac{B}{\Gamma(\frac{1}{2}S)} \cdot (RT)^{-\frac{1}{2}S} \int_{E_c}^{\infty} \frac{E}{RT^2} e^{-E/RT} \cdot E^{(\frac{1}{2}S-1)} dE \right\}$$
$$- \left\{ \frac{S}{2} \cdot \frac{B}{\Gamma(\frac{1}{2}S)} \cdot R^{-\frac{1}{2}S} \cdot T^{(-\frac{1}{2}S-1)} \int_{E_c}^{\infty} e^{-E/RT} \cdot E^{(\frac{1}{2}S-1)} dE \right\} \tag{62}$$

and

$$E_a = RT^2 \frac{d \ln k}{dT} = \frac{\displaystyle\int_{E_c}^{\infty} E \cdot e^{-E/RT} \cdot E^{(\frac{1}{2}S-1)} dE}{\displaystyle\int_{E_c}^{\infty} e^{-E/RT} \cdot E^{(\frac{1}{2}S-1)} dE} - \frac{1}{2}S\,RT \tag{63}$$

The quotient of the two integrals is the *average* energy of all species with energy greater than E_c, while $\frac{1}{2}SRT$ is the *average* energy of all molecules.

Another important consequence of activation in many degrees of freedom is that the rate coefficient predicted from (59) on the basis of an experimental activation energy may be considerably greater than that predicted by the simple collision theory. Using (60) to give a value for E_c in terms of E_a, the ratio of the two rate constants is:

$$e^{-(\frac{1}{2}S-1)} \cdot \frac{\left\{ \dfrac{E_a}{RT} + (\frac{1}{2}S-1) \right\}^{\frac{1}{2}S-1}}{(\frac{1}{2}S-1)!} \tag{64}$$

In a typical case, S might be 10 and E_a about 60,000 cal. mole^{-1}; RT at room temperature is about 600 cal. mole^{-1}, so that the rate coefficient calculated from the simple collision theory is between 10^5 and 10^6 times slower than that expected if energy in other degrees of freedom can help to overcome the activation barrier. This result can be looked upon as an explanation of probability factors greater than unity in the simple theory.

If energy possessed by various degrees of freedom *is* to overcome the activation barrier to a bimolecular reaction it must be able to appear in the place and form that it is needed *during the lifetime of a "collision"*. It might be expected, therefore, that activation in many degrees of freedom would be of greater importance in unimolecular reactions, since the species undergoing reaction retains its identity until it *does* react (although it can lose energy by quenching processes). Such experimental evidence as there is for the importance of activation in many degrees of freedom does, indeed, tend to concern unimolecular reactions, and discussion of the evidence will be deferred until section 5, where unimolecular reactions are considered in some detail.

3.6.4 Quantum mechanical tunnelling

Quantum mechanical tunnelling cannot properly be treated without reference to the forms of potential energy surfaces for reaction, and the subject is therefore discussed in section 4.5.6. For the present purposes, it is intended to investigate qualitatively any effect that tunnelling might have on the temperature dependence of A or E_a.

The picture of the activation barrier adopted so far is of a classical barrier through which no leakage may occur: the energy possessed by species is either sufficient to overcome the barrier or it is not. However, according to quantum mechanical reasoning there is a certain finite probability of a particle "tunnelling" through a barrier, and the probability depends, amongst other things, on the height and shape of the barrier, and is most important for particles of low mass such as electrons, and to a lesser extent protons and deuterons. Further, the probability of tunnelling increases with a decrease in temperature.

If tunnelling can occur, then reaction will proceed more rapidly than would be expected from the classical barrier height, and, in particular, the increase in reaction rate over that expected will become larger as the temperature is decreased. Fig. 2 shows the type of behaviour that might be expected for a $\log k - 1/T$ plot with and without tunnelling.

It is seen that the experimental activation energies will be temperature dependent, and will decrease with a decrease in temperature. Values of A calculated from $\log k - 1/T$ curves will also be affected by the deviation from linearity, since the extrapolation of experimental results obtained at finite temperatures to $1/T = 0$ will give a lower A than the "true" value of k at $1/T = 0$.

There appear to be a number of reactions in solution for which there are deviations from linearity of Arrhenius plots which can only reasonably be explained in terms of quantum mechanical tunnelling. These reactions will be discussed in the chapter on the Theory of Reactions in Solution. The evidence for the importance of tunnelling in gas-phase reactions is generally less convincing. The higher temperatures at which gas-phase reactions are usually performed tend to reduce the effect of tunnelling, while the accuracy of kinetic measurements in the gas phase is generally lower than that of solution phase measurements, and it is dif-

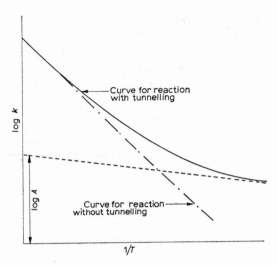

Fig. 2. Deviation from linearity of a log k *versus* $1/T$ plot for a reaction with quantum mechanical tunnelling.

ficult to demonstrate significant deviations of Arrhenius plots from linearity. Johnston *et al.*[30-32] have made a number of searches for experimental evidence of tunnelling, and have obtained some support for the view that tunnelling does occur. The most recent studies[33] of the $H+H_2$ reaction also show that the temperature dependence of reaction rate can only be explained reasonably on the basis of tunnelling.

3.7 SOPHISTICATED COLLISION THEORIES

One of the most obvious inadequacies of the simple collision theory is its treatment of collision partners as hard spheres. Attempts such as those described in section 3.6.3 to allow for activation in internal modes of excitation are also unrealistic in the sense that they do not differentiate between the effectiveness of energy in the various modes at overcoming activation barriers. To some extent, the Transition State Theory (to be developed in the next section) is less open to these criticisms, since it does allow for the internal motions of molecular species. However, the Transition State Theory assumes some kind of equilibrium between the reactants and an activated complex. In certain circumstances this assumption of equilibrium is not necessarily valid. Experiments using shock tubes or molecular beams, for example, do not even involve a Maxwell–Boltzmann distribution of velocities initially, while very fast reactions can cause depletion of the more energy rich species, and a consequent loss of thermal equilibrium. It is obvious that the postulation of equilibrium between reactants and activated complex loses its meaning for systems in which the various degrees of freedom of the reactants are not in thermal equilibrium. The most useful treatment for systems such as these may well be an accurate treatment using a collisional approach. Although it is not the primary purpose of this chapter to discuss specialised treatments of theoretical kinetics, it may nevertheless be of some interest to outline some of the methods which have been adopted to modify the collision theory.

The process of energy transfer is to be regarded as a special case of chemical reaction, and Widom and Bauer[34] have developed a collision theory for the energy transfer between vibrationally excited carbon dioxide and water which uses a Lennard-Jones "6–12" potential for the forces between "colliding" molecules. They calculate the probability of transition between various vibrational states on collision and allow for the possibility of the conversion of some energy into translation. Thus this theory is a first move away from the hard sphere model of collision.

Other approaches to modification of the collision theory seem to be divided into two categories. There are those treatments in which the potential energy surface for reaction plays no direct part in the formulation, while on the other hand there are the *kinematic* models in which the behaviour is examined of collision partners on a hypothetical reaction surface. A number of treatments of the first kind exist, amongst them those by Kramers[35], by Montroll and Schuler[36] and by Eliason and Hirschfelder[37]. Eliason and Hirschfelder divide a bimolecular reaction into a whole series of reactions

$$A_i + B_j \rightarrow C_k + D_l \tag{65}$$

where the subscripts refer to different internal quantum states of the reactants and products. The overall rate equation is then given by

$$-\frac{dn_a}{dt} = \sum_{ij} \sum_{kl} k_{ij}^{kl} n_{ai} n_{bj} \tag{66}$$

A formal kinetic theory for polyatomic molecules which takes into account transitions between the internal quantum states of the molecules has been developed by Wang-Chang and Uhlenbeck[38]. This theory forms the basis of Eliason and Hirschfelder's treatment of chemical reaction, since certain transitions may be taken to correspond to reaction. Eliason and Hirschfelder show that if a Boltzmann distribution among the internal energy states of the reactant molecules is assumed, then a rate equation of the usual form is obtained. However, perhaps the most interesting feature of Eliason and Hirschfelder's work is that they are able to show how the Eyring form of the rate expression, usually derived from the Transition State Theory (*cf.* section 4.3.2), may be derived from their collision theory if a number of rather drastic approximations are used. The forms of these approximations are naturally of interest, since they may throw light on the essential nature of the Eyring approach to the Transition State Theory. A rather similar approach has been used recently by Levine[39] in a discussion of *non-reactive* molecular encounters which might well be extended to include chemical reaction.

Kinematic calculations may be either classical or quantum mechanical in their treatment of the motions on a potential surface. Classical approximations may be valid for systems of the type where tunnelling is of only minor importance, although for small, low energy, species approaching a narrow energy barrier the use of quantum mechanical methods becomes essential if meaningful results are to be obtained. This limitation quite possibly applies to the $H + H_2$ reaction on which, in fact, many of the *classical* theories have been tested. Indeed, the problems of quantum mechanical treatment are only beginning to be tackled, and offer considerable difficulties.

Many of the kinematic calculations give potentially useful information about energy distributions in the products of reaction, and these predictions can be tested against experimental data to afford confirmation of the potential surfaces adopted. Research of this kind is of the greatest interest, since in the final event chemical kinetics turns on the potential surfaces for reaction. For the present purposes, however, we are concerned only with the rate at which reactants move over to products, and the topic of energy distribution in products is discussed in connection with potential energy surfaces in section 4.5.5.

The first kinematic calculations were those of Hirschfelder *et al.*[40] for the $H + H_2$ system, and much of this early work is described by Glasstone *et al.* (see Bibliography). The treatment employed a point by point calculation of the motion of a mass sliding on a potential surface (the surface was skewed in order that the motion should represent the relative motion of three co-linear hydrogen atoms: *cf.* section 4.5.1). However, the manual calculations did not allow analysis of a sufficient number of initial approach conditions to enable meaningful predictions to be made. The first workers to employ computers in kinematic calculations appear to have been Wall *et al.*[41,42], and many computer calculations have appeared subsequently. One of the most interesting of these is that of Karplus *et al.*[43], who used an LEP surface (see section 4.5.3) in a three-dimensional treatment of the $H + H_2$ reaction. The relative velocity of approach of the reactants and the vibrational (v) and rotational (J) quantum levels of the hydrogen molecule were varied systematically to obtain values for the total reaction cross-section. The reaction cross-section was found to be a monotonically increasing function of the relative velocity after a threshold value, and it also increased with increasing v and J. The threshold velocity was about 0.9×10^6 cm. sec^{-1} (corresponding to an energy of 5.69 kcal. mole^{-1}), and the reaction cross-section increased to a plateau at about twice this velocity. It was shown that the zero point energy of vibration contributes to the energy required for the threshold, while the rotational energy does not. The barrier height of the surface used by Karplus *et al.* was 9.13 kcal. mole^{-1}, while the zero point energy of hydrogen is about 6.20 kcal. mole^{-1}. It is immediately apparent that the calculated translational energy required for the threshold of reaction (5.7 kcal. mole^{-1}) is considerably in excess of the difference between the zero point energy and the barrier height. The exact computation in three dimensions of the amount of vibrational energy available was complicated, and this problem was discussed first in terms of the colinear model. The threshold energy for the colinear system *without zero point energy* is 9.4 kcal. mole^{-1}: that is, all but 0.3 kcal. mole^{-1} of the translational energy is available for reaction. With vibration, the threshold translational energy is 5.3 kcal. mole^{-1}, so that apparently about 4 kcal. mole^{-1} of the 6.20 kcal. mole^{-1} zero point energy can contribute to reaction.

Integration of the reaction cross-section over the velocity, vibrational and rotational distributions corresponding to temperatures between 300° K and 1000° K yielded rate coefficients which fitted an Arrhenius expression fairly well. The best fit for the points gave

$$k = 4.334 \times 10^{10} \exp(-7,435/RT) \tag{67}$$

in l. mole^{-1}. sec^{-1} units; the values of the pre-exponential factor and activation energy were compared with the best experimental estimates then available of 5.4×10^{10} l. mole^{-1}. sec^{-1} and 7.5 ± 1 kcal. mole^{-1} (see below). Detailed examination of the ln k *versus* $1/T$ plot led Karplus *et al.* to the conclusion that there was some slight deviation of their calculated rate coefficients from a simple Arrhenius relation, and the results were best fitted by an expression of the form

$$k = 7.867 \times 10^6 T^{1.1762} \exp(-6,234/RT) \tag{68}$$

Karplus *et al.* now performed the interesting exercise of calculating the expressions for the rate coefficients on the basis of the Transition State Theory *using the same potential energy surface* as they had employed for their collision calculations. They found that the rate coefficients (again in l. mole^{-1}. sec^{-1} units) could be expressed in the Arrhenius form

$$k = 7.413 \times 10^{10} \exp(-8,812/RT) \tag{69}$$

or in the temperature-dependent pre-exponential form

$$k = 2.163 \times 10^8 T^{0.797} \exp(-7,998/RT) \tag{70}$$

Thus the results of the two approaches give similar, but by no means identical,

TABLE 3

COMPARISON OF COLLISION THEORY AND TRANSITION STATE THEORY PREDICTED RATES[43]
FOR THE REACTION $H + H_2$ WITH EXPERIMENTAL RESULTS[33]

Temperature ($°K$)	$10^{-8}k$	$10^{-8}k$ ($l.\ mole^{-1}.\ sec^{-1}$)	$10^{-8}k$
	Collision	Transition State	Experimental
300	0.001845	0.000301	0.00118
400	0.03557	0.01112	0.0131
500	0.2213	0.0984	0.101
600	0.7795	0.4299	0.498
700	1.9742	1.260	1.18
800	4.0506	2.881	4.84
900	7.1965	5.584	11.1
1000	11.5280	9.640	22.0

Note: The experimental results were obtained over the approximate temperature range 300° K to 450° K, and should properly be compared with the predicted values over that range only.

values for the actual rate coefficients over the temperature range considered. The latest *experimental* results for the $H + H_2$ reaction are those of Schulz and LeRoy[33] who express their rate coefficients in the form

$$\log_{10} k = 12.45 - 3.49 \times \frac{10^3}{T} + 3.83 \times \frac{10^5}{T^2} \tag{71}$$

Table 3 lists the values of the rate coefficients calculated at various temperatures by Karplus *et al.* together with the experimental rate coefficients obtained by the use of relation (71). It is seen that the rate constants given by the kinematic calculations are consistently greater than those obtained by application of the Transition State Theory, the divergence between the two sets of figures being least at high temperatures. For the lowest temperatures the kinematic values appear to agree best with the experimental rate coefficient; less reliance can be placed on the "experimental" figures for temperatures above 450° K since they rely on large extrapolations.

Representative of the attempts to apply quantum mechanical principles to the problems of kinetics is the work of Golden *et al.*[44, 45]. In view of the great difficulty of solution of the wave equation for any realistic potential surface, a highly simplified surface (based on the LEP surface) was used for calculations on the $Br + H_2$ reaction. The formulation derived for the rate coefficient predicts a linear ln k *versus* $1/T$ relationship, and the calculated activation energy of 17.0 kcal. mole^{-1} is in fair agreement with the experimental value[46] of 18.1 ± 0.5 kcal. mole^{-1}. Bauer and Wu[47] have also performed simplified quantum mechanical calculations on the $Br + H_2$ reaction; their predicted rate coefficients are about

8×10^{-3} times lower than those predicted from the simple collision theory. The actual P factor seems to be rather less small than this result would indicate (*cf.* Table 1). A number of recent papers, of which one of the first is by Mortensen and Pitzer[48], describe attempts to use more realistic potential energy surfaces in quantum mechanical treatments, and the next few years should see the production of most useful results from this approach.

4. Transition state theory

4.1 INTRODUCTION

The collision theory, in its simpler forms, takes no account of the nature of the reacting species once they have collided except to consider whether or not they can carry sufficient energy to overcome an energy barrier to reaction. The transition state theory, on the other hand, concentrates on the configuration of the reactants just as they are about to pass over into products. This configuration is thought of as lying at the highest energy point of the potential energy surface describing the reaction, and is referred to as the *transition state* or *activated complex*.

The concept of the transition state has been used to develop theories of reaction kinetics in two equivalent ways. Both approaches start from an assumption that the reactants and activated complexes are in some kind of equilibrium, and they diverge according to whether they treat the equilibrium by the methods of classical thermodynamics or of statistical thermodynamics. The classical thermodynamic formulation is the more straightforward, and is discussed in section 4.2. The statistical thermodynamic formulation allows, in principle, the calculation of the reactant-activated complex equilibrium constant from a knowledge of partition functions for the reactants and activated complex. The partition functions are readily calculated from molecular parameters, although of course, in any real case the latter can be only rather roughly estimated for the activated complex. It will appear in section 4.3 that a knowledge of the potential surface for reaction should thus enable absolute calculation of the rate coefficient, and for this reason the theory is frequently rather hopefully known as *Absolute Rate Theory*. The major obstacle to the realisation of the hopes is the difficulty in the construction of sufficiently precise potential surfaces.

It seems appropriate, at this stage, to present a highly simplified derivation of the basic equation of the transition state theory. The formation of activated complexes, $AB^{\ddagger}$, is written as though reversible to reactants

$$A + B \rightleftarrows AB^{\ddagger} \tag{72}$$

and an "equilibrium" constant $K^{\ddagger}$ assigned to the process. Thus the concentration of $AB^{\ddagger}$ is given by

$$[AB^{\ddagger}] = K^{\ddagger}[A][B] \tag{73}$$

The rate of reaction is the rate at which the activated complexes pass over to products, and may be set equal to $v[AB^{\ddagger}]$ where v is the frequency of a critical "vibration" which results in the formation of products. Since a vibration of the kind envisaged will be totally excited, its frequency, v, will be given by

$$v = kT/h \tag{74}$$

(The derivation of equation (74) is given in section 4.3.2). Thus, using (73) and (74), the rate of reaction may be written

$$\text{Rate} = \frac{kT}{h} K^{\ddagger}[A][B] \tag{75}$$

Equation (75) is the basic form of the rate expression used in transition state calculations. A more rigorous justification for the introduction of the frequency, v, will be given in section 4.3.2, and the underlying assumption of equilibrium between reactants and activated complex will be examined.

It is usual to modify equation (75) to read

$$\text{Rate} = \kappa \frac{kT}{h} K^{\ddagger}[A][B] \tag{76}$$

where κ is known as the *transmission coefficient*. κ is introduced to allow for the possibility that not all activated complexes move over to the specific products under consideration. For most reactions κ is, in fact, near unity, although in certain circumstances it can be very much less. The factors influencing the magnitude of κ are discussed in section 4.4.7; in the general treatments κ will be taken to be unity for the sake of simplicity.

4.2 THERMODYNAMIC FORMULATIONS

The concept of "equilibrium" of the ordinary kind for process (72) is clearly not justified since one is dealing with a situation of maximum rather than minimum potential energy (and, presumably, free energy—see section 4.4.1). Nevertheless, even under these conditions it is possible to use the Van't Hoff relation

$$\Delta G_0^{\ddagger} = -RT \ln K_p^{\ddagger} \tag{77}$$

where $\Delta G_0^{\ddagger}$ is a "standard free energy of activation". Equation (75) is written in terms of concentration of reactants, and $K^{\ddagger}$ is the concentration equilibrium constant, $K_c^{\ddagger}$. For a gas-phase bimolecular reaction, as written in (72)

$$K_c^{\ddagger} = RT \cdot K_p^{\ddagger} \tag{78}$$

(and, in general, for a gas-phase reaction of molecularity n which produces a single activated complex)

$$K_c^{\ddagger} = (RT)^{n-1} K_p^{\ddagger} \tag{78a}$$

Equation (75) represents accurately the concentration dependence of rate of reaction, so that a rate coefficient for reaction, k, may be written in terms of $\Delta G_0^{\ddagger}$

$$k = \frac{kT}{h} \cdot e^{-\Delta G_0^{\ddagger}} \cdot RT \tag{79}$$

$\Delta G_0^{\ddagger}$ may be expressed in terms of standard[†] enthalpy and entropy changes in the usual way

$$\Delta G_0^{\ddagger} = \Delta H_0^{\ddagger} - T\Delta S_0^{\ddagger} \tag{80}$$

and the rate coefficient is then given by

$$k = \frac{kT}{h} e^{\Delta S_0^{\ddagger}/R} e^{-\Delta H_0^{\ddagger}/RT} \cdot RT \tag{81}$$

Equation (81) clearly bears some resemblance to the Arrhenius expression, and $\Delta H_0^{\ddagger}$ can be related to the experimental activation energy, viz.

$$E_a = RT^2 \frac{d \ln k}{dT} = \Delta H_0^{\ddagger} + 2RT \text{ (in general } = \Delta H_0^{\ddagger} + nRT) \tag{82}$$

It should be noted that the free energy of activation is sometimes defined by the relation

$$\Delta G_0^{*} = -RT \ln K^{\ddagger} \tag{83}$$

[†] The standard conditions chosen for ΔG, ΔH and ΔS depend upon the concentration units adopted for [A] and [B]. In the present chapter rates are usually quoted in l.mole^{-1}. sec^{-1}, so that the standard state is 1 mole.l^{-1}.

(we write here ΔG^* to distinguish this free energy change from $\Delta G^\ddagger$), and to define $\Delta H_0^* = \Delta H_0^\ddagger$ by the relation

$$\Delta H_0^* = RT^2 \cdot \frac{\mathrm{d} \ln K_p^\ddagger}{\mathrm{d}T} = RT^2 \cdot \frac{\mathrm{d} \ln K^\ddagger}{\mathrm{d}T} - RT \tag{84}$$

The entropy of activation, ΔS_0^*, is now defined by

$$\Delta S_0^* = \frac{\Delta H_0^* - \Delta G_0^*}{T} \tag{85}$$

and the rate coefficient may be written

$$k = \frac{kT}{h} \, e^{\Delta S_0^*/R} \, e^{-\Delta H_0^*/RT} \tag{86}$$

The relationship between experimental activation energy and heat of activation is still correctly represented by (82), and the difference between (81) and (86) is a result of the difference between $\Delta S_0^\ddagger$ and ΔS_0^*

$$\Delta S_0^* = \Delta S_0^\ddagger + R \ln RT \tag{87}$$

It is important to be sure which definition of $\Delta S_0^\ddagger$ is used since $R \ln RT$ (about 13 cal. degree^{-1} at room temperature) may be of the same order of magnitude as $\Delta S_0^\ddagger$ itself. The slightly greater simplicity of (86) compared with (81) commends this form of the expression, although the thermodynamic quantities have a slightly artificial flavour. The artificiality arises, of course, from using a Gibbs free energy in (83): perhaps it would be more justifiable to use a Helmholtz free energy, $\Delta A^\ddagger$, although the real problem is the nature of the equilibrium.

Substitution of the value of $\Delta H_0^\ddagger$ from (82) into (86) yields

$$k = \frac{kT}{h} \, e^{\Delta S_0^*/R} \cdot e^2 \cdot e^{-E_a/RT} \tag{88}$$

and the pre-exponential A factor in the Arrhenius equation is represented in (88) by $(kT/h) \, e^{\Delta S_0^*/R} \cdot e^2$. Furthermore, the collision theory equates A to $PZ/[A][B]$, and since at normal temperatures kT/h is about 10^{13} sec^{-1} and $Z/[A][B]$ is 10^{14} to 10^{15} cm^3. mole^{-1}. sec^{-1}, the probability factor is about unity for a zero entropy of activation (for the standard state of 1 mole. 1^{-1}). The actual dependence of $(kT/h) \, e^{\Delta S_0^*/R}$ on temperature will be related to the way in which ΔS_0^* itself varies with temperature.

In our discussion of the collision theory, it was pointed out that the "steric" component of the probability factor was equivalent to the operation of an entropy of activation. That this is so can be seen from the preceding discussion. In any process involving specialised geometrical orientations for reaction, the formation of the activated complex will involve a decrease in entropy, and the term $e^{\Delta S_0^*/R}$ will become smaller than unity. A positive entropy of activation suggests that, since the entropy of the activated complex is greater than that of the reactants, the complex must be very loosely bound: a decrease in entropy is, in fact, more usual than an increase. For some reactions ΔS_0^* is not very different from ΔS_0 for the overall reaction, a result which indicates that the entropy of the complex is similar to that of the products. This kind of behaviour is seen in reactions of the type

$$A + B \rightleftarrows AB^{\ddagger} \rightarrow AB \tag{89}$$

Entropies may, in principle, be calculated from calorimetric data. However, direct calorimetric calculation for the activated complex is clearly not possible, and it is more usual to obtain entropies of reactants and activated complex from statistical mechanical calculations using known or estimated partition functions for the various species. Prediction of entropies of activation in this way is tantamount to using the statistical mechanical formulation of the Transition State Theory, and is more conveniently developed as part of the statistical treatment.

4.3 STATISTICAL THERMODYNAMIC FORMULATION

4.3.1 The transition state

Before proceeding to derivations of the rate equation, it is perhaps of interest to examine the nature of the transition state. Fig. 3 is a representation of the potential energy of a system of reactants and products as a function of "reaction co-ordinate".

The reaction co-ordinate is looked upon as the lowest energy path between reactants and products over the potential energy surface linking them. The highest point on this path corresponds to a saddle-point on the surface, and species with the configuration described by the potential surface at that point are said to be activated complexes. The question arises whether the activated complex should more properly be defined as lying at a *free-energy* maximum: that is to say, would calculation of the free energy of activation indicate a different configuration of the activated complex from that given by the potential energy surface? Szwarc[49] has discussed the problems involved in considering free-energy maxima for reacting species. Free energy and entropy are in essence statistical concepts, while potential energy is a purely mechanical concept. Potential energy is therefore

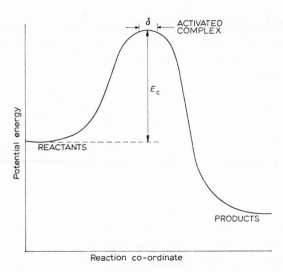

Fig. 3. Potential energy of a reaction system as a function of "reaction co-ordinate".

perfectly meaningful for a single set of reacting species while entropy and free energy are not. The potential energy of the system can be defined as a function of internal co-ordinates of the system, while the free energy cannot be defined until the potential energy is given as a function of those co-ordinates. Eyring[50] has suggested that even if the transition state were not at all well defined, the reaction rate could still be calculated by dividing into successful and unsuccessful paths all possible trajectories through the next most definitive gateway between reactants and products, and subsequent summation over the successful paths. Eyring goes on to say that for a well defined transition state the transmission coefficient could be accurately computed, so that the only problem is that of the difficulty of computation. There do, however, appear to be some cases where the identification of the transition state with a configuration of maximum free energy is necessary. Steel and Laidler[51] have pointed out that in the dissociation of species to produce two free radicals, although the transition state is not defined by a potential energy surface since the latter reaches a plateau rather than a maximum, the free energy passes through a maximum. The free energy decreases beyond its highest point as a result of the increase in entropy accompanying the loosening of vibrations and the development of internal rotations as dissociation sets in. The height of the maximum must in principle be lower than the potential energy plateau, and it should be possible for the activation energy of a dissociative process to be less than the endothermicity. Unfortunately there is little experimental evidence to confirm this conclusion; there are, however, some liquid phase experiments involving large increases in entropy for which the activation energy appears to be less than the endothermicity.

The actual choice of the reaction co-ordinate is of some moment in the accurate evaluation of rate coefficients. Precise determination of the lowest energy path requires a complete knowledge of the potential energy surface for reaction, and the problem is therefore discussed at greater length in connection with potential energy surfaces (section 4.5.8).

Finally, in this brief discussion of the nature of the transition state, some remarks should be offered about the stability of the activated complex. Certain theoretical potential surfaces possess a small potential well or "basin" at the configuration corresponding to the transition state. If the actual potential energy for the system does possess this basin then the activated complex could have an appreciable lifetime. On the other hand, if the potential basin arises only because of the approximations made in the calculations, then the activated complex has an existence only during the period of about one vibration at the top of the potential barrier. Experimental evidence about the stability of the activated complex tends to indicate that the complex is short-lived. Herschbach and collaborators (see, for example, references 52 and 53) have studied the angular scattering of alkali iodides formed in molecular beam experiments in the reaction of alkali metals with alkyl iodides. Most of the product molecules appear to recoil backwards along the direction of the relative velocity vector of the reactants. This result is only consistent with a short-lived complex. Again, Datz and Taylor[54] have shown that crossed molecular beams of $D + H_2$ show little angular scattering, and they attribute their results to the activated complex being short-lived (and linear).

4.3.2 The rate equation

The statistical mechanical formulation of the Transition State Theory treats the equilibrium of reaction (72) as a normal case, and derives a value for the equilibrium constant $K^{\ddagger}$ in terms of the partition functions of the reactants and activated complex and the height of the energy barrier (E_c of Fig. 3)

$$K^{\ddagger} = \frac{Q^0_{AB^{\ddagger}}}{Q^0_A Q^0_B} e^{-E_c/RT} \tag{90}$$

where Q^0 is a volume independent total partition function referred to a standard state. Reaction is then said to correspond to one vibration of the activated complex becoming, in the limit, a translation along the reaction co-ordinate. The partition function for this special vibration is therefore factorised out from $Q^0_{AB^{\ddagger}}$ either directly in the form of a vibrational partition function of near-zero frequency, or as a translational partition function. The frequency of the vibration which leads to reaction now needs only to be multiplied by the concentration of activated complex to give the reaction rate.

In this section we shall give two alternative derivations of the rate equation, and *allow the assumption of equilibrium* for the time being. We shall then examine the nature of the equilibrium assumption. It should be said, in anticipation of this examination, that the assumption of equilibrium appears to lead to no great errors. Non-equilibrium statistical treatments (for example, that of Present[55]) indicate that the rate given by an equilibrium treatment is only significantly in error for very fast reactions in which the more highly energetic reactants are removed more rapidly than equilibrium within the reactants can be regained. Further confirmation that the equilibrium assumption is a reasonable approximation may be drawn from the apparent invariance of rate coefficients as equilibrium for an *overall reaction* is approached. For overall equilibrium, reactants, products and activated complexes are all in equilibrium with each other, and if measured rate coefficients are identical under these conditions with those obtained where reactants and products are *not* in equilibrium, then it would seem that the assumption of equilibrium between reactants and activated complexes is always justified.

The factor to be extracted from the partition function $Q^0_{AB\ddagger}$ will be derived first as though it were a true vibrational factor. The partition function for a vibration of frequency v is (see section 4.3.3)

$$f^{\text{vib}} = \frac{e^{-\frac{1}{2}hv/kT}}{1-e^{-hv/kT}} \tag{91}$$

For any real vibration hv/kT is never small. However, for the "vibration" of the activated complex which leads to reaction, v may be taken to be very low and hv/kT taken to be $\ll 1$. Then (91) becomes

$$f^{\text{vib}} = \frac{kT}{hv} \tag{92}$$

and, using equation (90), the concentration of activated complexes is given by

$$[AB^\ddagger] = [A][B]K^\ddagger = \frac{kT}{hv}\frac{Q'^0_{AB\ddagger}}{Q^0_A Q^0_B} e^{-E_0/RT}[A][B] \tag{93}$$

where $Q'^0_{AB\ddagger}$ is the partition function for $AB^\ddagger$ from which the factor corresponding to the "reaction vibration" has been removed. The rate coefficient for reaction can now be written

$$k = \frac{v[AB^\ddagger]}{[A][B]} = \frac{kT}{h} \cdot \frac{Q'^0_{AB\ddagger}}{Q^0_A Q^0_B} \cdot e^{-E_0/RT} \tag{94}$$

The treatment as a vibration of the degree of freedom along the reaction coordinate is somewhat unsatisfactory, especially in view of the special assumptions

References pp. 298–301

that need to be made about the frequency. It is possible, however, to treat the degree of freedom as a translation, and no special conditions need be imposed. Let the distance along the reaction co-ordinate which corresponds to the activated complex be δ, as in Fig. 3. In this region the potential energy curve is flat, and motion corresponds to a translation, so we may extract from $Q^0_{AB\ddagger}$ the partition function for translation in one dimension for a particle mass $m^\ddagger$ of $(2\pi m^\ddagger kT)^{\frac{1}{2}}\delta/h$, viz.

$$Q^0_{AB\ddagger} = \frac{(2\pi m^\ddagger kT)^{\frac{1}{2}}\delta}{h} \cdot Q''^0_{AB\ddagger} \tag{95}$$

Hence

$$[AB^\ddagger] = [A][B] \cdot \frac{(2\pi m^\ddagger kT)^{\frac{1}{2}}\delta}{h} \cdot \frac{Q''^0_{AB\ddagger}}{Q^0_A Q^0_B} \cdot e^{-E_c/RT} \tag{96}$$

Now the kinetic theory of gases gives the average velocity at which particles move in *one direction* as

$$\bar{c} = \left(\frac{kT}{2\pi m^\ddagger}\right)^{\frac{1}{2}} \tag{97}$$

The frequency at which activated complexes pass over to products is the average velocity of particles moving from left to right divided by the distance δ (a frequency is the number of times a second that an oscillating system passes any given point in *one* direction). Thus

$$v = \left(\frac{kT}{2\pi m^\ddagger}\right)^{\frac{1}{2}} \frac{1}{\delta} \tag{98}$$

and the rate coefficient for reaction, k, is given by

$$k = \frac{v[AB^\ddagger]}{[A][B]} = \left(\frac{kT}{2\pi m^\ddagger}\right)^{\frac{1}{2}} \cdot \frac{1}{\delta} \cdot \frac{(2\pi m^\ddagger kT)^{\frac{1}{2}}\delta}{h} \frac{Q''^0_{AB\ddagger}}{Q^0_A Q^0_B} e^{-E_c/RT}$$

$$= \frac{kT}{h} \cdot \frac{Q''^0_{AB\ddagger}}{Q^0_A Q^0_B} \cdot e^{-E_c/RT} \tag{99}$$

This expression is identical to (94), since $Q'^0_{AB\ddagger}$ and $Q''^0_{AB\ddagger}$ are in fact the same.

The above derivation, given by Laidler (Bibliography), is equivalent to that given originally by Eyring[56], although Eyring worked in terms of the momentum and reduced mass of the activated complex. Evans and Polanyi[57], in their derivation, considered, in effect, the average velocity in both directions along the reaction co-ordinate, but also allowed for the fact that only one half of the complexes move towards products. Their final expression for the rate coefficient was therefore the same as Eyring's.

Laidler and Polanyi[58] have shown that the equilibrium treatment yields a valid result by the artifical device of considering a hypothetical system in which the potential energy curve rises again after the flat region (δ of Fig. 3). In this case the flat region corresponds to an ordinary state AB, and an equilibrium concentration of AB is established. By extracting the appropriate factor from the partition function for AB as before, the concentration of AB is given by

$$[AB] = [A][B] \cdot \frac{kT}{h\nu} \cdot \frac{Q'^0_{AB}}{Q^0_A Q^0_B} \cdot e^{-E_c/RT} \tag{100}$$

Since the state AB is in equilibrium with $A+B$, the rate at which species pass through the state from left to right is equal to the rate at which they pass through from right to left. The latter rate is equal to $\nu[AB]$ so that the rate at which species enter the region δ from the left is equal to

$$[A][B] \cdot \frac{kT}{h} \cdot \frac{Q'^0_{AB}}{Q^0_A Q^0_B} \cdot e^{-E_c/RT} \tag{101}$$

Since ν is independent of the subsequent form of the potential energy surface, (101) would also give the rate at which species entered the region δ even if the potential energy curve decreased again on the right of the flat region (as in Fig. 3). If every species that reaches the activated complex configuration moves on to products, then (101) gives the rate of reaction. This result is identical with the result normally obtained by applying the equilibrium assumption directly to the activated complex. What is important is that the *algebraic expression* is identical: a classical equilibrium between reactants and activated complexes does not exist (and, as Laidler and Polanyi[58] point out, if activated complexes were added to the system moving from left to right the "equilibrium" would not be disturbed).

4.3.3 Partition functions

The problem of evaluating rate coefficients from the statistical-thermodynamic form of the Transition State Theory requires calculation of partition functions for the reactants and activated complex. This section gives derivations of the partition functions for various degrees of freedom.

A total partition function per particle, Q, is defined by

$$Q = \sum_i g_i e^{-\varepsilon_i/kT} \tag{102}$$

where ε_i is the energy of the ith level of a quantised energy level system and g_i

its degeneracy. According to the usual approximation for separating energies, the total partition function can be factorised into separate components for the several degrees of freedom. The symbol f with appropriate superscripts is used here to represent the factorised components of the partition functions.

$$Q = f^{\text{trans}} f^{\text{vib}} f^{\text{rot}} f^{\text{elec}} \tag{103}$$

(Note that Q is volume dependent or not according to whether f^{trans} is divided by V or not: all Q in the rate equations are volume independent). A term f_0 is sometimes included to correspond to the zero point energy of the system, ε_0, and to give all the energy modes a common reference point. The nature of the summation differs in the various cases, and they are dealt with separately.

(a) Translational partition function

The quantised energy levels for a particle in a box of dimensions $a \times b \times c$ is given by

$$\varepsilon_i = \frac{h^2}{8m} \left(\frac{l_i^2}{a^2} + \frac{m_i^2}{b^2} + \frac{n_i^2}{c^2} \right) \tag{104}$$

where l, m, n are quantum numbers describing the motion in the x, y and z directions. Thus

$$f^{\text{trans}} = \sum_i g_i \exp \left\{ -\frac{h^2}{8mkT} \left(\frac{l_i^2}{a^2} + \frac{m_i^2}{b^2} + \frac{n_i^2}{c^2} \right) \right\} \tag{105}$$

The separation between the energy levels is very small for translational energy, and the summation of (105) over g_i-fold degenerate levels can be replaced by an integration over all values of l, m and n from 0 to ∞. The translational partition function is then given by

$$f^{\text{trans}} = \int\int\int\limits_0^{\infty} \exp \left\{ -\frac{h^2}{8mkT} \left(\frac{l^2}{a^2} + \frac{m^2}{b^2} + \frac{n^2}{c^2} \right) \right\} \, dl \, dm \, dn \tag{106}$$

The triple integral can be resolved into the product of three integrals of the form

$$\int_0^{\infty} e^{-\alpha^2 x^2} dx = \sqrt{\frac{\pi}{2\alpha}} \tag{107}$$

The partition function in three dimensions is, therefore, given by

$$f^{\text{trans}} = \left(\frac{2\pi mkT}{h^2}\right)^{\frac{3}{2}} \cdot abc = \left(\frac{2\pi mkT}{h^2}\right)^{\frac{3}{2}} V \tag{108}$$

where V is the volume. [The corresponding expression for translation in one dimension is clearly the result of a single integration

$$f^{\text{trans}} = \frac{(2\pi mkT)^{\frac{1}{2}}}{h} a, \qquad \text{etc.} \tag{109}]$$

The partition function derived is for a *single* particle. For a system of N non-interacting indistinguishable particles the total translational partition function is:

$$f_N^{\text{trans}} = \frac{(f^{\text{trans}})^N}{N!} \tag{110}$$

All further derivations are for single particles, and overall partition functions for assemblies of particles must be calculated according to (110).

(b) Vibrational partition function

The vibrational energy levels for a harmonic oscillator of frequency v are given by

$$\varepsilon_i = (v+\tfrac{1}{2})hv/kT \tag{111}$$

where v is the vibrational quantum number. Thus

$$f^{\text{vib}} = \sum_{v=0}^{\infty} e^{-(v+\frac{1}{2})hv/kT} = \frac{e^{-\frac{1}{2}hv/kT}}{1-e^{-hv/kT}} \tag{112}$$

The total vibrational partition function for a molecule will be the product of the partition functions for the $3N-6$ normal vibrations ($3N-5$ for a linear molecule). More complex summations are possible which take account of the anharmonicity of vibration.

If the partition function f_0 is included in the product of (103) then f'^{vib} is defined such that

$$f'^{\text{vib}} = (1-e^{-hv/kT})^{-1} \tag{113}$$

(c) Rotational partition function and statistical factors

The form of the rotational partition function depends on the rotational symmetry of the molecule. For a rigid linear rotator

$$\varepsilon_J = B_0 J(J+1) \tag{114}$$

and the degeneracy is given by

$$g_J = 2J+1 \tag{115}$$

It is customary to divide g_J by a "symmetry number" which is the number of equivalent rotational orientations that can be obtained for the molecules. However, several groups of workers have shown that the use of symmetry numbers in Transition State Theory formulations can give rise to erroneous rate coefficients. For this reason the symmetry number is omitted from the rotational partition functions derived in this section; an alternative method for obtaining the correct statistical factors to apply, due to Bishop and Laidler[59], is given at the end of the section.

The rotational constant, B_0, is equal to $h^2/8\pi^2 I$ where I is the moment of inertia of the molecule. Thus

$$f^{\text{rot}} = \sum_{J=0}^{\infty} (2J+1)e^{-J(J+1)h^2/8\pi^2 IkT} \tag{116}$$

For systems where only a few rotational levels are excited, it is probably best to carry out the summation. If, however, $h^2 \ll 8\pi^2 IkT$, then we can write f^{rot} as an integral

$$f^{\text{rot}} = \int_0^{\infty} e^{-J(J+1)h^2/8\pi^2 IkT} \cdot (2J+1)dJ \tag{117}$$

$$= \frac{8\pi^2 IkT}{h^2} \tag{118}$$

The problem of treating symmetric and asymmetric tops quantum mechanically is of some difficulty, and it is far easier to use a classical approximation equivalent to (117). For a polyatomic molecule with moments of inertia I_A, I_B and I_C about the principal axes

$$f^{\text{rot}} = \frac{(8\pi^2 kT)^{\frac{3}{2}}(\pi I_A I_B I_C)^{\frac{1}{2}}}{h^3} \tag{119}$$

In general, for substances in the gas phase at normal temperatures, the differences between quantum mechanical and classical rotational partition functions is small.

The problem of the correct statistical factors to be used in transition state formulations must now be considered. Schlag[60,61] has recently proposed treatments which use a group-theoretical approach, and has shown clearly that the use of symmetry numbers can yield erroneous results. Bishop and Laidler[59] have also presented a rather simpler treatment of the statistical factors, and, since the results seem to be in general identical with Schlag's, their method is described here.

Statistical factors s_f and s_r are defined for forward and reverse reactions in a process such as

$$A + B \rightleftarrows C + D \tag{120}$$

These factors are the numbers of different sets of products which could be formed if all identical atoms in the reactants were labelled. (It is noted that if both reactants are the same chemical species then the statistical factor is half the number of different sets.) An example may make the method clear. Consider the reaction

$$Cl + CH_4 \rightleftarrows HCl + CH_3 \tag{121}$$

The hydrogen atoms on the methane may be labelled H^1, H^2, H^3 and H^4 so that the following four sets of products may be formed: $H^1Cl + CH^2H^3H^4$, $H^2Cl + CH^1H^3H^4$, $H^3Cl + CH^1H^2H^4$ and $H^4Cl + CH^1H^2H^3$. Thus the statistical factor for the forward reaction, s_f, is four. The reverse process can be treated in a similar way, taking, say, $H^1Cl + CH^2H^3H^4$. The products are the two mirror image forms of $CH^1H^2H^3H^4 + Cl$, so that s_r is two. Bishop and Laidler show that for *any* reaction such as (120) the *ratio* s_f/s_r is the ratio of rotational symmetry factors for reactants and products.

The application of the method to the formulation of the rate equation is as follows. If the forward and reverse statistical factors for the process

$$A + B \rightleftarrows AB^{\ddagger} \tag{72}$$

are $s_f^{\ddagger}$ and $s_r^{\ddagger}$ respectively, then as a result of the property of quotients of symmetry numbers the concentration of $AB^{\ddagger}$ should be written

$$[AB^{\ddagger}] = [A][B] \cdot \frac{kT}{hv} \cdot \frac{s_f^{\ddagger}}{s_r^{\ddagger}} \cdot \frac{Q''^0_{AB^{\ddagger}}}{Q''^0_A Q''^0_B} \cdot e^{-E_c/RT} \tag{122}$$

where Q''^0 are total partition functions from which the symmetry numbers have been omitted. This omission will already be present in the rotational partition functions if the expressions (118) or (119) are used. It will be recalled that Laidler and Polanyi[58] obtained the rate at which reacting species entered the region δ in our Fig. 3 by consideration of the hypothetical equilibrium case where a stable species AB is formed rather than the unstable activated complex $AB^{\ddagger}$. It was shown that the rate is $v[AB]$ in the simple case, so it is $s_r v[AB]$ in the reaction system with statistical factor greater than unity. Thus

$$s_r v[AB] = [A][B] \frac{kT}{h} \cdot s_f \cdot \frac{Q''^0_{AB}}{Q''^0_A Q''^0_B} \cdot e^{-E_c/RT} \tag{123}$$

in the present case, and this would also be the rate at which reactants entered the transition state when the unstable $AB^{\ddagger}$ is formed. Hence the rate coefficient for reaction is

$$k = \frac{kT}{h} s_f^{\ddagger} \frac{Q''^0_{AB^{\ddagger}}}{Q''^0_A Q''^0_B} \cdot e^{-E_c/RT} \tag{124}$$

It is therefore apparent that the ordinary formulation using symmetry numbers in the rotational partition functions gives a rate coefficient that is too low by the factor $s_r^{\ddagger}$, and is in error for all reactions in which $s_r^{\ddagger} \neq 1$.

(d) Electronic partition function

The energy difference between the ground electronic state and the first excited state is frequently very large compared with kT, so that only the ground state makes an appreciable contribution to the electronic partition function. If the zero of energy is chosen to be the ground electronic state, then the electronic partition function is unity. However, in the general case we must write

$$f^{elec} = \sum_j g_j e^{-\varepsilon_j/kT} \tag{125}$$

where ε_j and g_j are the electronic energy and degeneracy of the jth electronic state. A number of species have relatively low lying excited electronic energy levels, while other are degenerate in their ground states [e.g. $O_2(^3\Sigma_g^-)$, $NO(^2\Pi)$, etc.], and for these species electronic partition functions must be calculated.

(e) Nuclear partition function

The energy of the ground nuclear state is taken to be unity, so that for species in their ground nuclear states the important contribution to the partition functions comes from the nuclear degeneracy $(2I+1)$ where I is the total nuclear spin of the species.

TABLE 4

PARTITION FUNCTIONS FOR THE SEVERAL DEGREES OF FREEDOM TOGETHER WITH
NUMERICAL ORDER OF MAGNITUDE

Energy form	Expression for partition function	Order of magnitude
Translation in 3-dimensions	$f'^{\,\text{trans}} = (2\pi mkT/h^2)^{\frac{3}{2}}$ (volume independent)	10^{24}–10^{27}
Vibration	$f^{\text{vib}} = e^{-\frac{1}{2}h\nu/kT}(1-e^{-h\nu/kT})^{-1}$ for each vibration. This includes zero-point energy.	1–10^2
Rotation (Classical forms)	$f^{\text{rot}} = (8\pi^2 kT)^{\frac{3}{2}}(\pi I_A I_B I_C)^{\frac{1}{2}}/h^3$ for three unequal moments of inertia. $f^{\text{rot}} = 8\pi^2 IkT/h^2$ for linear molecule	10–10^3
Electronic	$f^{\text{elec}} = \sum\limits_{j=0}^{j^*} g_n\, e^{-\varepsilon_j/kT}$ (j^* is highest populated state)	1
Nuclear	$f^{\text{nuc}} = (2I+1)$	1

To conclude this section a table (Table 4) is presented of the analytical expressions for partition functions, together with approximate numerical values obtained by substitution of reasonable magnitudes for the various molecular parameters needed.

4.3.4 Predicted pre-exponential factors

The pre-exponential factor for any reaction can now, in principle, be calculated from equation (94) or (99) and the appropriate values of the partition functions. Unfortunately, however, it is not possible to calculate accurate values for the partition functions of the activated complex unless the exact configuration of the activated complex and the form of the potential energy surface at the transition state are known. On the other hand, good estimates may be made; and rough guides may be obtained for pre-exponential factors just by using the orders of magnitude quoted in Table 4. Before carrying out the exercise of predicting numerical pre-exponential factors for a number of different types of reaction, it is of interest to see to what form the analytical expression for the rate coefficient reduces if the reactants are two atoms A and B (*i.e.* hard spheres). A and B will both have only translational partition functions, while the activated complex, $AB^{\ddagger}$, will behave as a diatomic molecule with a moment of inertia $I = \sigma_{AB}^2 \mu$ (σ, μ are respectively, the radius and reduced mass of the complex). The reaction co-ordinate is clearly the vibration along the line of centres, so no partition function need be written for this vibration. Substitution of the expressions for the partition functions into (94) then gives

$$
k = \frac{kT}{h} \frac{\left(\dfrac{\overline{2\pi m_A + m_B}\, kT}{h^2}\right)^{\frac{3}{2}} \dfrac{8\pi^2 \sigma_{AB}^2 \mu kT}{h^2}}{\left(\dfrac{2\pi m_A kT}{h^2}\right)^{\frac{3}{2}} \left(\dfrac{2\pi m_B kT}{h^2}\right)^{\frac{3}{2}}} \cdot e^{-E_c/RT}
$$

$$
= \sigma_{AB}^2 \left(\frac{8\pi kT}{\mu}\right)^{\frac{1}{2}} e^{-E_c/RT} \tag{126}
$$

which is identical to the value given by the collision theory (section 3.2). Thus the Transition State Theory and the Collision Theory are consistent if the hard sphere condition is imposed on the reactants.

The order of magnitude of the rate coefficient for the two atom reaction could also be obtained from the values quoted in Table 4. If the partition functions for the same degree of freedom are regarded as identical for the different species, then the rate coefficient could be written:

$$
k = \frac{kT}{h} \frac{f'^{\text{trans}}(f^{\text{rot}})^2}{(f'^{\text{trans}})^2} \cdot e^{-E_c/RT} = \frac{kT}{h} \frac{(f^{\text{rot}})^2}{f'^{\text{trans}}} \cdot e^{-E_c/RT} \tag{127}
$$

Taking $f'^{\text{trans}} = 10^{24}$ and $f^{\text{rot}} = 10$, and with $kT/h \sim 10^{13}$ sec^{-1}, we obtain a value for the frequency factor of about 10^{-9} cm^3. molec^{-1}. sec^{-1}, or about 10^{12} l. mole^{-1}. sec^{-1}. This is the same value as was calculated for the gas-kinetic collision rate coefficient, $Z_{AB}/n_A n_B$, in section 3.4. Comparison of the Transition State Theory and Collision Theory results may be extended by calculation of expected P factors for various types of reaction. For example, if the reaction between two non-linear molecules, containing n_1 and n_2 atoms respectively, proceeds via a non-linear complex, then the complex will possess translation in three dimensions, three rotations and $3(n_1+n_2)-6$ vibrations, of which one is along the reaction coordinate. The powers of the partition functions to be found in the pre-exponential part of the rate coefficient expression are, therefore,

$$
\frac{f'^{\text{trans}}(f^{\text{vib}})^{3(n_1+n_2)-7}(f^{\text{rot}})^3}{f'^{\text{trans}}(f^{\text{vib}})^{3n_1-6}(f^{\text{rot}})^3 \cdot f'^{\text{trans}}(f^{\text{vib}})^{3n_2-6}(f^{\text{rot}})^3} = \frac{(f^{\text{vib}})^5}{f'^{\text{trans}}(f^{\text{rot}})^3} \tag{128}
$$

The partition functions for the two atom reaction are in the ratio $(f^{\text{rot}})^2/(f'^{\text{trans}})$ so that the pre-exponential factor for the non-linear reaction will be in the ratio $(f^{\text{vib}}/f^{\text{rot}})^5$ to that for the hard sphere model, and this ratio is the probability factor, P. Using the orders of magnitude in Table 4, $P \sim 10^{-5}$, a result in keeping with the experimental data for the reaction rates of complex molecules. Probability factors for other reactions involving species of intermediate complexity may be calculated in a similar way, and, in general, they are similar to the experimentally determined values. The expression for P usually involves the ratio (f^{vib}/f^{rot}) raised to some power, and the form suggests an interpretation of low P factors in terms of the loss of rotational degrees of freedom of the reactants and their replacement by vibrations.

The temperature dependence of the pre-exponential term is clearly determined by the final ratio of partition functions, and will therefore be related both to the complexity of reaction and to the exact form adopted for the partition functions.

4.4 MORE GENERAL TREATMENTS OF THE TRANSITION STATE THEORY

During the last decade or so, a number of attempts have been made to formulate the Transition State Theory without recourse to all of the assumptions made by Eyring[56] or Evans and Polanyi[57]. A number of these have merely used the activated complex as a basis for statistical theory. For example, Light and his collaborators have produced statistical theories of reaction kinetics[62] which have now been extended to include reactions with activation energy[63]. The breakup of a strong-coupling complex is computed statistically, and the theory applied to a number of reactions (such as $K + HBr$, $H + Cl_2$, $H + HBr$, etc.) to determine rate coefficients, reaction cross sections and information about rotational, vibrational and electronic excitation of the products. In general the results are consistent with experimental data.

Of more direct concern to a discussion of the Transition State Theory are the attempts to investigate the effect of non-equilibrium distribution on calculated rates. Prigogine et al.[64, 65] made some of the earliest investigations of this kind by a derivation of the perturbation of the Boltzmann distribution as a result of reaction. Present[55] subsequently made a more reliable calculation which suggested that if the ratio E_c/RT is sufficiently large (so that the more energetic reactants are not too rapidly removed) the simple collision theory yields results which are not greatly in error. (For $E_c/RT = 5$, the collision theory is in error by 8 %.) Mahan[66] has performed similar calculations for radical combination processes (small activation energy). A collision cross section was calculated which indicated that translationally hot radicals combine more rapidly than average "thermal" radicals. The cross section was then used to calculate the disturbance of the velocity equilibrium in reactions of this kind, and it was concluded that the equilibrium assumption was justified in photochemical systems, where the radical concentration is low, although if the radical concentration approaches 0.1 mole fraction, the assumption may no longer be valid. Ross et al. [67, 68] have also attempted perturbation solutions of the Boltzmann equilibrium to obtain an equation for the rate of bimolecular reaction. Ross and Eu[69, 70] have, in addition, developed a theory of collisions to derive an activated complex theory of reaction rates.

Marcus[71] has given generalised forms of the Transition State Theory using both quantum mechanical and classical treatments. A rate expression is derived without introduction of the assumption that there is a Cartesian reaction co-ordinate, and it is shown that this expression reduces to the ordinary form if the Cartesian assumptions are subsequently made. The new equation for the transmission coefficient contains internal centrifugal terms, and a rotational interaction can be introduced into the formalism. The absence of the Cartesian assumption allows calculations to be made with a potential surface having separated variables, since the theory can deal with a separated wave equation for the reaction co-ordinate which is curvilinear (cf. ref. 72). A simpler and more general derivation of the final equations was subsequently given[73] which permitted the introduction of analytical mechanics. Marcus has also given[74] a quasiequilibrium expression which relates sums over reaction cross sections to properties of activated complexes. One of the most interesting features of the latter work is the comparison of the results with the kinematic calculations of Karplus et al.[43] (cf. section 3.7) on the $H + H_2$ reaction. The equations developed were used[75] to formulate a statistical dynamical model for the total chemical reaction cross section expressed as a function of the relative velocity and the vibrational and rotational states of the reactants. Encouraging agreement with the $H + H_2$ results of Karplus et al. was obtained[76] in the low to moderate velocity range, although at very high velocities the comparison between the predictions of Marcus' theory and the kinematic calculations indicated the occurrence of some vibrational non-adiabaticity.

4.5 POTENTIAL ENERGY SURFACES

4.5.1 General considerations

Some knowledge of the potential energy surface is needed in either collision or transition state theories of chemical kinetics. The simple collision theory requires only a value for the height of the energy barrier to be surmounted, while the transition state theory needs information also about the shape of the potential energy surface in the neighbourhood of the saddle point in order to calculate the partition functions for the activated complex. In principle, calculation of tunnelling corrections also presupposes a knowledge of the form of the surface near the saddle point, while the more sophisticated calculations of reaction rate, such as the kinematic treatments, need detailed information about the whole surface joining reactants and products. Direct calculation of a potential energy surface from experimental data is not, at present, possible, although experimental information of one kind or another can be used to test the validity of a hypothetical surface: this matter is discussed in section 4.5.5. The problem is therefore one of constructing the surface from quantum mechanical principles of systems involving several particles. Unfortunately, such approaches usually require drastic approximations which at once makes their validity questionable and reduces their general applicability. Empirical methods and semi-empirical methods are less satisfying intellectually, although they frequently have the advantage of greater tractability. For these reasons it seems useful to divide the treatment of potential energy surface calculations into three sections, 4.5.2–4, according to whether the approach is non-empirical, semi-empirical or empirical.

The potential energy of a many-particle system can only be represented as a function of co-ordinates of the particles on a many-dimensional hypersurface. However, many of the theoretical calculations to be discussed concern the three body problem connected with exchange reactions of the type $A + BC \rightarrow AB + C$ so that the atoms are confined to a single plane and their positions relative to each other can be expressed in terms of the distances between each of the atoms. If the angle of approach of A to BC is defined, it is now possible to define the relative positions by two variables, so that the potential energy surface may be represented in two dimensions by a contour map. For more complex systems contour maps may be drawn of sections through the potential energy hypersurface. Fig. 4 is a contour map for the reaction system formally represented by the formation of a linear complex in the reaction system $A + BC \rightarrow AB + C$ and the interatomic distances r_{AB}, r_{BC} define the relative positions of A, B and C. A section taken through the potential energy surface parallel to the r_{BC} axis and taken at large separation of A and BC (*i.e.* large r_{AB}) would be the ordinary potential energy

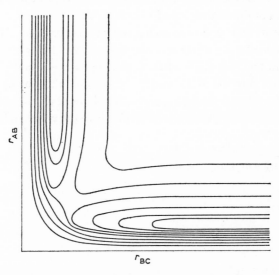

Fig. 4. Contour diagram for the potential energy of a linear system of three atoms A, B and C as a function of their relative positions defined by r_{AB} and r_{BC}.

curve for BC (and the similar section taken parallel to the r_{AB} axis would give the potential energy curve for AB).

Some comments by Marcus about the validity of expressing potential energies in the way described are put forward at the end of section 4.5.8.

The first kinematic calculations using potential energy surfaces (*cf.* section 3.7) were made by point by point calculation of the successive co-ordinates of a sliding mass point. In order that the relative kinetic energy of the three atoms should be represented by the kinetic energy of a single particle moving on the potential surface, it is necessary that the axes of the contour map be skewed. Although the concept of the sliding point mass is no longer especially valuable now that the calculations are performed by high speed computers, many contour maps have been presented in the past with skewed axes, and it is therefore proposed that the angle of skew required to satisfy the kinetic energy condition be derived. The calculations refer to a colinear system. Suppose three atoms A, B and C of mass m_A, m_B and m_C lie along a line so that the centres of A and C are separated by the distances r_{AB}, r_{BC} respectively from B. The total internal kinetic energy (T) with respect to the centre of mass of the system is given by:

$$T = \frac{1}{(m_A+m_B+m_C)} \{m_A(m_B+m_C)\dot{r}_{AB}^2 + 2m_A m_C \dot{r}_{AB}\dot{r}_{BC} + m_C(m_A+m_B)\dot{r}_{BC}^2\}$$

(129)

Now we require to choose the angle at which r_{AB} and r_{BC} must be plotted with respect to each other so that two conditions are filled:

(*i*) there must be no cross-terms if the kinetic energy is expressed in terms of the *perpendicular* axes of the system, x and y, and

(*ii*) the coefficients of x^2 and y^2 in such a system must be equal. That is, we wish to express T as $\frac{1}{2}m(x^2+y^2)$ where m is the mass of the sliding point. If θ be the angle between r_{AB} and the $r_{BC}(=x)$ axes, then $r_{AB} = x - y\tan\theta$ and $r_{BC} = cy$ $\sec\theta$ (c is a factor introduced to modify the scale of the sloping co-ordinate). Substitution of the expressions for r_{AB} and r_{BC} in terms of x, y, c and θ into (129), and application of the restrictions (*i*) and (*ii*) leads to the conclusions

$$\sin\theta = \left\{\frac{m_A m_C}{(m_A+m_B)(m_B+m_C)}\right\}^{\frac{1}{2}}$$

(130)

$$c = \frac{m_C}{m_A+m_B}$$

(131)

and

$$m = \frac{m_A(m_B+m_C)}{(m_A+m_B+m_C)}$$

(132)

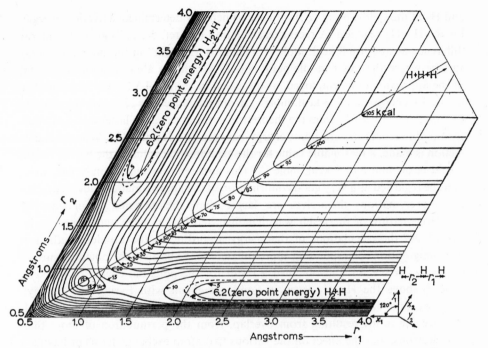

Fig. 5. Skewed axis contour map for the co-linear H–H–H system (after Eyring *et al.*[77]).

Fig. 5 shows a skewed potential energy contour map for the surface for co-linear H–H–H, taken from the paper by Eyring *et al.*[77].

4.5.2 Non-empirical treatments

London[78] made one of the earliest attempts to apply a simple valence-bond treatment to the H_3 system, and suggested that the energy of the system could be expressed in terms of the three Coulombic interaction integrals Q_A, Q_B and Q_C and the three exchange integrals α, β and γ, for the three pairs of atoms

$$E = Q_A + Q_B + Q_C \pm [\tfrac{1}{2}(\alpha - \beta)^2 + (\beta - \gamma)^2 + (\alpha - \gamma)^2]^{\frac{1}{2}} \tag{134}$$

(The contribution of overlap interaction was neglected, and the equation given without proof. Coolidge and James[79] showed that the derivation of (134) required that a number of approximations be made.)

Sugiura[80] had given the Coulombic and exchange integrals for the system, and Eyring and Polanyi[81] used these values to compute a potential energy surface for linear configurations of H_3 on the basis of (134). It should now be possible to compare the difference in the binding energy between the separated reactants and H_3 at the saddle point (E'_c) with experimentally determined activation energies for the $H + H_2$ reaction. A recent recalculation (see ref. 58, p. 7) gives this energy difference as about 8.8 kcal. mole^{-1}. (There is a "basin" in the potential energy surface where the exact saddle point should be: this is almost certainly a result of the approximations employed, as discussed in section 4.3.1.) The experimental activation energy (say 7–8 kcal. mole^{-1}, from reference 9 or 33) must be corrected for comparison with E'_c from the theoretical surface, since the latter does not allow for the differences in zero-point energy between $H + H_2$ and H_3. When this correction is made, a reasonable value for the experimental value of the barrier height would be around 8 or 9 kcal. mole^{-1}. This apparent agreement must not, however, be regarded as a very good confirmation of the approximations adopted, since the approximations themselves are larger than the energies calculated. The fortuitous nature of the result was demonstrated by Coolidge and James[79] who carried out the calculations of the energy according to the London method at the configuration corresponding approximately to the basin in the potential surface ($r_{AB} = r_{BC} = 0.9$ A) and obtained a result for the potential energy of 4.0 kcal. mole^{-1} above the $H + H_2$ level, in good agreement with Eyring and Polanyi. Coolidge and James then recalculated the potential energy after making allowance for terms resulting from overlap, from the permutation of more than two electrons, and the effect on the various two-atom exchange terms of having a third atom present. The potential energy at $r_{AB} = r_{BC} = 0.9$ A was now about 33 kcal. mole^{-1} above $H + H_2$, leaving no doubt that the more exact calculation gave a far worse estimate of the height of the energy barrier.

Quantum mechanical calculations, of the kind described above, for many bodies, are probably best solved by making use of the Variation Theorem. Hirschfelder et al.[82, 83] made the first attempts at a variational treatment of the linear H–H–H system. The "best" linear symmetrical configurations were based on homopolar, and then on homopolar and polar, combinations of three $1s$ atomic orbitals. Hirschfelder[84] later extended the treatment to include non-linear, but still symmetrical, configurations. The results of these variational calculations were rather unsatisfactory: the investigation of non-linear systems had suggested that the linear configuration would, in fact, give the lowest energy barrier. On the other hand, the lowest binding energy calculated by Hirschfelder et al.[82] was about 25 kcal. mole^{-1} calculated relative to $H + H_2$ (the $H + H_2$ level was calculated according to the same approximations).

The advent of high speed computers has allowed the use of more involved variational treatments, but the results have proved disappointing. Kimball and Trulio[85] set up an LCAO–MO wave function from five $1s$ orbitals for linear symmetric H_3; their rather complete variational treatment gave a value of 19.9 kcal. mole^{-1} binding energy relative to $H + H_2$ calculated by their treatment. It might be thought that this sophisticated treatment had brought the calculated binding energy considerably closer to the experimental value of 8–9 kcal. mole^{-1}. However, the $H + H_2$ energy calculated is itself in considerable error, and referred to the *experimental* $H + H_2$ energy, the binding energy is about 37 kcal. mole^{-1}. Boys and Shavitt[85a] have performed an even more involved variational calculation, and considered asymmetric and bent configurations. The linear configuration was found to be most stable, and the binding energy in the symmetric H_3 was 29.2 kcal. mole^{-1} referred to experimental values of $H + H_2$. Thus even the best calculated values of the barrier height are a factor of three higher than the values determined experimentally from reaction rates.

4.5.3 Semi-empirical surfaces

The rather unsatisfactory results obtained from non-empirical determinations of potential energy surfaces for even the simplest reaction systems such as $H + H_2$ have led to the parallel development of "semi-empirical" treatments in which the form of the surface is based on the quantum mechanical methods, but in which various parameters can be adjusted to fit experimental observations.

The most famous of the semi-empirical surfaces is the London–Eyring–Polanyi (LEP) surface. Eyring and Polanyi[81] first took the form of the potential energy curve for any H–H pair (obtainable from spectroscopic data) and equated the energy with the total binding energy for the pair (*i.e.* $Q_A + \alpha$, $Q_B + \beta$, $Q_C + \gamma$). Now Sugiura's values[80] for the integrals suggested that at internuclear separations greater than 0.8 A, the fraction of Coulombic to total energy

$$\rho_A = \frac{Q_A}{Q_A + \alpha} \tag{135}$$

was constant (and about 0.10 to 0.15). By taking a constant value of ρ in conjunction with the Morse form of the potential energy curve for H–H pairs, it is possible to calculate the Coulomb and exchange energies at any internuclear separation, and a potential energy surface may be constructed. This semi-empirical LEP surface predicts a basin at $r_{AB} = r_{BC} \sim 0.8$ A, although, as we have seen in section 4.5.1, experimental observations suggest that H_3 is not metastable. The depth of the basin increases with an increase in ρ. Table 5 shows the height of the energy barrier, E_c (referred to the LEP calculated $H + H_2$ energy), and the depth, E_d, of the basin at three values of ρ (*cf.* ref. 58).

Sato[86] has pointed out that not only does the LEP surface possess the basin at $r_{AB} = r_{BC} = 0.8$ A, but that it possesses a further basin at $r_{AB} = r_{BC} = 0.3$ A (which at $\rho = 0.12$, for example, is 75 kcal. mole^{-1} deep!), and he has proposed modifications to the LEP method to remove these features. Sato's technique was to obtain separated values for Coulombic and exchange components such as Q_A and α by taking the energy of the attractive H_2 state as $Q_A + \alpha$ and that of the

TABLE 5

HEIGHT OF ENERGY BARRIER (E_c) AND DEPTH OF BASIN (E_d) ON THE SEMI-EMPIRICAL
LEP SURFACE AT THREE RATIOS OF COULOMBIC TO TOTAL ENERGY

ρ	E_c	E_d
	(kcal. mole^{-1})	
0.07	20	0
0.14	14	2–3
0.20	7	5–6

repulsive state as $Q_A - \alpha$. Since the forms of the attractive and repulsive energy
curves are well known for H_2 it is possible to write out the London equation (134)
without reference to the arbitrary parameter ρ. The valence-bond energies for
the hydrogen molecule are

$$\frac{Q_A \pm \alpha}{1 + S_A^2} \qquad (136)$$

where S_A is the overlap integral for the pair of hydrogen atoms, and Sato suggested
that some allowance for the overlap integrals could be incorporated into the
London equation by writing

$$E = \frac{1}{1 + S^2} \{ Q_A + Q_B + Q_C \pm [\tfrac{1}{2}\{(\alpha - \beta)^2 + (\beta - \gamma)^2 + (\alpha - \gamma)^2\}^{\tfrac{1}{2}}]\} \qquad (137)$$

for the energy of the H_3 species. The parameter S is now to be adjusted to give
the best fit with experimental data. Table 6 shows the barrier heights calculated
by both Eyring and Sato procedures for four hydrogen atom abstraction reactions.
The figures are those given in Sato's paper[86], and the Sato calculations are based
on $S^2 = 0.18$. The experimental activation energies quoted should presumably

TABLE 6

EXPERIMENTAL AND CALCULATED ACTIVATION ENERGIES FOR SOME HYDROGEN AB-
STRACTION REACTIONS[86]

Reaction	Activation energy (kcal. mole^{-1})		
	Experimental	Eyring	Sato
$H + H_2 \rightarrow H_2 + H$	5–7	7.9	5.4
$H + HCl \rightarrow H_2 + Cl$	4.5	11.4	4.7
$H + HBr \rightarrow H_2 + Br$	1.2	10.4	0.7
$H + HI \rightarrow H_2 + I$	1.5	7.7	0.3

be corrected for the difference in zero-point energy between reactants and activated complex. For $S^2 = 0.18$ the potential surface has no basin at the configuration corresponding to the transition state, and in this respect is more in keeping with present experimental data than the LEP surface. However, Weston[87] has criticised the Sato surface on the grounds that there is no justification for using a constant value of S^2, especially when the value chosen is about one-third of the true value at the transition state configuration. Weston's further criticism that the Sato barrier is too thin and therefore predicts a high degree of quantum-mechanical tunnelling may be less cogent now that tunnelling effects have been recognised in the $H + H_2$ reaction[33].

Polanyi et al.[88] have introduced even more empiricism into the LEP calculations by allowing the three overlap integrals to be variable. They write their expression

$$E = \frac{Q_A}{x} + \frac{Q_B}{y} + \frac{Q_C}{z} + \left\{ \frac{\alpha^2}{x^2} + \frac{\beta^2}{y^2} + \frac{\gamma^2}{z^2} - \frac{\alpha\beta}{xy} - \frac{\beta\gamma}{yz} - \frac{\alpha\gamma}{xz} \right\}^{\frac{1}{2}} \qquad (138)$$

where x, y, z are $1 + S_A^2$, $1 + S_B^2$ and $1 + S_C^2$. The overlap parameters have been adjusted to give surfaces for kinematic calculation of energy distributions of products in exothermic reactions of the $A + BC \rightarrow AB + C$ type, and the predictions correlated with experimental results of the product energy distribution (see section 4.5.5).

The kinematic calculations of Karplus et al.[43] have already been described (section 3.7). The potential energy surface for H_3 used by these workers was based

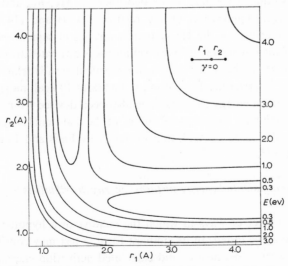

Fig. 6. Potential energy surface for linear symmetrical H_3 calculated by Porter and Karplus[89].

on a modified LEP surface devised by Porter and Karplus[89]. A combination of theoretical and semi-empirical treatments was used. The Coulombic term was taken to be $Q_A + Q_B + Q_C$ as before, but the exchange integrals were assumed to have the form $\alpha + \Delta$, where Δ is a residual term. A refined calculation from attractive and repulsive H_2 energy curves gave values for Q_A, Q_B, Q_C and α, β, γ. Overlap integrals and Δ were calculated from $1s$ atomic orbitals, as was a term for the exchange over the three atoms. The most favourable configuration for the activated complex was found to be linear and symmetrical with $r_{AB} = r_{BC} = 0.90$ A. The barrier height under these conditions is 8.58 kcal. mole^{-1}, in good agreement with experimental data. Fig. 6 shows the surface for linear, symmetric H_3 as calculated by Porter and Karplus[89].

4.5.4 Empirical potential energy surfaces

Johnston and Parr[90] have developed Johnston's Bond Energy Bond-Order (BEBO) method (see Johnston, Bibliography), which relies on two empirical rules relating bond length, r, and bond energy, D, for a bond of order n to the length r_s and energy D_s for the single bond. The relations are

$$r = r_s - 0.26 \ln n \tag{139}$$

and

$$D = D_s n^p \tag{140}$$

where p is a constant. For the $H + H_2$ reaction it is assumed that the *total* bond order is unity, so that the orders of the two bonds in H_3, n_{AB} and n_{BC}, are such that $n_{BC} = 1 - n_{AB}$. Application of the relations (139) and (140) now allows evaluation of the energy of the H_3 system at any set of separations (a term for the repulsion of the outer pair of hydrogen atoms is estimated from the repulsive energy curve for H_2). The result of this simple calculation is relatively good: for the H_3 system, the barrier height is found to be about 10 kcal. mole^{-1}.

Empirical potential energy surfaces have also been devised from manipulations of the Morse function for the potential energy of the diatomic H_2 system. Surfaces of this kind are described by Wall and Porter[91], by Blais and Bunker[92] and by Karplus and Raff[93].

4.5.5 Tests of potential energy surfaces: energy distributions in products

There are a number of ways in which the validity of a hypothetical potential energy surface may be tested. In its simplest form, a test may consist merely of a comparison between experimental and calculated activation energies and, as we have seen in section 4.5.2, the agreement even at this level between theory and experiment is not good. An alternative approach is to adjust the parameters of a semi-empirical surface to give a good agreement for barrier energy, and to test some other property. The barrier shape and thickness will affect the predicted rate of quantum mechanical tunnelling, while the form of the whole surface will affect the overall rate of reaction predicted by kinematic or generalised activated complex procedures. Unfortunately, however, the experimental results themselves are not sufficiently unequivocal to allow definite testing of the potential energy surface. On the other hand, investigations into the energy distribution in *product* species have recently proved most fruitful in attempts to test potential surfaces. Spectroscopic techniques used under proper conditions are able to determine accurately the energies of product molecules as they are formed, and velocity analysis techniques in molecular beam experiments can also yield precise information about the energy distribution of products. This section deals, therefore, with tests of potential energy surfaces based on determinations of excitation in products. The other methods whereby a potential energy surface may be tested are discussed implicitly in sections 3.7, 4.4 and 4.5.6.

The "atomic flame" reactions of sodium with the halogens were the earliest experiments in which the excitation of products was used to infer the nature of the potential energy surface for reaction. Evans and Polanyi[94] studied the chemiluminescent emission of sodium resonance radiation in the reaction of sodium vapour with iodine. The processes leading to excitation were suggested to be

$$Na + I_2 \rightarrow NaI + I \tag{141}$$

$$I + Na_2 \rightarrow NaI^\dagger + Na \tag{142}$$

$$NaI^\dagger + Na \rightarrow NaI + Na^* \tag{143}$$

($NaI^\dagger$ represents a molecule of *vibrationally* excited sodium iodide, while Na^* is an *electronically* excited, ^2P sodium atom). Although the nature of the energy transfer process (143) was not clear at the time (and has only recently been elucidated[95]) it was apparent that both (142) and (143) were highly efficient. Reaction (142) is only 1 kcal. mole^{-1} more exothermic than the energy needed to excite sodium resonance radiation. It follows, therefore, that not only must (143) occur without loss of energy, but also that all the energy liberated in (142) must go into vibration of the newly formed Na–I bond. In order that there shall be no loss of energy in rotational excitation, the activated complex must be a colinear system; and, further, there must be virtually no repulsion between the newly formed particles in the final state of the activated complex as otherwise translational excitation will result. Evans and Polanyi[94] showed what kinds of potential energy surface met these conditions, and a calculation of the surface for the reaction with atomic *chlorine* demonstrated the absence of repulsion as the activated complex moved over to products. More recently, crossed molecular beam experiments have confirmed this kind of behaviour in a more direct manner for the reaction of alkali metals with halogens. Datz and Minturn[96], for example, conclude from the scattering pattern from crossed beams of $Cs + Br_2$ that about 99 % of the energy released in the reaction goes into internal excitation of CsBr. The reaction is said to proceed by a "stripping" mechanism in which the Br atoms separate immediately after the Cs atom comes within its range of interaction; the momentum of the CsBr product is therefore just the sum of the initial momenta of the Cs and one of the Br atoms. It is not known how the internal energy is apportioned between vibration and rotation.

Polanyi *et al.*[88, 97] have categorised potential energy surfaces for exothermic exchange reactions of the type $A + BC \rightarrow AB + C$ into those which result in (*i*) *attractive* energy release (in which the atom A approaches BC at the ordinary BC separation), (*ii*) *repulsive* energy release (in which AB at its normal separation retreats from C) and (*iii*) *mixed* energy release (an intermediate case in which A approaches BC while BC extends). The reaction of alkali metals with the halogens clearly belong to the class of processes for which there is an attractive energy release. The degree of excitation has been put on a much more quantitative footing as a result of the kinematic computer calculations[88]. Eight potential energy hypersurfaces were considered covering the range of possibilities between purely repulsive and largely attractive behaviour. The surfaces themselves were derived from the modified LEPS surface expressed by equation (138). Product energy distributions (with respect to vibration, rotation and translation) were obtained as a function of initial position, impact parameter and kinetic energy for all eight combinations of light and heavy masses for A, B and C. Polanyi *et al.*[98–100] obtained experimental data on the vibrational excitation of product hydrogen chloride in the reaction

$$H + Cl_2 \rightarrow HCl^\dagger + Cl \tag{144}$$

from observation of the infrared chemiluminescence from the excited $HCl^\dagger$. The reaction is about 45 kcal. mole^{-1} exothermic and the activation energy is about 2.5 kcal. mole^{-1}, so that there is 47.5 kcal. mole^{-1} liberated after the barrier. This is very approximately the energy difference between the zero point and $v = 6$ for HCl. Excitation of levels up to the sixth does occur, and the probability of formation of a product molecule in any given level relative to $v = 3$ was measured[100]. The results are given in Table 7. Absolute intensity measurements suggest that about 7 % of the total energy available is converted into vibration. A high degree of rotational excitation was observed[99] and it would appear that the balance of the total energy is converted into rotation and translation. The type of behaviour observed for the $H + Cl_2$ reaction therefore corresponds to largely repulsive behaviour. Kuntz *et al.*[88] quote unpublished experimental results of Airey, Pacey and Polanyi for the $H + Br_2$ reaction which indicate a similar inefficiency for vibrational excitation in product HBr.

TABLE 7

RELATIVE PROBABILITY OF PRODUCT FORMATION IN DIFFERENT VIBRATIONAL LEVELS
FOR THE REACTION $H + Cl_2 \rightarrow HCl^\dagger + Cl$ (from reference 100)

Vibrational level	Probability
0	20
1	3
2	2
3	1
4	0.3
5	0.04
6	0.004

The experimental results quoted above indicate that a large fraction of the exothermicity of reaction is converted into internal excitation for the reactions of metals with halogens (e.g. $Na + I_2$, $Cs + Br_2$), and it would appear from the results of Evans and Polanyi[94] that this internal energy is largely vibrational (the molecular beam experiments[96] do not clearly show how much of the internal excitation is vibrational and how much rotational). On the other hand, only a small proportion of the available energy appears as vibration in the reaction of atomic hydrogen with halogens (e.g. $H + Cl_2$, $H + Br_2$). Kuntz et al.[88] showed that there was a tendency for the exchange reactions to give less vibrational excitation of products on surfaces with appreciable repulsive character if the attacking atom, A, were light. The greatest variation in the vibrational excitation with the nature of the potential energy surface was found for the attack of a light atom on a molecule BC in which both B and C were heavy. The final conclusion was that a surface with a large proportion of repulsive character provided the most likely qualitative explanation for the relatively low efficiency of vibrational excitation in the hydrogen atom–halogen systems.

A number of other reactions are known to give vibrationally excited products: for example, the processes

$$O(^1D) + O_3 \rightarrow O_2^\dagger + O_2 \qquad \text{(ref. 101)} \qquad\qquad (145)$$

$$Cl + NOCl \rightarrow Cl_2^\dagger + NO \qquad \text{(ref. 102)} \qquad\qquad (146)$$

$$H + O_3 \qquad \rightarrow OH^\dagger + O_2 \qquad \text{(ref. 103)} \qquad\qquad (147)$$

$$H + NOCl \rightarrow HCl^\dagger + NO \qquad \text{(ref. 104)} \qquad\qquad (148)$$

all give rise to vibrationally excited products, and in the latter two cases infrared chemiluminescence may be detected. The total yields of vibrationally excited products is in doubt, especially in the first two reactions, and detailed kinematic calculations for systems of the type $A + BCD \rightarrow AB + CD$ appear not to have been made so far. One general conclusion which seems to be valid for this kind of reaction is that such vibrational excitation as there is appears in the newly formed A–B bond, and not in the fragment CD.

The remaining tests of potential energy surfaces to be described here use experimental data from molecular beam studies. Reference has been made already (section 4.3.1) to the scattering observed in the $D + H_2$[54] and $M + RI$[52, 53] reactions, and to the inference drawn that the activated complex is short lived. The reaction of potassium with methyl iodide has also been the subject of theoretical calculation by both Blais and Bunker[92] and by Karplus and Raff[93, 105]. Although Blais and Bunker restricted their calculations to two dimensions, they were able to show that distributions of product scattering angle were peaked such that the product KI was emitted predominantly backward and the CH_3 forward with respect to the initial K direction. Furthermore, for an assumed potential energy surface of reasonable shape, most of the energy of reaction should go into internal vibrational and rotational excitation of KI. These results were found to be relatively insensitive to the presence or absence of rotation or vibration of CH_3I before collision and to minor changes in the assumed potential of interaction. The rotational energy distribution in KI corresponded closely to complete conversion of the initial angular momentum to product rotation. Karplus and Raff[93] subsequently extended the calculations to three dimensions and found that the potential assumed by Blais and Bunker, although giving results apparently in concordance with the observed energy distribution in the products, nevertheless led to significant disagreement with the experimental results for the total and differential reaction cross-section. It was found that the over-estimate of the long-range interaction between the reactants was the probable source of the error, and that introduction of an appropriate three-body attenuation term into the expression for the potential led to reasonable results. The three dimensional calculations have been followed up by Karplus and Raff[105] with the primary object of obtaining an understanding of the limitations on the form of the potential energy surface provided by the available experimental information. Four potential energy surfaces were examined: (i) a modified "Blais–Bunker" surface, (ii) a surface similar to (i) in which the three-body attenuation term was added, (iii) a surface similar to (i) in which the $K–CH_3$ repulsive term was replaced by an "exponential-six" interaction with the centre of mass of CH_3I as origin, and (iv) a special surface which was an artificial representation of the three-body potential for the (K, I, CH_3) system as a superposition of covalent and ionic structures. Comparisons were made of the total reaction cross-section, the rotational and vibrational state distribution of the product molecule, the reaction energy distribution, the differential cross-section and several further parameters. Surfaces (i) to (iii) all predicted that a large proportion of the energy of reaction would appear as internal excitation of the newly formed KI molecule, and that the centre-of-mass angular distribution for KI would be strongly peaked in the backward direction, although there were some quantitative differences between the results using the various surfaces. The artificial surface, (iv), disagreed with the other three in that it predicted that half of the energy of reaction should appear as kinetic energy. The first three surfaces gave results which suggested that the replacement of a potassium atom by an atom of a heavier alkali metal should result in the appearance of a larger fraction of the reaction energy in internal degrees of freedom, while a corresponding increase in the mass of the alkyl group would produce little or no kinematic effect.

The experimental results with which the predictions of Blais and Bunker, or Karplus and Raff, are to be compared are those of Herschbach and his collaborators. The results are summarised in reference 52 for a series of reactions $M + RI$ where $M = K$, Rb and Cs and $R = CH_3$, C_2H_5, etc. The general conclusions are very similar to the theoretical predictions outlined in the preceding paragraph, the main difference being that the experimental results suggest that the translational energy decreases with an increase in mass of the alkyl group (an effect not predicted by the kinematic considerations).

Two further exothermic reactions of atomic potassium have been studied by molecular beam techniques with the object of testing the results against hypothetical potential energy surfaces. These reactions are

$$K + HBr \rightarrow KBr + H \tag{149}$$

and

$$K + Br_2 \rightarrow KBr + Br \tag{150}$$

The internal energy of the products of the reaction with HBr was recently determined from a product recoil velocity analysis by Grosser et al.[106]. At two initial relative kinetic energies, it was found that the difference between initial and final relative kinetic energies was small, so that most of the heat of reaction must go into internal excitation of the newly formed KBr. The observed angular and velocity distribution of KBr implied that the reactive scattering in the centre-of-mass system was confined predominantly to low angles: that is, the product KBr "preferred" the direction of the incident K atom. However, further interpretation of the results was limited as a result of unfavourable kinematic factors. Experimental information for the $K + Br_2$ reaction is relatively extensive[107, 108]. Warnock et al.[109] have recently reported a computer analysis of the product velocity distribution experiments, and give references to previous work on the reaction. Studies have been carried out with both energy-selected[107] and thermal[108] beams, and suggest that most of the energy of reaction appears as internal excitation of the product. The technique used by Warnock et al. to compare the experimental velocity analyses with predictions based on calculations from hypothetical potential energy surfaces was to use a process of "inversion" of the data. Assumed functional forms for the differential reaction cross-section and product internal excitation function were used to compute contour maps of relative number density or flux; the experimental velocity analyses were then compared with cuts taken through the contour map at various laboratory scattering angles. The "best" trial distribution functions were obtained by iterative computation, and the final forms compared with theoretical distribution functions derived from trajectory analyses using semi-empirical potential energy surfaces. The results obtained were in fair accord with distribution functions obtained by Karplus and by Polanyi (unpublished) for certain hypothetical surfaces.

It would seem apparent that considerable advances will be made in the near future by studies of the kind outlined above. With sufficient refinement, studies of the product energy distributions may eventually allow rather complete description of the potential energy surface for an elementary chemical reaction.

4.5.6 Quantum mechanical tunnelling

In classical mechanics, the probability that a particle can overcome an energy barrier is zero if the particle has less energy than the barrier height, and unity if it has more. However, in quantum mechanics there is a finite probability that particles with energy less than the barrier height will appear on the other side of it. A wave reaching a barrier will be attenuated by the barrier; the wave will emerge on the far side of the barrier with its original period but with reduced amplitude. The largest deviations from classical behaviour are to be expected for particles of low mass, and it is well known that the motions of electrons cannot be treated classically. Bell suggested[110] that as a result of their relatively small mass, protons might also be subject to non-classical behaviour. At room temperature, protons moving with thermal velocities have De Broglie wavelengths of 0.1 to 1.0 A associated with them: barriers to chemical reaction probably have widths of the order of a few Angstroms so that proton tunnelling might be of importance. To a lesser extent, heavier particles may also tunnel through energy barriers although the expected effect becomes increasingly small as the particle mass increases.

The quantum mechanical versions of the trajectory studies (insofar as "trajectory" has a quantum mechanical meaning) presented in sections 3.7 and 4.4 make allowances for quantum mechanical tunnelling, and the rates of reaction predicted by such calculations need no further correction for non-classical behaviour. However, the calculations have been made only for a few, rather specific, reactions, and their validity depends on the form adopted for the potential energy surface on which the trajectory analysis is made. It has been more usual to apply a correction to the "classical" reaction rate which allows for tunnelling, and the form of this correction is what concerns us in this section. Most of the treatments which have been adopted involve a calculation of the tunnelling correction for a one-dimensional barrier of more or less artificial shape. A problem arises in this case, since there are a number of close-lying alternative paths on the reaction surface, each of which is probably curved, and tunnelling may occur from any of them. Both Johnston and Rapp[31] and Mortensen and Pitzer[48] have carried out multi-dimensional tunnelling studies which suggest that the one-dimensional approach is inadequate; unfortunately Mortensen and Pitzer give evidence that the one-dimensional treatment *under*estimates the extent of tunnelling, while Johnston and Rapp suggest that it *over*estimates the importance of tunnelling. (For some comments about multidimensional reaction paths by Marcus[72] see the final paragraph of section 4.5.8.) A further criticism may be levelled at the usual method by which correction factors are computed, and that is the assumption of a thermal equilibrium distribution of energies in the particles reaching the barrier. For those systems in which equilibrium cannot be assumed, the special statistical or kinematic treatments must be used.

If the "permeability" of the barrier (the probability that a particle will cross the barrier) is given the symbol $G(W)$ where the variable W is the energy of the particle, then for a system of particles possessing a Boltzmann energy distribution, the tunnel correction Q_t is given by[111]

$$Q_t = e^{-\varepsilon_c/kT} \int_0^\infty \frac{1}{kT} e^{-W/kT} \cdot G(W) dW \tag{151}$$

Q_t is to be multiplied by the expression for the reaction rate to give the true rate. Bell[112] has given a set of expressions for a wide range of parabolic barriers. An analytical expression is derived for a simple parabolic barrier of half-width a and height E_c: for a particle of mass m the tunnelling correction is given by

$$Q_t = \tfrac{1}{2}u_t/\sin \tfrac{1}{2}u_t, \quad \text{where} \quad u_t = \frac{h}{a\pi kT}\left(\frac{\varepsilon_c}{2m}\right)^{\frac{1}{2}} \tag{152}$$

If tunnelling were the *only* cause of temperature dependence of the Arrhenius parameters, A and E_a, then from (152) the relations

$$E_a = E_c - RT\{1 - \tfrac{1}{2}u_t \cot \tfrac{1}{2}u_t\} \tag{153}$$

and

$$A = A_c \cdot \frac{\tfrac{1}{2}u_t}{\sin \tfrac{1}{2}u_t} \exp \{\tfrac{1}{2}u_t \cot \tfrac{1}{2}u_t - 1\} \tag{154}$$

may be derived, where E_c is the (molar) classical barrier height and A_c the classical pre-exponential factor. Wigner[113] has also given an approximate expression for the tunnelling correction for a parabolic barrier in terms of the imaginary frequency, v_i, of vibration at the top of the barrier

$$Q_t = 1 + \frac{1}{24}\left(\frac{hv_i}{kT}\right)^2 \tag{155}$$

This is the first term of the more exact expression derived by Bell. Eckart[114], as long ago as 1930, devised an artificial barrier which may correspond more closely than the parabolic barrier to the actual shape of the potential barrier in a classical reaction. Fig. 7 shows the form of the Eckart barrier, and the potential energy, V, may be expressed by

$$V = \frac{Ce^u}{1+e^u} + \frac{De^u}{(1+e^u)^2}\,;\qquad u = 2\pi x/L \tag{156}$$

where

$$C = E_c - E_1\,;\qquad D = (E_c^{\frac{1}{2}} + E_1^{\frac{1}{2}})^2;\qquad L = \frac{2\sqrt{2\pi}}{E_c^{-\frac{1}{2}} + E_1^{-\frac{1}{2}}}\left(\frac{\partial^2 V}{\partial x^2}\right)^{-\frac{1}{2}}.$$

E_c and E_1 are identified in Fig. 7, and x is the reaction co-ordinate. $(\partial^2 V/\partial x^2)$ is calculable from the form of the potential surface adopted, and substitution of the Eckart expression for the potential energy into the Schrödinger equation allows values to be obtained for Q_t. Johnston et al.[31,115], Shavitt[116] and Shin[117] have given values of Q_t for a number of different parabolic and Eckart barriers.

The best-known example of a tunnelling process in the gas-phase is the proton tunnelling in ammonia[118], which gives rise to the inversion doubling observed in some of the lines of the ammonia vibrational spectrum. A double-minimum potential field exists, with a barrier through which tunnelling can occur to cause splitting of the vibrational levels. The accuracy with which energies can be measured by spectroscopic means is, of course, far greater than the secondhand evidence that can be obtained from rates of chemical reactions. In any case, the complete lack

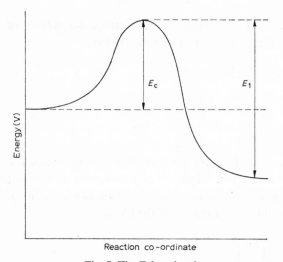

Fig. 7. The Eckart barrier.

of reliance that can be placed on theoretically predicted barrier heights shows that it is not possible to test a tunnelling correction by comparison of a measured rate and the "classical" predication. Experimental tests of tunnelling theories therefore tend to fall into two classes: experiments in which a search is made for curvature in an Arrhenius plot, and experiments in which the rates of processes involving similar reactants are compared.

The general behaviour of an Arrhenius plot for a reaction in which there is quantum mechanical tunnelling was discussed in section 3.6.4 in terms of Fig. 2. All treatments of tunnelling give a correction factor whose importance increases at low temperatures [*e.g.* equations (152) and (155)], and it is at low temperatures, therefore, that curvature of the Arrhenius plot should be most apparent. Curvature of Arrhenius plots which must be ascribed to tunnelling is fairly well established in solution processes; the greater accuracy of many condensed phase observations and the lower temperatures at which reaction may be carried out both help to make the identification of tunnelling effects relatively unambiguous. On the other hand, curvature of Arrhenius plots for gas-phase experiments is much less easy to demonstrate. The *ortho–para* hydrogen conversion, $H+H_2 \rightarrow H_2+H$, is a reaction in which tunnelling would be expected to play a significant part since it involves the transfer of hydrogen atoms. However, until quite recently, the experimental data available showed no evidence of deviation from linear Arrhenius behaviour. Both Weston[87] and Shavitt[116] reviewed the evidence available in 1959. For example, in Weston's work, a parabola was fitted to the profile of a Sato-type potential energy surface, and the tunnelling correction calculated by Bell's method. The calculated correction term ranged from 52.6 at 295 °K to 1.23 at 1000 °K. Tunnelling of this magnitude would certainly give rise to curvature of the Arrhenius plot, and since the experimental results showed little signs of curvature it was concluded that either the parabolic correction terms gave an overestimate of the tunnelling, or that the barrier profile given by the Sato surface was too thin. In view of the claim by Mortensen and Pitzer[48] that the one-dimensional treatment underestimates the amount of tunnelling, the second explanation seemed more reasonable. The situation has altered to some extent, however, as a result of the recent re-examination of the $H+H_2$ kinetics by Schulz and Le-Roy[33]. In the temperature range 300 °K to 444 °K, these workers find a marked curvature of the Arrhenius plot, and they were able to fit the temperature dependence of the rate constant by assuming quantum mechanical tunnelling through a parabolic barrier of imaginary frequency, v_i, equal to 1213 cm^{-1} (more properly 1213 i cm^{-1}). The height of the barrier was 9.7 kcal. mole^{-1} measured between zero-point energy levels of reactants and activated complex, and if the tunnelling treatment is valid, it would seem that this value, corrected for the zero point energies, is the one to compare with predicted values of the barrier height E_c. One interesting point that Schulz and LeRoy make is that assumption of Eckart-type barriers could lead to linear Arrhenius behaviour even in the presence of significant tunnelling, so that linearity of an Arrhenius plot does not necessarily indicate that tunnelling is unimportant. (Conversely, of course, deviation from Arrhenius behaviour does not unequivocally show the presence of tunnelling, since a number of other causes—for example, anharmonicity or vibration–rotation interaction—could be responsible for the deviations.)

The second procedure that may be used to demonstrate the presence of tunnel effects is to compare the rates of similar reactions. In particular, the use of isotopically substituted reactants can lead to changes in reaction rate which can be associated in part with tunnelling. *Kinetic isotope effects* arise from a number of causes[119]. First, the zero-point energies of both reactants and activated complex will be changed by the substitution: except in rare cases these changes will not cancel out, so the barrier height will be altered by the substitution. Secondly, the component terms of the partition functions in the pre-exponential function will be changed. To some extent, since the changes in translational and rotational partition functions are in the same direction for both reactants and activated complex, the main effect on the pre-exponential term results from changes in the vibrational partition function. Finally, the extent to which tunnelling occurs will depend on the masses of the particles reacting. In all respects, hydrogen–deuterium substitution tends to have the largest effect, especially since tunnelling is likely to be most important in reactions involving light particles. The diagnostic tests for tunnelling are, qualitatively: (*i*) that the ratio of rate coefficients, say k_H/k_D, for the two processes is greater than the value calculated from classical theory, and it should be temperature dependent; (*ii*) the difference in experimental activation energies, $E_a^D - E_a^H$, is greater than the difference predicted on zero-point energy considerations alone; (*iii*) the ratio of experimental pre-exponential factors, A_D/A_H, should be much greater than unity—this is, perhaps, the best test for tunnelling since classical theory would usually give a value close to unity for the ratio.

Johnston *et al.*[32,120] have studied the reactivity of CF_3 with a number of hydrocarbons in an attempt to establish the validity of both potential energy surfaces and tunnelling corrections. For example, Sharp and Johnston[32] studied the pair of reactions

$$CF_3 + CHD_3 \rightarrow CF_3H + CD_3 \tag{157}$$

and

$$CF_3 + CHD_3 \rightarrow CF_3D + CHD_2 \tag{158}$$

A Sato surface was constructed, and the value of S^2 [equation (137)] was estimated by fitting it to give the correct rate for the exchange reaction

$$CH_3 + CH_4 \rightarrow CH_4 + CH_3 \tag{159}$$

The variation of the ratio of rate coefficients for (157) and (158) was found to be quite different from that expected in the absence of tunnelling. By using the Johnston and Rapp method[31] for calculation of the tunnelling correction, good agreement was obtained with the experimental ratio of rate coefficients, on condition that a complete vibrational analysis of the activated complex was employed which involved the variation of potential energy with more co-ordinates than just the reaction co-ordinate. One-dimensional calculations did not give such good agreement with experiment as the multidimensional ones.

The reactions of halogen atoms with hydrogen or deuterium have also been used in studies of tunnelling in gas-phase reactions[121-125]. In the latest series of experiments[124] tritium has also been used, and the rates of reaction of Cl atoms with H_2, HD, D_2, HT, DT and T_2 were measured over the temperature range $-30\ °C$ to $+70\ °C$. The isotope effects observed could only be explained if tunnelling were present, and it was found that the best agreement (within 15 %) of the calculated values with experiment was obtained by the use of a generalised Sato model for the potential together with the Johnston–Rapp treatment[31] of tunnelling through an asymmetrical barrier. Empirical sets of four force constants describing the transition state H–H–Cl were also given, and their use in the calculation of isotope effects gave excellent agreement with experimental values. In most cases a multidimensional treatment of tunnelling gives better correlation with experimental results than does the one-dimensional calculation. However, in experiments on the pair of reactions[125]

$$*Cl + HCl \rightarrow H*Cl + Cl \tag{160}$$

$$*Cl + DCl \rightarrow D*Cl + Cl \tag{161}$$

(*Cl is a radioactive chlorine atom), although tunnelling was important at the low temperatures used, the Johnston–Rapp treatment was unable to give corrections which accounted for the experimental results. On the other hand, the Bell correction gave satisfactory correction terms for a number of possible barrier profiles.

The discussion of this section will have made it apparent that what are sometimes known as "Tests of the Transition State Theory" may in fact be testing the potential energy surface used, or the form of the tunnelling corrections applied. Orders of magnitude of frequency factors have been calculated a number of times, and, if the vibration frequencies and geometries of activated complexes are sufficiently carefully chosen, good agreement is obtained with experiment (see, for example, the calculations of Herschbach et al.[126]). On the other hand, attempts to calculate activation energies and the exact form of the variation of rate coefficient with temperature have been rather less successful.

4.5.7 The transmission coefficient, κ

The transmission coefficient, κ, is the probability that a system which has reached the transition state will continue to the final state. It is unity if systems cross a classical barrier only once in their passage from initial to final states; in the general case, some of the systems which cross the transition state will cross it again in the opposite direction. In classical theory, κ is near unity at low temperatures as it is very improbable that a system with barely enough energy to cross the barrier will find its way back[127]. At higher temperatures, however, there may be a decrease in the magnitude of κ. Departures of κ from unity are to be expected as a result of quantum mechanical effects. Quantum mechanical tunnelling can give rise to a situation in which the rate of reaction is *apparently* greater than the rate at which activated complexes are formed at the top of the classical energy barrier. For systems in which tunnelling is important, therefore, a transmission coefficient greater than unity can be obtained. The correction Q_t introduced in section 4.5.6 allows for tunnelling effects, so for the purposes of the present section, κ refers to a transmission coefficient which excludes tunnelling corrections. A particle which has wave properties associated with it may be reflected both on its way up and on its way down a potential barrier separating final and initial states, and κ can be less than unity from this cause. The generalised treatments of reaction rates (sections 3.7 and 4.4) should allow for these reflections, and the rate coefficients predicted from the most complete of these treatments need no further adjustment. For the ordinary statistical-thermodynamic formulation of the equilibrium transition state theory the magnitude of κ is still required. Hirschfelder and Wigner[128] first showed that the transmission coefficient is a rapidly fluctuating function of the energy of the system, so that if the thermal energy spread is sufficiently great to cover several periods of the fluctuation, an average transmission coefficient may be defined which agrees closely with the classical value. For a one-dimensional barrier the corrections are small, although in a multidimensional system the transmission coefficient is affected by the interchange of vibrational and translational energy. It was concluded that for a large number of reactions involving species in their ground electronic states, κ is close to unity.

In section 3.4, two special causes of small P factors were discussed. The reactions concerned were (*i*) the bimolecular recombination of atoms and (*ii*) non-adiabatic reactions, and the small P factor can be identified in each case with a low probability of the activated complex reaching the final state, rather than a low probability that the activated complex is formed. Examples of reactions were given which were electronically or vibrationally non-adiabatic. It is proposed to discuss further some considerations respecting electronically non-adiabatic processes. The confusion about terminology pointed out in section 3.4 will be avoided here by not using the term "non-adiabatic", and instead referring to the actual conditions which give rise to high or low values of κ.

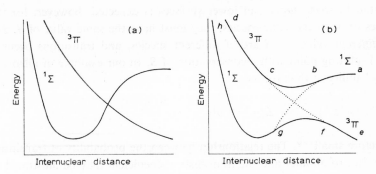

Fig. 8. Crossing potential energy curves.

The representation of the "crossing" of surfaces is difficult in two dimensions, and an example is given here of the "crossing" of two potential energy curves (which may be a section through a surface). Fig. 8 shows a pair of curves: they are designated $^3\Pi$ and $^1\Sigma$, and in Fig. 8(a) they are represented as zero-order crossing curves, while in Fig. 8(b) they are represented as a pair of higher order non-crossing curves. The important point is that if the curves are regarded as non-crossing, then the spectroscopic designation of the curves changes at the point which corresponds with crossing in Fig. 8(a).

An actual example of a system with curves similar to those in Fig. 8 is nitrous oxide. Ground state $N_2O(^1\Sigma^+)$ correlates with ground state $N_2(^1\Sigma_g^+)$ and excited atomic oxygen, $O(^1D)$, while the $^3\Pi$ repulsive state of N_2O correlates with $O(^3P)$ and $N_2(^1\Sigma_g^+)$.

The extent of the separation of the two curves at the "crossing" point in Fig. 8(b) depends on the symmetry properties of the two states. If the states are of the same "species" (e.g. they have the same values of Λ and S), then the resonance separation, equal to the matrix element $\oint \psi_1 H \psi_2 d\tau$ where ψ_1, ψ_2 are the wavefunctions for the two states, is large (several kcal. mole^{-1}). On the other hand, if the states are of different species, as they are in Figure 8, then the resonance splitting is small. For a small resonance splitting, the probability of transition in either direction between upper and lower surfaces is near unity, and the curves behave as though they really crossed. Conversely, the probability of transition between curves separated by a large resonance splitting is small. It is found that for states of different symmetry species, the resonance splitting depends on spin–spin and spin–orbital interactions, and it is possible to derive selection rules for transitions between two crossing curves or surfaces[129]. For example, where the quantum numbers Λ and S have a meaning, $\Delta\Lambda = 0, \pm 1$ and $\Delta S = 0$ are selection rules for radiationless transitions. One aspect of these considerations is that reaction systems for which there are "crossing" potential energy surfaces tend to resist any change in S: this is the Wigner spin-conservation rule[19]. If the two potential surfaces are of the same species the resonance splitting is large and no transition between upper and lower surfaces is expected; however, for the two surfaces to be of the same species, they must have the same value of S. Surfaces with *different* values of S are of different species, and transitions occur at the (small) splitting point with a conservation of S. In our example of nitrous oxide, the probability of the process

$$N_2O(^1\Sigma^+) \rightarrow N_2(^1\Sigma_g^+) + O(^3P) \tag{162}$$

is therefore small[130]. The relationship between the probability of transition from one surface to another and the transmission coefficient may be illustrated for the quenching process

$$O(^1D) + N_2(^1\Sigma_g^+) \rightarrow O(^3P) + N_2(^1\Sigma_g^+) \tag{163}$$

If we take Fig. 8 to refer to nitrous oxide, then there are two ways in which (163) can occur *via* N_2O. The point a in Fig. 8(b) corresponds to the reactants, and the point e to the products. Point e can be reached (*i*) by the system remaining on *abcd* during a vibration and crossing from c to f in a subsequent vibration, or (*ii*) by direct crossing from b to g and subsequent passing to e. If the probability of crossing is P, then the overall probability of (*i*) is the probability of first remaining on *abcd*, $(1-P)$, multiplied by the probability of crossing, P. Similarly, the probability of (*ii*) is also $P(1-P)$, so the overall probability of the quenching process (163) is $2P(1-P)$, and

$$\kappa = 2P(1-P) \tag{164}$$

An interesting inference to be drawn from the form of (164) is that quenching is unlikely either for very high or for very low values of P, and κ has a maximum of $\frac{1}{2}$ when $P = \frac{1}{2}$. The more complex relationships between κ and transition probabilities in systems where there are several potential energy surfaces have been discussed by Laidler[131] and by Magee[132].

The magnitude of P has been given by Landau[133] and Zener[134] for crossing between states of different species as

$$P = \exp[-4\pi^2\varepsilon_{12}/hV|s_1 - s_2|] \tag{165}$$

ε_{12} is the resonance splitting, V the velocity with which the system passes the point of closest approach of the surface, and $|s_1 - s_2|$ the magnitude of the difference of slopes with which the curves would cross in the zero-order approximation (*i.e.* the difference of slopes of cf and hg). No treatment seems to have been given for transitions between curves of the same species. The Landau–Zener equation, (165), indicates that $P = 1$ when $\varepsilon_{12} = 0$, and for systems in which ε_{12} is small, the probability that the system will *remain* on a given curve, P' $(= 1-P)$, is given by

$$P' = \frac{4\pi^2 \varepsilon_{12}^2}{hV(s_1 - s_2)} \tag{166}$$

The value of κ calculated[130] for reaction (162) is 10^{-4}.

4.5.8 The reaction co-ordinate

The use of a one-dimensional representation for the path of the reacting species over an energy barrier, such as the representation in Fig. 3, suggests that the reaction system may be treated as confined to motion along just the one path which has been called the reaction co-ordinate. Some of the discussion of earlier sections has indicated that the one-dimensional treatment is inadequate, particularly when the one-dimensional energy barriers are used to calculate tunnelling corrections. The nature of the reaction co-ordinate is required in the ordinary formulation of the transition state theory, since it determines which vibrational term is to be omitted from the vibrational partition function for the activated complex. For the purpose of obtaining a rough estimate of reaction rates the exact choice of reaction co-ordinate may not be of great importance, since vibrational stretching frequencies are mostly of the same order of magnitude. (Laidler and Polanyi[58] note that in *cis–trans* isomerizations, where the angle of twist is the reaction co-ordinate, the extraction of a normal vibrational frequency from the partition function calculation will give an erroneous rate coefficient.) The correct procedure for obtaining accurate values for rate coefficients is to attempt to identify the reaction co-ordinate with a normal mode of vibration: normal-mode analysis of the vibrations will then indicate the frequency of the vibration to be omitted. This situation exists, for example, in the $H + H_2$ reaction where linear H–H–H is thought to be the activated complex configuration. The H_3 complex has four normal modes of vibration: symmetric and asymmetric stretching and the doubly degenerate bending modes. The symmetrical nature of any reasonable potential energy surface for H_3 suggests at once that the reaction co-ordinate may be identified with the normal-mode asymmetric stretching vibration, ν_2. As the reactant H atom approaches colinearly with H_2, the most distant H atom of the molecule recedes. The actual frequencies of ν_3 and ν_1 can then be calculated by using force constants estimated from the shape of the potential energy surface around the top of the barrier along the chosen reaction co-ordinate. In the case of a polyatomic activated complex, if the reaction co-ordinate corresponds to a linear combination of several normal modes, then the frequencies of the vibrations to remain in the partition function are obtained from a normal-mode analysis with the combination held fixed. A rather different approach may be needed in cases where the reaction co-ordinate cannot be identified with a normal mode. For example, the "repulsive" energy surfaces, believed to be of importance in the reactions of atomic hydrogen with halogens (see section 4.5.6), are characterised by behaviour in which for a reaction $A+BC \rightarrow AB+C$, AB at its normal separation retreats from C. The reaction co-ordinate is therefore the B–C distance, and the analysis of frequencies must be carried out with the B–C distance held fixed. The remaining vibrations are the two bending modes and the motion of A along the line of centres in B–C.

The assumption that the potential energy is such as to permit separation of the variables r_{AB} and r_{BC} from all the others is an approximation. Marcus[72] has shown, however, that *in the immediate vicinity* of a potential energy maximum or of a saddle point, the major topographical features of a "non-separable" potential energy surface can be imitated by those of a surface permitting separation of variables. According to this local approximation, an orthogonal curvilinear co-ordinate system is introduced which permits separation of the variables, the origin of the co-ordinate system being selected so that there exists a pair of curvilinear co-ordinate curves intersecting, and cotangential with the paths of steepest ascent and descent, at the saddle point. The theory of reaction rates developed by Marcus allows the use of curvilinear reaction paths. The curvature of the separable curvilinear reaction co-ordinate is found to be less than that of the path of steepest ascent. In classical terms, this means that a trajectory passing through the saddle point proceeds along the sides of the valley, rather than following the path of steepest ascent, as a result of the centrifugal force experienced by the particle as it follows the curved path. (The centrifugal force is least at the saddle point where the velocity is smallest.) The effect occurs also in quantum mechanical formulations, and one of the consequences of the curvature of the reaction path is that the potential barrier is "thicker" than that predicted from the paths of steepest ascent, so some correction may be needed to tunnelling terms.

5. Unimolecular reactions

5.1 INTRODUCTION

A unimolecular reaction may be represented by the elementary process

$$A \rightarrow \text{Products} \tag{167}$$

and the rate of reaction in (167) will obey a first-order kinetic law

$$\frac{-d[A]}{dt} = k[A] \tag{168}$$

There are, however, a large number of reactions which are kinetically first-order under suitable experimental conditions but which are not elementary processes in the sense that only a single step is involved in the reaction. These reactions involve a unimolecular transformation in their final stage, and are always treated together with any discussion of unimolecular reactions.

For a unimolecular reaction to take place, the molecule which undergoes reaction must have sufficient energy to allow the necessary rearrangement of the chemical bonds. The required energy can derive from several sources: it may be energy received by the absorption of light or ionizing radiation (as in photochemical or radiochemical processes), or it may be the excess energy with which the molecule has been formed in an exothermic chemical reaction ("chemical activation"). The source of the energy in thermal reactions must be molecular collision. Lindemann[135] made the original suggestion that unimolecular reaction could take place under the influence of collisions, and Hinshelwood[136] showed that if molecular collisions could both activate and deactivate a reactant molecule, then collisional activation could be reconciled with first-order kinetic behaviour. This first theory of unimolecular reactions explained the more general kinetic features of experimental observations, although it did not allow for the varying rate of the unimolecular step with differing degrees of excitation of the reactant. Subsequent developments of the theory have been made in order to account for the more detailed behaviour of unimolecular reactions.

Kassel[137] and Rice and Ramsperger[138] extended the Lindemann–Hinshelwood theory for activated molecules of differing levels of excitation on the assumption that energy in the energised molecule could flow freely between some or all of the normal modes of excitation in the molecule. The initials of the proponents are often used to designate this theory—RRK (sometimes RRKH to include the parentage of the theory from Hinshelwood's earlier ideas). Slater[139] proposed an alternative theory of unimolecular reactions in which energy flow in the activated molecule was forbidden. Instead, the intrinsic rate of decomposition of activated molecules was predicted by superposition of the oscillator amplitudes on the assumption that the oscillators were strictly harmonic. That is to say, the activated molecule decomposes only when the phase relationships lead to sufficient distortion of the bond to allow the chemical transformation. The application of the transition state theory to unimolecular processes is subject to two major considerations. First, the equilibrium assumption may be quite invalid under low pressure conditions where the rate of reaction of activated molecules may be greater than the rate at which they can be activated by the collision process. Secondly, it is difficult to see exactly what is the transition state of the system. The potential energy of interaction of the two fragments of a decomposing molecule will not have a well-defined maximum (see, however, the discussion of *free-energy* maxima in section 4.3.1). Marcus[140, 141] has developed a quantum theory of unimolecular reactions which uses the methods of the transition state theory to extend the RRK approach.

5.2 THE LINDEMANN–HINSHELWOOD THEORY

The principle steps of activation, deactivation and unimolecular transformation of the Lindemann–Hinshelwood theory[135, 136] are

$$A + A \xrightarrow{k_1} A^* + A \qquad \text{(activation)} \tag{169}$$

$$A^* + A \xrightarrow{k_2} A + A \qquad \text{(deactivation)} \tag{170}$$

$$A^* \xrightarrow{k_3} \text{products} \qquad \text{(unimolecular reaction)} \tag{171}$$

Solution of the steady-state equations for $[A^*]$ gives

$$[A^*] = \frac{k_1[A]^2}{k_2[A] + k_3} \tag{172}$$

so that the rate of reaction is given by

$$-\frac{d[A]}{dt} = k_3[A^*] = \frac{k_1 k_3[A]^2}{k_2[A] + k_3} \tag{173}$$

At sufficiently high pressures, $k_2[A] \gg k_3$, and equation (173) becomes, approximately

$$-\frac{d[A]}{dt} = \frac{k_1 k_3}{k_2}[A] \tag{174}$$

The Lindemann–Hinshelwood mechanism therefore predicts first-order kinetics for thermal unimolecular reactions under conditions where the concentration of activated molecules is controlled by the gas-phase bimolecular activation–deactivation processes (169) and (170), rather than by the rate at which A* is removed in (171).

At low pressures, the concentration of A* will be determined by the rate of activation and the rate of unimolecular removal of A*. That is, $k_2[A] \ll k_3$, and

$$-\frac{d[A]}{dt} = k_1[A]^2 \tag{175}$$

Thus, under low pressure conditions, the rate of reaction is kinetically second-order in $[A]$, even though the step leading to loss of A is unimolecular. The explanation of first-order behaviour at high pressures, and the turning over to second-order kinetics at low pressures was one of the main achievements of the Lindemann–Hinshelwood theory. Experimental results were found to behave approximately in the manner predicted, and by observing the pressure at which first-order gave way to second-order behaviour, some estimate could be made of the relative magnitudes of k_2 and k_3. A "first-order" rate coefficient, k', may be written for the processes, such that

$$k' = -\frac{1}{[A]}\frac{d[A]}{dt} = \frac{k_1 k_3 [A]}{k_2 [A] + k_3} \tag{176}$$

although k' will be a function of [A]. At low pressures, indeed, k' will show an approximately linear dependence on [A], while at high pressures it will be nearly independent of [A]. As an example of the way in which experimental observations qualitatively support the predictions of the Lindemann–Hinshelwood scheme, Fig. 9 shows the variation of k' with pressure for some results obtained[142] for the reaction

$$CD_3 NC \rightarrow CD_3 CN \tag{177}$$

It is seen that the general features of the pressure dependence of k' are as anticipated, and that k' tends towards a limiting value at high pressures (the actual experimental results extend to pressures of several atmospheres, and k' remains virtually constant). At high pressures, therefore, the reaction is very nearly first-order.

The pre-exponential factors of k_1 and k_2 might be expected to be approximately dentical, so that the high pressure value of k' (which we will refer to as k_∞)

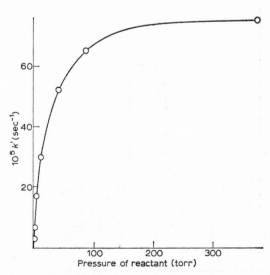

Fig. 9. The variation of k' with pressure in the isomerization of CD_3NC to CD_3CN[142].

should be consistent with an Arrhenius-type expression

$$k' = Ae^{-E_a/RT} \tag{178}$$

where A is the pre-exponential factor of k_3, and E_a a composite activation energy for k_1, k_2 and k_3. Although, in general, experimental rate determinations show, to a greater or lesser extent, deviations from the simple Arrhenius law, they can be fitted sufficiently closely to a linear plot to yield values of A. It has frequently been pointed out that the pre-exponential factor for unimolecular reaction has a tendency to lie near $10^{13.5}$ sec^{-1}, and frequency distribution curves or histograms have often been presented to bear out this contention. Fig. 10 is a histogram of numbers of reactions known to have A factors lying within various ranges of log $A = 0.5$, taken from recent tables of experimental A factors[143] for isomerization and fission reactions.

The pre-exponential factor for unimolecular reaction has a high probability of having a value close to the vibration frequency of the bond involved in the reaction (*i.e.* 10^{13}–10^{14} sec^{-1}), and this result is the one expected on the basis of either RRK or Slater theories, as will appear in due course. (The vibration frequency does not arise as such in the Lindemann–Hinshelwood formulation, although it was, of course, understood that k_3 had the dimensions of frequency.) The existence of reactions possessing A factors considerably in excess of vibration frequencies—there are, for example, 6 reactions with $A > 10^{17}$ sec^{-1} given by Benson and DeMore[143]—requires consideration. The explanation frequently suggested is that the ratio of A factors for k_1 and k_3 is *not* unity. The treatment

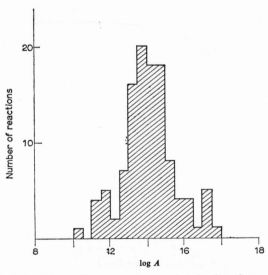

Fig. 10. Histogram of numbers of unimolecular reactions having given pre-exponential factors.

given in section 3.6.3 for activation in many degrees of freedom indicates that if a total energy greater than E_c can lead to excitation of A*, no matter where that energy is stored in A, then the rate of excitation will be

$$\frac{e^{-E_c/RT}\left(\dfrac{E_c}{RT}\right)^{\frac{1}{2}S-1}}{(\frac{1}{2}S-1)!} \cdot Z_{AA} \qquad (59)$$

Since the collision rate coefficients for (169) and (170) will be identical, the ratio of pre-exponential factors, A_1/A_2 will be given by

$$\frac{A_1}{A_2} = \frac{\left(\dfrac{E_c}{RT}\right)^{\frac{1}{2}S-1}}{(\frac{1}{2}S-1)!} \qquad (179)$$

As was pointed out in section 3.6.3 for a reaction with a reasonable activation energy (related to E_c), this ratio could reach 10^5 or 10^6 for 10 "square terms" (S). The absolute value of k_1 may be calculated from experimental results by inversion of (176), *viz.*

$$\frac{1}{k'} = \frac{k_2}{k_1 k_3} + \frac{1}{k_1[A]} \qquad (180)$$

A plot of $1/k'$ against $1/[A]$ should be linear and have a slope of k_1. In an early experiment on the isomerization of cyclopropane, Chambers and Kistiakowsky[144] calculated that to fit their values of k_1, S must be about 26 (*i.e.* 13 vibrations must be involved) on the basis of a collision rate estimated from a molecular diameter of 3.9 A for cyclopropane. Cyclopropane has 9 atoms in the molecule, which means that there are 21 vibrational modes. Between one-half and two-thirds of the total available number of vibrations must be excited, therefore, in order to account for the value of the observed rate coefficient, k_1. Although Chambers and Kistiakowsky fitted their results at different temperatures to a linear Arrhenius plot, later experiments on various unimolecular processes have been fitted to the temperature dependence of E_a suggested by equation (60) in section 3.6.3, using S as a variable parameter. In most cases the best agreement with experiment was obtained with $S/2$ set equal to between one-half and two-thirds of the number of vibrational degrees of freedom, thus bearing out in a general way the conclusions of Chambers and Kistiakowsky.

If, in fact, equation (180) represented adequately the variation of k' with [A], the value of $1/k'$ at $1/[A] = 0$ would give the ratio $k_2/k_3 k_1$, and since the value of k_1 is also known, a value for the relative efficiencies of quenching or reaction, k_2/k_3, could be obtained. Estimates of the quenching rate might be made from the

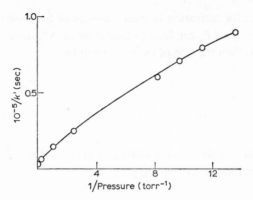

Fig. 11. Plot of $1/k'$ against inverse pressure for the isomerization of CD_3NC [142].

results of energy transfer experiments, and each rate coefficient would be known absolutely. How far equation (180) is an adequate representation is shown in Fig. 11.

Fig. 11 is a plot of $1/k'$ against 1/pressure for some of the experimental results[142] on the isomerization of CD_3NC to CD_3CN. Although over any small region, especially at high values of $1/p$, a straight line might be drawn through the points, it is seen that the plot is really curved. The curvature can be taken to mean that the value of the intercept for $1/p = 0$ varies with pressure, and is greatest at low pressures. This means that if k_2/k_1 is independent of pressure, k_3 *is a function of pressure*. The pressure dependence of k_3 reveals the weakness of the Lindemann–Hinshelwood theory. According to the reaction scheme (169) to (171), there are only two kinds of A molecule: those with and those without enough energy to undergo unimolecular reaction (171). No provision is made for a molecule with a considerable amount of energy beyond the critical amount to react more rapidly than a molecule with just barely the necessary energy. The inference to be drawn from Fig. 11, that k_3 is lower at lower pressures, is in accordance with an energy dependence of k_3. A more complete theory of unimolecular reactions will therefore differentiate between all levels of excitation of the reactant, and ascribe a separate value of k_3 to the reactivity of molecules in each level. In addition, the values of k_1 and k_2 should now be separated out for the production and deactivation of each level of excitation of A.

It is possible to derive an expression for the rate of reaction, using the idea that k_3 is energy dependent, which shows the same pressure dependence of kinetic order as that predicted by the simple Hinshelwood scheme. Let us suppose that there exists an active state of A, A_a^i, such that A_a^i has a non-zero chance of being converted into a "reactive" species A_i^* on collision. Then the rate of *formation* of A_i^* is given by

$$+ \frac{d[A_i^*]}{dt} = k_c[A][A_a^i] \tag{181a}$$

[k_c is here the "collision rate coefficient" $= Z_{AA_a}/[A][A_a] = 4\sigma^2(RT/\pi M)^{\frac{1}{2}}$].
If the cross-section, σ^2, for activation and deactivation is identical, then the rate
at which A_i^* is deactivated in reactions of type (170), $(-d[A_i^*]/dt)_d$, is given by

$$\left(-\frac{d[A_i^*]}{dt}\right)_d = k_c[A_i^*][A] \tag{181b}$$

while the rate at which A_i^* is removed by unimolecular reaction, $(-d[A_i^*]/dt)_r$,
is equal to

$$\left(-\frac{d[A_i^*]}{dt}\right)_r = k_3^i[A_i^*] \tag{182}$$

k_3 now being split into k_3^i for each energy level, E_i, of A_i^*. Solution of the steady
state equations for $[A_i^*]$ gives

$$[A_i^*] = \frac{k_c[A][A_c^i]}{k_c[A]+k_3^i} \tag{183}$$

so that the overall rate of reaction can be expressed by

$$-\frac{d[A]}{dt} = \sum_i k_3^i[A_i^*] = \sum_i \frac{k_c k_3^i[A][A_c^i]}{k_c[A]+k_3^i} \tag{184}$$

The value of $[A_c^i]$ may be calculated from the Maxwell–Boltzmann law if thermal
equilibrium may be assumed, viz.

$$[A_c^i] = [A]\frac{g_i e^{-E_i/RT}}{\sum_j g_j e^{-E_j/RT}} \tag{185}$$

For the high pressure condition, $k_3^i \ll k_c[A]$, equation (184) may therefore be
rewritten

$$-\frac{d[A]}{dt} = [A]\frac{\sum_i k_3^i g_i e^{-E_i/RT}}{\sum_j g_j e^{-E_j/RT}} \tag{186}$$

Similarly, at low pressures, where $k_3^i \gg k_c[A]$

$$-\frac{d[A]}{dt} = k_c[A]^2 \cdot \frac{\sum_i g_i e^{-E_i/RT}}{\sum_j g_j e^{-E_j/RT}} \tag{187}$$

The kinetic orders of (186) and (187) are thus the same as those of equations (174) and (175).

It is of interest to calculate the "experimental" activation energy for the high pressure case. We may write

$$
E_a = RT^2 \frac{d \ln k_\infty}{dT} = \frac{\sum_i E_i k_3^i g_i e^{-E_i/RT}}{\sum_i k_3^i g_i e^{-E_i/RT}} - \frac{\sum_j E_j g_j e^{-E_j/RT}}{\sum_j g_j e^{-E_j/RT}}
$$

$$
= \begin{pmatrix} \text{average energy of} \\ \text{molecules with } E > E_i \end{pmatrix} - \begin{pmatrix} \text{average energy of} \\ \text{all molecules} \end{pmatrix} \quad (188)
$$

The equation should be compared with equation (63) of section 3.6.3.

One very important observation must be made about the general derivation of the kinetic orders. Although the assumption of thermal equilibrium may be justified in the high pressure case, it may very well *not* be valid at low pressures. Indeed, the derivation depends on the fact that the concentration of A_i^* is determined by removal processes not involving collisional deactivation, and if k_3^i is greater for more energetic molecules, the rate of depletion of energetic molecules may be greater than the rate of production needed to maintain the equilibrium concentration.

Before moving on to discuss the application of the transition state theory to unimolecular reactions, and to more general treatments, it seems appropriate to point out the dangers of assuming that all first-order processes involve either a true unimolecular step or an activation–deactivation mechanism. A chain mechanism, not involving activation, can give rise to first-order kinetics. The first-order decomposition of dinitrogen pentoxide was, for a long time, regarded as a "unimolecular" process, although it became apparent that the kinetics were of the first order down to the lowest pressures at which the experiments were conducted. Ogg[145] and Mills and Johnston[146] have shown that, in fact, the reaction proceeds *via* a chain mechanism involving the radical NO_3

$$
N_2O_5 \underset{k_5}{\overset{k_4}{\rightleftarrows}} NO_2 + NO_3 \tag{189}
$$

$$
NO_2 + NO_3 \overset{k_6}{\rightarrow} NO + O_2 + NO_2 \tag{190}
$$

$$
NO + NO_3 \overset{k_7}{\rightarrow} 2NO_2 \tag{191}
$$

Solution of the steady state equation for $[NO_3]$ leads to an expression

$$
-\frac{d[N_2O_5]}{dt} = \frac{2k_6 k_4 [N_2O_5]}{k_5 + 2k_6} \tag{192}
$$

which is first-order at all pressures. A considerable number of thermal reactions, formerly thought to be unimolecular, has been shown to proceed *via* a chain mechanism. The nitric oxide inhibition experiments devised by Hinshelwood[147] appear to show that a number of pyrolyses proceed *via* both a chain and a unimolecular mechanism. Even the thermal decomposition of azomethane, which was for many years regarded as a good example of a unimolecular reaction, has been shown[148] to involve short chains initiated and terminated at the surface of the reaction vessel.

5.3 APPLICATION OF THE TRANSITION STATE THEORY TO UNIMOLECULAR REACTIONS

There are two main obstacles to the use of the transition state model for unimolecular reactions. First, there is the difficulty in defining the transition state for a unimolecular process, and secondly there is the question of the validity of the equilibrium assumption (although for the high pressure condition such an assumption may be applicable). In the introduction to this discussion of unimolecular reactions, section 5.1, and in section 4.3.1, it was pointed out that the transition state for dissociative processes may be defined by a maximum in a free-energy surface. An alternative suggestion is that the effective energy of the reaction system should be taken to be the sum of the potential energy and the rotational energy, *i.e.*

$$\varepsilon_{\text{eff}} = \varepsilon + \varepsilon_J \tag{193}$$

The rotational energy for any value of J depends on the moments of inertia, and thus ultimately on the internuclear distance between a pair of atoms which separate during a unimolecular dissociation. The rotational contribution to the total energy therefore rises to a maximum and then finally disappears as dissociation occurs. If, in fact, it is the rotational energy maximum that defines the transition state, then it is clearly of importance to know the value of ε_J under these conditions.

Substitution of the approximate classical rotational partition function (119) for three unequal moments of inertia into the usual expression for rate coefficient (99) yields

$$k = \kappa \frac{kT(I_A^{\ddagger} I_B^{\ddagger} I_C^{\ddagger})^{\frac{1}{2}} f_{\text{vib}}^{\prime\ddagger}}{h(I_A I_B I_C)^{\frac{1}{2}} f_{\text{vib}}^{\prime}} e^{-E_c/RT} \tag{194}$$

The relative simplicity of this equation is a direct result of the nature of a unimolecular process. The masses of the initial molecule and the activated state are identical, and since the translational partition functions are raised to the first power in both numerator and denominator of the expression they cancel, as do the constants in the rotational partition functions.

A further simplification may be introduced if only one of the moments of inertia, I_A say, is increased significantly on stretching of the bond to be broken. If the expression for the vibrational partition functions (113) is now inserted into (194), the equation becomes

$$k = \kappa \frac{kT}{h} \left(\frac{I_A^{\ddagger}}{I_A}\right)^{\frac{1}{2}} \frac{\prod\limits^{3N-7} [1 - e^{-h\nu^{\ddagger}/kT}]}{\prod\limits^{3N-6} [1 - e^{-h\nu/kT}]} e^{-E_c/RT} \tag{195}$$

For normal bonds at room temperatures, $h\nu \gg kT$ for both initial and activated states, and the vibrational partition functions are unity. Thus

$$k = \kappa \cdot \frac{kT}{h} \left(\frac{I_A^{\ddagger}}{I_A}\right)^{\frac{1}{2}} e^{-E_c/RT} \tag{196}$$

Although the calculation of the exact value of $I_A^{\ddagger}$ may be difficult, it is probable that $I_A^{\ddagger}$ is of the same order of magnitude as I_A, and the pre-exponential factor becomes $\kappa(kT/h)$. At room temperature kT/h is about 10^{13} sec^{-1}, so that for $\kappa = 1$, the "normal" frequency factor is obtained (see section 5.2).

If the vibration frequency of the bond to be broken is abnormally low, then just the first term in the expansion for $e^{-h\nu/kT}$ need be taken in evaluating the partition function. Under these circumstances, (195) with $I_A^{\ddagger} \sim I_A$ simplifies to

$$k = \kappa \nu_b e^{-E_c/RT} \tag{197}$$

where ν_b is the frequency of the breaking bond. For a temperature around 300 °K, the form of the expression given in (197) may be valid for $\nu_b < 10^{12}$ sec^{-1}.

The transition state theory does not give any explanation of abnormally large A factors in unimolecular reactions. Some increase in the pre-exponential factor above 10^{13} sec^{-1} can be achieved if $I_A^{\ddagger} > I_A$, although it is hardly possible to find a *reasonable* value for $I_A^{\ddagger}$ which will give a factor of 10^{16} or 10^{17}.

One of the most valuable applications of the transition state theory to unimolecular reactions has been made by Marcus[140, 141]. The theory is in principle a quantum mechanical transition state reformulation of the RRK free-flow theory, and includes both quantum and anharmonic effects. In view of the importance of Marcus' theory, and its assumption of a free flow of energy, it seems that the theory merits a short section of its own which should follow the discussion of free-flow theories. The Marcus approach is therefore developed in section 5.5.

Another important treatment of an equilibrium theory of unimolecular reactions has been given by Giddings and Eyring[149] for both dissociation and isomerization processes; the values to be expected for κ in the unimolecular step are discussed. For example, they point out that since κ is small in a bimolecular association process (see section 3.4), the transmission coefficient in the reverse, unimolecular dissociation, process will also be small. Following the presentation of Giddings and Eyring, we may write the general reaction scheme as

$$A+A \underset{k_{-i}}{\overset{k_i}{\rightleftarrows}} A_\varepsilon^* + A \tag{198}$$

$$A^* \overset{k_\varepsilon}{\rightarrow} A^\ddagger \rightarrow \text{products} \tag{199}$$

A_ε^* is a molecule of energy ε about to undergo unimolecular reaction *via* a transition state $A^\ddagger$. If $K_\varepsilon = k_i/k_{-i}$, then the fraction of molecules, F_ε, with energy ε per unit energy range is given by

$$F_\varepsilon = a_\varepsilon K_\varepsilon \tag{200}$$

where a_ε is a non-equilibrium constant, *viz.*

$$a_\varepsilon = \frac{k_{-i}[A]}{k_{-i}[A]+k_\varepsilon} \tag{201}$$

The overall rate coefficient for reaction is then given by

$$k' = \int_{\varepsilon=0}^{\infty} F_\varepsilon k_\varepsilon d\varepsilon = \int_0^{\infty} \frac{k_\varepsilon k_i[A]}{k_\varepsilon + k_{-i}[A]} d\varepsilon \tag{202}$$

This expression possesses the correct dependence of k' on [A]. It is now possible to write a transition state formulation of k' in which the non-equilibrium effects can be included in the value of the transmission coefficient. If κ is the ordinary transmission coefficient, then the pre-exponential part of the expression must be multiplied by a factor γ given by

$$\gamma = \kappa \int_{\varepsilon=0}^{\infty} k_\varepsilon a_\varepsilon K_\varepsilon d\varepsilon \Big/ \int_{\varepsilon=0}^{\infty} k_\varepsilon K_\varepsilon d\varepsilon \tag{203}$$

The evaluation of k_ε by the transition state theory has been considered by Rosenstock *et al.*[150]. They have shown that for any system in which $\kappa = 1$

$$k_\varepsilon = \frac{1}{h} \int_{\varepsilon=0}^{\varepsilon-\varepsilon_c} \frac{\rho^\ddagger(\varepsilon, \varepsilon_c, \varepsilon_t)}{\rho(\varepsilon)} d\varepsilon_t \tag{204}$$

where $\rho, \rho^\ddagger$ are energy densities, and the additional energy variable, ε_t, is the translational energy of the system passing across the saddle point of the transition state. If every vibration in A_ε^* and $A^\ddagger$ is a classical harmonic oscillator, then (204) may be written in the form

$$k_\varepsilon = \frac{\prod\limits_{s-1}^{s} v_i}{\prod v_i^\ddagger} \left(\frac{\varepsilon-\varepsilon_c}{\varepsilon}\right)^{s-1} \tag{205}$$

where s is here the *number of oscillators*, and v_i are the vibrational frequencies.

The final result for the case at low pressure, where thermal equilibrium is not maintained, is obtained by formulating the rate coefficients k_i and k_{-i} in terms of the transition state theory and summing over all states. Giddings and Eyring[149] give the expression

$$k' = \frac{1}{hQ_A} \int_0^\infty a_\varepsilon \rho(\varepsilon) e^{-\varepsilon/kT} d\varepsilon \int_0^{\varepsilon-\varepsilon_c} \frac{\rho^\ddagger(\varepsilon, \varepsilon_c, \varepsilon_t)}{\rho(\varepsilon)} d\varepsilon_t \tag{206}$$

(where Q_A is the partition function for A) for the rate coefficient, k', at any pressure.

The relation of the expressions obtained by the transition state theory to the results of the RRK free-flow theory will be shown in section 5.4.

5.4 FREE FLOW THEORIES

The interpretation of the kinetics of unimolecular processes in terms of the free flow of energy between the various degrees of freedom in the molecule was considered first by Kassel[137,151] and Rice and Ramsperger[138,152]. Kassel has summarised the work in his book (see Bibliography). For the present purposes, the form of the theory used by Kassel is adopted. Let there be s oscillators, each with the same frequency v; the energy of a state defined by the quantum number i is then taken to be $\varepsilon = ihv$.

It is now necessary to calculate the probability, p_i, that of this energy an amount $\varepsilon_c (= mhv)$ shall be concentrated in any one oscillator. The coupling between the oscillators is regarded as sufficiently weak to allow each oscillator to be independent, but at the same time sufficiently strong that the total energy may be freely distributed about the various degrees of freedom. It may be shown that

$$p_i = \frac{i!(i-m+s-1)!}{(i-m)!(i+s-1)!} \tag{207}$$

Kassel then assumed that the rate coefficient for the unimolecular decomposition process (171), k_3^i, is proportional to p_i. It may be shown that

$$p_i \rightarrow (1 - m/i)^{s-1} \quad \text{as} \quad i \rightarrow \infty \tag{208}$$

and since the ratio m/i may be expressed in terms of the molar energies E and E_c ($\varepsilon_c/\varepsilon = E_c/E = ih\nu/mh\nu$), then

$$k_3^i = A(1 - E_c/E)^{s-1} \tag{209}$$

(where A is the constant of proportionality) if the system behaves as a system of non quantised oscillators (*i.e.* if i is very large).

An expression has already been obtained for the rate of a thermal first order reaction in terms of k_3^i (section 5.2, equations 184 to 187). In the high pressure case

$$k' = k_\infty = \frac{\sum_i k_3^i g_i e^{-E_i/RT}}{\sum_j g_j e^{-E_j/RT}} \tag{210}$$

The statistical weight g_i is given by

$$g_i = \frac{(i+s-1)!}{i!(s-1)!} \tag{211}$$

(g_j is given by a similar expression with i replaced by j). Substitution of g_i, g_j and k_3^i into (210) gives

$$k_\infty = A e^{-E_c/RT} \tag{212}$$

without use of the approximation (208).

To obtain the rate coefficient, k', at a general pressure, the expression derived from (184) must be used, *viz.*

$$k' = \sum \left(\frac{k_c k_3^i [A_c^i]}{k_c[A] + k_3^i} \right) \tag{213}$$

$[A_c^i]$ may be obtained by a classical statistical procedure for excitation in several degrees of freedom, and the function of $[A_c^i]$ is represented by equation (57) of section 3.6.3 (with $S/2$ replaced by s because of the new definition of s). If the differential form of (61) with $\Gamma(s) = (s-1)!$ is used together with the classical value of k_3^i given in equation (209), and the summation of (213) replaced by integration, we obtain

$$k' = \frac{A}{(s-1)!(RT)^s} \int_{E_c}^{\infty} \frac{(E-E_c)^{s-1}e^{-E/RT}dE}{1+(A/k_c[A]) \cdot (1-E_c/E)^{s-1}} \tag{214}$$

It may be shown that at high pressures, the value obtained (214) is identical with the result of the non-approximate calculation (212). At low pressures, (214) becomes, on performing the integration

$$k' = \frac{(E_c/RT)^{s-1}}{(s-1)!} \cdot k_c[A] \cdot \left[1 + \frac{(s-1)}{E_c/RT} + \frac{(s-1)(s-2)}{(E_c/RT)^2}\right.$$

$$\left. + \cdots + \frac{(s-1)!}{(E_c/RT)^{s-1}}\right] e^{-E_c/RT} \tag{215}$$

Thus it can be seen that if E_c/RT is large, the experimental activation energy is given by

$$E_a = RT^2 \frac{d \ln k'}{dT} = E_c - (s-1)RT \tag{216}$$

This expression is the same as the activation energy obtained by Hinshelwood for activation in many degrees of freedom (see equation (60), bearing in mind that $S/2$ is called s in the Kassel notation). However, the Hinshelwood theory predicts the same activation energy at all pressures, while the Kassel theory suggests that the measured activation energy will vary from E_c at high pressures (equation (212)) to the lower value predicted by (216) at low pressures. The lowering of the effective activation energy at low pressures indicates that the predicted *pressure* dependence of k' will be different for the two theories, a sharper change of k' with pressure being expected on the basis of the Kassel theory.

   Kassel's classical theory[137] was followed by two quantum-mechanical versions[151], which give substantially the same results as the classical theory. For example, in one of the theories the s classical oscillators are replaced by quantal oscillators of frequency v. The form of k_3^i is retained as

$$k_3^i = Ap_i \tag{217}$$

and the high pressure rate coefficient $k' = k_\infty$ is obtained in the quantum mechanical case by writing

$$[A_c^i] = [A](1-\alpha)^s \alpha^i p_i \tag{218}$$

where $\alpha = e^{-hv/kT}$. Thus, using the high pressure form of (184) together with (218) and summing over i from m to infinity, we obtain

$$k_\infty = A\alpha^m = Ae^{-E_c/RT} \qquad (219)$$

The general rate coefficient is obviously given by the general form of (184) after substitution of $[A_c^i]$ from (218), and k_3^i, p_i from (217) and (207).

The discussion of this section has been concerned with a model allowing a free-flow of energy, and Kassel's treatment has been used as an illustration. Rice and Ramsperger[138, 152] developed their theory at about the same time as Kassel produced his formulation, and their theory differs from Kassel's only in that it requires the energy ε_c to accumulate in one degree of freedom rather than in one oscillator (= two degrees of freedom), for dissociation to occur. The outcome of the difference is that k_∞ is not quite of the form (212), although the variation of k' with pressure expressed as k'/k_∞ is very similar to the result obtained from the Kassel theory. The actual comparison of experimental results with the various theoretical predictions will be presented, after the Slater and Marcus theories have been discussed, in section 5.7.

It may finally be of interest to compare the RRK result for the unimolecular rate coefficient, k_3^i, with the result given by Giddings and Eyring[149] in their application of the transition state theory to unimolecular reactions. The transition state formula for k_ε (205) and the RRK formula for k_3^i (209) may be identified, and since $(\varepsilon - \varepsilon_c)/\varepsilon = (E - E_c)/E$, it may be shown that

$$A = \frac{\prod\limits^{s} v_i}{\prod\limits^{s-1} v_i^\ddagger} \qquad (220)$$

5.5 MARCUS' QUANTUM MECHANICAL TRANSITION STATE FORMULATION

The errors inherent in any theory of unimolecular reactions which treats the vibrations as classical and which neglects the importance of anharmonicity in the coupling of vibrations led Marcus to develop[140, 141] a formulation of the RRK theory using the transition state method. Two specific advantages of the Marcus formulation are (i) that the curve-fitting parameter often used in the RRK formulation is absent, and (ii) that the theory is applicable under conditions where anharmonicity or quantum effects are important. This latter point is of particular significance, since it covers just those situations where the Slater theory to be developed in section 5.6 is not applicable.

A recent extension[153] of Marcus' theory has allowed both for centrifugal effects and for reaction path degeneracy. For the present purposes, however, it is intended to set out the results of only the earlier work. The major changes effected by the modifications are first the inclusion of detailed rotational energies in the energy-balance equation, and secondly the introduction of a term for reaction-path degeneracy which in some places requires the use of symmetry numbers rather than rotational partition functions.

The original derivation given by Marcus[141] uses a reaction scheme

$$A + M \underset{k_2}{\overset{k_1}{\rightleftarrows}} A^* + M \tag{221}$$

$$A^* \xrightarrow{k_a} A^+ \tag{222}$$

$$A^+ \rightarrow products \tag{223}$$

The notation is that used by Marcus, and the subscripted rate coefficients should not be confused with those used previously, although they correspond roughly with those used for equations (169) and (170). A^* is an "active" molecule of A, and A^+ is the transition state; activation and deactivation is by means of a third body, M, which may or may not be A itself. The rate coefficients are all functions of the energy of the initial A^* molecule. It is assumed that each active molecule is randomly distributed about all configurations subject to energy and angular momentum conservation, and the degrees of freedom of the active molecule are classified as "active", "adiabatic", or "inactive", according to their role in intermolecular energy transfer. The "adiabatic" degrees of freedom are taken to remain in a given quantum state during the decomposition of the molecule, and do not, therefore, contribute much energy to the breaking bond; it was shown from an argument involving angular momentum conservation that the two rotational degrees of freedom involving the larger moments of inertia were approximately adiabatic. Energy in the "inactive" degrees of freedom is assumed to be transferred to the breaking bond only when the molecule is essentially in the transition state, and A^* must possess a "non-fixed" energy E^* such that E^* is greater than the sum of the activation energy and the "non-fixed" energy of the "inactive" modes (a "non-fixed" energy is energy which does not contain the energy "fixed" as zero-point energy).

The probability that A^* is in some critical configuration A^+ is now calculated by the methods of statistical mechanics. The vibrational energy levels of A^+ are treated quantum mechanically, as are the levels of A which are *not* levels of A^*: for simplicity of calculation, it was assumed that the number of energy levels for a given energy range in A^* could be computed classically (since the energies involved are relatively large). A weighted frequency factor, describing passage over the potential energy surface characterising the activated complex, is now multiplied by the probability that A^* was A^+ to give the rate coefficient k_a as a function of the energy of the active molecule. The result is

$$k_a = \frac{P_1^+}{P_1} \frac{P_R^+}{h} \frac{1}{(r/2)!} \sum_{E_v^+ \leq E^+} \left\{ \frac{[(E^+ - E_v^+)/RT]^{r/2} P(E_v^+)}{N^*(E_a + E^+ + E_0)} \right\} \tag{224}$$

This is the original result[141], but expressed using the more recent nomenclature, viz.

P_1^+, P_1 are the partition functions of the *adiabatic* rotations of A^+ and A.

P_R^+ is the partition function of the *r active* rotations of A^+.

E^+ is the "non-fixed" energy of A^+ (see above).

E_v^+ is the difference between the zero-point energy of A^+ and the energy of the vth vibrational level in A^+; it is $P(E_v^+)$ fold degenerate.

E_0 is the zero-point energy of A.

E_a is the difference between the energy of A in its lowest state (all modes) and A^+ in its lowest state.

N^* (energy) is the number of energy states per unit energy of the active modes of A^* at the stated energy.

Treatment of $[A^*]$ by steady-state methods gives a value for the "unimolecular" rate coefficient, k', if it is assumed that the fraction of energetic molecules having *insufficient* energy to react is the equilibrium value at all pressures (an assumption which is not always justifiable). The equation derived for k' is

$$k' = \frac{kT}{h} \frac{P_1^+ P_R^+}{P_1 P_2} \frac{e^{-E_a/RT}}{(r/2)!} \int_0^\infty \sum_{E_v^+ \leq E^+} \frac{[(E^+ - E_v^+)/RT]^{r/2} P(E_v) e^{-E^+/RT}}{1 + k_a/k_2 P} \, \mathrm{d}(E^+/RT) \tag{225}$$

and it is shown that the high pressure value of k', k_∞, is given by

$$k_\infty = \frac{kT}{h} \cdot \frac{P_1^+ P_2^+}{P_1 P_2} \cdot e^{-E_a/RT} \tag{226}$$

where P_2^+ and P_2 are the partition functions for *all* active modes of A^+ and A, respectively. This equation is equivalent to (194) given in section 5.3. The quantum mechanical treatment of energy levels manifests itself in equation (225) and (226) by the presence of the quantum mechanical partition functions for A and A^+, P_2 and P_2^+, and the degeneracy $P(E^+)$.

Calculation of a unimolecular rate coefficient at any pressure P now requires that a value for $N^*(E_a + E^+ + E_0)$ should be obtained, and that a systematic method for enumerating the number of energy levels be evolved to evaluate $\Sigma[(E^+ - E_v^+)/RT]^{r/2} P(E_v^+)$. If t rotations of A^* are active as well as the vibrations, then any non-fixed energy u may have a non-fixed rotational energy x separated from it, and $N^*(u+E_0)$ may be expressed as

$$N^*(u+E_0) = \int_0^{u+E_0} N_v^*(u-x+E_0)N_r^*(x)\mathrm{d}x \tag{227}$$

where

$$N_r^*(x) = \frac{(x/RT)^{\frac{1}{2}t-1}P_R}{\Gamma(\frac{1}{2}t)RT} \tag{228}$$

(P_R is the partition function for the t active rotations of A*). A classical treatment of the number of energy states per unit energy is much simpler than the exceedingly laborious quantum treatments, but generally gives a gross overestimate of $N_v^*(u-x+E_0)$. Marcus and Rice[140] have shown that the semi-classical expression

$$N_v^*(u-x+E_0) = (u-x+E_0)^{s-1}/\Gamma(s)\prod_{i=1}^s hv_i^* \tag{229}$$

for a system of s oscillators of frequencies v_i^*, is a reasonable approximation. Substitution of (229) and (228) into (227) gives

$$N^*(u+E_0) = \frac{P_R(u+E_0)^{s+\frac{1}{2}t-1}}{(RT)^{\frac{1}{2}t}\Gamma(s+\frac{1}{2}t)\prod_{i=1}^s hv_i^*} \tag{230}$$

It is suggested[154] that when some vibration frequencies of A* are very high in relation to the others, a better approximation than (229) is obtained by separating the frequencies into two classes, and computing the degeneracies of the high frequency group by a quantum mechanical treatment.

A number of modifications have been made to these expressions for N^*. For example Rabinovitch and Diesen[155] propose a corrected form

$$N_{corrected}^*(u-x+E_0) = (u-x+aE_0)^{s-1}/\Gamma(s)\prod_{i=1}^s hv_i^* \tag{231}$$

where a is an empirical correction factor. Whitten and Rabinovitch[156] have obtained exact sums of harmonic vibrational energy level densities for 45 molecules at various energies by the use of a digital computer. The results have been used to characterise the correction factor a with regard to its magnitude, its energy dependence and its dependence on the nature of the molecular frequency pattern concerned. The use of the approximation has the great advantage of being very rapid and of not requiring any calculational detail; the actual molecular frequencies are employed. An extension of the method has been given to obtain an approximation for rotation–vibration energy level sums[157]. It is shown that the approximation used by Marcus[141], equivalent to equation (227), for the separation of vibrational and rotational sums leads to an error of less than 5 % for molecules having $s < 40$ and that the error decreases with increasing energy.

Wieder and Marcus[154] have reduced the quantum form of the theory to its classical limit, and show that the equations (224) and (225) reduce to the classical Kassel form (see section 5.4). Two important particular cases may be quoted: the ratio of rate coefficients on classical and quantum theories at the high pressure limit and as $p \to 0$. At the high pressure limit

$$\frac{k_\infty^c}{k_\infty^q} = \left(\frac{Q_2^+}{Q_2}\right) \bigg/ \left(\frac{Q_2^+}{Q_2}\right)^q \tag{232}$$

(The superscripts c and q refer to classical and quantum quantities, and Q_2^+, Q_2 correspond to P_2^+, P_2, but are measured from the potential energy minimum rather than ground vibrational state). Using an approximation for the low pressure rate coefficients, it is shown that the ratio of the leading terms in expansions of k_0^c and k_0^q is given by

$$\frac{k_0^c}{k_0^q} = \frac{Q_2^q}{Q_2^c} \cdot \frac{e^{-E_0^+/RT}}{(1+E_0^+/U)^{s+\frac{1}{2}t-1}} \tag{233}$$

where E_0^+ is the zero point energy of A^+, and U is the difference in potential energy of the *least unstable* configuration of A^+ and the *most stable* configuration of A.

Although the value of k_0^q may be calculated without a knowledge of the quantitative properties of A^+, the more general test of the RRKM theory—a plot of k'/k_∞ against pressure—does require this knowledge. The properties of A^+ are in general deduced from a knowledge of the high-pressure frequency factor, guided by certain considerations about the type of activated complex. Each reaction should be treated individually, and the reader is referred to the several papers which present RRKM treatments for specific chemical processes. The predicted RRKM rate coefficients for some reactions are quoted (but not derived) in the comparison of experimental results with the various theories presented in section 5.7.

5.6 SLATER'S THEORY

Slater[139, 158, 159] has suggested a theory of unimolecular reactions whose basic assumptions are entirely different from those of the RRK theory. The postulates of the theory and some of the main conclusions are given in this section. (The "Slater Theory" discussed here should not be confused with "Slater's New

Approach" which will be touched upon briefly in section 5.8).

Collisional energisation of a molecule is assumed to result in a statistical distribution of the energy between the degrees of freedom: *energy does not flow between the degrees of freedom during subsequent vibrations of the molecule.* A molecule undergoes reaction when a certain critical co-ordinate undergoes a sufficient extension: that is to say, reaction occurs if and when the phase relationships of the various vibrations allow the critical bond to be broken. Let the critical co-ordinate be given the symbol q_1, and the normal co-ordinates be Q_i, then if α_{1i} is the transformation matrix of q_1 to Q_i

$$q_1 = \sum_i \alpha_{1i} Q_i \tag{234}$$

We may write for a system in which n' normal vibrations take part in activation

$$q_1 = \sum_{i=1}^{n'} \alpha_{1i} \varepsilon_i^{\frac{1}{2}} \cos 2\pi(\nu_i t + \psi_i) \tag{235}$$

where ε_i and ν_i are the energy and frequency of the ith vibration, and ψ_i is a phase angle. The energies and phases can change only on collision, and the time t is calculated from the moment of the last collision. It follows that, on Slater's theory, not only must the total energy in the molecule be greater than a critical amount E_c, but the critical co-ordinate must reach a value q_c corresponding to E_c, so that

$$|\alpha_{1i}| \varepsilon_i^{\frac{1}{2}} \geqq q_c \tag{236}$$

The probability of unimolecular conversion therefore depends on the frequency of molecular collisions and the properties of the function (235). Slater's theory shows[139] that at the high pressure limit

$$k_\infty = \bar{\nu} e^{-E_c/RT} \tag{237}$$

where $\bar{\nu}$ is a weighted r.m.s. frequency of normal vibrations of the molecule given by

$$\bar{\nu} = \left[\frac{\sum \alpha_{1i}^2 \nu_i^2}{\sum \alpha_{1i}^2} \right]^{\frac{1}{2}} \tag{238}$$

Equation (237) is very similar to equation (212) of the Kassel theory with the frequency factor A replaced by the weighted mean frequency $\bar{\nu}$. Indeed, Slater points out that his theory gives a value for $k_3(E)$, the rate coefficient of unimolecular conversion of an energised species, energy E, such that

$$k_3(E) = \bar{v} \left(\frac{E - E_c}{E} \right)^{n'-1} \tag{239}$$

The equation is formally identical to Kassel's [equation (209)] with A set equal to $\bar{v}$. However, A is an arbitrary constant which represents the frequency of interchange of energy between oscillators, a process not allowed in the Slater theory. Thus the two relations are different in concept, although the mathematical similarity ensures the same forms (237) and (212) for the limiting rate at high pressures. Slater derives (239) from an equilibrium distribution of energies, and the value of $k_3(E)$ may not be applicable at low pressures. Slater has[160], however, used the theory to obtain a value for the general rate. The approximate formula derived is

$$k' = \frac{\bar{v}e^{-b}}{m!} \int_0^\infty \frac{x^m e^{-x} dx}{1 + x^m \theta^{-1}} \tag{240}$$

where

$$b = \varepsilon_c/kT$$

$$x = \varepsilon_i/kT$$

$$m = \tfrac{1}{2}(n'-1)$$

$$\theta = m! \omega / \bar{v} (4\pi b)^m \prod_{i=1}^{n'} \mu_i$$

$$\mu_i = \alpha_i / (\sum_i \alpha_i^2)^{\frac{1}{2}}$$

$$\omega = k_c p$$

At high pressures, $x^m \theta^{-1} \ll 1$ and $k_\infty \to \bar{v} e^{-b}$ which is identical with the exact result (237). At low pressures

$$k' = k_c p (4\pi b)^m \prod_{i=1}^{n'} \mu_i \tag{241}$$

approximately, which indicates that the overall reaction is kinetically of second order. One of the consequences of the form of (241) is that the activation energy at low pressures is given by

$$E_a = RT^2 \frac{d \ln k'}{dT} = E_c - \tfrac{1}{2}(n'-1)RT \tag{242}$$

Comparison of (242) with the expression (216) for E_a on the Kassel theory shows that the two activation energies are identical only if

$$n' = 2s - 1 \qquad (243)$$

Slater has suggested that the non-correspondence of n' and s results from the different criteria of a critical energy and critical vibration amplitude in the Kassel and Slater theories.

The variation of the ratio of general to limiting high pressure rate coefficients with pressure is frequently used as a test of theories of unimolecular reaction. From (237) and (240) it is seen that

$$I_m(\theta) = (k'/k_\infty)_n = \frac{1}{m!} \int_0^\infty \frac{x^m e^{-x} dx}{1 + x^m \theta^{-1}} \qquad (244)$$

A curious result of the Slater theory is obtained for a diatomic species which has $n = 1$ and thus $m = 0$. The *exact* form of k'/k_∞ from (244) is given by

$$\frac{k_\infty}{k'} = 1 + \frac{\bar{v}}{\omega} \qquad (245)$$

Since, in gases, $\bar{v} \gg \omega$, k' cannot become a first order rate coefficient, and a diatomic molecule (represented by simple oscillators) will not undergo unimolecular decomposition.

Slater[160] has considered the possibility that although the assumption of no energy flow might be valid at moderately low pressures, at *very* low pressures the RRK predictions might nevertheless be valid. This situation would arise if there were a *slow* flow of energy between normal modes in the reactant, which could only make a significant contribution to energization of the critical bond if the time between collisions were sufficiently long.

5.7 EXPERIMENTAL TESTS OF THEORIES OF UNIMOLECULAR REACTION

In principle, the validity of the main theories of unimolecular reaction can be tested by means of "experimental" (*i.e.* computer) studies of trajectories for reaction. Such studies have been made, and will be mentioned very briefly in section 5.8. The tests of the theories with which we are concerned here involve laboratory chemical experiments.

There are, of course, several levels at which the tests might be made. Predicted and experimental high pressure first-order rate coefficients may be compared in the first instance. These comparisons may be extended to a low pressure case, and finally the entire shape of the "fall-off" curve (usually plotted as $\log(k'/k_\infty)$ *versus* pressure) may be compared with experiment. Fig. 11 is, in effect, a fall-off test of the Lindemann–Hinshelwood theory, and the shortcomings of the theory are well revealed. In point of fact, the experimental data is taken from work of Schneider and Rabinovitch[142, 161, 162] on the isomerization of isocyanides carried out to test the RRKM theory. The reactions which have been studied are

$$CH_3NC \rightarrow CH_3CN \tag{246}$$

$$CD_3NC \rightarrow CD_3CN \tag{247}$$

$$CH_2DNC \rightarrow CH_2DCN \tag{248}$$

Two tests are implicit in this work, and they will be examined in reverse order. Curve fitting of the fall-off gave the best fit for the Slater theory with $n = 5.5$, 6.1 and 5.6 for reactions (246), (247) and (248) respectively. The Slater theory predicts that there should be no change of n on perdeuteration of CH_3NC since the original symmetry is retained. On the other hand, monodeuteration removes the degeneracy present in some vibrational modes, and it is shown[162] that n should increase by *at least* one for CH_2DNC. The experimental results suggest that n is virtually constant, and argue against the validity of the Slater theory. Tests of the RRKM theory may now be made. First, the variation of k'/k_∞ with pressure may be examined for each isocyanide. Fig. 12 shows the results obtained by

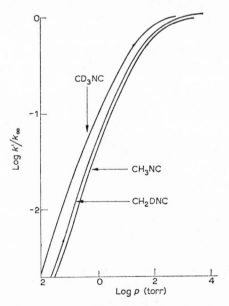

Fig. 12. Logarithmic "fall-off" plots for the isomerization of methyl isocyanides (from Schneider and Rabinovitch[162]).

Schneider and Rabinovitch[162] plotted on a logarithmic scale against pressure. The fall-off curves shift to progressively lower pressures with increasing deuteration, as expected on the basis of the RRKM theory. A more exact test could be made of the rates of isomerization of either the mono- or the perdeuterated compound *relative* to the rate of isomerization of CH_3NC.

At high pressures, the isotopic rate ratio, called k_H/k_D, was found to be slightly greater than unity, in agreement with the predicted rates. A non-equilibrium inverse isotope effect is predicted by the RRKM theory for low pressures, and the experimental value of k_H/k_D decreases at low pressures in the manner expected. The inversion of the isotope effect is a result of opposing statistical isotope effects which become more important than the primary isotope effect at low pressures[163].

Another reaction system in which the RRK and Slater theories have been differentiated by means of behaviour on isotopic substitution is the thermal isomerization of cyclopropanes[164]. The value of n determined from fall-off experiments in cyclopropane and cyclopropane-d_2 is 13–14 in both cases. In Slater's theory, the highest value n can take for light cyclopropane (which has seven doubly degenerate vibrational modes) is 14; for cyclopropane-d_2 there are no degenerate vibrations, and n could reach 20 or 21. The actual invariance of n as determined from the fall-off is evidence against the postulate of Slater's theory.

An important series of fall-off tests of the RRKM theory has been made by Wieder and Marcus[154] for unimolecular reactions of cyclopropane, cyclobutane, methyl cyclopropane, but-2-ene, ethyl chloride, nitrous oxide and nitrogen pentoxide. The tests were made with energy densities calculated by the approximate formula (229), and by the grouping procedure involving quantum mechanical treatment of the high frequency groups. Fig. 13 shows the results obtained for the isomerization of methylcyclopropane to butenes (several butenes are produced, but the frequency factors and activation energies are similar in the various cases). The

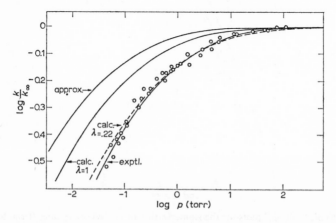

Fig. 13. Fall-off test of the thermal isomerization of methylcyclopropane (Wieder and Marcus[154]).

figure shows curves calculated for the two counting procedures as "approx." and "calc.". A better fit to the experimental results is obtained if the collisional deactivation efficiency, λ, is set equal to 0.22 rather than unity.

Application of the Slater theory to the prediction of unimolecular rates necessitates a complete vibrational analysis of the reacting molecule. Slater himself undertook such an analysis for cyclopropane[165], in the first complete test of his theory. It is, of course, important to decide which co-ordinate is the critical one, and the decision rests on choosing a correct model for the mechanism of reaction. In the main part of Slater's test, it was assumed that isomerization occurred when the vibrations carried any hydrogen atom too near to a carbon of another methylene group. On this model, the frequency factor found for the high pressure rate coefficient was 4×10^{14} sec^{-1} compared with an early experimental value of Chambers and Kistiakowsky[166] of about 15×10^{14} sec^{-1}. Adoption of a second model in which isomerization followed the over-stretching of a carbon–carbon bond yielded a high pressure frequency factor of only 6×10^{13} sec^{-1}. The shape of the predicted $\log(k'/k_{\infty})$ versus $\log p$ curve for the first reaction model was compared with the experimental curve obtained by Pritchard et al.[167]. Although the theoretical curve was of the correct shape, it was displaced by a factor of $3\frac{1}{2}$ times on the pressure scale from the experimental curve. However, the absolute relation between θ and pressure was calculated from the activation energy of Chambers and Kistiakowsky, and assuming unity collisional deactivation efficiency. Error in either of these factors could lead to appreciable error in the absolute position of the theoretical curve.

Gill and Laidler[168] have examined the experimental data for the low-pressure second-order rate coefficients of a number of unimolecular processes in the light of the RRK and Slater theories. This work was complementary to a complete vibrational analysis of the hydrogen peroxide[169] and ozone[170] molecules. For hydrogen peroxide, the choice of the O–O distance as the critical co-ordinate leads to a rate of energization on the Slater theory which is about 10^{-5} of the experimental value, and a maximum possible rate of reaction which is 10^{-2} of the experimental rate. On the other hand, for ozone the Slater rates of energization appear to be of the correct order of magnitude, while the RRK rates are too high unless fewer degrees of freedom are employed than are actually in the molecule. The Slater theory was found to be consistent in a similar way with the experimental results for the reactions of nitrogen pentoxide, cyclopropane and ethyl chloride, while the RRK theory predicted the rates more correctly for ethane and nitrous oxide decomposition as well as for the hydrogen peroxide reaction. The results were explained on the hypothesis that the Slater theory is correct as far as the breakdown of A* is concerned, but not always correct with regard to the rate of formation of A*. If flow of energy between normal modes can take place, a molecule A′ energised in the RRK sense, but not having the distribution of energy needed to be energised in the Slater sense, may become an A*. Gill and Laidler[168] propose an extension of the Lindemann–Hinshelwood scheme which allows of this process

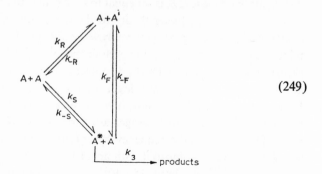

$$\tag{249}$$

k_R, k_{-R} are RRK activation, deactivation rate coefficients, k_S, k_{-S} the Slater rate coefficients, and k_F, k_{-F} the rate coefficients for flow of energy between A' and A*. A steady state treatment shows that if k_F, k_{-F} are small, the ordinary Slater situation obtains. Further, if the pressure is high

$$- \frac{d[A]}{dt} = \frac{k_3 k_S [A]^2}{k_3 + k_{-S}[A]} \tag{250}$$

under all conditions, even if k_F is large. However, at low pressures

$$- \frac{d[A]}{dt} = (k_S + k_R)[A]^2 \tag{251}$$

and since, probably, $k_R \gg k_S$, this expression reduces to

$$- \frac{d[A]}{dt} = k_R [A]^2 \tag{252}$$

Thus at low pressures the rate should be given by the RRK formula for the rate of energization. The experimental results suggest, on this modified theory, that k_F, k_{-F} are more rapid for N_2O, H_2O_2 and C_2H_6 than for O_3, N_2O_5, C_3H_6 and C_2H_5Cl. Gill and Laidler advance reasons for believing that in small molecules the flow of energy will tend to be more rapid than in large ones.

Contribution to the activation and deactivation processes can be made by an added gas which is otherwise inert chemically with respect to the reaction system. Experimental results show qualitatively that addition of an unreactive gas can restore first order kinetic behaviour to a system which was previously at a low enough pressure to display second order behaviour. Quantitative data obtained from experiments of this kind tends, however, to give information about the relative energy transfer efficiencies for the added gas and reactant rather than provide tests of anything more than the basic Lindemann activation–deactivation hypothesis. There do exist, of course, a number of techniques whereby the absolute efficiency for vibrational energy transfer of a gas may be determined, and a discussion of this topic appears elsewhere in *Comprehensive Chemical Kinetics* (see last Chapter, Vol. 3). In instances where these efficiencies are known unequivocally, a more direct test of unimolecular reaction theory is possible. The main "use" of added gases appears, however, to be in allowing rate determinations of unimolecular processes to be made at high pressures. For example, Schneider and Rabinovitch[162] conserved CH_2DNC by using a high pressure of chemically inert gas to simulate the high pressure conditions.

Lyon[171] has recently suggested a simple test of unimolecular reaction theory. If a molecule undergoes sequential unimolecular isomerization at high pressures, then the isomerization will become direct at sufficiently low pressures. A typical reaction scheme might be

$$A + M \rightleftarrows A^* + M \tag{253}$$

$$A^* \rightarrow B^* \tag{254}$$

$$B^* + M \rightleftarrows B + M \tag{255}$$

$$B^* \rightarrow C \tag{256}$$

Since the lifetime of B^* is inversely proportional to pressure in (255) and pressure independent in (256), the overall reaction selectivity will shift with decreasing pressure from $A \rightarrow B$ to $A \rightarrow C$. This shift may be calculated from RRK theory without any empirical information or assumptions concerning the molecules or transition states if A, B and C are virtually identical molecules. As an example of such a system, Lyon has studied the isomerization of 1-proto-perdeuterocyclopentadiene as a function of pressure. The observed low-pressure selectivity shift is in good agreement with the prediction.

Tests of unimolecular reaction theory based on fall-off curves suffer from the disadvantage that they test simultaneously a whole series of assumptions. For example, the curves of Fig. 13 are a test of the RRKM theory, but at the same time they test the state counting procedures, and, since λ is introduced as a variable, the efficiency of collisional processes. One solution to the problem might be to start with a molecule in an initially well-defined activated state. Ideally the species would be monoenergetic and the energy would be distributed in a known fashion about the various degrees of freedom, so that the rate coefficient k_3^i could be predicted on any theory of unimolecular reaction and compared with experiment. A technique which produces activated molecules which come some of the way to meeting these requirements is that of "chemical activation".

Chemical activation studies have yielded some valuable information about unimolecular reaction theory, and a few examples of the many applications of the technique are described here. A review of chemical activation processes has been given by Rabinovitch and Flowers[172] and part of the discussion is devoted to the relation of chemical activation studies to reaction theory.

Rabinovitch et al.[173] reported one of the first chemical activation studies of a unimolecular reaction. Methylene (CH_2) was produced by the photolysis of ketene at two wavelengths, and was added to trans-ethylene-d_2. The addition was stereospecifically cis-, and the "hot" cyclopropane molecules suffered geometric and structural isomerizations at rates depending on the wavelength of photolysis of ketene. According to RRK or Slater theories

$$k_3^i = A(1 - E_c/E)^{r-1} \tag{257}$$

where r is s or n in the Kassel or Slater models [cf. equations (209) and (239)]. Agreement between (257) and the observed experimental rate coefficient could be obtained only if quite unrealistic values were adopted for the energy, E, of the hot cyclopropane. Rabinovitch et al. write a form of the Marcus expression, using the "semi-classical" counting procedure, which allows direct comparison with (257)

$$k_3^i = \text{constant} \left(\frac{E - E_c + E_0^+}{E + E_0^*}\right)^{r-1} \tag{258}$$

where E_0^+, E_0^* are vibrational zero point energies of A^+, A^* as before. The exponent r is now the total number of vibrational modes which interact with the bond breaking co-ordinate, degenerate modes being counted fully. The values of k_3^i predicted from (258) were found to be satisfactory with the constant set equal to the observed high pressure frequency factor, and with reasonable values for E, E_c and the zero-point energies. Chemical activation systems such as this one may be the most suitable cases in which to apply the Marcus semi-classical approximation, since the requirement that $E \gg E_c$ may be met in these instances.

An experimental distinction between free-flow and Slater mechanisms was made by Butler and Kistiakowsky[174]. Methylene was produced by the photolysis of either ketene or diazomethane at a number of wavelengths, and allowed to react with either cyclopropane or propylene

$$CH_2 + \triangle \longrightarrow CH_3 \triangleleft{}^* \tag{259}$$

$$CH_2 + CH_2 = CH.CH_3 \longrightarrow CH_3 \triangleleft{}^* \tag{260}$$

The hot methylcyclopropane produced in (259) or (260) rearranges by a unimolecular process into a mixture of butenes unless it is stabilised by collision. The *lifetime* of the methylcyclopropane depends greatly on whether it is produced from cyclopropane or propylene and on the source of methylene, while the *composition* of the butenes does not depend on any of these factors. This latter result is taken to mean that the energy of the "hot" molecules migrates freely among the normal modes of vibration of methylcyclopropane molecules, and is therefore evidence strongly in favour of a free-flow rather than Slater theory. A similar result was obtained by Frey[175] and by Setser and Rabinovitch[176] for the lifetimes of "hot" dimethylcyclopropanes formed by the addition of methylene from different sources to *cis-* and *trans-*but-2-ene. It was shown[176] that the activated molecules were formed in a distribution which was nearly monoenergetic.

Rabinovitch and his collaborators have subsequently studied the unimolecular conversions of a large number of chemically activated species, and, in general, the results give good agreement with RRKM theory for a suitable choice of critical configuration, and with all degrees of freedom active. The technique typically involves analysis of decomposition (D) and stabilisation (S) products of the activated species. An average rate coefficient for unimolecular reaction, k_a, is then defined by

$$k_a = \omega \frac{[D]}{[S]} \tag{261}$$

where ω is the collision frequency for deactivation ($k_c p$) with an assumed efficiency, λ, for deactivation of unity. Since the chemically activated species still possesses some energy spread, represented by a distribution function $f(E)$, (261) may be rewritten

$$\frac{k_a}{\omega} = \frac{[D]}{[S]} = \frac{\int_{E_c}^{\infty} \frac{k_3(E)}{k_3(E)+\omega} f(E)dE}{\int_{E_c}^{\infty} \frac{\omega}{k_3(E)+\omega} f(E)dE} \tag{262}$$

The unimolecular decomposition of "hot" secondary butyl radicals provides a good example of the type of study described. References to the original papers are given in the review by Rabinovitch and Flowers[172]. Activated secondary butyl radicals may be produced with varying degrees of excitation by the addition of atomic hydrogen to (*i*) but-1-ene, (*ii*) *cis*-but-2-ene and (*iii*) *trans*-but-2-ene, *viz.*

$$H + C_4H_8 \rightarrow sec.\text{-}C_4H_9^* \tag{263}$$

Furthermore, since the activation energy for (263) is low (about 2 kcal. mole^{-1}), the reaction can be carried out at reduced temperatures (-78 °C in the studies reported) in order to reduce the average energy of the *sec*-butyl radicals as well as their energy spread. The energy possessed by the radicals produced in (263) is only just that required for the reverse dissociation to butene and atomic hydrogen, and the reverse of (263) is thought to be unimportant. Thus the important reactions open to hot *sec*-butyl radicals are

$$sec\text{-}C_4H_9^* \rightarrow CH_3 + C_3H_6 \text{ (D)} \tag{264}$$

$$sec\text{-}C_4H_9^* + M \rightarrow sec\text{-}C_4H_9 + M \text{ (S)} \tag{265}$$

and measurement of suitable reaction products gives the ratio [D]/[S]. A similar series of hot deutero-*sec*-butyl radicals may also be formed by the addition of atomic deuterium to the three butenes. The known thermochemistry of the systems can now be used to calculate the average energy of the radicals produced at reduced or room temperatures. The ratio of the limiting high and low pressure forms of k_a, *viz.* $k_{a\infty}/k_{a0}$, provides a measure of the energy spread described by the distribution $f(E)$. For monoenergetic systems, the ratio is, of course, unity, and for the butyl radical system under consideration, the ratio was found to be less than 2 at room temperature and less than 1.3 at $-78\ ^\circ C$. The ratio is orders of magnitude smaller than the equivalent ratio for thermal activation. A comparison of the experimental values of k_a with RRKM theory for the twelve different average energies is given by Rabinovitch *et al.*[177] and the agreement is shown to be good. Another noteworthy feature of the unimolecular reaction of butyl radicals is that the hot radicals formed from but-1-ene and but-2-ene decompose at comparable rates. In the first case the butyl radical formed decomposes by rupture of the C–C bond furthest from the point at which H added, and in the second case the rupture is of a bond attached to the carbon to which H added. The apparent lack of importance of the initial location of excess energy suggests strongly that there is a free flow of energy.

Competitive unimolecular decomposition processes have also been studied with chemically activated systems[178]. A general formulation of the rates was derived, together with various relationships between the rates and between the energetics of the competing processes. The experimental system employed was activated 3-hexyl radicals from the addition of atomic hydrogen to *trans*-hex-3-ene at 300 $^\circ C$. Two unimolecular decomposition reactions occur

$$C\text{-}C\text{-}C\text{-}\dot{C}\text{-}C\text{-}C^* \rightarrow CH_3 + 1\text{-}C_5H_{10} \tag{266}$$

$$\rightarrow C_2H_5 + 1\text{-}C_4H_8 \tag{267}$$

and the relative and absolute rates of the two processes were determined. A quantum statistical model was used to predict the rate coefficients of (266) and (267) as a function of energy, for use in the general relations. Good agreement between the theoretical and experimental data was obtained over the pressure range 0.004 to 90 torr.

Hassler and Setser[179] have carried out a very extensive test of the RRKM theory for the HCl elimination reactions of hot C_2H_5Cl, $1,1\text{-}C_2H_4Cl_2$ and $1,2\text{-}C_2H_4Cl_2$. The experimental rates for both thermal and chemically activated systems were compared with rates predicted from RRKM calculations employing several models for the transition state. Chemically activated mono- and di-chloroethanes were formed by the various recombination reactions of methyl and chloromethyl radicals with about 90 kcal. mole^{-1} of excess energy. The distribution function is so narrow that less than 10 % spread in k_a exists between high and low pressure values at 300 °K. Table 8 shows the experimental and calculated values of the Arrhenius A factors for thermal activation at 800 °K, and the k_a values for 300 °K in the chemically activated system. The model of the transition state adopted to give the calculated results is an asymmetric four centre complex with torsional motion in the molecule treated as a vibration.

TABLE 8

EXPERIMENTAL AND CALCULATED VALUES OF ARRHENIUS A FACTORS (THERMAL, 800 °K) AND k_a (CHEMICAL ACTIVATION, 300 °K) IN HCl ELIMINATION REACTIONS[179]

Halide	Average excess energy in chemically activated system (kcal. mole^{-1})	$10^{-13}A$		$10^{-9}k_a$	
		exptl. (sec^{-1})	calc.	exptl. (sec^{-1})	calc.
C_2H_5Cl	91.5	0.97	1.14	2.6	1.7
$1,1\text{-}C_2H_4Cl_2$	92.3	1.24	2.75	11	4.9
$1,2\text{-}C_2H_4Cl_2$	88.7	1.5	2.69	0.18	0.82

Sufficient examples of the use of chemically activated systems in tests of reaction theory have been given to show the power of the technique. It is apparent that chemical activation studies can also give valuable evidence about collisional de-activation. This topic lies just outside the scope of the present section, and the reader is therefore referred to the review by Rabinovitch and Flowers[172] for a brief summary of this application of activation experiments.

5.8 CONCLUSION

The summary of unimolecular reaction theory which has been given has intentionally concentrated on those theories which have found favour in comparisons with experimental work. The discussion of section 5.7 suggests that, so far as the theories considered are concerned, free-flow theories are more consistent with the experimental results than are purely harmonic theories, and of these theories, the quantum-statistical model of Marcus appears to give the most reliable calculated rates. However, a number of more or less formal unimolecular reaction theories have appeared over the last decade or so, and it seems appropriate to mention some of them.

Slater[180] first questioned the hypothesis of "random incidence of dissociation" [see Chapter 9, of Slater's book (Bibliography)]. The theories described hitherto have assumed that the dissociation of individual "interesting" molecules is random (although the *average* lifetime before dissociation is precise). In the "new approach to rate theory", Slater points out that if the precise classical state of an interesting molecule were known, then it would be possible to calculate the precise time, s say, it required to reach the dissociation configuration. At a concentration where the collision frequency is ω, the absolute chance of dissociation (*i.e.* of *no* collision) would be $e^{-\omega s}$, and it should be possible to calculate a reaction rate constant as the integral of this probability over the collisional rate of production of interesting molecules in the various relevant states. The mathematical equipment for handling the $3N$ co-ordinates and $3N$ momenta of a system of N particles requires the use of the *phase-space* concept, and much of the more recent work is expressed in terms of phase-space boundaries and gaps. Bunker (Bibliography, p. 51 *et seq.* of his book) gives a useful summary of the methods used. Slater uses the technique to re-derive the rate equations for the harmonic oscillator model, and finds that at high and low concentrations the results are identical with those derived previously assuming random behaviour. Thiele[181] has also re-examined the Slater, Kassel and Giddings–Eyring theories in the light of Slater's "new approach" and made some suggestions about further development of a non-linear theory. A quantum model adopted by Wilson and Thiele[182] has been elaborated by Thiele[183] to provide a formal quantum theory for unimolecular reactions.

A second assumption made in all the theories of activation and deactivation is the "strong-collision" assumption. There is some reason to suppose[184] that collisions may be inefficient in certain cases, so that the rate of activation controls the overall rate of the unimolecular process. Discussions of weak collisions have been given by Buff and Wilson[185] and Hoare[186]. Using Landau–Teller probabilities to obtain rates of collisional energy transfer Buff and Wilson investigated the "weak collision" unimolecular rates and found that the fall-off region was more extended than for strong collisions.

Two further theories of unimolecular reaction have appeared recently. One is to be found in a series of papers by Levine[187], who discusses quantum mechanical dissociation rates in terms of Green's operator formalism. Valence and Schlag[188] have obtained theoretical unimolecular rate coefficients for many-level systems, and examined the case where any or all of the reactant levels have non-equilibrium effects.

Finally, brief mention must be made of computer kinematic or trajectory studies for unimolecular reactions. Several such studies have been made[189–193]:

all show that the normal-mode approximation used is a bad one, and that an-harmonicity has a very important influence on the vibrational motion and on the coupling of rotation to it. There seems, therefore, to be no real reason for retaining Slater's strictly harmonic model. In general, the computer values for $k_3(E)$ are smaller than the RRKM ones; part of the difference can be ascribed to a larger value of the energy density of states, N^*, obtained in the trajectory studies. Although there is some evidence that non-random lifetimes may arise under some circumstances, the trajectory tests suggest that non-randomness will be unimportant for large molecules.

6. Termolecular and third-order processes

The last class of chemical process with which we shall deal is that of third-order reactions. Termolecular reactions are relatively rare, and the best authenticated examples are the reactions of nitric oxide with hydrogen, deuterium, oxygen, bromine and chlorine, and the recombination of atoms or small radicals. Examples of the two groups of reactions are

$$2NO + O_2 \rightarrow 2NO_2 \tag{268}$$

$$H + H + M \rightarrow H_2 + M \tag{269}$$

In view of the great importance of the recombination reactions in a variety of gas phase processes, it is necessary to examine briefly elementary reactions which are kinetically third order.

Bimolecular recombination of two atoms or radicals naturally produces a species which initially possesses the entire energy of the newly formed bond. In the sense described in section 5.7, this species is therefore "chemically activated", and can undergo spontaneous unimolecular dissociation unless it is stabilised. So far as the formation of diatomic molecules in processes such as (269) is concerned, this energy must remain in the only bond present, and dissociation is likely to occur in about one vibration, unless a collision with a "third-body" takes place during this vibrational lifetime. Stabilization of a vibrationally excited molecule by a vibrational radiative process is extremely unlikely (and impossible if the diatomic molecule is homonuclear). Even if the molecule first formed is *electronically* as well as vibrationally excited, electronic emission processes will not usually help to stabilise the molecule, since at the very best the electronic radiative lifetime will not be much shorter than 10^{-8} sec, as against a vibration period of about 10^{-13} sec. Collisional stabilization is therefore the most probable way in which the newly formed molecule can be stabilized. If, however, the species formed in a recombination reaction possesses a large number of bonds, then a different situation arises. The discussion of section 5 suggests that vibrational energy can flow freely between the various vibrational degrees of freedom in a polyatomic molecule, so that the energy concentrated initially in the newly formed bond can begin to distribute itself about several bonds during the course of a vibrational period. It is true that the molecule is still able to undergo unimolecular dissociation, but since the energy available is only just equal to the critical energy for dissociation and is distributed about many degrees of freedom, the lifetime of the molecule will be considerable. In this case, the molecule will suffer many deactivating collisions, and the rate controlling process will be the rate of recombination rather than recombination plus deactivation, and the kinetics of recombination will be second order. At sufficiently low pressures, of course, such a process will ultimately become third order if *wall* deactivation steps are not admitted.

For recombination reactions of species larger than, say, ethyl radicals, the process seems to be adequately described for all experimental conditions as bimolecular, while the recombination of atoms, as in (269) is certainly termolecular. The recombination of methyl radicals affords an example of intermediate behaviour, and the molecularity of the process has often been discussed. The number of bonds available is not sufficient to ensure a dissociation lifetime greater than the collisional lifetime at all pressures. Kistiakowsky and Roberts[194] give the effective rate coefficient for the recombination of methyl radicals in the presence of acetone at 180 °C

$$CH_3 + CH_3 \rightarrow C_2H_6 \tag{270}$$

as

$$k_{eff} = k''/(1+4.6/p) \tag{270a}$$

where k'' is a second order rate coefficient, and p the pressure of acetone in torr. At pressures above a few cm Hg, therefore, $k_{eff} \approx k''$, although at low pressures $k_{eff} \approx k''p/4.6$ and the process is third order.

One requirement of any third body in a recombination process is that it shall be able to remove vibrational energy efficiently on collision. The details of vibrational energy transfer are not within the scope of this chapter (see last Chapter, Volume 3), but two general points should be made. Both vibration–vibration and vibration–translation energy exchange processes can occur, and the exchange appears to become more efficient if the energy difference between exchanging states is small. Secondly, the probability that quenching will occur on collision by a vibration–vibration exchange process is increased by an increase in the number of vibrational degrees of freedom in the quenching molecule. The validity of these considerations may be seen qualitatively from Table 9 which shows the third order recombination rate coefficients for the reaction

$$O + O_2 + M \rightarrow O_3 + M \tag{271}$$

TABLE 9

THIRD ORDER RATE COEFFICIENTS FOR THE GAS-PHASE RECOMBINATION OF O WITH O_2
(from Kaufman and Kelso[195])

Third body	$10^{34} \times$ rate coefficient (cm^6. $molec^{-2}$. sec^{-1})
He	4
Ar	4
N_2	5.6
O_2	6.5
CO_2	15
N_2O	15
CF_4	16
SF_6	34
H_2O	60

with a number of different third bodies[195]. The increasing rate coefficients with increasing atomicity of M reflect the greater efficiency of the polyatomic molecules in vibrational quenching processes.

The problem now arises as to what exactly constitutes a termolecular process. The examples given so far appear to be complex reactions which display third-order kinetics. For instance, the recombination of atomic hydrogen might be written formally

$$H+H \rightarrow H_2^* \tag{272}$$

$$H_2^* + M \rightarrow H_2 + M \tag{273}$$

It seems fairly well established that the recombination of iodine atoms in the presence of certain third bodies (e.g. aromatic hydrocarbons[196]) involves the formation of a complex

$$I+M \rightarrow IM \tag{274}$$

$$IM + I \rightarrow I_2 + M \tag{275}$$

and several complex schemes have been put forward for the nitric oxide reactions. One such scheme involves "sticky" collisions which may be represented for the $NO + Cl_2$ reaction as

$$NO + Cl_2 \rightarrow NO–Cl–Cl \tag{276}$$

$$NO–Cl–Cl + NO \rightarrow 2NOCl \tag{277}$$

In every case, the third order behaviour depends on competition between the reverse (dissociation) of the first reaction of the pair and the second reaction.

The concept of a termolecular reaction is itself untenable on a hard-sphere collision theory, since the chance of simultaneous contact of the surfaces of three spherical molecules is zero. Several authors have discussed three-body collision rates. For example, Tolman[197] assumed that the approach of rigid spheres to within an arbitrary distance of each other constituted a collision, and introduced a rather ill-defined distance parameter, δ. The collision rate calculated, Z_{ABC}, for three-body collision was given by

$$Z_{ABC}^2 = n_A n_B n_C 4\pi\sigma_{AB}^2 \sigma_{BC}^2 \delta \sqrt{8\pi kT} \left(\frac{1}{\sqrt{\mu_{AB}}} + \frac{1}{\sqrt{\mu_{BC}}} \right) \tag{278}$$

where the symbols have the meanings associated with equation (9). An alternative approach was adopted by Steiner[198], who assumed that unstable binary complexes of lifetime τ were formed, and these then collided with the third species. The result for Z_{ABC} is

$$Z_{ABC} = 54\sqrt{3}\,\pi \frac{kT}{m}\, d^4\tau \cdot n_A n_B n_C \tag{279}$$

where d and m are mean molecular diameters and masses for A, B and C. The lifetime τ was estimated from the time spent by one particle close to another when they are moving freely (i.e. d/v where v is the mean relative velocity of the two particles). Smith[199,200] has developed new equations for the rates of three-body collisions, using a symmetric representation. The collisions are assumed to occur with attractive potentials with a sharp cut-off, and the potential is represented in a 6-dimensional normalised centre of mass system. Simplification of the basic equations leads to the result

$$Z_{ABC} = n_A n_B n_C \cdot \left(\frac{15\pi}{8} \right) \left(\frac{kT}{2\pi\mu} \right)^{\frac{1}{2}} \sigma_I \tag{280}$$

where

$$\mu = \left[\frac{m_A m_B m_C}{(m_A + m_B + m_C)} \right]^{\frac{1}{2}}$$

and σ_I is the 3-body reaction cross-section. In principle, σ_I is temperature dependent, although in this work a less realistic temperature independent cross-section was employed. Smith[200] gives methods for the calculation of σ, and quotes results for collision rate coefficients ($Z_{ABC}/n_A n_B n_C$) for the recombination of iodine atoms in the presence of various third bodies. For M = Ne, Ar or Xe, the frequency factor is about 1.3×10^{-30} cm^6. molec^{-2}. sec^{-1} at 293 °K. Using lighter masses of around 30–40 units, and molecular diameters of 3 Å, formulae (278) and (279) give frequency factors of 10^{-32} to 10^{-31} cm^6. molec^{-2}. sec^{-1} at 300 °K with $\delta \sim 1$ A or $\tau \sim 10^{-12}$ sec. Comparison of these values with the bimolecular collision frequency derived under the same conditions (section 3.4) of about 10^{-10} cm^3. molec^{-1}. sec^{-1} indicates that the ratio of termolecular to bimolecular frequency factors is about $10^{-22} \times$ (concentration in molec. cm^{-3}). Thus up to about 3×10^4 torr, the termolecular frequency factor will always be less than the bimolecular factor. At ordinary pressures it seems reasonable to suppose that only those termolecular processes with a low activation energy can be of importance in relation to possible bimolecular mechanisms. Smith[200] used his calculated collision rates to obtain an overall efficiency of the iodine atom recombination at 293 °K by comparison with the experimental results of Christie et al.[201]. The efficiency was found to increase with increasing mass of the inert gas, being 1/720 for helium as third body and 1/44 for xenon.

Porter[202] has discussed a series of mechanisms for third-order recombination reactions. These are, formally, processes of the types (272–273) and (274–275) together with

$$R + M \rightarrow RM^* \tag{281}$$

$$RM^* + M \rightarrow RM + M \tag{282}$$

$$RM + R \rightarrow R_2 + M \tag{283}$$

The three mechanisms all explain two notable features of third-order atomic recombination processes: (1) the temperature coefficients are negative, and (2) the efficiency of different third bodies varies over a factor of at least 10^3. The recombination of iodine has been studied in great detail, and the temperature dependence has been obtained by the variable temperature flash photolysis experiments of Porter and Smith[196] for a number of third bodies. Porter[202]shows that, at any rate for the iodine recombination, the energy transfer theory is less successful in predicting the absolute rates and dependence on the nature of M than are the radical–molecule complex mechanisms. One particularly interesting aspect of the experimental work on iodine recombination is the variation of the temperature dependence of rate with M. Porter and Smith[196] plot their results in the form

$$_lk_R = A_2 \, Te^{-E_2/RT} \tag{284}$$

as well as in the ordinary Arrhenius form. Table 10 shows the values obtained for k_R at 20 °C, and A_2 and E_2. The correlation that is immediately apparent is the increase in $-E_2$ with increasing efficiency of the third body (as reflected in larger values of k_R). The very high efficiencies of aromatic hydrocarbons have been ascribed to the formation of charge-transfer IM complexes, and in the case of iodine itself, it is believed that the species I_3 is formed, *viz.*

$$I + I_2 \rightarrow I_3 \tag{285}$$

$$I + I_3 \rightarrow 2I_2 \tag{286}$$

The formation of a complex in processes such as (274) or in (281) with R = I, can be defined in terms of an equilibrium constant, K, and the overall rate of recombination is given by

$$-\frac{d[I]}{dt} = PZ_2 K \tag{287}$$

where P and Z_2 are the probability factor and collision rate for the last process of the sequence (275) or (283). Porter and Smith[196] obtained the constant K by a statistical thermodynamic calculation, using a binding energy for the complex estimated from the observed temperature coefficient of the overall reaction, and hence calculated rate coefficients for the various third bodies. The last column of Table 10 shows these rate coefficients calculated on the basis of the rate equation (284), and it is seen that at least the relative values are satisfyingly close to the experimental ones. A rather similar treatment has been given by Bunker and Davidson[203], except that they were able to calculate the temperature dependence of k_R since they estimated the binding energy of IM from independent sources. They found good agreement with the experimental results, even to the extent that the calculations predicted an increased temperature dependence with increased efficiency of M. Kim[204] has extended the calculations using the radical–molecule complex mechanism, and obtained an expression for the rate coefficient for the iodine recombination reaction as a function of temperature. The equilibrium

TABLE 10

RATE COEFFICIENTS AND MODIFIED ARRHENIUS FACTORS FOR THE REACTION
$I + I + M \rightarrow I_2 + M$ (from Porter and Smith[196])

Third body	$10^{32}k_R$ (20° C) (cm^6. $molec^{-2}$. sec^{-1})	$10^{36}A_2$ (cm^6. $molec^{-2}$. sec^{-1})	E_2 ($kcal$. $mole^{-1}$)	$10^{32}k_{calc}$ (cm^6. $molec^{-2}$. sec^{-1})
He	0.84	2.6	−1.4	3.0
Ar	1.64	1.8	−2.0	8.1
O_2	3.72	2.9	−2.2	11
CO_2	7.41	4.1	−2.4	17
benzene	43.9	24	−2.4	12
toluene	107	11	−3.4	86
ethyl iodide	144	24	−3.1	54
mesitylene	223	2.0	−4.8	1500
I_2	760	2.9	−5.3	2100

constant K is calculated by a method using the Lennard-Jones (6–12) potential, and the rate coefficient for the process equivalent to (275) was estimated from a collision rate assuming a Sutherland potential between I and IM. The temperature dependence of the overall termolecular rate coefficient is expressed in terms of a transcendental function of the temperature, although it is admitted that the theoretical expression is approximated equally well by the Arrhenius equation or by an inverse power dependence of the temperature over the temperature range of most experiments. The actual calculated rate coefficients are in good agreement for argon as third body with all experimental data tested. The best fit was obtained with a Lennard-Jones potential whose depth was 700 k (k is here the Boltzmann constant).

The preceding discussion of third order processes has been given in terms of the collision theory of reaction. It is also possible to use the transition state theory to predict the rates of termolecular processes. Gershinowitz and Eyring[205] applied the transition state formulation to the termolecular reactions of nitric oxide. For example, they investigated the process

$$2NO + O_2 \rightarrow 2NO_2 \tag{288}$$

and assumed a form

for the activated complex. From an analysis of the experimental results, it was shown that the energy, E_c, in the exponential of the transition state expression was small, and that most of the observed *negative* temperature dependence arose from a high negative order temperature term in the pre-exponential part of the expression. The calculated and observed rate coefficients were within a factor of about two over the temperature range 80 °K to 662 °K. Eyring *et al.*[77] subsequently gave a general formulation of the transition state theory for termolecular reactions of the type

$$A + B + C \rightarrow AB + C \tag{289}$$

It seems more appropriate at the present time to suggest orders of magnitude for rate coefficients predicted by the transition state theory for certain types of termolecular reaction, rather than to present detailed analyses of specific systems. The approach to be adopted is identical to that used in section 4.3.4, and the orders of magnitude used for f'^{trans}, f^{vib} and f^{rot} will be the lower of those in Table 4. For a reaction of three atoms, as in (289) with a linear transition state, the rate coefficient will be given by

$$k = \frac{kT}{h} \cdot \frac{(f'^{\,\text{trans}})(f^{\text{rot}})^2(f^{\text{vib}})^3}{(f'^{\,\text{trans}})^3} e^{-E_c/RT} \approx 10^{-33} e^{-E_c/RT} \text{ cm}^6. \text{ molec}^{-2}. \text{ sec}^{-1}$$

$$(290)$$

In a more complex system, say the recombination reaction of two atoms with a diatomic molecule as third-body, proceeding *via* a linear complex, *viz.*

$$A + A + B_2 \rightarrow A_2 + B_2 \tag{291}$$

the rate coefficient will be

$$k = \frac{kT}{h} \frac{(f'^{\,\text{trans}})(f^{\text{rot}})^2(f^{\text{vib}})^6}{(f'^{\,\text{trans}})^3(f^{\text{rot}})^2(f^{\text{vib}})} e^{-E_c/RT} \approx 10^{-35} e^{-E_c/RT} \tag{292}$$

Thus it is seen that the pre-exponential factors are roughly of the same order of magnitude as those observed, although the estimate is crude. Perhaps of greater importance is the form of the cancelled expression $kT/h(f^{\text{vib}})^5/(f'^{\,\text{trans}})^2$, since if vibrations are not excited the temperature dependence of the pre-exponential factor will be affected by the temperature dependence of $(f'^{\,\text{trans}})^2$ alone (*i.e.* the dependence will be T^{-1}). For more complex systems, even higher negative temperature exponents can be obtained. These results may well explain the observed negative temperature coefficients observed for termolecular processes, although it should be remembered that *experimentally* the results can well be fitted to an Arrhenius expression with a negative activation energy. In any case, the transition state formulation should be used with care for termolecular reactions which are thought to occur *via* a complex mechanism.

Some theoretical investigations of termolecular processes have been made which follow the general lines of the statistical trajectory studies described in other sections. Of these studies, that of Keck[206] may be given as an example. Rate coefficients were obtained as a function of temperature for the $I + I + Ar$ recombination, although the results were not in very good accord with experiment. Considerably more attention has been paid by theorists to the reverse process of bimolecular dissociation, *viz.*

$$M + AB \rightarrow M + A + B \tag{293}$$

both along the lines of the trajectory studies (*e.g.* Keck[207]) and in the direction indicated by Montroll and Shuler[36] in their "stochastic" treatment. For the present, however, it would appear that the most satisfactory treatment of termolecular reactions lies in the collision theory, and the important object is to elucidate the exact mechanism of the processes.

BIBLIOGRAPHY

BELL, R. P., *The Proton in Chemistry*, Methuen, London, 1959.
BENSON, S. W., *The Foundations of Chemical Kinetics*, McGraw-Hill, New York, 1960.
BUNKER, D. L., *Theory of Elementary Gas Reaction Rates*, Pergamon Press, Oxford, 1966.
EYRING, H. AND EYRING, E. M., *Modern Chemical Kinetics*, Chapman and Hall, London, 1965.
FROST, A. A. AND PEARSON, R. G., *Kinetics and Mechanism*, Wiley, New York, 1961.
GLASSTONE, S., LAIDLER, K. J. AND EYRING, H., *The Theory of Rate Processes*, McGraw-Hill, New York, 1941.
HINSHELWOOD, C. N., *The Kinetics of Chemical Change*, O.U.P., Oxford, 1940.
JOHNSTON, H. S., *Gas Phase Reaction Rate Theory*, The Ronald Press, New York, 1966.
KASSEL, L. S., *The Kinetics of Homogeneous Gas Reactions*, Chemical Catalog Co. Inc., New York, 1932.
KONDRATIEV, V. N., *Chemical Kinetics of Gas Reactions*, Pergamon Press, Oxford, 1964.
LAIDLER, K. J., *Chemical Kinetics*, McGraw-Hill, New York, 1965.
LAIDLER, K. J., *The Chemical Kinetics of Excited States*, O.U.P., Oxford, 1955.
MELANDER, L., *Isotope Effects on Reaction Rates*, The Ronald Press, New York, 1959.
RICE, O. K., *Statistical Mechanics, Thermodynamics and Kinetics*, Freeman, San Francisco, 1967.
SLATER, N. B., *Theory of Unimolecular Reactions*, Methuen, London, 1959.
TROTMAN-DICKENSON, A. F., *Gas Kinetics*, Butterworths, London, 1955.

REFERENCES

1 S. ARRHENIUS, Z. Physik. Chem., 4 (1889) 226.
2 C. N. HINSHELWOOD, The Kinetics of Chemical Change, O.U.P., Oxford, 1940, pp. 36–39.
3 M. BODENSTEIN, Z. Physik. Chem., 13 (1894) 22; 22 (1897) 1; 29 (1899) 295.
4 W. C. McC. LEWIS, J. Chem. Soc., (1918) 471.
5 J. H. SULLIVAN, J. Chem. Phys., 46 (1967) 73.
6 A. B. NALBANDYAN AND V. V. VOEVODSKII, Mekhanizm Okisleniya i Goreniya Vodoroda, Acad. Sci. USSR, Moscow, 1949.
7 L. I. AVRAMENKO AND V. N. KONDRATIEV, Zh. Fiz. Khim., 24 (1950) 207.
8 P. G. ASHMORE AND J. CHANMUGAM, Trans. Faraday Soc., 49 (1953) 254.
9 M. VAN MEERSCHE, Bull. Soc. Chim. Belg., 60 (1951) 99.
10 C. F. GOODEVE AND A. W. TAYLOR, Proc. Roy. Soc. (London), Ser. A, 152 (1935) 221; 154 (1936) 181.
11 H. S. JOHNSTON AND D. M. YOST, J. Chem. Phys., 17 (1949) 386.
12 P. G. ASHMORE AND J. CHANMUGAM, Trans. Faraday Soc., 49 (1953) 270.
13 M. R. BERLIE AND D. J. LEROY, J. Chem. Phys., 20 (1952) 200.
14 M. A. A. CLYNE, B. A. THRUSH AND R. P. WAYNE, Trans. Faraday Soc., 60 (1964) 359.
15 L. MANDELCORN AND E. W. R. STEACIE, Can. J. Chem., 30 (1952) 800.
16 A. F. TROTMAN-DICKENSON, J. R. BIRCHARD AND E. W. R. STEACIE, J. Chem. Phys., 19 (1951) 163.
17 D. ROWLEY AND H. STEINER, Discussions Faraday Soc., 10 (1951) 198.
18 C. C. SCHUBERT AND R. N. PEASE, J. Chem. Phys., 24 (1956) 919.
19 E. P. WIGNER, Gött. Nachr., 4 (1927) 375.
20 A. B. CALLEAR, Chapter 7 in Photochemistry and Reaction Kinetics, P. G. ASHMORE, F. S. DAINTON AND T. M. SUGDEN (Eds.), Cambridge University Press, 1967.
21 J. HIRSCHFELDER, J. Chem. Phys., 9 (1941) 645.
22 M. G. EVANS AND M. POLANYI, Trans. Faraday Soc., 34 (1938) 11.
23 E. T. BUTLER AND M. POLANYI, Trans. Faraday Soc., 39 (1943) 19.
24 N. N. SEMENOV, Chapter 1 in Some Problems in Chemical Kinetics and Reactivity, Pergamon Press, Oxford, 1958.
25 M. BODENSTEIN AND H. LÜTKEMEYER, Z. Physik. Chem., 114 (1925) 208.
26 D. J. LEROY, Discussions Faraday Soc., 14 (1953) 120.
27 H. O. PRITCHARD, J. B. PYKE AND A. F. TROTMAN-DICKENSON, J. Am. Chem. Soc., 76 (1954) 1201.
28 G. B. KISTIAKOWSKY AND E. R. VAN ARTSDALEN, J. Chem. Phys., 12 (1944) 469.
29 E. H. TAYLOR AND S. DATZ, J. Chem. Phys., 23 (1955) 1711.
30 H. S. JOHNSTON, Advan. Chem. Phys., 3 (1960) 131.
31 H. S. JOHNSTON AND D. RAPP, J. Am. Chem. Soc., 83 (1961) 1.
32 T. E. SHARP AND H. S. JOHNSTON, J. Chem. Phys., 37 (1962) 1541.
33 W. R. SCHULZ AND D. J. LEROY, J. Chem. Phys., 42 (1965) 3869.
34 B. WIDOM AND S. H. BAUER, J. Chem. Phys., 21 (1953) 1670.
35 H. A. KRAMERS, Physica, 7 (1940) 284.
36 E. W. MONTROLL AND K. E. SCHULER, Advan. Chem. Phys., 1 (1958) 361.
37 M. A. ELIASON AND J. O. HIRSCHFELDER, J. Chem. Phys., 30 (1959) 1426.
38 WANG-CHANG AND G. E. UHLENBECK, Transport Phenomena in Polyatomic Molecules, University of Michigan, 1951.
39 R. D. LEVINE, J. Chem. Phys., 46 (1967) 331.
40 J. O. HIRSCHFELDER, H. EYRING AND B. TOPLEY, J. Chem. Phys., 4 (1936) 170.
41 F. T. WALL, L. A. HILLER AND J. MAZUR, J. Chem. Phys., 29 (1958) 255.
42 F. T. WALL, L. A. HILLER AND J. MAZUR, J. Chem. Phys., 35 (1961) 1284.
43 M. KARPLUS, R. N. PORTER AND R. D. SHARMA, J. Chem. Phys., 43 (1965) 3259.
44 S. GOLDEN, J. Chem. Phys., 17 (1949) 620.
45 S. GOLDEN AND A. M. PEISER, J. Chem. Phys., 17 (1949) 630.
46 G. C. FETTIS AND J. H. KNOX, Progr. Reaction Kinetics, 2 (1964) 1.
47 E. BAUER AND T. Y. WU, J. Chem. Phys., 21 (1953) 726.

48 E. M. MORTENSEN AND K. S. PITZER, *Chem. Soc., Special Publ.*, 16 (1962) 57.
49 M. SZWARC, *Chem. Soc., Special Publ.*, 16 (1962) 25.
50 H. EYRING, *Chem. Soc., Special Publ.*, 16 (1962) 27.
51 C. STEEL AND K. J. LAIDLER, *J. Chem. Phys.*, 34 (1961) 1827.
52 D. R. HERSCHBACH, *Discussions Faraday Soc.*, 33 (1962) 149.
53 D. R. HERSCHBACH, *Advan. Chem. Phys.*, 10 (1966) 319.
54 S. DATZ AND E. H. TAYLOR, *J. Chem. Phys.*, 39 (1963) 1896.
55 R. D. PRESENT, *J. Chem. Phys.*, 31 (1959) 747.
56 H. EYRING, *J. Chem. Phys.*, 3 (1935) 107.
57 M. G. EVANS AND M. POLANYI, *Trans. Faraday Soc.*, 31 (1935) 875.
58 K. J. LAIDLER AND J. C. POLANYI, *Progr. Reaction Kinetics*, 3 (1964) 3.
59 D. M. BISHOP AND K. J. LAIDLER, *J. Chem. Phys.*, 42 (1965) 1688.
60 E. W. SCHLAG, *J. Chem. Phys.*, 38 (1963) 2480.
61 E. W. SCHLAG, *J. Chem. Phys.*, 42 (1965) 584.
62 J. C. LIGHT, *J. Chem. Phys.*, 40 (1964) 3221.
63 J. LIN AND J. C. LIGHT, *J. Chem. Phys.*, 45 (1966) 2545.
64 I. PRIGOGINE AND E. XHROUET, *Physica*, 15 (1949) 913.
65 I. PRIGOGINE AND M. MAHIEU, *Physica*, 16 (1950) 51.
66 B. H. MAHAN, *J. Chem. Phys.*, 32 (1960) 362.
67 J. ROSS AND P. MAZUR, *J. Chem. Phys.*, 35 (1961) 19.
68 C. W. PYNN AND J. ROSS, *J. Chem. Phys.*, 40 (1964) 2572.
69 H. C. EU AND J. ROSS, *J. Chem. Phys.*, 44 (1966) 2467.
70 H. C. EU AND J. ROSS, *J. Chem. Phys.*, 46 (1967) 411.
71 R. A. MARCUS, *J. Chem. Phys.*, 41 (1964) 2614, 2624.
72 R. A. MARCUS, *J. Chem. Phys.*, 41 (1964) 610.
73 R. A. MARCUS, *J. Chem. Phys.*, 43 (1965) 1598.
74 R. A. MARCUS, *J. Chem. Phys.*, 45 (1966) 2138.
75 R. A. MARCUS, *J. Chem. Phys.*, 45 (1966) 2630.
76 R. A. MARCUS, *J. Chem. Phys.*, 46 (1967) 959.
77 H. EYRING, H. GERSHINOWITZ AND C. E. SUN, *J. Chem. Phys.*, 3 (1935) 786.
78 F. LONDON, *Z. Elektrochem.*, 35 (1929) 552.
79 A. S. COOLIDGE AND H. M. JAMES, *J. Chem. Phys.*, 2 (1934) 811.
80 Y. SUGIURA, *Z. Physik*, 45 (1927) 484.
81 H. EYRING AND M. POLANYI, *Z. Physik. Chem.*, B12 (1931) 279.
82 J. O. HIRSCHFELDER, H. EYRING AND N. ROSEN, *J. Chem. Phys.*, 4 (1936) 121.
83 J. O. HIRSCHFELDER, H. DIAMOND AND H. EYRING, *J. Chem. Phys.*, 5 (1937) 695.
84 J. O. HIRSCHFELDER, *J. Chem. Phys.*, 6 (1938) 795.
85 G. E. KIMBALL AND J. G. TRULIO, *J. Chem. Phys.*, 28 (1958) 493.
85a S. F. BOYS AND I. SHAVITT, University of Wisconsin Naval Res. Lab. Tech. Rep. WIS-AF-13 (1959).
86 S. SATO, *J. Chem. Phys.*, 23 (1955) 592, 2465.
87 R. E. WESTON, *J. Chem. Phys.*, 31 (1959) 892.
88 P. J. KUNTZ, E. M. NEMETH, J. C. POLANYI, S. D. ROSNER AND C. E. YOUNG, *J. Chem. Phys.*, 44 (1966) 1168.
89 R. N. PORTER AND M. KARPLUS, *J. Chem. Phys.*, 40 (1964) 1105.
90 H. S. JOHNSTON AND C. PARR, *J. Am. Chem. Soc.*, 85 (1963) 2544.
91 F. T. WALL AND R. N. PORTER, *J. Chem. Phys.*, 36 (1962) 3256.
92 N. C. BLAIS AND D. L. BUNKER, *J. Chem. Phys.*, 37 (1962) 2713.
93 M. KARPLUS AND L. M. RAFF, *J. Chem. Phys.*, 41 (1964) 1267.
94 M. G. EVANS AND M. POLANYI, *Trans. Faraday Soc.*, 35 (1939) 178.
95 M. C. MOULTON AND D. R. HERSCHBACH, *J. Chem. Phys.*, 44 (1966) 3010.
96 S. DATZ AND R. E. MINTURN, *J. Chem. Phys.*, 41 (1964) 1153.
97 J. C. POLANYI AND S. D. ROSNER, *J. Chem. Phys.*, 38 (1963) 1028.
98 J. K. CASHION AND J. C. POLANYI, *Proc. Roy. Soc. (London), Ser. A*, 258 (1960) 570.
99 P. E. CHARTERS AND J. C. POLANYI, *Discussions Faraday Soc.*, 33 (1962) 107.
100 J. R. AIREY, R. R. GETTY, J. C. POLANYI AND D. R. SNELLING, *J. Chem. Phys.*, 41 (1964) 3255.

101 W. D. McGrath and R. G. W. Norrish, *Proc. Roy. Soc. (London), Ser. A*, 242 (1957) 165.
102 N. Basco and R. G. W. Norrish, *Proc. Roy. Soc. (London), Ser. A*, 268 (1962) 291.
103 D. Garvin and M. Boudart, *J. Chem. Phys.*, 23 (1955) 784.
104 J. K. Cashion and J. C. Polanyi, *J. Chem. Phys.*, 35 (1961) 600.
105 L. M. Raff and M. Karplus, *J. Chem. Phys.*, 44 (1966) 1212.
106 A. E. Grosser, A. R. Blythe and R. B. Bernstein, *J. Chem. Phys.*, 42 (1965) 1268.
107 A. E. Grosser and R. B. Bernstein, *J. Chem. Phys.*, 43 (1965) 1140.
108 J. H. Birely and D. R. Herschbach, *J. Chem. Phys.*, 44 (1966) 1690.
109 T. T. Warnock, R. B. Bernstein and A. E. Grosser, *J. Chem. Phys.*, 46 (1967) 1685.
110 R. P. Bell, *Proc. Roy. Soc. (London), Ser. A*, 139 (1933) 466.
111 R. P. Bell, *The Proton in Chemistry*, Cornell University Press, Ithaca, N.Y., 1959, p. 205.
112 R. P. Bell, *Trans. Faraday Soc.*, 55 (1959) 1.
113 E. P. Wigner, *Z. Physik. Chem.*, B19 (1932) 203.
114 C. Eckart, *Phys. Rev.*, 35 (1930) 1303.
115 H. S. Johnston and J. Heicklen, *J. Phys. Chem.*, 66 (1962) 532.
116 I. Shavitt, *J. Chem. Phys.*, 31 (1959) 1359.
117 H. Shin, *J. Chem. Phys.*, 39 (1963) 2934.
118 G. Herzberg, *Infrared and Raman Spectra*, Van Nostrand, Princeton, New Jersey, 1945, p. 221.
119 J. Bigeleisen and M. Wolfsberg, *Advan. Chem. Phys.*, 1 (1958) 15.
120 H. Carmichael and H. S. Johnston, *J. Chem. Phys.*, 41 (1964) 1975.
121 J. Bigeleisen, F. S. Klein, R. E. Weston and M. Wolfsberg, *J. Chem. Phys.*, 30 (1959) 1340
122 R. B. Timmons and R. E. Weston, *J. Chem. Phys.*, 41 (1964) 1654.
123 J. H. Sullivan, *J. Chem. Phys.*, 39 (1963) 3001.
124 A. Persky and F. S. Klein, *J. Chem. Phys.*, 44 (1966) 3617.
125 F. S. Klein, A. Persky and R. E. Weston, *J. Chem. Phys.*, 41 (1964) 1799.
126 D. R. Herschbach, H. S. Johnston, K. S. Pitzer and R. E. Powell, *J. Chem. Phys.*, 25 (1956) 736.
127 E. Wigner, *Trans. Faraday Soc.*, 34 (1938) 29.
128 J. O. Hirschfelder and E. Wigner, *J. Chem. Phys.*, 7 (1939) 616.
129 R. de L. Kronig, *Z. Physik*, 43 (1927) 524.
130 A. E. Stern and H. Eyring, *J. Chem. Phys.*, 3 (1935) 778.
131 K. J. Laidler, *J. Chem. Phys.*, 10 (1942) 34.
132 J. L. Magee, *J. Chem. Phys.*, 8 (1940) 677.
133 L. Landau, *Physik. Z. Sowjetunion*, 1 (1932) 88.
134 C. Zener, *Proc. Roy. Soc. (London), Ser. A*, 137 (1932) 696; 140 (1933) 666.
135 F. A. Lindemann, *Trans. Faraday Soc.*, 17 (1922) 598.
136 C. N. Hinshelwood, *Proc. Roy. Soc. (London), Ser. A*, 113 (1927) 230.
137 L. S. Kassel, *J. Phys. Chem.*, 32 (1928) 225.
138 O. K. Rice and H. C. Ramsperger, *J. Am. Chem. Soc.*, 49 (1927) 1617.
139 N. B. Slater, *Proc. Roy. Soc. (London), Ser. A*, 194 (1948) 112.
140 R. A. Marcus and O. K. Rice, *J. Phys. and Colloid Chem.*, 55 (1951) 894.
141 R. A. Marcus, *J. Chem. Phys.*, 20 (1952) 359.
142 F. W. Schneider and B. S. Rabinovitch, *J. Am. Chem. Soc.*, 85 (1963) 2365.
143 S. W. Benson and W. B. DeMore, *Ann. Rev. Phys. Chem.*, 16 (1965) 397.
144 T. S. Chambers and G. B. Kistiakowsky, *J. Am. Chem. Soc.*, 56 (1934) 399.
145 R. A. Ogg, *J. Chem. Phys.*, 15 (1947) 337.
146 R. L. Mills and H. S. Johnston, *J. Am. Chem. Soc.*, 73 (1951) 938.
147 L. A. K. Staveley and C. N. Hinshelwood, *Nature*, 137 (1936) 29.
148 C. Steel and A. F. Trotman-Dickenson, *J. Chem. Soc.*, (1959) 975.
149 J. C. Giddings and H. Eyring, *J. Chem. Phys.*, 22 (1954) 538.
150 H. M. Rosenstock, M. B. Wallenstein, A. L. Wahrhaftig and H. Eyring, *Proc. Natl. Acad. Sci. U.S.*, 38 (1952) 667.
151 L. S. Kassel, *J. Phys. Chem.*, 32 (1928) 1065.
152 O. K. Rice and H. C. Ramsperger, *J. Am. Chem. Soc.*, 50 (1928) 617.
153 R. A. Marcus, *J. Chem. Phys.*, 43 (1965) 2658.

154 G. M. WIEDER AND R. A. MARCUS, *J. Chem. Phys.*, 37 (1962) 1835.
155 B. S. RABINOVITCH AND R. W. DIESEN, *J. Chem. Phys.*, 30 (1959) 735.
156 G. Z. WHITTEN AND B. S. RABINOVITCH, *J. Chem. Phys.*, 38 (1963) 2466.
157 G. Z. WHITTEN AND B. S. RABINOVITCH, *J. Chem. Phys.*, 41 (1964) 1883.
158 N. B. SLATER, *Proc. Cambridge Phil. Soc.*, 35 (1939) 56.
159 N. B. SLATER, *Nature*, 159 (1947) 264.
160 N. B. SLATER, *Phil. Trans. Roy. Soc. (London), Ser. A*, 246 (1953) 57.
161 F. W. SCHNEIDER AND B. S. RABINOVITCH, *J. Am. Chem. Soc.*, 84 (1962) 4215.
162 F. W. SCHNEIDER AND B. S. RABINOVITCH, *J. Am. Chem. Soc.*, 87 (1965) 158.
163 B. S. RABINOVITCH, D. W. SETSER AND F. W. SCHNEIDER, *Can. J. Chem.*, 39 (1961) 2609.
164 E. W. SCHLAG AND B. S. RABINOVITCH, *J. Am. Chem. Soc.*, 82 (1960) 5996.
165 N. B. SLATER, *Proc. Roy. Soc. (London), Ser. A*, 218 (1953) 224.
166 T. S. CHAMBERS AND G. B. KISTIAKOWSKY, *J. Amer. Chem. Soc.*, 56 (1934) 399.
167 H. O. PRITCHARD, R. G. SOWDEN AND A. F. TROTMAN-DICKENSON, *Proc. Roy. Soc. (London) Ser. A*, 217 (1953) 563.
168 E. K. GILL AND K. J. LAIDLER, *Proc. Roy. Soc. (London), Ser. A*, 250 (1959) 121.
169 E. K. GILL AND K. J. LAIDLER, *Proc. Roy. Soc. (London), Ser. A*, 251 (1959) 66.
170 E. K. GILL AND K. J. LAIDLER, *Trans. Faraday Soc.*, 55 (1959) 753.
171 R. K. LYON, *J. Chem. Phys.*, 46 (1967) 4504.
172 B. S. RABINOVITCH AND M. C. FLOWERS, *Quart. Rev.*, 18 (1964) 122.
173 B. S. RABINOVITCH, E. TSCHNIKOW-ROUX AND E. W. SCHLAG, *J. Am. Chem. Soc.*, 81 (1959) 1081.
174 J. N. BUTLER AND G. B. KISTIAKOWSKY, *J. Am. Chem. Soc.*, 82 (1960) 759.
175 H. M. FREY, *Proc. Roy. Soc. (London), Ser. A*, 251 (1959) 575.
176 D. W. SETSER AND B. S. RABINOVITCH, *Can. J. Chem.*, 40 (1962) 1425.
177 B. S. RABINOVITCH, R. F. KUBIN AND R. E. HARRINGTON, *J. Chem. Phys.*, 38 (1963) 405.
178 D. C. TARDY, B. S. RABINOVITCH AND C. W. LARSON, *J. Chem. Phys.*, 45 (1966) 1163.
179 J. C. HASSLER AND D. W. SETSER, *J. Chem. Phys.*, 45 (1966) 3246.
180 N. B. SLATER, *J. Chem. Phys.*, 24 (1956) 1256.
181 E. THIELE, *J. Chem. Phys.*, 36 (1962) 1466.
182 D. J. WILSON AND E. THIELE, *J. Chem. Phys.*, 40 (1964) 3425.
183 E. THIELE, *J. Chem. Phys.*, 45 (1966) 491.
184 B. H. MAHAN, *J. Phys. Chem.*, 62 (1958) 100.
185 F. P. BUFF AND D. J. WILSON, *J. Chem. Phys.*, 32 (1960) 677.
186 M. HOARE, *J. Chem. Phys.*, 38 (1963) 1630.
187 R. D. LEVINE, *J. Chem. Phys.*, 44 (1966) 1567, 2029, 2035, 2046, 3597.
188 W. G. VALENCE AND E. W. SCHLAG, *J. Chem. Phys.*, 45 (1966) 216, 4280.
189 E. THIELE AND D. J. WILSON, *J. Chem. Phys.*, 35 (1961) 1256.
190 N. C. HUNG AND D. J. WILSON, *J. Chem. Phys.*, 38 (1963) 828.
191 R. J. HARTER, E. B. ALTERMAN AND D. J. WILSON, *J. Chem. Phys.*, 40 (1964) 2137.
192 E. THIELE, *J. Chem. Phys.*, 38 (1963) 1959.
193 D. L. BUNKER, *J. Chem. Phys.*, 40 (1964) 1946.
194 G. B. KISTIAKOWSKY AND E. K. ROBERTS, *J. Chem. Phys.*, 21 (1953) 1637.
195 F. KAUFMAN AND J. R. KELSO, *J. Chem. Phys.*, 46 (1967) 4541.
196 G. PORTER AND J. A. SMITH, *Proc. Roy. Soc. (London), Ser. A*, 261 (1961) 28.
197 R. C. TOLMAN, *Statistical Mechanics*, Chemical Catalogue Co., New York, 1921, p. 245.
198 W. STEINER, *Z. Physik. Chem.*, B15 (1932) 249.
199 F. T. SMITH, *J. Chem. Phys.*, 36 (1962) 248.
200 F. T. SMITH, *Discussions Faraday Soc.*, 33 (1962) 183.
201 M. I. CHRISTIE, A. J. HARRISON, R. G. W. NORRISH AND G. PORTER, *Proc. Roy. Soc. (London), Ser. A*, 231 (1955) 446.
202 G. PORTER, *Discussions Faraday Soc.*, 33 (1962) 198.
203 D. L. BUNKER AND N. DAVIDSON, *J. Am. Chem. Soc.*, 80 (1958) 5090.
204 S. K. KIM, *J. Chem. Phys.*, 46 (1967) 123.
205 H. GERSHINOWITZ AND H. EYRING, *J. Am. Chem. Soc.*, 57 (1935) 985.
206 J. KECK, *J. Chem. Phys.*, 40 (1964) 1166.
207 J. KECK, *Discussions Faraday Soc.*, 33 (1962) 173.

Chapter 4

The Theory of Elementary Reactions in Solution

I. D. CLARK AND R. P. WAYNE

1. General characteristics of solution reactions

In a system containing two reactant species at concentrations of about 0.02 moles per litre the reactant molecules have an average separation of about 10 molecular diameters. This corresponds roughly to the situation in the gas phase at a pressure of one atmosphere, and the kinetic behaviour of such a gaseous system has been treated in the previous chapter. The present chapter will be concerned with the effect that "filling up the empty space" with solvent molecules has on the kinetic behaviour of the reactants. The formation of an aqueous solution from the gaseous system of reactants requires the addition of about 1000 times their number of water molecules. The mean free paths of the reactant molecules are then reduced about 1000-fold to become of the order of a molecular diameter. Consequently a reactant molecule is, to a first approximation, constantly in contact with solvent molecules and the resulting electrical interactions between reactant and solvent may be strong enough to stabilize a *charge* on the reactant.

The production of ions from neutral species in the gas phase demands considerable amounts of energy (a few hundred kcal. mole^{-1}), and unless this energy is made available (*e.g.* by nuclear or short wavelength UV radiation), ionic species are unlikely to appear in gaseous reactions. In particular, at normal temperatures the *thermal* reactions of neutral species will probably proceed *via* a lower energy path involving uncharged intermediates rather than the higher energy path involving ionic species. In solution, however, the interaction energy between solvent and ions may be of the same order of magnitude as the energy required for the formation of ions from the neutral precursor, and the participation of ion intermediates in low-temperature thermal reactions may be expected in many condensed phase systems.

The stabilization of charge is not the only way in which the presence of solvent molecules may influence the reaction. The first section of this chapter is devoted to a consideration of how these general *non electrical* liquid phase properties affect the kinetic behaviour of the reactants. Section 2 deals with the *electrical* properties of solutions and current treatments of reactions which involve ions and dipoles; the section includes a discussion of the important and relatively new field of *electron transfer* processes. Section 3 is concerned with the kinetics of *proton transfer* reactions, which constitute the most widely studied class of solution reactions. The use of *free energy relationships* to describe kinetic behaviour in terms of the structures of reactants is discussed briefly in the final section.

1.1 COLLISIONS IN SOLUTION

As a result of the small centre-to-centre distances of nearest neighbours in solution, any one molecule is always interacting with many surrounding molecules at the same time. The definition of a bimolecular *collision* in solution therefore becomes somewhat arbitrary. It requires a choice of a collision diameter, r_{AB}, such that two particles are considered to be in a state of collision if their centres are closer than r_{AB}.

The most common choice of r_{AB} for theories of collisions is the value at which the interaction energy becomes positive with respect to the energy at infinite separation, so that a strong, short range, *repulsive* interaction results for separations less than r_{AB}. The situation is illustrated in Fig. 1. Defined in this way the resulting collision number (collisions per second per mole in 1 litre) will be equal to that for hard spheres the sum of whose radii is equal to r_{AB}. Since the free volume for a liquid is less than the molecular volume, the collision diameter will, on the average, be less than the distance between adjacent non-colliding molecules and the passage of a specified "reference" molecule between any two neighbours will be restricted by repulsive forces. To escape from the "*cage*" formed by the "nearest-neighbour" molecules the reference molecule will be forced to surmount the energy barrier presented by the repulsive forces. Consequently the reference molecule will undergo many collisions with each of its nearest neighbours before it escapes from the cage.

Rabinowitch and Wood[10] demonstrated this *cage effect* by the use of a mechanical model. They recorded electrically the collisions of a reference sphere in a tray containing many spheres which were agitated to simulate random molecular motion. For a high density of spheres (corresponding to liquid phase) it was found that the collisions occurred in *sets*. Such sets of repeated collisions in the solution phase are commonly called *encounters*, and if the reference molecule and one of its

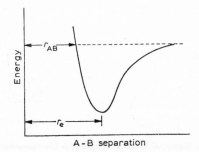

Fig. 1. Definition of a collision diameter, r_{AB}.

nearest neighbours are reactant species then the pair of reactant molecules is known as an *encounter pair*. It will be demonstrated that the encounter pair concept distinguishes the solution from the gas phase not only by its effect on very fast (diffusion controlled) reactions but on slow reactions as well. The distinction arises because the formation of encounter pairs is predicted to increase the number of collisions per second of reactant species over that for reactants of the same concentration in the gas phase.

The number of collisions experienced by a reference molecule in each encounter can be estimated from a lattice-like model for the liquid phase, and this approach has been used by Frenkel[11] and by North[1]. Because of the very short mean free path of the reference molecule inside a cage, its collisional motion can be considered to be a series of very rapid reflections or *vibrations*. The mean vibration time can be estimated from the velocity of the reference molecule and the collision duration. Using gas phase values of $\sqrt{kT/m}$ and 3×10^{-13} sec for the velocity and collision duration[12] respectively, and a path length, δ_1, of a few Ångstroms, the mean vibration period, τ_0, is of the order of 10^{-13} to 10^{-12} seconds. The *total* time, τ, spent by the reference molecule inside the cage (the duration of the encounter), can be estimated by regarding the potential barrier for escape as a square well (Fig. 2).

From the Boltzmann relation, the relative probabilities of finding the centre of the molecule in region 2 (the top of the barrier) and in region 1 (the well, or cage) are given by

$$\frac{p_2}{p_1} = \frac{\delta_2}{\delta_1} e^{-W/kT} \tag{1}$$

where δ_1 and δ_2 denote the widths of the well and the barrier respectively. The molecule passes over the barrier with the average thermal velocity $v = \sqrt{kT/m}$ (independent of its potential energy) so that the time spent in state 2 is $\delta_2/\sqrt{kT/m}$. The state probabilities are proportional to the times spent in the states and the

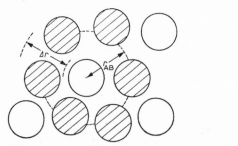
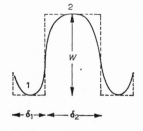

Fig.2. Diagrammatic representation of the cage effect and the potential barrier for escape of the reference molecule from the cage.

time in the lower state is therefore given by

$$\tau = \frac{\delta_1}{kT/m} \, e^{W/kT} \tag{2}$$

However, at moderate temperatures, $\delta_1/\sqrt{kT/m}$ approximately equals τ_0, the period of vibration of a particle about its equilibrium position so that

$$\tau = \tau_0 \, e^{W/kT} \tag{3}$$

For most common liquids at room temperature the height of the potential barrier, W, is thought[1] to be of the order of 5 kT, so that encounter durations will be in the order of 10^2 to $10^4 \, \tau_0$. The number of collisions between two molecules of an encounter pair would be $(2/Z)(\tau/\tau_0)$, where Z is the co-ordination number (equal to 12 for a hexagonally close packed configuration). It can therefore be expected that in most solvents at room temperature two reactant molecules in an encounter pair collide from 10 to 10^3 times before they separate. The height of the barrier depends on solvent viscosity and molecular separations, so that high viscosities, low temperatures and high pressures would favour a large number of collisions per encounter.

It is apparent that a quantitative application of collision theory to kinetic processes in solution requires an accurate knowledge of both the rate of encounter pair formation and the average number of collisions per encounter. A detailed account of a collision theory treatment of solution kinetics may be found in the monograph by North[1].

1.2 DIFFUSION EFFECTS IN SOLUTION REACTIONS

1.2.1 General considerations

The three processes involved in any chemical reaction are:
1. diffusion of reactants toward each other;
2. the actual reaction between the species;
3. diffusion of the products apart.

For slow reactions in solution, step 2 is rate controlling so that the barrier to diffusion presented by the solvent molecules does not affect the overall reaction rate. However for fast reactions, such as the quenching of fluorescence or reactions of free radicals, the probability of reaction at each encounter approaches unity and the observed reaction rate will be limited by step 1. Such a reaction is referred to as *diffusion controlled*; its observed rate is equal to the rate of encounter pair formation and is unrelated to the overall collision rate. The encounter rate of two species in solution can be calculated, on the basis of certain assumptions, using Fick's two Laws of Diffusion.

Fick's First Law relates the flux, J (the amount of material passing across a reference plane of unit area in unit time), to the concentration gradient at the plane

$$J = -\mathscr{D}\frac{dc}{dx} \tag{4}$$

$\mathscr{D}$ is the diffusion constant with units of *area × time*, and is often a function of the concentration of the diffusing material at the reference plane. For steady state conditions in solution only the first law (4) is required, and diffusion considerations can be incorporated into the kinetic equations using a first order differential equation.

However, the calculation of the time dependence of the concentration in a particular volume of material requires application of *Fick's Second Law*

$$\frac{\partial c}{\partial t} = \frac{\partial}{\partial x}\left(\mathscr{D}\frac{\partial c}{\partial x}\right) \tag{5}$$

If $\mathscr{D}$ is independent of concentration the Second Law has the form

$$\frac{\partial c}{\partial t} = \mathscr{D}\frac{\partial^2 c}{\partial x^2} \tag{6}$$

since $\mathscr{D}$ will also be independent of x. Equations (5) and (6) are second order differential equations and their solutions have been found for various boundary conditions[13,14].

1.2.2 Calculation of diffusion controlled rates

The calculation of rate coefficients for reactions where diffusion is important has involved three different approaches, which have been discussed in some detail by North[1,15] and Noyes[16].

(a) Simple Fick's Law approach

Since Fick's Laws apply to isotropic continuous media their application to the calculation of encounter rates in molecular media is not strictly valid, especially in the region where encountering partners are separated by only a few solvent molecules. The Fick's Law approach assumes that the solvent can be approximated by an isotropic continuum even at distances just outside the collision radius of a reactant.

If such a model is assumed, Fick's Laws are applicable and a steady state solution can be obtained from (4) by considering a reference molecule A to be surrounded by a spherically symmetrical gradient of B molecules diffusing towards it with a net average flux J. At a distance r from the reference molecule the area of the spherical surface is $4\pi r^2$. Application of Fick's First Law at this surface yields

$$J = 4\pi r^2 \mathscr{D}_{AB} \left(\frac{dc_B}{dr}\right)_r \tag{7}$$

where $\mathscr{D}_{AB}$ is the coefficient of relative diffusion of species A and B and has been shown[17] to be equal to $\mathscr{D}_A + \mathscr{D}_B$. The rate coefficient for a diffusion controlled reaction can now be calculated, using a boundary condition (first proposed by Smoluchowski[18]) which corresponds to a reaction occurring at each encounter. As soon as molecule B reaches the encounter distance r_{AB} it will be removed by reaction, so that $c_B = 0$ for $r \leqq r_{AB}$. Integration of (7), with $c_B = 0$ at $r = r_{AB}$ and $c_B = c_B^0$ at $r = \infty$, gives $c_B^0 = J/4\pi r_{AB}\mathscr{D}_{AB}$. The observed rate coefficient will be

$$k_{obs} = -\frac{dc_A}{dt} / c_A^0 c_B^0 \tag{8}$$

But $-dc_A/dt = Jc_A$, and substituting $c_B^0 = J/4\pi r_{AB}\mathscr{D}_{AB}$ yields

$$k_{obs} = 4\pi r_{AB}\mathscr{D}_{AB} \ (cm^3.molec^{-1}.sec^{-1}) \tag{9}$$

$$= \frac{4\pi N_0}{1000} r_{AB}\mathscr{D}_{AB} \ (l.mole^{-1}.sec^{-1}) \tag{10}$$

where N_0 is Avogadro's number.

The Smoluchowski diffusion controlled rate coefficient given by (10) is thus equal to the rate of encounter pair formation. Typical values of $r_{AB} = 4$ A and $\mathscr{D}_{AB} = 2 \times 10^{-5}$ cm^2. sec suggest an encounter frequency of about 6×10^9 l. mole^{-1}. sec^{-1}, which is about 30 times smaller than typical collision frequencies in the gas phase.

Equation (7) may be solved for a more general case than reaction at every encounter in the following way[1]. If c_B^* is the average concentration of B molecules just undergoing an encounter with the reference A molecule, then the flux of B towards the reference molecule will be

$$J = kc_B^* \tag{11}$$

where k is a rate coefficient equal to that which would be observed if the diffusion was *infinitely fast* thus keeping c_B^* equal to the average value, c_B^0. Integration of (7) between the limits $c_B = c_B^0$ at $r = \infty$ and $c_B = c_B^*$ at $r = r_{AB}$ yields

$$c_B^* = c_B^0 - \frac{J}{4\pi r_{AB} \mathscr{D}_{AB}} \tag{12}$$

Substitution of $J = k c_B^*$ and rearrangement gives

$$c_B^* = \frac{c_B^0}{1 + \dfrac{k}{4\pi r_{AB} \mathscr{D}_{AB}}} \tag{13}$$

As in (8), $k_{obs} = J/c_B^0$, and therefore

$$k_{obs} = \frac{k c_B^*}{c_B^0} = \frac{k}{1 + \dfrac{k}{4\pi r_{AB} \mathscr{D}_{AB}}} = \frac{4\pi r_{AB} \mathscr{D}_{AB}}{1 + 4\pi r_{AB} \mathscr{D}_{AB}/k} \tag{14}$$

For the elementary reaction $2A \rightarrow A: A \rightarrow$ products, the rate of reaction is

$$-2dc_A/dt = k_{obs} c_A^2 = J c_A$$

so that $J = 2k c_B^*$, which yields

$$k_{obs} = \frac{2\pi r_{AB} \mathscr{D}_{AB}}{1 + 2\pi r_{AB} \mathscr{D}_{AB}/k} \tag{15}$$

It can be seen from (14) that the simple Smoluchowski equation (9) will only be valid for the condition $k \gg 4\pi r_{AB} \mathscr{D}_{AB}$, which in some cases may not be so even if the reaction occurs at every collision. At the other extreme (14) shows that k_{obs} and k are equal (and that diffusion considerations can be ignored) only if $k \ll 4\pi r_{AB} \mathscr{D}_{AB}$. The typical values adopted for r_{AB} and $\mathscr{D}_{AB}$ suggest that k_{obs} will be within 1 % of k only if k is less than about 10^7 l. mole^{-1}. sec^{-1}.

North[1] shows how the simple Fick's Law treatment can be extended to the initial non-stationary parts of a reaction by application of the Second Law. If the A molecules were formed instantaneously at random positions in solution then c_B^* would be equal initially to c_B^0 before it fell rapidly to the steady state value (12). The time dependence of the material flux is analogous to that of the heat flux in problems of thermal conduction and can be shown to be given by

$$
J = \frac{4\pi r_{AB} \mathscr{D}_{AB} c_B^0}{1 + 4\pi r_{AB} \mathscr{D}_{AB}/k} \left\{ 1 + \frac{k \exp(\alpha^2) \operatorname{erfc}(\alpha)}{4\pi r_{AB} \mathscr{D}_{AB}} \right\}
\tag{16}
$$

where $\alpha = (\mathscr{D}_{AB} t)^{\frac{1}{2}}(1 + k/4\pi r_{AB} \mathscr{D}_{AB})/r_{AB}$ and erfc is the error function. Using only the first terms of the power series expansion for $\exp(\alpha^2)\operatorname{erfc}(\alpha)$ the observed rate coefficient $k_{obs}(= J/c_B^0)$ is given by (17)

$$
k_{obs} = \frac{4\pi r_{AB} \mathscr{D}_{AB}}{1 + 4\pi r_{AB} \mathscr{D}_{AB}/k} \left\{ 1 + \frac{r_{AB}}{(\pi \mathscr{D}_{AB} t)^{\frac{1}{2}}[1 + 4\pi r_{AB} \mathscr{D}_{AB}/k]} \right\}
\tag{17}
$$

The power series approximation used in (17) is accurate to within 1 per cent of $\exp(\alpha^2)\operatorname{erfc}(\alpha)$ for typical values of $r_{AB} \mathscr{D}_{AB}$ and k for times greater than 10^{-11} sec after the initiation of the reaction. At times greater than 10^{-7} sec, the steady state equation (14) will be valid.

(b) The diffusion and pair probability approach

The second method of treating rate processes where diffusion is important involves consideration of the probability distribution of molecular pairs in the solution. The detailed development of the theory may be found in the discussions of Collins and Kimball[19] and Waite[20]. The same parameters are employed as in the previous simple treatment with the addition of Δr, the thickness of the cage surrounding the reference molecule as shown in Fig. 2. The chemical rate coefficient for reaction of the two molecules in an encounter pair is found from analogy with gas phase collision theory to be $vPe^{-E/RT}$ where v is the average frequency of A–B collision, P is the "probability" factor and E is the chemical activation energy. The resultant expression for the overall steady state rate coefficient is

$$
k_{obs} = \frac{4\pi r_{AB} \mathscr{D}_{AB}}{1 + \mathscr{D}_{AB}/(r_{AB}\Delta r vPe^{-E/RT})}
\tag{18}
$$

This expression reduces to the Smoluchowski equation for the diffusion controlled limiting rate in the same way as equation (14), under those conditions where the rate of reaction during an encounter is much greater than the rate of diffusion of reactants together, i.e. where $r_{AB}\Delta r vPe^{-E/RT} \gg \mathscr{D}_{AB}$.

(c) Molecular pair approach[16]

The third treatment does not employ Fick's Laws at all but considers the potential reactivity of isolated molecular pairs. This approach does not involve any of the approximations inherent in the application of laws for isotropic continua to actual systems containing discrete molecules. However, some of the parameters involved in the molecular pair treatment are difficult to relate to experimental observables without additional assumptions, which are in some ways equivalent to those inherent in the diffusion equation.

Noyes[16] has compared some of the methods used to compute the diffusion controlled rate and has discussed the assumptions contained in each approach.

1.3 COMPARISON OF REACTION RATES IN SOLUTION AND THE GAS PHASE

1.3.1 Collision theory considerations

Collision theory applied to an ideal gas identifies the pre-exponential term A of the calculated bimolecular rate coefficient as a collision number, the number of bimolecular collisions in unit time of the reactants at unit concentration. Typical values for molecular masses give A as 10^{10}–10^{11} l. mole^{-1}. sec^{-1} at room temperature, and since experimentally determined rate coefficients often can have pre-exponential factors of this magnitude, 10^{10}–10^{11} l. mole^{-1}. sec^{-1} is considered to be the "normal" gas phase frequency factor. North[1] suggests that the common practice of assuming this number to be a criterion of "normal" behaviour in *solution* reactions is mistaken, and he shows how liquid phase collision theory predicts an A factor of at least an order of magnitude greater.

The liquid phase collision number is equal to the molar concentration of encounter pairs multiplied by the frequency of collision of two reactant species in an encounter pair and divided by the reactant concentrations. In section 1.1 it was seen that the frequency of reactant species collisions in an encounter pair is approximately $2/Z\tau_0$ where τ_0 is the period of "vibration" of the reference molecule about its equilibrium point caused by collisions with its Z nearest neighbours. The solution phase collision number is therefore given approximately by

$$A = \frac{c_{AB}}{c_A c_B} \frac{2}{Z\tau_0} \tag{19}$$

The concentration of encounter pairs, c_{AB}, can be calculated from a consideration of the cages around the reference, A, molecules as shown in Fig. 2. The probability of a reactant, B, molecule being adjacent to an A molecule is simply equal to the probability of B being in a cage of average thickness $\overline{\Delta r}$, $4\pi r_{AB}^2 \overline{\Delta r} c_B$, multiplied by the concentration of cages, c_A, so that

$$c_{AB} = 4\pi r_{AB}^2 \overline{\Delta r} c_A c_B \tag{20}$$

Substitution of this value for c_{AB} into (19) and conversion to l. mole^{-1}. sec^{-1}

units yields

$$A = \frac{N_0}{1000} 4\pi \overline{r_{AB}^2} \overline{\Delta r} \frac{2}{Z\tau_0} \quad \text{l.mole}^{-1}.\text{sec}^{-1} \tag{21}$$

Substitution of reasonable values for $\overline{r_{AB}}$ and $\overline{\Delta r}$ of 5 A, for Z of 12 and for τ_0 of 10^{-13} to 10^{-12} sec, as discussed in Section 1.1, yields a value for A lying in the range of 10^{11} to 10^{12} l. mole^{-1}. sec^{-1}.

In addition North shows how liquid phase collision theory predicts a more favourable exponential factor than that for the gas phase. The effect is ascribed to the high degree of vibrational motion between encounter pairs which is possible in solution: the vibration makes available more collisional energy and therefore provides a better chance for two colliding reactant molecules to overcome the energy barrier for reaction at each collision. In the language of classical statistical mechanics, when a pair of colliding molecules is situated in an ordered lattice, each translational degree of freedom (one "square term") can be replaced by a vibrational degree of freedom (two "square terms"), the change being associated with the potential energy of the intermolecular forces. Because the liquid phase is not a completely ordered lattice it is difficult to estimate quantitatively the magnitude of this effect. However, when both the exponential and the pre-exponential terms are included in the bimolecular rate coefficient, the ratio of rate coefficients for reactions occurring in both solution and the gas phase would be predicted to be: $k_{(s)}/k_{(g)} \approx 10\text{--}10^2$.

1.3.2 Transition state theory considerations

In the transition state theory (see Chapter 3, Section 4) the rate coefficient for the process, $A + B \rightarrow AB^{\ddagger} \rightarrow$ products, is obtained by writing the rate as $vc_{\ddagger}$, where v is a frequency (which can be shown to be equal to kT/h) and $c_{\ddagger}$ the concentration of the activated complex, $AB^{\ddagger}$. Thus for a gas-phase reaction the rate is given by

$$k_{(g)} = v \frac{c_{\ddagger}}{c_A c_B} = vK^{\ddagger} = \frac{kT}{h} K^{\ddagger} \tag{22}$$

$K^{\ddagger}$ is an equilibrium constant defined here in terms of concentrations. Properly speaking, a true equilibrium constant should be defined in terms of activities, a_i, rather than concentrations, c_i. Although it is frequently possible to write $a_i \approx c_i$ for gases at low pressures, in solution the approximation is no longer valid, and an activity coefficient, f_i, must be introduced, defined by the relation

$$a_i = f_i c_i \tag{23}$$

The nature of the "equilibrium" has been discussed in Chapter 3, and the strictures there make questionable the propriety of constructing an "equilibrium" constant from activities. However, many of the discussions of the application of transition state theory to condensed phases have been concerned, *inter alia*, with the magnitude of the activity coefficients, and the reasonably good agreement obtained between predicted and experimental rate coefficients justifies a description of the theories. To conform with normal practice, the free-energy of activation is also discussed as though it were a *Gibbs* free energy, $\Delta G^{\ddagger}$, rather than a *Helmholtz* free energy, $\Delta A^{\ddagger}$. The equilibrium constant, $K^{\ddagger}$, is meant to be a *concentration* equilibrium constant, $K_c^{\ddagger}$, and, even though the pressure may normally be regarded as constant, the Helmholtz free energy should really be employed. (For a further discussion of the difficulties that arise, see pp. 216–217 of this volume).

In the generally accepted treatments, the rate coefficient, $k_{(s)}$, for a particular solvent is related to that in a different solvent (or in the gas phase) by formulating the equilibrium constant for the formation of activated complex in terms of activities

$$K^{\ddagger} = \frac{a_{\ddagger}}{a_A a_B} = \frac{c_{\ddagger}}{c_A c_B} \cdot \frac{f_{\ddagger}}{f_A f_B} \qquad (24)$$

Substitution of (24) in (22) yields the result

$$\frac{k_{(s)}}{k_{(g)}} = \frac{K_{(s)}^{\ddagger}}{K_{(g)}^{\ddagger}} \cdot \frac{f_A f_B}{f_{\ddagger}} \qquad (25)$$

if the gas-phase $K^{\ddagger}$, *i.e.* $K_{(g)}^{\ddagger}$, is written directly in terms of concentration, and (25) relates the rate coefficient of a reaction in any solution to that in the gas phase. The rate coefficient, $k_{(s)}$, may also be expressed in terms of the rate coefficient, k_0, for an ideal solution where $f_A = f_B = f_{\ddagger} = 1$, *viz.*

$$k_{(s)} = k_0 \cdot \frac{f_A f_B}{f_{\ddagger}} \qquad (26)$$

Equation (26) is known as the *Brønsted–Bjerrum equation*. The precise activity coefficients, f, to be used, naturally depend on the standard state for which k_0 is defined. It is usual to give k_0 as the rate coefficient for an ideal solution referred to infinite dilution.

A roughly quantitative comparison of solution and gas phase reaction rates has been made from equation (25) in the following way by Benson[2]. An ideal solution is assumed so that (25) becomes

$$\frac{k_{(s)}}{k_{(g)}} = \frac{K^{\ddagger}_{(s)}}{K^{\ddagger}_{(g)}} \tag{27}$$

and Raoult's Law, $p_A = x_A p^0_A$, is assumed to hold at all concentrations (p^0_A is the vapour pressure of pure reactant A and x_A is its mole fraction in solution). For ideal gases $p_A = c_{A(g)}RT$ and for dilute solutions $x_A \approx c_{A(s)}/c_s \approx c_{A(s)}V_s$ where V_s is the molar volume of the solution. Substitution into (27) yields

$$\frac{k_{(s)}}{k_{(g)}} = \frac{c_{\ddagger(s)}c_{A(g)}c_{B(g)}\cdots}{c_{\ddagger(g)}c_{A(s)}c_{B(s)}\cdots} \tag{28}$$

$$= \frac{p^0_A p^0_B \cdots}{p^0_{\ddagger}}\left(\frac{V_s}{RT}\right)^{n-1}$$

where n is the molecularity of the reaction. The integrated form of the Clausius–Clapeyron equation yields an expression for the reactant vapour pressures

$$p^0 = e^{\Delta S^0/R}e^{-\Delta H^0/RT} \tag{29}$$

where ΔS^0 and ΔH^0 are the entropy and enthalpy of vaporization of the reactant referred to a standard state of one atmosphere. From the free volume theory of liquids $\Delta S^0/R$ can be replaced by RT/V_f where V_f is the free volume per mole. Thus (28) becomes

$$\frac{k_{(s)}}{k_{(g)}} = \frac{V_s}{V_{fA}} \cdot \frac{V_s}{V_{fB}} \cdots \frac{V_{f\ddagger}}{V_s}e^{\Delta H_{vap}^0/RT} \tag{30}$$

where ΔH^0_{vap} is the *difference* in heat of vaporization between the activated complex and the reactants and is related to the differences in energies of vaporization by $\Delta H^0_{vap} = \Delta E^0_{vap} - (n-1)RT$.

For reactions in which the reactants form a loosely bound activated complex the energy of vaporization for the complex is approximated by the sum of the energies of vaporization of the reactants, so that $\Delta E_{vap} = 0$ and $\Delta H^0_{vap} = -(n-1)RT$. For most liquids the free volumes are about one per cent of their molar volumes. Application of two final assumptions, that the molar volumes of the solvent are about equal to those of the reactants and that the molar volume of the activated complex is equal to the sum of those of the reactants, means that

$$\frac{V_s}{V_{fA}} \sim \frac{V_s}{V_{fB}} \sim \cdots \sim 100, \qquad \frac{V_{f\ddagger}}{V_s} \sim \frac{n}{100} \tag{31}$$

so that (30) becomes

$$\frac{k_{(s)}}{k_{(g)}} = \frac{n10^{2n-2}}{e^{n-1}} \qquad (32)$$

where n is the molecularity of the reaction. The values of $k_{(s)}/k_{(g)}$ predicted from (32) for unimolecular, bimolecular and termolecular reactions ($n = 1, 2, 3$) are 1, 86 and 4050 respectively. The prediction that a unimolecular reaction would proceed at about the same rate in both solution and gas phase has been supported experimentally for the few reactions that have been studied in both phases. The predicted effect of the presence of a solvent on the rate of a bimolecular reaction agrees qualitatively with the increase predicted by collision theory in section 1.3.1. However, experimental results for bimolecular reactions, which are thought to occur by the same mechanism in both phases, are too sparse to provide a test of these predictions; for termolecular reactions data is virtually nonexistent.

1.4 EFFECT OF SOLVENT ON NON-IONIC REACTIONS

Collision theory indicates that the rate of a reaction depends on the relative concentration of encounter pairs A: B in the solution. If the solvent molecules are larger than the reactant molecules then the relevant parameters $\overline{r_{AB}^2}$ and $\overline{\Delta r}$ in equation (20), (and therefore the encounter pair concentration) will depend on the nature of the solvent molecules. Such very simple considerations predict that the bimolecular rate constant which varies as $c_{AB}/c_A c_B$ will be higher for solvents of high molar volume. On the other hand if the solvent molecules are smaller than the reactant molecules, $\overline{r_{AB}^2}$ and $\overline{\Delta r}$ will be governed primarily by the nature of the reactants and the molecular size of the solvent species would have little effect on the rate.

A second solvent property that is predicted by collision theory to have an important effect on reaction rates is *viscosity*. Since the collision theory indicates that the cage effect increases the collision frequency (section 1.3), more viscous solvents should enhance this effect by making the cage a more rigid environment for the vibratory motion of the encounter pair. An increase in solvent viscosity would therefore be expected to *retard* the rate of diffusion controlled reactions by reducing the rate of encounter pair formation, but to *increase* the rate of non-diffusion controlled reactions as a result of the increased collision number.

A prediction of the solvent effects on reaction rates due to differences in the degree of *solvation* of reactant species can be made by extending the ideal solution treatment in 1.3.2.

Even for dilute solutions ($< 0.1\ M$), there may be significant deviations from Raoult's Law as a result of solvation of the solute. A Raoult's Law activity coefficient may be defined by

$$p = xfp^0 \tag{33}$$

and used in the derivation of (30) from (25) rather than (27). The expression

$$\frac{k_{(s)}}{k_{(g)}} = \frac{f_A f_B}{f_{\ddagger}} \cdots \frac{V_s}{V_{fA}} \cdot \frac{V_s}{V_{fB}} \cdots \frac{V_{f\ddagger}}{V_s} \cdot e^{\Delta H_{vap}^0/RT} \tag{34}$$

results. The effect of solvation will be observed in the activity coefficients and in the exponential term of (34), since ΔH_{vap}^0 is the difference in heats of vaporization of the activated complex and the reactant species. For a reaction in which the activated complex is more highly solvated than the reactants the exponential term will increase, but the increase will be partially compensated by a decrease in the pre-exponential term due to the change in the activity coefficients. In most cases the exponential term dominates and, consequently, such reactions have higher rates in solvents of high than in those of low solvating power.

In solutions where solvation and association of reactant species is nonexistent the solvent effects are less pronounced, and due only to differences in the activity coefficients, f, see equation (26). A solution in which association and solvation effects are negligible, so that the molecular distribution is completely random, is known as a *regular solution*[21], and is often closely approximated by solutions containing only neutral species with negligible dipole moments. In these systems Coulombic forces are unimportant and the forces between molecules are primarily of the van der Waals type. For separations greater than the collision diameter the forces are attractive and lead to a cohesive potential energy in liquids that manifests itself in the heat of vaporization. Hildebrand and Wood[22] have demonstrated that the activity coefficients of a nonpolar reactant species in solution can be expressed in terms of the relative strengths of the van der Waals attractive forces characteristic of the reactants and of the solvent. The characteristic attractive potential energy per mole, E_v, for a particular species (experimentally measurable as the molar heat of vaporization in the pure liquid state) is approximately equal to a/V where a is the van der Waals attractive constant and V is the molar volume. The quantity E_v/V, or a/V^2, therefore represents the *cohesive energy density* of the species. Since it has the dimensions of force per unit area, E_v/V is often referred to as the *internal pressure* of the particular substance.

In their derivation, Hildebrand and Wood assumed a van der Waals attractive potential which varied as the sixth power of the distance between two molecules. The resulting expression which relates the activity coefficient of a species to the difference in internal pressures between it and the solvent species, S, is given by

$$\ln f_A' = \frac{V_A}{RT} \left(\frac{x_S V_S}{x_A V_A + x_S V_S} \right) \left[\left(\frac{E_{vA}}{V_A} \right)^{\frac{1}{2}} - \left(\frac{E_{vS}}{V_S} \right)^{\frac{1}{2}} \right]^2 \tag{35}$$

where x_A, x_S are the mole fractions of reactant A and solvent respectively. It should be noted that f'_A is an activity coefficient referred to a standard state of pure A. If a parameter Δ is defined by

$$\Delta_A = \left[\left(\frac{E_{vA}}{V_A}\right)^{\frac{1}{2}} - \left(\frac{E_{vS}}{V_S}\right)^{\frac{1}{2}}\right]^2$$

and the solution is sufficiently dilute that $x_A \ll x_S$, then (35) becomes

$$\ln f'_A = \frac{V_A}{RT} \cdot \Delta_A \tag{36}$$

If values for f'_A, f'_B and $f'_{\ddagger}$ are given by expressions such as (36) and substituted into a modified Brønsted–Bjerrum equation (26) (where k'_0 is a rate coefficient corresponding to the new definition of ideality), an expression

$$\ln k = \ln k'_0 \left(\frac{V_A}{RT} \Delta_A + \frac{V_B}{RT} \Delta_B - \frac{V_{\ddagger}}{RT} \Delta_{\ddagger}\right) \tag{37}$$

may be obtained. The molar volumes in (37) are almost independent of the identity of the solvent so that the variations in rate for a given reaction in different solvents depend on the Δ terms and thus on the internal pressures of the solvents. Since V and Δ for the activated complex cannot be measured, a quantitative calculation of k/k_0 is not possible. However, qualitative predictions of solvent effects can be made if it is assumed that the properties of the complex resemble those of the products. It is then apparent that if the products and reactants have similar internal pressures the solvent will have little effect on the rate. If, on the other hand, the products have higher heats of vaporization than the reactants the reaction will be favoured by solvents of high internal pressure. Such a conclusion has received experimental support and indeed was proposed as an empirical relation as early as 1929[23].

1.5 THE EFFECT OF PRESSURE ON REACTION RATE[24,25]

Externally applied pressure may affect all three of the processes involved in a solution reaction: the diffusion of reactants together, the chemical reaction itself and the diffusion of the products apart. The viscosities of organic solvents exhibit an approximately exponential increase with increasing pressure, and pressure may either increase or decrease the rate of the chemical step of the reaction. It is therefore possible for the rate determining step to change as the pressure is increased so that a reaction, which is normally "slow", may be diffusion controlled at high pressures and *vice versa*.

The way in which the chemical step of a reaction is influenced by pressure is best understood from the transition state theory formulation. If the transmission coefficient is unity the rate coefficient can be expressed in thermodynamic quantities as

$$k = \frac{kT}{h} \cdot K^{\ddagger} \cdot \frac{f_A f_B \cdots}{f_{\ddagger}} \tag{38}$$

For dilute solutions ($< 0.1 \, M$) the activities rarely differ significantly from the concentrations, so that the activity coefficients drop out of (38) to give

$$k = \frac{kT}{h} \cdot K^{\ddagger} \tag{39}$$

The thermodynamic relations

$$\Delta G^{\ddagger} = -RT \ln K^{\ddagger} \quad \text{and} \quad \frac{\partial \Delta G^{\ddagger}}{\partial P} = \Delta V^{\ddagger} \tag{40}$$

can be substituted into the differential form of equation (39) to give

$$\frac{\partial \ln k}{\partial P} = -\frac{\Delta V^{\ddagger}}{RT} \tag{41}$$

Rigorously, $K^{\ddagger}$ should be based on mole fraction concentrations in which case (41) would include terms containing coefficients of compressibility and thermal expansion. However, in practice, these terms can usually be ignored. To produce significant changes in ln k pressures of several hundred atmospheres are commonly used. In many cases a plot of ln k against pressure is linear[25] indicating that for these reactions the quantity $\Delta V^{\ddagger}$ is independent of pressure. $\Delta V^{\ddagger}$ is known as the *volume of activation* and is the difference between the volume of the activated complex including its molecules of solvation and the volumes of the reactant molecules with their associated solvent molecules. It is convenient to consider $\Delta V^{\ddagger}$ as composed of $\Delta V_1^{\ddagger}$, the change in volume of the reactant molecules themselves in forming the activated complex, and $\Delta V_2^{\ddagger}$, the change in volume of the solvated solvent molecules. For strongly dipolar or ionic species the degree of solvation may be extensive and the $\Delta V_2^{\ddagger}$ term often dominates.

In the reaction $A \rightleftarrows X^{\ddagger} \rightleftarrows C + D$ a bond (or bonds) will be stretched to form $X^{\ddagger}$ and stretched further to form $C + D$, so that $\Delta V_1^{\ddagger}$ would be expected to be positive for a unimolecular decomposition and negative for the reverse bimolecular association. For reactions $A + B \rightleftarrows X^{\ddagger} \rightleftarrows C + D$ the sign of $\Delta V_1^{\ddagger}$ will depend on the specific reaction and is usually fairly small.

The $\Delta V_2^{\ddagger}$ contribution is caused by the rearrangement of the solvent molecules due to steric requirements of the reaction and to changes in electric field surrounding the solvated species. The latter effect predominates in reactions of ions. An increase in the charge on species in solution increases the electrostatic attractive forces between the species and the permanent or induced dipoles in the solvating molecules. This leads to a reduction in volume (*electrostriction*) of the solvated complex. For a sphere the extent of electrostriction varies as the *square* of the charge on the sphere so that the association of molecules with like charges will increase the electrostriction in the solvent and $\Delta V_2^{\ddagger}$ will be large and negative, while association of two oppositely charged reactants will be accompanied by a large positive $\Delta V_2^{\ddagger}$.

These considerations would suggest a correlation between $\Delta V^{\ddagger}$ and the entropy of activation, $\Delta S^{\ddagger}$, which is in turn related to the freedom of motion within the solvated complex. A roughly linear correlation has been demonstrated for ionic reactions in aqueous solution[26]. A relation could therefore be expected between the direction of the pressure effect, which depends on the sign of $\Delta V^{\ddagger}$, and the rate of a chemical reaction, which contains $\Delta S^{\ddagger}$ in its frequency factor term. This was proposed by Perrin[27] in 1938, who observed that reactions fell into three broad classes according to their behaviour under applied pressure. Reactions with "normal" ($\sim 10^{12}$ l. mole^{-1}. sec^{-1}) frequency factors exhibited only a slight increase, slow reactions a large increase, and fast reactions (along with unimolecular decompositions) a large decrease in reaction rate when the pressure was increased

2. Reactions of ions and dipoles

2.1 ENERGETIC CONSIDERATIONS OF IONS IN SOLUTION

In nonionic solutions molecules can often be considered to be independent nonassociated species whose energy is not affected (other than thermally by collision) by other species in the solution. In this respect the behaviour of molecules in solution is very similar to that in the gas phase. Since the van der Waals forces are short range, dropping off as $1/r^7$, even at nearest neighbour distance the energy of interaction is only of the order of kT, the thermal energy of the molecule (equivalent to 0.59 kcal. mole^{-1} at 25 °C).

However, in an ionic solution with a polar solvent the reactant species can be considered neither as independent nor as unassociated entities. Since the main forces between ions are Coulombic the energy of interaction falls off as $1/r$, and a charged particle many molecular diameters away may contribute significantly to the energy of an ion. It is the strength of these ion–ion interactions that precludes any sizeable equilibrium concentration of ions in the gas phase. The major difference between reactions in solution and in the gas phase arises from the fact that the associative electrostatic interaction between individual ions and the solvent molecules can be strong enough to stabilize an ionic charge and thus to make the existence of ions in solution energetically possible.

To gain a very rough idea of the magnitude of such electrostatic interactions we can consider point charges in a vacuum. Two points of atomic charge, $\varepsilon = 4.8 \times 10^{-10}$ esu, and separated by a distance r of 10 A will have a potential energy of interaction of $U(r) = \varepsilon^2/r = 3.8 \times 10^{-12}$ erg or about 100 kT at room temperature. The energy of interaction between an ion and a (neutral) dipole falls off as $1/r^2$ and depends on the relative orientation of the dipole. However, for nearest neighbour interaction distances its magnitude can also be considerably greater than kT. If a point dipole having a dipole moment, μ, equal to that possessed by a water molecule (1.85×10^{-18} esu. cm) is situated 3 A away from a point charge and has its dipole moment oriented along the lines of centres then the energy of interaction, $U(r) = \varepsilon\mu/r^2$, is about 20 kT.

Even the interaction of ions with molecules not possessing permanent dipoles can involve considerable energies at small separations. For a point charge and a spherically symmetric molecule this energy is $U(r) = \alpha\varepsilon^2/2r^4$, so that for a typical value of molecular polarizability, $\alpha = 5 \times 10^{-24}$ cm^3, and of r, 3 A, the interaction energy is again about 20 kT. Polarization also occurs in dipoles and ions and for small separations a correction term may be required. However, since the energy falls off as $1/r^4$ it is not important at distances much greater than nearest neighbour separations.

In practice, the electrostatic interactions discussed above are reduced by a factor of $1/D$, where D is the dielectric constant of the medium. Although macroscopic dielectric constants are considerably greater than 1 (D $\approx$ 80 for H_2O), the effective D may be much smaller than this for interaction distances less than a few Ångstroms, as discussed in section 2.2.1.

The treatment of ionic reactions in this chapter is restricted to those having a significant free energy of activation, *i.e.* reactions involving the making and breaking of *covalent* bonds and electron transfer reactions whose solvent reorganization requirements give rise to a fairly large activation energy. Simple recombinations such as the reaction of silver and chloride ions to form an insoluble salt have negligible activation energies. These reactions are very rapid and must be treated as diffusion controlled processes.

Scatchard[28] and Moelwyn-Hughes[29] have treated ionic reactions using collision theory, by modifying the equations to deal with the electrostatic interactions. This results in P factors which are greater than unity for reactions of oppositely charged species and which are correspondingly reduced for reactions of ions with the same sign. In the transition state approach used in this chapter the rate depends upon the free energy of formation of the activated complex, according to

$$k = \frac{kT}{h} e^{-\Delta G^{\ddagger}/RT} \tag{42}$$

where $\Delta G^{\ddagger}$ is the free energy of activation per mole.

2.2 FREE ENERGY OF ACTIVATION IN IONIC REACTIONS

For reactions in ionic solution $\Delta G^{\ddagger}$, the free energy of activation for the formation of an activated complex from two ions, can be separated into three parts:

(1) $\Delta G_0^{\ddagger}$, the free energy of activation in the absence of all electrostatic effects.

(2) $\Delta G_{es}^{\ddagger}$, the free energy change associated with the electrostatic forces *between the two reactants* as they are brought together.

(3) $\Delta G_{\mu}^{\ddagger}$, the free energy contribution from electrostatic interactions *between the reactants and other charged species* in the solution.

Thus

$$\Delta G^{\ddagger} = \Delta G_0^{\ddagger} + \Delta G_{es}^{\ddagger} + \Delta G_{\mu}^{\ddagger} \tag{43}$$

Although both $\Delta G_{es}^{\ddagger}$ and $\Delta G_{\mu}^{\ddagger}$ result from electrostatic interactions in solution the term "electrostatic contribution to the free energy of activation" will be reserved for $\Delta G_{es}^{\ddagger}$; $\Delta G_{\mu}^{\ddagger}$ will be referred to as the ion atmosphere contribution.

The nonelectrostatic contribution, $\Delta G_0^{\ddagger}$, arises mainly from the chemical forces between the reactant species in the formation of the activated complex. Although it does include changes in the van der Waals and other nonelectrostatic interactions of the separate reactant species with the solvent molecules, $\Delta G_0^{\ddagger}$ is primarily a characteristic of the reaction itself and not of the solvent. On the other hand, both $\Delta G_{es}^{\ddagger}$ and $\Delta G_{\mu}^{\ddagger}$ are strongly affected by the electrical properties of the medium, $\Delta G_{es}^{\ddagger}$ by the dielectric constant and $\Delta G_{\mu}^{\ddagger}$ by both the dielectric constant of the solvent and by the ionic strength of the solution. The nonelectrostatic $\Delta G_0^{\ddagger}$ term can be thought of as the free energy of activation for the reaction in a medium of infinite dielectric constant where all electrostatic intermolecular forces (and therefore $\Delta G_{es}^{\ddagger}$ and $\Delta G_{\mu}^{\ddagger}$) vanish.

In the treatment that follows it will be convenient to consider the free energy contributions in terms of single molecules. Since the definition of free energy contains an entropy term which is statistical and therefore macroscopic in nature, the concept of free energy per molecule is not strictly rigorous, and it is possibly better to consider this quantity as the free energy per mole divided by Avogadro's number, N_0. No symbolic distinction will be made between molar and molecular expressions for free energy or for the bimolecular rate coefficient, k. However, the appropriate units will always be indicated by the nature of the expression if it is remembered that the gas constant R is related to the Boltzmann constant k by $R = N_0 k$. Therefore the molar energy or entropy values will simply be greater than the corresponding molecular values by the factor N_0, and the bimolecular rate coefficient in units of l. mole^{-1}. sec^{-1} will be greater than the value in units of cm^3. molecule^{-1}. sec^{-1} by the factor $N_0/1000$.

The problems concerning the evaluation of $\Delta G_0^{\ddagger}$ for ordinary chemical reactions are similar to those encountered in gas-phase studies, and except for the special case of electron transfer reactions, will not be treated further. The electrostatic and ion atmosphere contributions to $\Delta G^{\ddagger}$ are peculiar to solution reactions, and they are each discussed in the following sections.

2.2.1 The electrostatic contribution, $\Delta G^{\ddagger}_{es}$

The simplest model for the calculation of $\Delta G^{\ddagger}_{es}$ is that of two conducting spheres of charge $z_A \varepsilon$ and $z_B \varepsilon$ and of radius r_A and r_B which are brought together to a distance of $r_{\ddagger}$ in an isotropic continuous dielectric medium without distortion of the original charge. At a separation x the force between the two ions is $-z_A z_B \varepsilon^2/Dx^2$, and the electrostatic contribution to the free energy per molecule is simply the *work* required to bring the ions together from infinity to $r_{\ddagger}$, viz.

$$\Delta G^{\ddagger}_{es} = W = -\int_{\infty}^{r_{\ddagger}} \frac{z_A z_B \varepsilon^2}{Dx^2} \, \mathrm{d}x = \frac{z_A z_B \varepsilon^2}{Dr_{\ddagger}} \tag{44}$$

From this equation it is seen that the model predicts an electrostatic contribution to the free energy of activation which is inversely proportional to the dielectric constant of the solvent. The sign of this contribution depends on whether the reactants are similarly or oppositely charged. For conditions where $\Delta G^{\ddagger}_{\mu} = 0$ (either for very dilute solutions where the ionic strength approaches zero or for rate coefficients that have been extrapolated to zero ionic strength) equations (42), (43) and (44) can be arranged to give

$$\ln k = \ln \frac{kT}{h} - \frac{\Delta G^{\ddagger}_0}{kT} - \frac{\Delta G^{\ddagger}_{es}}{kT} \tag{45}$$

$$= \ln k_0 - \frac{z_A z_B \varepsilon^2}{kTDr_{\ddagger}} \tag{46}$$

where k_0 is the rate coefficient for the reaction where electrostatic forces are non-existent (*i.e.* in a medium of infinite dielectric constant).

The model predicts that a plot of logarithm of rate coefficient against the reciprocal of dielectric constant should be linear (see equation 46). In practice the situation is complicated by the fact that a change of the dielectric constant involves changes in the nature of the solvent which may affect both $\Delta G^{\ddagger}_0$ and $r_{\ddagger}$. Despite these complications, Fig. 3 shows that, except at low values of D, the predicted relationship is obeyed for the reactions[30,31]

(a) $S_2O_3^= + BrCH_2COO^- \overset{k_a}{\rightleftarrows} S_2O_3CH_2COO^= + Br^-$

(b) $N_2(COO)_2^= + H_3O^+ \overset{k_b}{\rightleftarrows} HN_2(COO)_2^- + H_2O$

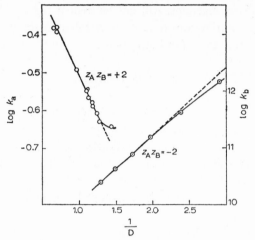

Fig. 3. The influence of solvent dielectric constant on the rate of reaction between ionic species of like (○)$^{V}_{A}$and opposite (○) charges (from Laidler and Eyring[30] and King and Josephs [31]).

The measured slopes yield values for $r_{\ddagger}$ for the activated complexes in the two reactions of 5.1 A and 3.4 A respectively, values which are certainly of the right order of magnitude. Fig. 3 shows that marked deviations from the predicted relation occur at low values of dielectric constant. This is thought to be a result of *ion pairing* of reactant ions with ions of opposite charge present in the solution. In the case of like-charged reactants such pairing would reduce the repulsive electrostatic forces between the reactants and thus effectively increase the rate; the reverse effect may be expected for oppositely charged reactants.

Other complications, such as *selective solvation* and *dielectric saturation*, result from the failure of the "structureless continuum" model for the solvent to describe the molecular effects of the solvent in the region of an ion. In solvent mixtures, the ion–dipole forces would tend to attract selectively the molecules with higher dipole moments to the immediate vicinity of the ion. For this reason the behaviour of a reaction in a solvent pair mixture would tend towards that exhibited in the pure high-dielectric solvent.

Dielectric saturation is due to the "freezing" of solvent molecules into a rigid pattern in the vicinity of an ion which is caused by the extremely high field strengths in the region near the ionic charge. That such "freezing" reduces the dielectric constant of the solvent in the vicinity of an ion can be seen from the Debye equation[32]

$$\frac{D-1}{D+2} \cdot \frac{M}{\rho} = \mathscr{P} \tag{47}$$

where M and ρ are the molecular weight and density of the solvent and $\mathscr{P}$ is the polarization. Since the largest component of the polarization in polar liquids is that due to orientation of the solvent dipoles, the freezing of the dipoles into line with the force field will drastically reduce the overall polarization. The dielectric constant is thus reduced to a low value (equal to the square of the refractive index) characteristic of nonpolar solvents. For an aqueous solution, Laidler[33] has calculated that although D is equal to its macroscopic value of 78.5 at a distance of 8 A away from a singly charged ion it is reduced to its limiting saturation value of 1.78 at a distance of 2 A.

2.2.2 The ion atmosphere contribution, $\Delta G_\mu^\ddagger$

The alterations in the rate of an ionic reaction brought about by the addition of non-reacting ions to the solution are known as *salt effects*. The changes in rate result either from changes in $\Delta G_\mu^\ddagger$ or from changes in the concentrations of ionic reactant species present. The phenomenon associated with changes in $\Delta G_\mu^\ddagger$ is known as the *primary* salt effect (*cf.* section 3.3.3) and its basis is discussed in this section. As a result of the long-range nature of Coulombic forces, interactions with all the surrounding ions will affect the free energy of any charged particle in solution by contributing a term G_μ. Since both the reactant ions and the activated complex (if it is charged) will be surrounded by such an "atmosphere" of ions the ion atmosphere contribution to the activation energy for the formation of the activated complex is

$$\Delta G_\mu^\ddagger = G_\mu^\ddagger - G_\mu^A - G_\mu^B \tag{48}$$

The first theory to account successfully for the interactions of ions in (dilute) solutions was formulated by Debye and Hückel. The theory gives an expression for the spherically symmetrical electrical potential ψ_j at a distance r from a reference ion j. A detailed treatment and discussion of the theory can be found in treatises on electrolyte solutions[34, 35] and only a brief derivation will be presented here. The basis of the formulation is to combine the Poisson equation relating the potential to the charge density, ρ, with the Boltzmann distribution law. The Poisson equation is given in spherical co-ordinates by

$$\frac{1}{r^2} \frac{d}{dr} \left(r^2 \frac{d\psi_j}{dr} \right) = \frac{4\pi}{D} \rho \tag{49}$$

and the Boltzmann distribution law for the ion system is

$$n_i = n_0 e^{-z_i \varepsilon \psi_j / DkT} \tag{50}$$

The charge density around the reference ion j is found from (50) by summation of the product of number density and charge for each of the ionic species

$$\rho_j = \sum_i n_i z_i \varepsilon e^{-z\varepsilon\psi_j/DkT} \tag{51}$$

Expansion of the exponential term gives

$$\rho_j = \sum_i n_i z_i \varepsilon - \sum_i n_i z_i \varepsilon \left(\frac{z_i\varepsilon\psi_j}{DkT}\right) + \sum_i \frac{n_i z_i \varepsilon}{2!} \left(\frac{z_i\varepsilon\psi_j}{DkT}\right)^2 \dots \tag{52}$$

The first term of (52) is zero owing to the electrical neutrality of the solution. Further, for high dilutions where the electrical potential energy of the ions will be small compared to their thermal energy (*i.e.* $z\varepsilon\psi_j/0 \ll kT$) the charge density may be approximated by the linear term only to give

$$\rho_j = \sum_i \frac{n_i z_i^2 \varepsilon^2 \psi_j}{DkT} \tag{53}$$

It can be seen, that for the special case of the solution containing only symmetrical electrolytes, the second order term in (52) vanishes and the approximation (53) becomes much better.

Substitution of (53) into the Poisson equation (49) yields

$$\frac{1}{r^2} \frac{d}{dr} \left(r^2 \frac{d\psi_j}{dr}\right) = \kappa^2 \psi_j \tag{54}$$

where κ is the "Debye kappa" given by

$$\kappa = \left(\frac{4\pi\varepsilon^2}{DkT} \sum_i n_i z_i^2\right)^{\frac{1}{2}} \tag{55}$$

Equation (54) is a straightforward second order linear differential equation, and using the boundary condition $\psi_j \to 0$ as $r \to \infty$ the solution is found to be

$$\psi_j = \frac{\alpha e^{-\kappa r}}{r} \tag{56}$$

The evaluation of the constant α requires the boundary condition for electrical neutrality, which is that the total charge outside a sphere of diameter a_j, equal to the distance of closest approach for the ion j, be equal and opposite to its own charge, *i.e.*

$$\int_{a_j}^{\infty} 4\pi r^2 \rho_j dr = -z_j \varepsilon \tag{57}$$

Substitution of (53) and (56) into this equation yields

$$4\pi\alpha \sum_i \frac{n_i z_i^2 \varepsilon^2}{DkT} \int_{a_j}^{\infty} re^{-\kappa r}\mathrm{d}r = \alpha\kappa^2 \sum_i \int_{a_j}^{\infty} re^{-\kappa r}\mathrm{d}r = z_j\varepsilon \tag{58}$$

and integration by parts gives

$$\alpha = \frac{z_j\varepsilon}{D} \cdot \frac{e^{\kappa a_j}}{1+\kappa a_j} \tag{59}$$

so that

$$\psi_j = \frac{z_j\varepsilon}{D} \cdot \frac{e^{\kappa a_j}}{1+\kappa a_j} \cdot \frac{e^{-\kappa r}}{r} \tag{60}$$

This is the total potential in the vicinity of the ion j, and for a_j and $r < 1/\kappa$, the potential ψ_j consists of two parts, $\psi_j' = z_j\varepsilon/Dr$, the Coulombic potential due to its own charge, and ψ_j'' due to the *ion atmosphere* of surrounding ions. Thus

$$\psi_j'' = \frac{z_j\varepsilon}{Dr} \left(\frac{e^{\kappa a_j}e^{-\kappa r}}{1+\kappa a_j} -1 \right) \tag{61}$$

For high dilution, where κa_j and $\kappa r \ll 1$, the exponential terms in (60) can be expanded to give

$$\psi_j'' = - \frac{z_j\varepsilon\kappa}{D(1+\kappa a_j)} \tag{62}$$

For $a_j \ll 1/\kappa$ we see that $\psi_j'' = -z_j\varepsilon/D(1/\kappa)$, so that ψ_j'' can be thought of as originating from a spherical shell, radius $1/\kappa$, of evenly distributed charge, $-z_j\varepsilon$, surrounding the central ion.

The contribution of the ion atmosphere to the electrical energy of the central ion is simply the product of the potential at the ion, ψ_j'', and the ion's charge, $z_j\varepsilon$. To avoid counting each interaction twice a factor of $\frac{1}{2}$ is required, so that the ion atmosphere contribution to the free energy per mole is

$$G_\mu = \tfrac{1}{2}\psi_j'' z_j\varepsilon \tag{63}$$

and, from (62)

$$G_\mu = - \frac{z_j^2\varepsilon^2}{2D} \cdot \frac{\kappa}{1+a_j\kappa} \tag{64}$$

From the definition of *ionic strength*, $\mu = \frac{1}{2}\Sigma_i c_i z_i^2$, and the relation $n_i = N_0 c_i/1000$ where N_0 is Avogadro's number, (55) becomes

$$\kappa = \left(\frac{8\pi N_0 \varepsilon}{1000DkT}\right)^{\frac{1}{2}} \mu^{\frac{1}{2}} = \mathcal{A}\mu^{\frac{1}{2}} \tag{65}$$

If the deviations from ideality are due entirely to the electrical interactions, then the activity coefficient, f_μ, and G_μ are related by $\ln f_\mu = G_\mu/kT$, and f_μ can thus be evaluated from equation (64). At high dilutions where the ionic strengths approach zero, the result is known as the *Debye–Hückel limiting law*

$$\ln f_\mu = -\frac{z_j^2 \varepsilon^2 \mathcal{A}}{2kTD}\frac{\mu^{\frac{1}{2}}}{1+\mathcal{A}a_j\mu^{\frac{1}{2}}} \xrightarrow{\mu\to 0} -\frac{z_j^2 \varepsilon^2 \mathcal{A}}{2kTD} \tag{66}$$

Although expressions (62) and (64) were formulated by considering a_j as the distance of closest approach of ion j, there is uncertainty as to its actual significance and it is customary to regard it as an overall empirical parameter rather than a characteristic of a particular species. For this reason the subscript j is usually dropped.

Finally, application of (64) to (48) using a without the subscript, yields the required expression for the ion atmosphere contribution to the free energy of activation

$$\Delta G_\mu^\ddagger = [(z_A+z_B)^2 - z_A^2 - z_B^2]\frac{\varepsilon^2 \mathcal{A}}{2D}\frac{\mu^{\frac{1}{2}}}{1+\mathcal{A}a\mu^{\frac{1}{2}}} = -\frac{z_A z_B \varepsilon^2 \mathcal{A}}{D}\cdot\frac{\mu^{\frac{1}{2}}}{(1+\mathcal{A}\mu^{\frac{1}{2}})} \tag{67}$$

It is seen that the sign $\Delta G_\mu^\ddagger$ is opposite to that of $\Delta G_{es}^\ddagger$, so that in solutions of high ionic strength the ion atmosphere effect partially compensates for the inhibiting (or enhancing) effect of the repulsive (or attractive) forces between the charged reactant species.

If $\Delta G_0^\ddagger$ and $\Delta G_{es}^\ddagger$ are included with this value for $\Delta G_\mu^\ddagger$ in the overall expression for $\Delta G^\ddagger$, on substitution in (42) the following equation for the rate coefficient of an ionic reaction is obtained

$$\ln k = \ln\frac{kT}{h} - \frac{\Delta G_0^\ddagger}{kT} - \frac{z_A z_B \varepsilon^2}{kTDr_\ddagger} + \frac{z_A z_B \varepsilon^2 \mathcal{A}\mu^{\frac{1}{2}}}{kTD(1+\mathcal{A}a\mu^{\frac{1}{2}})} \tag{68}$$

The expression is often known as the *Brønsted–Christiansen–Scatchard equation*. If the first three terms in (68) are taken to comprise $\ln k_0$ where k_0 is the rate coefficient at zero ionic strength and dielectric constant D then, for dilute solutions where $\mathscr{A}a\mu^{\frac{1}{2}} \ll 1$, we obtain

$$\ln \frac{k}{k_0} = z_A z_B \frac{\varepsilon^2 \mathscr{A}}{kTD} \mu^{\frac{1}{2}} \tag{69}$$

Equation (69) was first derived by Brønsted[36] using the Brønsted–Bjerrum equation (26) together with the Debye–Hückel limiting law (64) and has been the subject of many investigations[37]. For water at 25 °C, $\varepsilon^2 \mathscr{A}/2.303kTD = 1.02$, so that

$$\log \frac{k}{k_0} = 1.02 z_A z_B \mu^{\frac{1}{2}} \tag{70}$$

Thus the effect of increasing the ionic strength of the solution is to decrease the rate for oppositely charged reactants and increase it for like charged ones. Fig. 4 shows how closely some measured reaction rates follow the predicted variation (indicated by the solid lines).

Although equation (69) is obeyed remarkably well for the reactions included in Fig. 4 at low ionic strengths, some deviations do occur, especially for high valence ions[38]. *Ion association* is a major complicating factor with such ions of high valency and for concentrated solutions where association reduces the true ionic strength of the solution. The effect will obviously be most pronounced in media of low

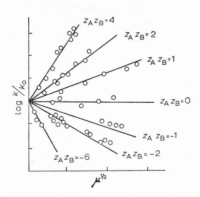

Fig. 4. The influence of ionic strength, μ, of the medium on the rate of reaction between ionic species with various charges z_A and z_B (from Benson[2])

$z_A z_B$	Reaction
+4	$[Co(NH_3)_5Br]^{++} + Hg^{++} + H_2O \rightleftarrows Co(NH_3)_5(H_2O)^{3+} + (HgBr)^+$
+2	$S_2O_8^= + 2I^- \rightleftarrows 2\,SO_4^= + I_2$
+1	$NO_2: N \cdot COOC_2H_5^- + OH^- \rightleftarrows N_2O + CO_3^= + C_2H_5OH$
0	cane sugar $+ OH^- \rightleftarrows$ invert sugar
−1	$2H_3O^+ + 2Br^- + H_2O_2 \rightleftarrows 4H_2O + Br_2$
−2	$[Co(NH_3)_5Br]^{++} + OH^- \rightleftarrows [Co(NH_3)_5(OH)]^{++} + Br^-$
−6	$Fe^{++} + Co(C_2O_4)_3^{3-} \rightleftarrows Fe^{+++} + [Co(C_2O_4)_3]^{4-}$

dielectric constant, and the resulting associated species can range from covalently bonded complexes (for weak electrolytes) to purely electrostatically bonded complexes (for strong electrolytes) which are known as *Bjerrum ion pairs*. In his treatment of ion pairs Bjerrum[39] suggested that any pair of ions whose electrostatic interaction energy is greater than $2kT$ should be regarded as an ion pair and not as independent ions describable by the Debye–Hückel treatment. The critical separation of two oppositely charged ions is then $d = z_A z_B \varepsilon / 2kTD$ which for divalent ions in aqueous solution is about 20 A (*cf.* the average separation of ions in $0.1M$ solution of about 25 A). Adoption of $2kT$ as the critical energy in the calculation of the relative numbers of paired and nonpaired ions has been shown by Kraus and Fuoss[40] to produce values for dissociation constants of strong electrolytes that agree well with conductivity data. Although the critical separation is seen to depend on D, the dielectric constant is not the only solvent property which influences the extent of ion pair formation, and specific interactions may be important. For example the dissociation constant of $[o\text{-ClC}_6\text{H}_4\text{N}(\text{CH}_3)_3]^+$ ClO_4^- is ten times greater in CH_3CHCl_2 than in $\text{CH}_2\text{ClCH}_2\text{Cl}$, even though both solvents have the same dielectric constant[41].

2.3 ENERGY AND ENTROPY OF ACTIVATION IN IONIC REACTIONS

It is sometimes convenient to divide the free energy of activation into the enthalpy and entropy terms of which it is composed. In this way, the experimental activation energy may be compared with an activation energy derived from the enthalpy term (*cf.* p. 216). For solution reactions, it is usually assumed that $K_c^\ddagger = K_p^\ddagger = K^\ddagger$, so that to a first approximation

$$\Delta H^\ddagger = RT^2 \cdot \frac{\text{d} \ln K^\ddagger}{\text{d}T} \tag{71a}$$

The proper definition of activation energy is

$$E = RT^2 \cdot \frac{\text{d} \ln k}{\text{d}T} \tag{71b}$$

so that, in fact

$$E = \Delta H^\ddagger + RT \tag{71c}$$

The thermodynamic relations between ΔG, ΔS and ΔH are

$$\Delta G = \Delta H - T\Delta S \tag{72a}$$

and

$$\Delta S = -\frac{\partial \Delta G}{\partial T} \tag{72b}$$

The entropy and energy of activation, $\Delta S^{\ddagger}$ and E, can now be separated into non-electrical, electrostatic and ion-atmosphere contributions in a manner analogous to that employed for the free energy, $\Delta G^{\ddagger}$

$$\Delta S^{\ddagger} = \Delta S_0^{\ddagger} + \Delta S_{es}^{\ddagger} + \Delta S_{\mu}^{\ddagger} \tag{73}$$

$$E = E_0 + E_{es} + E_{\mu} \tag{74}$$

Since the non-electrical term E_0 makes the largest contribution to the total energy of activation, in the following treatment relating the components of energy to free energy of activation the RT term will be grouped with E_0. If this is done we have

$$E_0 = \Delta G_0^{\ddagger} + T\Delta S_0^{\ddagger} + RT \tag{75a}$$

$$E_{es} = \Delta G_{es}^{\ddagger} + T\Delta S_{es}^{\ddagger} \tag{75b}$$

$$E_{\mu} = \Delta G_{\mu}^{\ddagger} + T\Delta S_{\mu}^{\ddagger} \tag{75c}$$

where the entropy of activation terms are calculated from the relation (72b).

In the next two sections the electrostatic and the ion atmosphere components of energy and entropy of activation will be dealt with separately; an estimate of the relative magnitudes of the various contributions will then be made from experimental data for a particular reaction.

2.3.1 The electrostatic contributions

The expression derived in section 2.2.1 for the electrostatic contribution to the free energy of activation per mole was

$$\Delta G_{es}^{\ddagger} = \frac{N_0 z_A z_B \varepsilon^2}{D r_{\ddagger}} \tag{76}$$

Using relations (72b) and (75b), this expression yields

$$\Delta S_{es}^{\ddagger} = \frac{\Delta G_{es}^{\ddagger}}{T} \cdot \frac{\partial \ln D}{\partial \ln T} \tag{77}$$

$$E_{es} = \Delta G_{es}^{\ddagger} \left(1 + \frac{\partial \ln D}{\partial \ln T}\right) \tag{78}$$

For aqueous solution, $\partial \ln D/\partial T = (1/T)\, \partial \ln D/\partial \ln T$ remains constant at -0.0046 over a large temperature range, so that for a value of $r_{\ddagger}$ of 2 A and D of 80, $\Delta S_{es}^{\ddagger}$ is roughly $-10\, z_A z_B$ cal. deg^{-1}. According to the Arrhenius equation and equation (71) the logarithm of the A factor depends on $\Delta S^{\ddagger}/R$ according to

$$\ln A = \ln \frac{kT}{h} + \frac{\Delta S^{\ddagger}}{R} + 1 \tag{79}$$

Thus, if $\Delta S_{es}^{\ddagger}$ provides the major contribution to the total entropy of activation, $\Delta S^{\ddagger}$, the frequency factor would be expected to be higher than the "normal" value of 10^{12} l. mole^{-1}. sec^{-1} for reactions of oppositely charged ions and lower for reactions of like charge, and the increment in $\ln A$ would be proportional to the product of the charges on the reacting ions. For many ionic reactions the experimental results are roughly consistent with these expectations[3].

The entropy change can be thought of as resulting from the electrostrictive "freezing" of solvent molecules around a charged particle in solution. It was mentioned in section 1.4 that because the extent of electrostriction depends roughly on the square of the charge on a body, the formation of an activated complex from two like charges will be accompanied by a net increase in electrostriction and thus a decrease in entropy.

For many solvents $1 + \partial \ln D/\partial \ln T$ is negative and in these cases (77) and (78) show that the rate of an ionic reaction will be affected more by the electrostatic entropy of activation than the electrostatic energy of activation and the two effects will act in opposite directions. This can be illustrated by a comparison of rates for charged and uncharged reactants of similar structure. For example, Bell and Lindars[42] found that the overall rate of alkaline hydrolysis of the positively charged ester $(C_2H_5)_3N^+CH_2COOC_2H_5$ was 200 times greater than for the neutral acetate. The increase in rate on changing from the charged acetate to the neutral ester was due to the $e^{\Delta S^{\ddagger}/R}$ factor, since it increased 500 fold while the $e^{-E/RT}$ term decreased by a factor of less than three.

Since the dielectric constant of the solvent changes with temperature, a simple measurement of the temperature coefficient of rate to find the activation energy of an ionic reaction cannot distinguish between the change in rate due to the "Arrhenius temperature effect" and that due merely to the *change in dielectric constant*. Svirbely and Warner[43] first defined a "true" activation energy, E_D, as that measured in a system where the dielectric constant can be held fixed as the temperature is changed. The difference between E_D measured in isodielectric solvents and E measured at constant solvent composition at infinite dilution was given as

$$E - E_D = RT^2 \frac{\partial \ln k}{\partial D} \cdot \frac{dD}{dT} \qquad (\mu = 0) \tag{80}$$

This relation was found to be in excellent agreement with measurements of E_D for the ammonium cyanate–urea conversion, D being held constant by use of a methanol–water solvent mixture.

A comparable expression for $E - E_D$ can be derived from (74). Since E_0 is independent of D, at zero ionic strength the difference $E - E_D$ depends only on E_{es}. In isodielectric solvents $\partial \ln D / \partial \ln T = 0$ so that from (78), $(E_{es})_D = \Delta G_{es}^{\ddagger}$, and therefore at zero ionic strength

$$E - E_D = \Delta G_{es}^{\ddagger} \frac{\partial \ln D}{\partial \ln T} \qquad (\mu = 0) \tag{81}$$

This equation predicts that E_D will be larger than E for reactions of like charged ions but less than E for oppositely charged ions. In addition, for any given reaction, (81) states that $E - E_D = (\text{constant}) \times T/D$ which agrees with the experimental findings of Svirbely and Warner[43]. A further consequence of the condition $\partial \ln D / \partial \ln T = 0$ in isodielectric solvents is that both $\Delta S_{es}^{\ddagger}$ and $\Delta S_{\mu}^{\ddagger}$ would vanish if results were extrapolated to zero ionic strength. $\Delta S^{\ddagger}$ should therefore be equal to $\Delta S_0^{\ddagger}$ and the frequency factor measured in this way would be expected to be comparable with that of a similar reaction involving uncharged reactants.

2.3.2 The ion atmosphere contribution

Substitution in the relations (72b) and (75c) of the expression

$$\Delta G_{\mu}^{\ddagger} = -\frac{N_0 z_A z_B \varepsilon^2}{2D} \kappa \tag{82}$$

(where κ is the Debye–Hückel kappa, equation (55), and is proportional to $(DTV)^{-\frac{1}{2}}$) yields

$$\Delta S_{\mu}^{\ddagger} = \frac{\Delta G_{\mu}^{\ddagger}}{T} \left(\frac{1}{2} + \frac{3}{2} \frac{\partial \ln D}{\partial \ln T} + \frac{1}{2} \frac{\partial \ln V}{\partial \ln T} \right) \tag{83}$$

$$E_{\mu} = \Delta G_{\mu}^{\ddagger} \left(\frac{3}{2} + \frac{3}{2} \frac{\partial \ln D}{\partial \ln T} + \frac{1}{2} \frac{\partial \ln V}{\partial \ln T} \right) \tag{84}$$

For a given temperature and solvent dielectric constant, $\Delta S_0^{\ddagger}$, E_0, $\Delta S_{es}^{\ddagger}$, and E_{es} are constant so that both the total $\Delta S^{\ddagger}$ and E should depend only on $\Delta G_{\mu}^{\ddagger}$. For dilute solutions at constant T and D, $\Delta G_{\mu}^{\ddagger}$ varies directly as $z_A z_B \mu^{\frac{1}{2}}$. Amis and La Mer[44] have shown that plots of both E and $\ln A$ against $\mu^{\frac{1}{2}}$ have the slopes predicted by (83) and (84) for the tetrabromophenolsulphonphthalein–hydroxide ion reaction at high dilution.

2.3.3 Comparison of the magnitudes of $\Delta G^{\ddagger}$, E, and $\Delta S^{\ddagger}$

It is of interest to compare the magnitudes of the contributions to $\Delta G^{\ddagger}$, $\Delta S^{\ddagger}$ and E from nonelectrical, electrostatic and ion atmosphere effects by treatment of experimental data for an ionic reaction using the relations derived in the previous sections. The particular reaction to be used as an example is the conversion of ammonium cyanate to urea, as it was the first to be extensively studied in iso-dielectric media[43, 45]. More recent study[46, 47] has suggested that the reaction may not be the simple ionic process considered by the original workers, but for the purpose of this example we shall consider that the only important step is

$$NH_4^+ + NCO^- \rightleftarrows CO(NH_2)_2$$

Table 1 shows the isodielectric activation energies and $\log_{10}A$ extrapolated to $\mu = 0$, measured at five different solvent dielectric constants by Svirbely and Schramm[45]. In an isodielectric solvent $\partial \ln D / \partial \ln T = 0$ so that $\Delta S_{es}^{\ddagger} = 0$, and, since $\mu = 0$, $(\Delta S^{\ddagger})_D$ is equal to $\Delta S_0^{\ddagger}$. It can be seen that the measured $(\Delta S^{\ddagger})_D$ changes very little with changing D which supports the assumption that $\Delta S_0^{\ddagger}$

TABLE 1

DATA FOR THE REACTION $NH_4^+ + NCO^- \rightleftarrows CO(NH_2)_2$ FOR $\mu = 0$

D	$(\log_{10}A)_D$	$(\Delta S^{\ddagger})_D (cal.\ deg^{-1})$	$E_D (kcal.\ mole^{-1})$
63.5	11.77	−4.5	18.84
55	11.80	−4.4	18.50
50	11.71	−4.7	18.81
45	11.75	−4.6	17.84
40	11.78	−4.5	17.59

TABLE 2

THE COMPONENTS OF $\Delta G^{\ddagger}$, E AND $\Delta S^{\ddagger}$ FOR THE REACTION $NH_4^+ + NCO^- \rightleftarrows CO(NH_2)_2$ IN WATER AT 25° C AND IONIC STRENGTH OF 0.025

	$\Delta G^{\ddagger}$ $(kcal.\ mole^{-1})$	E $(kcal.\ mole^{-1})$	$T\Delta S^{\ddagger}$ $(kcal.\ mole^{-1})$	$\Delta S^{\ddagger}$ $(cal.\ deg^{-1})$
Total	20.6	21.6	0.46	1.6
"non-electrical" component	21.9	21.2	−1.34	−4.5
"electrostatic" component	−1.92	0.71	2.63	8.9
"ion atmospheric" component	0.56	−0.27	−0.83	−2.8

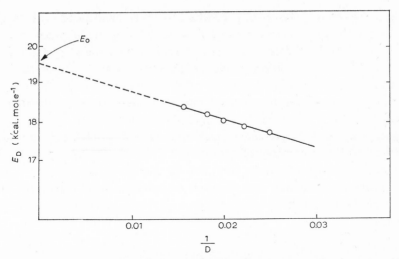

Fig. 5. Extrapolation of isodielectric constant activation energy results for the $NH_4^+ + NCO^- \rightleftharpoons$ $CO(NH_2)_2$ reaction to infinite dielectric constant where electrostatic effects disappear (data from Svirbely and Warner[43]).

is primarily a characteristic of the reaction and not of the solvent.

Since $\partial \ln D / \partial \ln T = 0$, from (78) we see that $E_{es} = \Delta G_{es}^{\ddagger} = N_0 z_A z_B \varepsilon^2 / D r_{\ddagger}$. E_{μ} is zero and E_0 is assumed to be independent of D, so that E_0 should be calculable by means of an extrapolation of the measured total E_D to infinite dielectric constant. The plot of the measured E_D against $1/D$ is shown in Fig. 5. The intercept gives $E_0 = 21.2$ kcal. mole^{-1} and the slope of the line yields the reasonable value of 2.2 A for $r_{\ddagger}$. From (75a) $\Delta G_0^{\ddagger}$ is seen to be equal to $E_0 - T\Delta S_0^{\ddagger} - RT$, *i.e.* 21.9 kcal. mole^{-1} for this reaction.

By using the value of $\partial \ln D / \partial \ln T = -1.37$, D = 78.5 for water and 2.2 A for $r_{\ddagger}$, the values of $\Delta G_{es}^{\ddagger}$, $\Delta S_{es}^{\ddagger}$, and E_{es} can be calculated from (76), (77), (78) for the reaction in aqueous solution at 25 °C. Warner and Stitt[48] showed that the Debye–Hückel limiting law is obeyed by this system for values of $\mu^{\ddagger}$ up to 0.16. Taking[49] $\partial \ln D / \partial \ln T = -1.37$ and $\partial \ln V / \partial \ln T = 0.076$, $\Delta G_{\mu}^{\ddagger}$, $\Delta S_{\mu}^{\ddagger}$, and E_{μ} can be calculated at 25 °C and ionic strength of 0.025 from relations (82), (83) and (84). The resulting values are recorded in Table 2.

2.4 REACTIONS INVOLVING DIPOLES

In the previous section it was seen that for ion–ion reactions the contribution to the free energy of activation from electrostatic effects is of the order of a few kcal per mole. For such reactions the small changes in $\Delta G_0^{\ddagger}$ from solvent to solvent are usually overshadowed by changes in $\Delta G_{es}^{\ddagger}$ and $\Delta G_{\mu}^{\ddagger}$ so that the solvent effects can usually be accounted for by the variation in solvent dielectric constant.

However, the electrostatic forces between ions and dipoles, and especially between two dipoles, are much smaller than those between two ions. In fact, they may be as small as the van der Waals forces between nearest neighbours. The variation of $\Delta G_{es}^{\ddagger}$ and $\Delta G_{\mu}^{\ddagger}$ with the dielectric constant of the solvent may thus be of the same magnitude as the variation in $\Delta G_0^{\ddagger}$, or, in other words, the solvent effect due to the dielectric constant may be no more important than specific solvent and structural effects for which it may be extremely difficult to make allowances.

Expressions will now be presented which attempt to describe the electrostatic contributions to the free energy of activation for ion–dipole and dipole–dipole reactions. However, because these contributions are small, the changes in the total free energy of activation with dielectric constant are small compared with other solvent effects and the theoretical expressions are usually difficult to verify experimentally.

2.4.1 The electrostatic contribution, $\Delta G_{es}^{\ddagger}$

A treatment analogous to that for ion–ion reactions may be used to derive very simple expressions for $\Delta G_{es}^{\ddagger}$ by equating $\Delta G_{es}^{\ddagger}$ to the electrostatic energy, W, required to bring the reactant species from infinity to a separation $r_{\ddagger}$. In the case of *point* charges and *point* dipoles the expressions are

$$|W| = \frac{z\varepsilon\mu \cos\theta}{Dr_{\ddagger}^2} \tag{85}$$

$$|W| = \frac{2\mu_1\mu_2 \cos\theta_1 \cos\theta_2}{Dr_{\ddagger}^3} + \frac{\mu_1\mu_2 \sin\theta_1 \sin\theta_2}{Dr_{\ddagger}^3} \tag{86}$$

where μ is the dipole moment of the dipolar species and the θ values represent the *acute* angle between the line of the particular dipole and the line of approach. For "in line" (or "head-on") reactions the cosine terms are unity and the sine terms vanish so that the $\Delta G_{es}^{\ddagger}$ expressions become

$$\Delta G_{es}^{\ddagger} = -\frac{z\varepsilon\mu}{Dr_{\ddagger}^2} \qquad \text{ion–dipole reaction} \tag{87}$$

$$\Delta G_{es}^{\ddagger} = -\frac{2\mu_1\mu_2}{Dr_{\ddagger}^3} \qquad \text{dipole–dipole reaction} \tag{88}$$

The electrostatic effects will be negative as in (87) and (88) (*i.e.* will lower the total free energy of activation) only if the reaction site is at the favourable end of the dipolar molecule. In this case the head-on reaction path will be energetically the most favourable and the assumption of $\cos\theta = 1$ is reasonable. If, however, the reaction is between an ion and a like charged end of a dipolar molecule, or between two like charged ends of dipolar molecules, the $\Delta G_{es}^{\ddagger}$ contributions will be positive. The assumption of a head-on line of approach for these reactions is unreasonable unless special steric factors are important (as may be the case in some Walden inversion reactions) and, in general, expressions for $\Delta G_{es}^{\ddagger}$ will contain angular terms as in (85) and (86).

Substitution of (87) and (88) for $\Delta G_{es}^{\ddagger}$ in (45) yields

$$\ln k = \ln k_0 + \frac{z\varepsilon\mu}{kTDr_{\ddagger}^2} \tag{89}$$

$$\ln k = \ln k_0 + \frac{2\mu_1\mu_2}{kTDr_{\ddagger}} \tag{90}$$

where, as before, k_0 is the rate coefficient for the reaction when electrostatic forces are non-existent. Both (89) and (90) predict that a plot of log k against 1/D will be linear and have a positive slope when the reacting species are favourably oriented. Good linearity has been found for log k against 1/D plots for several ion–dipole reactions. In the reactions of chloride ion with N-chloropropionan-ilide[30] and hydroxide ion with diacetone alcohol[50] the values of $r_{\ddagger}$ obtained from (89), 3.4 A and 5.5 A respectively, certainly have the right order of magni-tude. However, the fact that the slope of the log k against 1/D plot is positive for the first reaction but negative for the second is difficult to explain on purely electrostatic grounds. In these reactions solvation effects are undoubtedly im-portant so that the reaction having a transition state less polar than the reactants is favoured by low dielectric (less aqueous) media and *vice versa*.

If the two reactants in *ion–ion* reactions have significant dipole moments the ion–dipole and dipole–dipole interaction contributions to $\Delta G_{es}^{\ddagger}$ should not be neglected. All three types of interaction are incorporated in the Laidler–Lands-kroener equation[51]. Kirkwood's general equation[52] for the free energy of charging a sphere is used to derive expression (91) for the free energy required to transfer a sphere of charge, $z\varepsilon$, and dipole moment, μ, from a medium of unit dielectric constant to one of dielectric constant D, *viz.*

$$\Delta G_{es} = -\frac{z^2\varepsilon^2}{2r}\left(1-\frac{1}{D}\right) - \frac{3}{8}\frac{\mu^2}{r^3}\left(\frac{D-1}{D+1}\right) \tag{91}$$

The electrostatic free energy for the two reactants, A and B, and the activated complex, $X^{\ddagger}$, may be computed from equation (91), and the electrostatic contribution to the free energy of activation becomes

$$\Delta G_{es}^{\ddagger} = \Delta G_{esD=1}^{\ddagger} + \frac{\varepsilon^2}{2}\left(1-\frac{1}{D}\right)\left[\frac{z_A^2}{r_A} + \frac{z_B^2}{r_B} - \frac{(z_A+z_B)^2}{r_{\ddagger}}\right]$$

$$+ \frac{3}{8}\left(\frac{D-1}{D+1}\right)\left[\frac{\mu_A^2}{r_A^3} + \frac{\mu_B^2}{r_B^3} - \frac{\mu_{\ddagger}^2}{r_{\ddagger}^3}\right] \quad (92)$$

The dependence of the rate coefficient on $\Delta G_{es}^{\ddagger}$ implicit in equation (45) may be used together with the approximation that for high D, $(D-1)/(D+1) \approx 1-(2/D)$, to obtain an expression for the rate coefficient, k, in a medium of dielectric constant D, referred to the rate constant for a medium with $D = 1$, viz.

$$\ln k = \ln k_{D=1} - \frac{\varepsilon^2}{2kT}\left(1-\frac{1}{D}\right)\left[\frac{z_A^2}{r_A} + \frac{z_B^2}{r_B} - \frac{(z_A+z_B)^2}{r_{\ddagger}}\right]$$

$$- \frac{3}{8kT}\left(1-\frac{2}{D}\right)\left[\frac{\mu_A^2}{r_A^3} + \frac{\mu_B^2}{r_B^3} - \frac{\mu_{\ddagger}^2}{r_{\ddagger}^3}\right] \quad (93)$$

The first two terms are seen to be equivalent to expression (45) for the reaction between two ions, if $r_{\ddagger} = r_A = r_B$. The third term in (93) can be added to (45) if the ionic reactants have large dipole moments.

For ion–dipole reactions both terms in (93) must be used, the second term being reduced to

$$- \frac{z_A^2\varepsilon^2}{2kT}\left(1-\frac{1}{D}\right)\left(\frac{1}{r_A} - \frac{1}{r_{\ddagger}}\right) \quad (94)$$

where z_A is the charge on the ion and therefore on the activated complex as well. Since $r_{\ddagger}$ will be larger than r_A, (94) predicts that the rate should be reduced in media of high dielectric constant, and a plot of log k against $1/D$ will be linear and have a positive slope. It can be seen that equation (93) is unable to account for the fact that the electrostatic effects for reactions with unfavourable dipole orientations may impede rather than facilitate the reaction and thus make the slope of the log k versus $1/D$ plot negative rather than positive.

For reactions of uncharged dipolar molecules the second term in equation (93) disappears and the third term alone describes the electrostatic solvent effects. When the activated complex is significantly more polar than the reactant molecules (as when an ionic salt is formed from the reaction of two neutral molecules) the third term of (93) predicts a higher reaction rate the greater the dielectric constant of the solvent. In such a reaction the electrostriction of the solvent molecules will be greater for the activated complex than for the separate reactants and a large negative entropy of activation would be expected.

Other theoretical expressions, such as those by Laidler and Eyring[30] and Amis and Jaffe[53], have been proposed for ion–dipole and dipole–dipole reactions, but their success has been limited. These treatments and their application to experimental data have been discussed in detail by Amis[5].

2.4.2 The ion atmosphere contribution

It can be seen from equation (68) that the simple Debye–Hückel treatment of dilute solutions predicts that if one of the reactants is uncharged, a reaction should show no primary salt effect. The plot of $\ln k/k_0$ against $\mu^{\frac{1}{2}}$ in Fig. 4 shows that the prediction is correct at very low concentrations. For higher ionic strengths, however, the ion atmosphere effects might be expected to influence the free energy of both ions and dipoles in a way not predicted by the original Debye–Hückel theory.

On theoretical grounds Hückel[54] and Debye and McAulay[55] proposed that terms depending on the first power of the ionic strength be introduced into the expressions for the logarithm of activity coefficient for ions

$$\ln \gamma = -z_i^2 \mathscr{A} \sqrt{\mu} + b\mu \tag{95}$$

and for neutral molecules

$$\ln \gamma = b'\mu \tag{96}$$

For reactions of an ion and a neutral molecule the activated complex, $X^{\ddagger}$, has the same charge as the ion, A, so that the terms involving the square root of the ionic strength in (95) for A and $X^{\ddagger}$ cancel and

$$\ln k = \ln k_0 + \ln \frac{\gamma_A \gamma_B}{\gamma^{\ddagger}} = \ln k_0 + (b_A + b'_B - b_{\ddagger})\mu \tag{97}$$

The logarithmic dependence of the reaction rate on the first power of ionic strength suggested by (97) for ion–dipole reactions has been demonstrated for the hydrolysis of acetals by hydroxide ion by Brønsted and Wynne-Jones[56].

The $b\mu$ term for ions was originally introduced by Hückel to account for dielectric saturation effects and Robinson and Stokes[34] have presented a detailed treatment of the term. However, both b for ions and b' for molecules remain largely empirical, and it is difficult to predict even the sign of the primary salt effect for reactions involving dipoles.

2.5 ELECTRON TRANSFER REACTIONS

An electron transfer reaction is a type of oxidation–reduction process in which the overall reaction consists of the migration of one or more electrons between two ions. Examples are

$$Cr^{2+} + Co^{3+} \rightarrow Cr^{3+} + Co^{2+} \tag{98}$$

$$Fe^{2+} + {}^*Fe^{3+} \rightarrow Fe^{3+} + {}^*Fe^{2+} \tag{99}$$

where the species may be complex ions with a co-ordination shell of ligands. When the electron transfer under consideration is between two oxidation states of the same species, as in (99), isotopic labelling techniques may be used to follow the reaction; the process is often called an *electron exchange* reaction. The field of electron transfer reactions has been one of the most active areas in solution kinetics during the last twenty years and both experimental and theoretical progress has been extensive. The various aspects of the field have been well reviewed[7, 57–59], and space permits in this section only an outline of the principles currently used to treat electron transfer reactions.

2.5.1 Mechanisms of electron transfer

It now appears that most electron transfer reactions occur by direct mechanisms with an activated complex that incorporates both the oxidant and reductant. An ion in solution can be considered to be surrounded by two spheres of molecules. In the case of a transition metal ion the inner sphere is a co-ordinated shell of ligands and the outer one is a solvation sphere consisting of less rigidly aligned solvent molecules. The extent of involvement of the ligand and solvent molecules in the activated complex is the major factor in determining the detailed mechanism of the transfer of the electron. Two broad classes of mechanism can be distinguished depending on whether the inner or outer spheres of the reactants "overlap" in the activated complex.

(a) Inner sphere reactions

In this type of reaction the electron transfer is preceded by substitution of a ligand from one reactant into the co-ordination sphere of the other to form a bridged activated complex where the inner spheres for both ions share the substituted ligand. The reduction of a Co(III) complex ion by Cr^{2+} aq. provides an example of this type of process[60] and is believed to proceed *via* the mechanism

$$Co^{III}(NH_3)_5Cl^{2+} + Cr^{2+}(aq) + 5H^+ \rightarrow [(NH_3)_5Co^{III} \ldots Cl \ldots$$

$$Cr^{II}(H_2O)_5]^{4+} \rightarrow Co^{2+}(aq) + Cr^{III}Cl(H_2O)_5^{2+} + 5NH_4^+ \quad (100)$$

The bridging group is thought to assist electron transfer by reducing the repulsive forces between the two ions, and by acting as a conductor of electric charge through the provision of orbitals able to interact with and delocalize the metal ion orbitals. Inner sphere reactions can occur with or without group transfer. In reaction (100) for example, whether or not the Cl "group" is transferred depends only on the relative strengths of the Co^{III}–Cl and Cr^{III}–Cl bonds formed in the activated complex. When group transfer does occur (as in reaction (100)) an inner sphere mechanism can be demonstrated by showing that relevant reactant and product complexes (in this case Co^{III} and Cr^{III}) are substitution inert (*i.e.* that they have ligand substitution rates which are considerably slower than the measured electron transfer rate).

(b) Outer sphere reactions

In this class of reactions the inner co-ordination shells of the reactants remain intact and the reaction proceeds through an extended activated complex with "overlapping" of the outer spheres. This type of mechanism can be established by showing either that the rate law corresponds to an activated complex which contains the ions with all their ligands, or that electron transfer is faster than the rate of substitution of ligands into the co-ordination shell of both reactants. Both these are difficult to demonstrate for aquo ions, and for this reason it can rarely be conclusively shown that such ions react *via* an outer sphere mechanism.

Bridging is also possible in the outer sphere with the bridging group between the ligand shells assisting the reaction in the same way that the bridging group facilitates inner sphere reactions. When bridging does not occur it is common to think of the two inner co-ordinating spheres of the reactants as approaching within van der Waals distances during the formation of the activated complex.

Because the outer sphere mechanism does not involve displacement of a ligand the fastest electron transfer reactions are probably outer sphere reactions. However, the mechanism for many electron transfer reactions is still uncertain and other mechanistic variations, such as doubly bridged intermediates, and hydrogen atom transfer, have been proposed. A comprehensive presentation of the evidence for the various possible mechanisms has been made by Halpern[57], and most of the known electron transfer reactions are discussed.

2.5.2 Adiabatic and non-adiabatic transfer reactions

A discussion of the present theoretical treatments of electron transfer reactions can most easily be approached through consideration of potential energy surfaces[64]. If there is no electronic interaction between two reacting species the reactants and the products can be considered as separate systems describable by the wave functions $\Psi_{reactants}$ and $\Psi_{products}$. The potential energy of each of these systems may be plotted against N atomic co-ordinates involving all the atoms in

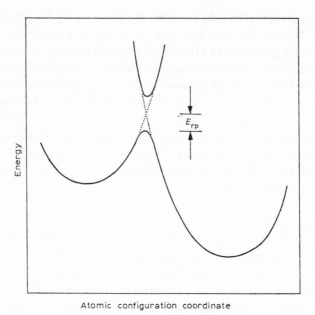

Atomic configuration coordinate

Fig. 6. Potential energy surfaces of reactant and product wave functions in an electron transfer reaction (after Marcus[64]).

the inner and outer spheres of the ions. The surfaces for the two wave functions intersect along an $(N-1)$ dimensional "line" or surface. Fig. 6 indicates a profile of the two surfaces showing the intersection and two minima representing the configurations of lowest energy of reactants and products. The intersection region for two non interacting states is indicated by the dotted lines in Fig. 6. Electronic interaction between the two states removes the degeneracy at the intersection and leads to a splitting of the two surfaces (indicated by the solid lines). The extent of the separation between the upper and lower states in the intersection region is equal to twice the interaction energy (or resonance energy) between the two states $E_{rp} = \langle \Psi_{react} | \mathcal{H} | \Psi_{prod} \rangle$, where $\mathcal{H}$ is the relevant Hamiltonian operator.

The magnitude of E_{rp} leads to the distinction between *adiabatic* and *non-adiabatic* reactions. This matter was discussed in section 4.5 of Chapter 3, and it was pointed out that the probability of crossing between two surfaces decreases rapidly with an increase in the separation between them. For extremely weak interactions, the probability of crossing is high, so that the probability of the system remaining on the solid (adiabatic) line during a passage through the intersection region is small. The transmission coefficient, κ, for the reaction, defined by the relation

$$k = \kappa \cdot \frac{kT}{h} \cdot e^{-\Delta G^{\ddagger}/RT} \tag{101}$$

will be small for such processes occurring adiabatically. It was pointed out in Chapter 3 that processes, characterized by a value of κ that is much less than unity for the above reasons, are often erroneously called *non-adiabatic*. Conversely, for larger (although still "weak") interaction energies in the order of 1 kcal. mole^{-1}, κ is near unity and the reaction is termed *adiabatic*. Any interactions in ordinary chemical reactions normally involve large energies and therefore are also adiabatic.

Attempts have been made to calculate the reaction rates for non-adiabatic electron transfers by calculating the probability of an electron tunnelling through a potential barrier of some assumed shape[61, 62]. These treatments have been very approximate at best, and the error in assuming that the tunnelling probability is equivalent to κ in equation (101) has been discussed by Marcus[59]. In any case, it is difficult to justify the assumption that a reaction has a weak enough energy of interaction to be treated as non-adiabatic, and it now appears that the great majority of electron transfer reactions are adiabatic[7].

The most successful theoretical treatments of electron transfer reactions have been those which assume that the interaction between reactants is strong enough for the reaction to be adiabatic, and yet weak enough to have a negligible effect on the free energy of activation, $\Delta G^{\ddagger}$. This assumption is probably valid for most outer sphere reactions, and most of the recent theoretical work in the field has been directed toward electron transfer reactions of this type.

2.5.3 Components of the free energy of activation

Because electron transfer reactions involve ions, the two Coulombic terms $\Delta G_{es}^{\ddagger}$ and $\Delta G_{\mu}^{\ddagger}$ in equation (43) will contribute to the free energy of activation. In the ordinary ionic reactions treated by equation (43), $\Delta G_0^{\ddagger}$ arises mainly from the making and breaking of bonds in forming the activated complex and was considered as a constant for a given reaction and independent of the electrical properties of the solvent. Although $\Delta G_0^{\ddagger}$ was the largest contribution to $\Delta G^{\ddagger}$ it could not be calculated explicitly for these reactions. This is because in any reactions where electronic interaction between the reactants is strong enough to produce changes in covalent bonding, the quantum mechanical treatment which would be required for calculation of the absolute potential energies of the activated complex is prohibitively difficult. However, in outer sphere electron transfer reactions where covalent bonds are not made and broken the electronic interaction between reactants is small enough to be neglected and $\Delta G_0^{\ddagger}$ can be treated by quasi-classical methods and *absolute rate calculations* can be attempted.

Although the free energy contribution from covalent bonding in electron transfer reactions is negligible compared with that expected in ordinary reactions, the very nature of the weak interaction between the reactants in electron transfer reactions introduces a special component of $\Delta G^{\ddagger}$ required by the Franck–Condon principle. Electronic transitions occur very rapidly compared with the rates of nuclear motions, and the Franck–Condon principle requires that, to a first approximation, electronic transitions must occur without change in internuclear distances. The principle is therefore sometimes called the "Principle of Vertical Transitions", and in the present system it requires that the transition between the two states (Fig. 6) can only take place in the vicinity of the intersection region where the states have equivalent configuration and energy. For this reason the formation of an activated complex incurs the expenditure of a free energy, $\Delta G_r^\ddagger$, in the *reorganization* of the two reactants to form a configuration in the activated complex equivalent to that of the products. Since this will require reorganization of both the inner co-ordinate shells of ligands and the outer sphere of solvated molecules, $\Delta G_r^\ddagger$ can be separated into $\Delta G_{ri}^\ddagger$ and $\Delta G_{ro}^\ddagger$ for reorganization of the inner and outer spheres respectively.

These reorganizational free energies, along with the usual Coulombic terms, constitute the largest part of the free energy of activation for electron transfer reactions. The total free energy of activation can thus be written

$$\Delta G^\ddagger = \Delta G_{es}^\ddagger + \Delta G_\mu^\ddagger + \Delta G_{ri}^\ddagger + \Delta G_{ro}^\ddagger + \text{other small terms} \tag{102}$$

Each component of (102) will now be treated in turn.

(a) $\Delta G_{es}^\ddagger$ and $\Delta G_\mu^\ddagger$

These terms are identical to those discussed in sec. 2.2. They represent the Coulombic energy required to bring the reactants together to a centre-to-centre separation of the metallic ions of $r_\ddagger$. The $\Delta G_{es}^\ddagger$ term is much larger than $\Delta G_\mu^\ddagger$ for moderate ionic strengths. For the usual case of electron transfer reactions involving two positively charged ions, $\Delta G_{es}^\ddagger$ represents a *repulsive* term (*i.e.* it makes a positive contribution to the free energy of activation).

The remaining components of the free energy of activation have been treated by Hush[63] and by Marcus[64]. The two theories have been shown to give equivalent terms for $\Delta G_{ri}^\ddagger$ and $\Delta G_{ro}^\ddagger$ in special cases[59]. Since the Marcus theory has been developed further and appears slightly easier to apply, only the expressions derived from this theory will be considered here.

(b) $\Delta G_{ro}^\ddagger$

The contribution to the free energy of activation due to outer sphere (solvent molecule) rearrangement is given by Marcus as

$$\Delta G_{ro}^\ddagger = m^2 \lambda_{ro} \tag{103}$$

where

$$m = -\tfrac{1}{2} - \frac{\Delta G^0 + (\Delta G_{es}^\ddagger + \Delta G_\mu^\ddagger - \Delta G_{es}^{\ddagger\prime} - \Delta G_\mu^{\ddagger\prime})}{2(\lambda_{ro} + \lambda_{ri})} \tag{104}$$

$$\lambda_{ro} = (\Delta z)^2 \varepsilon^2 \left(\frac{1}{2r_A^{\ddagger}} + \frac{1}{2r_B^{\ddagger}} - \frac{1}{r_{\ddagger}}\right)\left(\frac{1}{D_{op}} - \frac{1}{D}\right) \tag{105}$$

The meaning of the various symbols used is discussed below.

ΔG^0 refers to the overall free energy change for the electron transfer step, although it is not usually identical with the standard free energy change for the step. The term $(\Delta G_{es}^{\ddagger} + \Delta G_{\mu}^{\ddagger} - \Delta G_{es}^{\ddagger\prime} - \Delta G_{\mu}^{\ddagger\prime})$ is the difference in Coulombic energy between the reactants and products in the activated complex configuration. Since $r_{\ddagger}$ is the same for both reactants and products this term will be zero for isotopic exchange reactions and all electron transfers where the product of charges on the reactants is the same as that on the products. Therefore, because ΔG^0 is also zero for isotopic exchange reactions, m will be equal to $-\frac{1}{2}$ for electron transfers between isotopic ions of the same metal. Δz is the number of electrons transferred, $r_A^{\ddagger}$ and $r_B^{\ddagger}$ are the radii of the reactants in the activated complex and $r_{\ddagger}$ is the distance between the metal ion centres. λ_{ri} is a contribution from the inner sphere and is discussed in the next section. D is the usual static dielectric constant and D_{op} is the optical dielectric constant equal to the square of the refractive index (η) of the solvent. For most solvents $1/D$ is only a few per cent of $1/\eta^2$ so that $\Delta G_{ro}^{\ddagger}$ is relatively unaffected by any dielectric saturation in the outer sphere.

Insertion of the conditions for isotopic exchange reactions ($m = -\frac{1}{2}$ and $r_A = r_B$) into equation (103) yields

$$\Delta G_{ro}^{\ddagger} = \frac{(\Delta z)^2 \varepsilon^2}{8r_A^{\ddagger}} \left(\frac{1}{\eta^2} - \frac{1}{D}\right) \tag{106}$$

if $r_{\ddagger}$ is taken to be equal to $r_A^{\ddagger} + r_B^{\ddagger}$. In water $\Delta G_{ro}^{\ddagger}$ is equal to 22.7 $(\Delta z)^2/r_A^{\ddagger}$ kcal. mole^{-1}, where $r_A^{\ddagger}$ is measured in Ångstrom units. If $r_A^{\ddagger}$ is taken to be the radius of the metal ion with its co-ordinated shell of ligands, then it equals $r_{ion} + 2r_{ligand}$, and is about 3.5 A for most metallic ions co-ordinated with water.

(c) $\Delta G_{ri}^{\ddagger}$

The inner sphere reorganization contribution is due to the changes in ligand geometry in the formation of a suitable activated complex. Evaluation will therefore require a knowledge of the force constants of the ligand–ion vibrations. The Marcus expression is

$$\Delta G_{ri}^{\ddagger} = m^2 \lambda_{ri} \tag{107}$$

where

$$\lambda_{ri} = \sum_j \frac{f_j f_j^p}{(f_j + f_j^p)} (\Delta r_j^0)^2 \tag{108}$$

and m is the same as in (104). f_j and f_j^p refer to the force constants for vibration of the jth co-ordinate in the reactant and its corresponding co-ordinate in the product. Δr_j^0 represent the changes in bond length and bond angles to form the activated complex from the reactants.

A simplified treatment by Sutin[58] gives an expression for $\Delta G_{ri}^{\ddagger}$ for a one electron isotope exchange reaction. It assumes that the only motion in the inner co-ordination shell is the symmetric and harmonic "breathing" vibration of the ligands. To satisfy the Franck–Condon principle the metal ligand equilibrium distances, d, for the two reactants must be adjusted to a common value $d^{\ddagger}$ in the activated complex. The energy required for this process is

$$\Delta G_{ri}^{\ddagger} = \frac{nf_A}{2}(d_A - d^{\ddagger})^2 + \frac{nf_B}{2}(d_B - d^{\ddagger})^2 \tag{109}$$

where f_A and f_B are the force constant for the metal–ligand bonds associated with the reactants A and B and n is the number of co-ordinated ligands. The value of $d^{\ddagger}$ required for the smallest $\Delta G_{ri}^{\ddagger}$ is found by differentiating $\Delta G_{ri}^{\ddagger}$ with respect to $d^{\ddagger}$ and is equal to $(f_A d_A + f_B d_B)/(f_A + f_B)$. Substitution of this value for $d^{\ddagger}$ into (109) gives

$$\Delta G_{ri}^{\ddagger} = \frac{nf_A f_B(d_A - d_B)^2}{2(f_A + f_B)} \tag{110}$$

The expression of Basolo and Pearson[65] to calculate the force constants has been used by Sutin[58] to calculate $\Delta G_{ri}^{\ddagger}$ for the $Fe^{+2} + {}^*Fe^{+3}$ reaction in water. Taking values of 1.49×10^5 dyne. cm^{-1}, 4.16×10^5 dyne. cm^{-1} and 2.21 A, 2.05 A for the force constants and Fe–O distances in hexaquo iron(II) and (III) respectively, the value of $\Delta G_{ri}^{\ddagger}$ from (110) is found to be 12.1 kcal. $mole^{-1}$. The result is about twice the magnitude of $\Delta G_{ro}^{\ddagger} = 6.4$ kcal. $mole^{-1}$ for the same reaction, calculated from (106) using $r_A^{\ddagger}$ as the mean value between $r_A = 3.59$ A for $Fe(H_2O)_6^{+2}$ and $r_B = 3.43$ A for $Fe(H_2O)_6^{+3}$. These values for $\Delta G_{ro}^{\ddagger}$ and $\Delta G_{ri}^{\ddagger}$ can be added to $\Delta G_{es}^{\ddagger} = 3.6$ kcal. $mole^{-1}$ for the reaction obtained using $r_{\ddagger} = 2r_A^{\ddagger} = r_A + r_B$ and $D = 78.5$, and give the sum of $\Delta G_{es}^{\ddagger}$, $\Delta G_{ri}^{\ddagger}$ and $\Delta G_{ro}^{\ddagger}$ as 22.1 kcal. $mole^{-1}$ at zero ionic strength. With the addition of the $RT \ln (kT/\rho Zh)$ factor discussed below, this yields an absolute calculation of $\Delta G^{\ddagger} = 23.1$ kcal. $mole^{-1}$, compared with the measured[66] total free energy of activation for the reaction of 16.6 kcal. $mole^{-1}$.

(d) Additional contributions to $\Delta G^{\ddagger}$

It should be noted at this point that the Marcus formulation of the theory does not use equation (101) for the rate coefficient but rather[59]

$$k = \kappa \rho Z e^{-\Delta G^*/RT} \tag{111}$$

where κ is the same transmission coefficient as in (101), ρ is a dimensionless statistical factor having a value close to unity and Z is the bimolecular collision frequency of two uncharged species in solution when they are at unit concentration. ΔG^* is not equivalent to the usual free energy of activation, $\Delta G^{\ddagger}$, but refers to the free energy of activation for a hypothetical complex whose configurational distribution is referred to as the "equivalent equilibrium distribution"[64b]. For this reason the force constants in (108) should really be the effective force constants characteristic of the equivalent equilibrium distribution. However the differences are small and may be neglected in a simple treatment. Comparison of (111) and (101) indicates that ΔG^* is related to the conventional free energy of activation by

$$\Delta G^{\ddagger} = \Delta G^* + RT \ln \frac{kT}{\rho Z h} \tag{112}$$

The Marcus theory therefore requires that the term $RT \ln kT/\rho Z h$ be included in equation (102). ρ is usually close to unity and if a "normal" solution collision number of 10^{12} l. mole^{-1}. sec^{-1} is used, then in molar units Z becomes 10^{12} sec^{-1}, and $RT \ln (kT/\rho Z h)$ is equal to 1.0 kcal. mole^{-1}.

Further small terms in (102) can be added to account for the omission of the resonance energy and for possible changes of electron multiplicity, M, in forming the activated complex. The latter term is given by $\Delta G_M^{\ddagger} = RT \ln (M^{\ddagger}/M_{reactants})$ and is never a large contribution.

A final factor that could affect the free energy of activation is *nuclear tunnelling*. Since the motions of ligand nuclei required to reorganize themselves to satisfy the Franck–Condon principle are small and the energy barrier presented by $\Delta G_{ri}^{\ddagger}$ is fairly high, there exists a possibility that nuclear re-organization could take place by tunnelling. If such nuclear tunnelling did occur it would reduce the effective value of $\Delta G_{ri}^{\ddagger}$, and to a much lesser extent that of $\Delta G_{ro}^{\ddagger}$. Sutin and Wolfsberg[67] have calculated that nuclear tunnelling would reduce $\Delta G_{ri}^{\ddagger}$ for the aqueous Fe^{+2}–Fe^{+3} reaction by 2.2 kcal. mole^{-1}.

3. Proton transfer reactions

3.1 ACIDS AND BASES[8]

The transfer of a proton between two chemical species occurs as at least one step in a very large number of reactions in solution. The unique properties of the proton are primarily a result of the absence of any surrounding electron shell so that it has a radius about 10^5 times smaller than other cations. The resulting lack of any need for a special geometric orientation in reactions of the proton, combined with its small mass, give it an extremely high mobility in solution. For these reasons, solution reactions in which hydrogen is transferred between two species tend to involve transfer of the hydrogen nucleus rather than the atom itself.

Its extremely small size relative to its charge makes the proton a very strong polarizing agent, and in solution it is always highly solvated. The primary solvation process in aqueous media involves the formation of a covalent bond with the oxygen atom of a water molecule to form the hydronium ion, H_3O^+. The independent existence of the hydronium ion and the symmetrical distribution of its positive charge has been demonstrated by nuclear magnetic resonance[68]. Secondary solvation in the form of three additional water molecules joined by hydrogen bridges to the hydronium ion produces the relatively stable complex $H_9O_4^+$.

The ease of proton transfer is responsible for the importance of the acid–base concept in solution chemistry, or more specifically, for the Brønsted concept of acids and bases. Brønsted[69] defined an acid as any species which tends to lose a proton and a base as any species which tends to gain one. It follows that any proton transfer process having the general form

$$HX + Y \rightarrow X + HY \tag{113}$$

(where any species may be charged) can be considered as an *acid–base reaction* between a Brønsted acid HX and Bronsted base Y to form a conjugate base X and acid HY.

Lewis[70] has offered a different definition which is applicable in aprotic systems. A Lewis acid or electrophile is a species which tends to gain an electron pair and a Lewis base or nucleophile is a species tending to lose an electron pair. The Brønsted concept of acids and bases is obviously more relevant to a discussion of proton transfer reactions and it will be used in this section.

3.2 ELEMENTARY PROTON TRANSFER PROCESSES

The high mobility of the proton makes many proton transfer processes very rapid, and, indeed, the reaction between strong acids and bases (*i.e.* $H_3O^+ + OH^-$) was long thought to be instantaneous. However, modern relaxation techniques have recently provided a means for the direct measurement of even the fastest acid–base reactions. The most successful ideas on the nature of proton transfer processes have been put forward by Eigen and his school[71] and the treatment presented here will be based on their approach.

The mechanism for acid–base (proton transfer) reactions can be separated into three steps

$$HX + Y \underset{}{\overset{1}{\rightleftarrows}} (XH, Y) \overset{2}{\rightleftarrows} (X, HY) \overset{3}{\rightleftarrows} X + HY \tag{114}$$

The first step represents the diffusion of the reactants towards one another to a distance close enough to facilitate hydrogen bridge formation between the donor and acceptor. Step 2 is the transfer process itself and includes the formation of the hydrogen bridge. The third step is the separation of the products by diffusion. Using the steady state approximation for the concentration of the encounter pair species (HX, Y) and (X, HY), the general forward rate coefficient for the proton transfer process becomes

$$k = \frac{k_1 k_2 k_3}{k_{-1} k_3 + k_2 k_3 + k_{-1} k_{-2}} \tag{115}$$

It is possible that the hydrogen bridge may involve one or more H_2O molecules, and the proton transfer may occur by a Grotthuss mechanism where the proton is transferred along a chain of water molecules in a series of very rapid jumps. The question of whether the solvent molecules are involved in the hydrogen bridge is the subject of a recent review by Albery[72]. Such considerations will not significantly affect our discussion since in most cases the presence or absence of H_2O molecules in the hydrogen bridge would not greatly alter the relative ability of different species to form the bridges. The ease of formation of a hydrogen bridge, and thus of the proton transfer, is one of the most important factors which influence the rate of step 2. The ease of bridge formation is in the order

$$OH \cdot \cdot O > OH \cdot \cdot N,\ NH \cdot \cdot O > NH \cdot \cdot N \gg CH \cdot \cdot O,\ OH \cdot \cdot C \tag{116}$$

and the rates of step 2 for reactions involving the various species will therefore decrease from $OH \cdot \cdot O$ to $OH \cdot \cdot C$ if other factors are equal. In addition, if an unperturbed hydrogen bridge can be formed, very little rearrangement of nuclei will be necessary in a proton transfer reaction and Franck–Condon type restrictions will be less important than for electron transfers (section 2.5.3).

The effect of a difference in efficiency of hydrogen bridge formation on the rate of step 2 can be visualized by reference to Fig. 7. The relative heights of the $HX + Y$ and the $X + HY$ equilibrium positions are related to the overall free energy difference between HX and HY and therefore to the pK difference, ΔpK, where

$$pK_{HX} = -\log \frac{[H^+][X^-]}{[HX]} \quad \text{(and similarly for } pK_{HY}) \tag{117}$$

$$\Delta pK = pK_{HY} - pK_{HX} \tag{118}$$

The intersecting dotted lines represent a situation where there is no electronic interaction between the reactant, $HX + Y$, and the product $X + HY$ wave func-

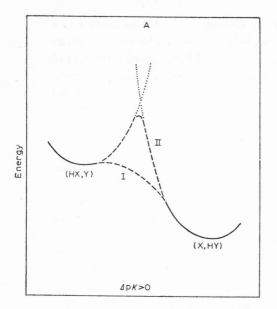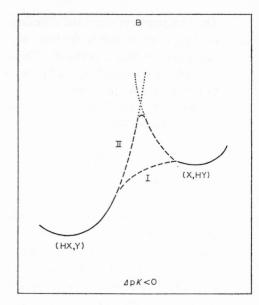

Reaction coordinate

Fig. 7. Potential energy curves for the proton transfer reactions $HX+Y \rightleftarrows X+HY$ with efficient (I) and inefficient (II) hydrogen bridge formation for reactions in which $\Delta pK > 0$ and $\Delta pK < 0$ where $\Delta pK = pK_{HY} - pK_{HX}$.

tions (*cf.* Fig. 6). On the other hand, when a symmetrical unperturbed hydrogen bridge can be formed between HX and Y the resonance interaction is very large and, as indicated by curve I, the activation barrier can be almost zero. In such a case, proton transfer in the downhill direction (A, curve I) will be unhindered and extremely rapid. The transfer will take place and the products will be formed before thermal motion is able to separate the reactants in the encounter complex. This corresponds to $k_{-1} \ll k_2$ and k_3 so that $k = k_1$ and the overall proton transfer rate coefficient will be equal to the rate coefficient for the diffusion controlled formation of the (HX, Y) encounter pair and will be independent of ΔpK. This diffusion controlled limit is about 10^{11} l. mole^{-1}. sec^{-1}, and is found to apply to reactions of strong acids and bases where hydrogen bridge formation is efficient and where the pK difference is large in the favourable direction.

An expression for the diffusion controlled rate coefficient k_1 is given by (119), which is simply the Smoluchowski equation (10) with the addition of an electrostatic term[73], which has no effect unless both acid and base are charged or have strong dipole moments, *viz.*

$$k_1 = \frac{4\pi N_0}{1000} \cdot r_{(HX, Y)} \cdot D_{HX, Y} \cdot \frac{\Delta G_{es}/kT}{(e^{\Delta G_{es}/kT} - 1)} \tag{119}$$

ΔG_{es} is the electrostatic free energy of the encounter pair given by equations

(44, 87 or 88). Thus, for proton transfer reactions where the encounter pair formation is rate controlling, the reaction rate depends only on the charge and diffusion coefficients of the donor and acceptor, the encounter pair separation, and the dielectric constant of the medium.

The activation energy of many proton transfer reactions is, however, appreciable, so that there is an energy barrier to surmount even though the overall reaction is "downhill". The potential curve for such a reaction is illustrated by curve II in Fig. 7A, and step 2 may be the rate determining process even when a highly favourable pK difference exists. Curve II is typical of reactions of CH acids where hydrogen bridge formation probably does not occur to a significant extent and also of reactions where specific effects hinder bridge formation. An example of the latter case can be seen in reaction of OH^- with

which is 10^5 times slower than the diffusion controlled rate because the internal hydrogen bond must be broken before the hydrated OH^- is able to form a hydrogen bridge to the proton.

For reactions in which the pK difference is unfavourable (see Fig. 7B), the standard free energy difference is included in the free energy of activation. Step 2 will then be rate determining for both case BI and case BII and the overall reaction rate will depend on ΔpK.

In the case of efficient hydrogen bridge formation with an unfavourable ΔpK (curve I, Fig. 7B) the free energy of activation $\Delta G^{\ddagger}$ is equal to the standard free energy change ΔG^0. Since log K varies linearly with ΔG^0 and log k with $\Delta G^{\ddagger}$ we have

$$\log k_2 = \text{const.} + \Delta pK \tag{120}$$

If for the variation of the overall rate coefficient k with ΔpK

$$\alpha = \frac{d \log k}{d\Delta pK} \tag{121}$$

then for the ideal case (curve BI) if step 2 is rate controlling, $\alpha = 1$ for all ΔpK less than zero. At $\Delta pK = 0$ there will be a rapid transition to $\alpha = 0$, where even at small positive values of ΔpK the reaction becomes diffusion controlled and independent of ΔpK. The behaviour of the ideal case (curve I) of zero activation energy is shown in Fig. 8I. A similar curve is also given for the reverse reaction of (114) for which d log $k/d\Delta pK$ is denoted β.

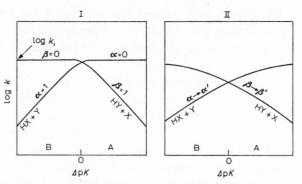

Fig. 8. Theoretical log k *versus* ΔpK plots for the reactions $HX+Y \rightarrow X+HY$ (α curves) and $HX+Y \leftarrow X+HY$ (β curves) for the cases of efficient (I) and inefficient (II) hydrogen bridge formation. The regions A and B correspond to favourable ΔpK differences for the $HX+Y \rightarrow X+HY$ transfer as shown in Fig. 7.

For most reactions of interest in solution, however, perfect hydrogen bridging is not possible and a plot of log k against ΔpK is not as simple as that in Fig. 8I. Even when the pK difference is large and favourable the diffusion controlled limit may not be reached (*e.g.* curve II, Fig. 7). For unfavourable pK difference (curve BII) the $\Delta G^{\ddagger}$ will always be greater than ΔG^0, and α will approach unity only at very large negative ΔpK. In the wide intermediate region α', defined by

$$\alpha' = \frac{d \log k_2}{d\Delta pK} \tag{122}$$

will not be constant, but will depend on ΔpK and will range between 0 and 1. This situation is represented by Fig. 8II.

The value of α for the overall reaction can be found by differentiating the logarithm of (115) with respect to ΔpK to yield

$$\alpha = \frac{k_{-1}k_3\alpha' + k_{-1}k_{-2}}{k_{-1}k_3 + k_2k_3 + k_{-1}k_{-2}} \tag{123}$$

The three limiting cases for the reaction (114) can easily be seen from (123) to be (*a*) the forward reaction is diffusion controlled

$$k_{-2}, k_{-1} \ll k_2, k_3; \text{ then } \alpha \rightarrow 0.$$

(*b*) the reverse reaction is diffusion controlled

$$k_2, k_3 \ll k_{-2}, k_{-1}; \text{ then } \alpha \rightarrow 1.$$

(*c*) the proton transfer step is rate determining in both directions

$$k_2 \ll k_{-1} \text{ and } k_{-2} \ll k_3; \text{ then } \alpha \to \alpha' \qquad (0 < \alpha' < 1).$$

If the reactants HX and Y have opposite charges then the forward reaction diffusion limited rate coefficient, k_1, will be greater than the reverse direction diffusion limited rate coefficient, k_{-3}, for the reaction of neutral X and HY. In such a case the α curves would be displaced slightly higher than the β curves in Fig. 8.

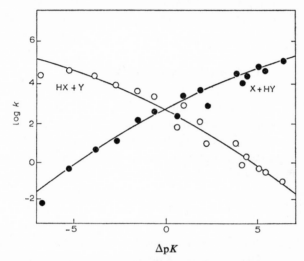

Fig. 9. The ΔpK dependence of log k for proton transfer from HX to Y ($\bullet$) and HY to X (○) where

X = the enolate form of acetylacetone (HX)
Y = (from right to left) : OH$^-$, glycerol (anion), mannose (anion), glucose (anion), trimethylphenoxide, phenoxide, pyrophosphate, chlorophenoxide, *p*-nitrophenoxide, HS$^-$, dimedone (enolate), pyridine, acetate, benzoate, hydroxyacetate and chloroacetate (from Eigen[71a]).

Fig. 9 illustrates an experimental plot of log k against ΔpK for the ionization of the carbon–hydrogen bond of various ketonic donors with different acceptors. As expected from (116) the transfer is preceded by *inefficient* formation of the hydrogen bridge and the curves approximate to those in Fig. 8II.

3.3 ACID–BASE CATALYSIS

Acid and base catalyzed reactions have for many years provided the major means of studying proton transfer processes. A catalyst is generally defined as a species that accelerates a particular reaction without itself being consumed in the overall process[†]. The role of the catalyst is usually to provide a different mechanism involving a series of steps whose overall free energy of activation is less than that for the uncatalyzed reaction path. An acid or base catalyzed reaction must have at least one step consisting of an *acid–base reaction (proton transfer) between the catalyst and the substrate*. The protonated or deprotonated substrate then proceeds to a further reaction such as decomposition or reaction with another species in solution. For an acid catalyzed reaction the substrate is Y in equation (114) and HX is the catalyst, while for base catalysis HY is the substrate and X the catalyst. In almost all cases the observed acid or base catalyzed reaction cannot proceed in the absence of a proton transfer between substrate and "catalyst", so that the function of the catalyst (which may be the solvent) is not merely to accelerate the reaction but to allow it to occur at all.

3.3.1 General and specific catalysis

It was originally thought that in aqueous solution only H_3O^+ or OH^- could affect the rate of an acid or base catalyzed reaction. It is now apparent, however, that, because of the nature of acid and base catalyzed reactions, the catalyst can be any species able to undergo a proton transfer with the substrate, and for many reactions the rate may be written

$$\text{rate} = v_0 + k_{H^+}[H_3O^+] + k_{OH^-}[OH^-] + k_{HX_i}[HX_i] + k_{X_j}[X_j] \tag{124}$$

where HX_i and X_j represent all the acid and base species present in the solution apart from H_3O^+ and OH^-. In general the extent to which each acid and base species in (124) affect the rate will be determined by the specific nature of the reaction and the equilibria that exist in the solution. Reactions for which only $[H_3O^+]$ or $[OH^-]$ appear in the overall rate equation are considered to be examples of *specific catalysis* while those involving $[HX]$ or $[X]$ are classified as *general catalysis*.

The various possible cases for acid and base catalyzed reactions are treated exhaustively in other books on the subject[4, 9] and only the most common examples will be mentioned here. Since the treatments of acid and base catalysis are so closely analogous, attention is confined to acid catalysis.

A reaction in which a substrate, Y, undergoes rapid protonation in an equilibrium step, followed by a rate determining unimolecular reaction of the protonated substrate, is subject to specific catalysis. The reaction sequence may be written as

---

[†] Since a reactant or product may also be a catalyst, Bell[9] defines a catalyst as a substance which appears in the rate law to a power higher than that to which it appears in the stoichiometric equation.

follows

$$HX + Y \underset{k_r}{\overset{k_f}{\rightleftarrows}} X + HY$$

$$HY \overset{k'}{\rightarrow} \text{products}$$

and assuming HY is maintained at a steady state concentration

$$[HY] = \frac{k_f[HX][Y]}{k' + k_r[X]}$$

Thus

$$\text{rate} = k'[HY] = \frac{k'k_f[HX][Y]}{k' + k_r[X]}$$

For an equilibrium in which $k_r[X] \gg k'$

$$\text{rate} = \frac{k'k_f[H^+][Y]}{K_{HX}k_r}, \qquad \text{where } K_{HX} = \frac{[H^+][X^{(-)}]}{[HX]}$$

and clearly the rate depends specifically on the concentration of hydrogen ions. The same, of course, is true if the reaction of HY is bimolecular provided that bases do not in general participate in the reaction.

On the other hand a reaction in which a proton transfer step is rate determining is subject to general acid catalysis. An example of this is provided by the previous reaction when $k' \gg k_r[X]$, *i.e.* the step

$$HX + Y \overset{k_f}{\rightarrow} X + HY$$

is rate determining and

$$\text{rate} = k_f[HX][Y]$$

Now both the proton and the conjugate base of the acid are present in the transition state and it is the concentration of the molecular acid which appears in the rate equation. The same is true when a slow proton transfer is preceded by an equilibrium protonation of the substrate. In this case the anion of the acid acts as a *base* in the rate determining step, as for example in the acid catalyzed enolisation of ketones[8]. The reaction may be written generally as follows

$$HX + Y \underset{k_r}{\overset{k_f}{\rightleftarrows}} X + HY$$

$$X + HY \overset{k''}{\rightarrow} products$$

and, if there is a stationary concentration of HY

$$rate = \frac{k''k_f[HX][Y]}{k_r + k''}$$

The same kinetic law is followed independently of the relative magnitudes of k_{-1} and k_2.

The observation of general catalysis cannot always be taken as a criterion of rate determining proton transfers. A common modification of the last example occurs when the anion of the acid participates not as a base but as a nucleophile. Clearly, the kinetic behaviour simulates that of general catalysis. However, when a series of acids is compared it is found that their relative effectiveness as acids correlates not with their pK's but with the nucleophilicities of their conjugate bases. "Nucleophilic catalysis" of this type is commonly observed in the addition reactions of carbonyl compounds[74].

3.3.2 The Brønsted relation

In 1923 Brønsted and Pederson[75] suggested the following equations to relate the effectiveness of the catalyst to its acid or base strength

$$k_{HX} = G_{HX} K_{HX}^{\alpha} \qquad k_X = G_X K_X^{\beta} \tag{125}$$

K_{HX} is the usual acid constant defined by (117) and K_X is the base dissociation constant which can be taken to be equal to the reciprocal of the acid dissociation constant for the conjugate acid corresponding to that particular base, viz.

$$K_X = \frac{1}{K_{HX}} \tag{126}$$

A statistical correction is often made to (125) to allow for acids which have a number, p, of equivalent dissociable protons, their respective conjugate bases having q equivalent positions where a proton can be accepted. Equation (125) is then modified to read

$$\frac{k_{HX}}{p} = G_{HX} \left[\left(\frac{q}{p} \right) K_{HX} \right]^{\alpha} \qquad \frac{k_X}{q} = G_X \left[\left(\frac{p}{q} \right) K_X \right]^{\beta} \qquad (127)$$

In practice the effect of the statistical correction is usually insignificant, since it changes the rate typically by a factor of less than two, which is small compared to the large differences of several powers of ten commonly covered in Brønsted plots.

Equations (125) or (127) can be written in the form

$$\log k_{HX} = \alpha(-pK_{HX}) + \text{const.} \qquad (128)$$

so that the Brønsted α is identical to that defined by (121). From the considerations of the previous section it is apparent as Brønsted himself suggested[75], that α could not be expected to remain constant over large variations in catalyst pK. From the definition of α (121) it is seen that an expectation of a linear dependence of log k over a limited range of pK is equivalent to the retention of only the first term of a Taylor series expansion for d log k/d ΔpK. Nevertheless, when the transition range from the limiting cases of $\alpha = 0$ to $\alpha = 1$ is very wide as is the case especially for proton transfers to or from carbon, α will appear constant at some intermediate value over a fairly large ΔpK range. A striking example of this seems to be the acid catalyzed hydrogen exchange in trimethoxybenzene[76], the Brønsted plot of which is shown in Fig. 10. Similar behaviour has been observed for proton transfer between oxygens as in the case of the hydration of acetaldehyde[77], where the log k against pK plot is linear over a range of 11 pK units. However, this behaviour is sufficiently unusual for reactions involving simple

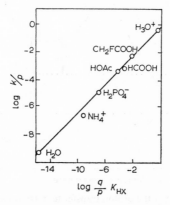

Fig. 10. A Brønsted plot for acid catalyzed aromatic hydrogen exchange in 1,3,5-trimethoxybenzene (from Kresge and Chiang[76]).

proton transfers from an oxygen atom that it has been considered as evidence for a more complex reaction mechanism[73].

When the rates are measured for different substrates (*i.e.* pK_{HY} is varied) as well as different catalysts, a wider range of ΔpK is accessible, and Fig. 9 shows that the predicted curvature is found.

The Brønsted equations are an example of a class of relations known collectively as *linear free energy relationships* which will be discussed in greater detail in section 4. A diagrammatic interpretation of the Brønsted equation indicating the relation between free energies is shown in Fig. 11.

Fig. 11 applies to the type of acid–base reaction shown on curve BII of Fig. 7, where interaction between the HX + Y and the X + HY curves is minimal and their intersection provides a good approximation to the activation energy barrier. The dashed curve (Fig. 11) represents the potential energy of a different catalyst HX′ having a similar structure to HX but a slightly greater pK. From Fig. 11, it is apparent that if the HX′ curve is identical in shape to that of HX and if all three curves can be approximated by straight lines in the intersection region, then the difference in activation energy will be proportional to the difference in standard energy, and we can write

$$E_{HX'} - E_{HX} = \alpha'(E^0_{HX'} - E^0_{HX}) \tag{129}$$

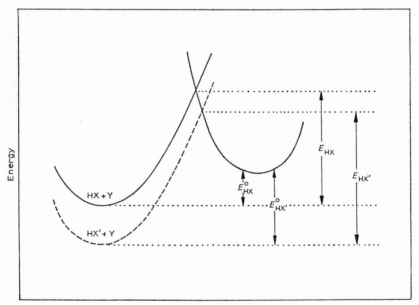

Fig. 11. Potential energy curves for the proton transfer to Y from two different acid catalysts HX and HX′. E_{HX}, $E_{HX'}$ and E^0_{HX}, $E^0_{HX'}$ represent the activation energies and standard free energies, respectively, for proton transfer.

If the entropy changes for the reactions of HX and HX′ are the same then the energies in (129) can be replaced by free energies. Now, from the relations

$$\Delta G_{HX'}^{\ddagger}-\Delta G_{HX}^{\ddagger} = RT \ln \frac{k_{HX}}{k_{HX'}} \quad \text{and} \quad \Delta G_{HX'}^{0}-\Delta G_{HX}^{0} = RT \ln \frac{K_{HX}}{K_{HX'}} \quad (130)$$

we can write

$$\log k_{HX'} - \log k_{HX} = \alpha'(\log K_{HX'} - \log K_{HX}) \quad (131)$$

Equation (131) is seen to be identical with the definition of α', *viz.* (d log k_2/d ΔpK), in equation (122), and is identical also with the Brønsted relation if $\alpha' = \alpha$, which is the case if the actual proton transfer is the rate determining step.

In practice the curves in Fig. 11 are not straight lines in the intersection region so that even if the catalyst potential curves were identical in shape the curvature would be expected to produce a smooth variation of α over any finite ΔpK range. If the shape of a potential curve of, say, HX″ differs markedly from that of HX and HX′, large deviations from the Brønsted plot would be expected.

The solvent also may have an important effect on Brønsted plots. Bell and Wilson[78] have shown that the Brønsted plots for the amine catalyzed decomposition of nitramide are in the form of parallel lines corresponding to the primary and tertiary amines. The separation between these two lines is found to be much greater for reaction in water than in the non-polar solvent anisole. This difference in behaviour is attributed to hydrogen bonding affecting the amine dissociation constant in water. Such hydrogen bonding is not possible in anisole and the solvation of the cations has a much smaller effect on the dissociation constant.

3.3.3 Salt effects

The term *salt effects* refers to the influences that the presence of charged species in solution exert on a reaction. The effects are classified as primary and secondary, and are both caused by the perturbation of the activities of species in solutions by an ion atmosphere. The ion atmosphere effects for dilute solutions can be treated by the Debye–Hückel theory (section 2.2.2) and are given by equation (66).

The *primary salt effects* are due to the change in activity coefficients of the *reactant* species and the transition state. They were shown in section 2.2.2 to be reflected in the free energy of activation, $\Delta G^{\ddagger}$, and they have a *kinetic* effect on the reaction rate. The effect can be expressed in terms of activity coefficients by the Brønsted–Bjerrum equation (26), and for acid catalyzed reactions has the form

$$k = k_{\mu=0} \frac{f_Y f_{HX}}{f_{\ddagger}} \quad (132)$$

Although it is often possible to determine f_Y experimentally, the ratio $f_{HX}/f_{\ddagger}$ cannot be evaluated directly and must be determined by extrapolating the rate coefficient to zero ionic strength, $\mu = 0$.

The *secondary salt effect*, on the other hand, influences the reaction rate through the purely *thermodynamic* effect of altering the concentration of dissociable reactant species. These concentration changes are due to a perturbation of the dissociation constant, K, expressed by

$$K = K_{\mu=0}\frac{f_{H^+}f_{X^-}}{f_{HX}} \tag{133}$$

The Debye–Hückel Limiting Law (66) combined with (133) predicts a variation of K with the square root of ionic strength, and this has been demonstrated for many reactions[79]. The magnitude and direction of the secondary salt effect on reaction rate will depend on which concentration terms appear in the overall rate expression.

3.3.4 Acidity functions[80, 82]

From the discussion in section 3.2 it is apparent that the effectiveness of acid catalysis depends on the *proton donating power* of the acid. In order to study acid catalyzed reactions of extremely weak bases, very strongly acidic media are required. In dilute solutions of strong acids a good measure of the proton donating power of the medium is provided by pH measurements, which are based on hydronium ion concentration. In very concentrated solutions, salt and other effects cause the activity coefficients, especially those of H^+, to differ greatly from unity so that hydrogen ion or hydronium ion concentration is no longer a valid measure of the proton donating power (acidity) of the solution. As the concentration of acid is increased, the acidity of the medium, as measured by its ability to protonate indicator bases and by its effect on reaction rates, increases much faster than the H^+ or H_3O^+ concentration.

In 1932 Hammett and Deyrup[81] introduced an experimentally determinable quantity, H_0, designed to provide a measure of proton donating power to electrically neutral entities of a medium at all acid concentrations[†]. The function H_0 is determined by the ability of the medium to protonate an indicator base, and is defined in such a way that it becomes equivalent to pH at low concentrations, *viz.*

$$H_0 \equiv pK_{BH^+} - \frac{\log[BH^+]}{[B]} \tag{134}$$

[†] There have to be different H functions for proton donation to electrically charged species.

It can be seen that H_0 will be a unique property of the medium only if equation (134) gives the same result when any indicator base is used. This has been found to be very nearly the case for bases of similar structure. The indicator base must be chosen so that the ratio $[BH^+]/[B]$ is measurable spectrophotometrically when BH^+ and B are in low enough concentrations to have a negligible effect on the reaction being studied. The pK of the indicator base is a thermodynamic quantity, independent of the medium, which can be determined using other appropriate systems.

Substituting $K_{BH^+} = a_B a_H / a_{BH^+}$ into (134) yields

$$H_0 = -\log \frac{a_H f_B}{f_{BH^+}} \equiv -\log h_0 \tag{135}$$

It can be seen that at infinite dilution, where activities approach concentrations and activity coefficients approach unity, $H_0 \rightarrow$ pH and $h_0 \rightarrow [H^+]$.

For some time it was thought that a very wide range of weakly basic indicators gave nearly the same value for an acidity function[80]. However, it has since become clear that the condition that f_B/f_{BH^+} be independent of the nature of the base is true only for structurally related groups of compounds, and different acidity functions have now been measured for several such groups, including secondary and tertiary amines, amides, indoles and olefins.

A principal use of acidity functions has been the correlation of the acidity dependences of the rates of acid catalyzed reactions[82]. On the basis of a tentative suggestion by Zucker and Hammett[83], reactions were classified as to whether the acidity dependence of their rate was better described by using the Hammett function or the stoichiometric acid concentration. The two types of behaviour were thought to correspond to the presence or absence of a water molecule acting as a nucleophile or base in the transition state of the reaction, and acidity functions were regarded as providing a useful criterion of reaction mechanism. More recently it became apparent that this distinction lacks experimental and theoretical support. Clearly, for example, it depends on the existence of a unique acidity function. Nonetheless, comparisons of acidity dependences for reaction rates and equilibria have continued and recently have been formulated within the framework of a linear free energy relation[84]. Where the acidity dependence of the rate and equilibrium of a reaction can be studied, this comparison leads to a parameter comparable to the exponent in the Brønsted relation. At present, however, the kinetic significance of acidity functions cannot be considered to be fully established.

3.4 ISOTOPE EFFECTS ON PROTON TRANSFER PROCESSES[8]

The reactivity of most chemical species depends primarily on the properties of the electron shells which surround the nuclei and is therefore relatively unaffected by isotopic substitution. The proton, however, has no electron cloud and the properties of proton transfer reactions attributable to the proton will depend on the properties of a bare nucleus. The most obviously important of these properties is the nuclear mass, and as a result the different isotopes of hydrogen, protium, deuterium and tritium, undergo transfer reactions at measurably different rates. The study of these isotope effects on reaction rates can help to elucidate the nature of the proton transfer process. In particular, isotopic studies have been used to provide information about the structure of the transition state and about the nature and extent of tunnelling in proton transfer reactions. In order to discuss the latter it will first be necessary to examine the isotopic effects to be expected if tunnelling were not present.

3.4.1 Isotope effects in the absence of tunnelling

If the possibility of tunnelling is ignored, the proton transfer process can be discussed in terms of the usual transition state formalism. The reaction will be characterized by a temperature independent activation energy E, and a pre-exponential factor A. The isotope effect on each of these factors will be considered separately.

(a) Effect on energy of activation

The electronic shell in both protium and deuterium is the same and to the accuracy of the Born–Oppenheimer approximation the electronic distributions for bonds involving either isotope are identical. If such is the case, both a proton and a deuteron transfer can be represented by the same potential curve as shown in Fig. 12 taken from Bell[8].

Because of the existence of a zero point energy, $\frac{1}{2}h\nu_0$, which is mass dependent, the unreacted XH and XD systems do not have equal energy but are separated by a zero point energy difference, ΔE_0. A transfer reaction would therefore be expected to have different activation energies with the various hydrogen isotopes. The fact that the measured difference in activation energy, ΔE, is usually found to be less than ΔE_0 probably indicates that the transition state also has quantized levels and therefore a zero point energy depending on isotopic mass as shown in Fig. 12. Thus the magnitude of the difference between ΔE and ΔE_0 gives as indication of the magnitude of the transition state zero point energy and in this way provides a possible insight into the vibrational modes, and ultimately the structure, of the transition state. Although data is still limited an interpretation of such differences in terms of transition state structure has been given recently by Bell[85].

The existence of a zero point energy in the transition state was originally thought to be due to the vibration of the "incompletely broken" $X \cdot \cdot H$ bond. It has since been demonstrated by Westheimer[86] that such a view, which is equivalent to assigning a real frequency to the stretching mode, $\overleftarrow{X} \cdot \cdot \overrightarrow{H} \cdot \cdot \overleftarrow{Y}$, corresponding to

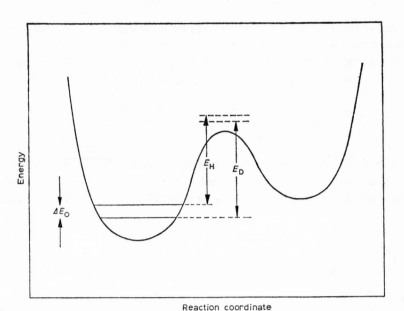

Reaction coordinate

Fig. 12. Potential energy curve for proton or deuteron transfer showing the zero point energies in the initial and the transition states (from Bell[8]).

the reaction co-ordinate, is not consistent with the definition of the transition state. However, Westheimer does point out that the second stretching mode for the transition state $\overset{\leftarrow}{X} \cdots H \cdots \vec{Y}$ (where the motion of H is unspecified) does possess zero point energy and that the sensitivity of this zero point energy to isotopic substitution depends upon the force constants of the partial bonds to the hydrogen. For a *symmetrical* transition state where these force constants are the same, the hydrogen is not in motion and the frequency is independent of the hydrogen isotope. In this case there is no zero point energy in the transition state to offset that in the ground state and a maximum isotope effect should be observed. It is assumed of course that the effect of hydrogen bonding vibrations is largely cancelled between the reactants and the transition state and that the use of such a simple model for discussion of the stretching vibrations is justified by the low mass and largely uncoupled motion of the hydrogen. In practice it seems likely that the force constants to hydrogen in the transition state will depend largely on the *relative base strengths* of X and Y. Indeed, a plot of log k_H/k_D against ΔpK ($= pK_{HX} - pK_{HY}$) for proton and deuteron transfer reactions from C–H bonds of twenty different compounds leads to a fairly smooth curve with a maximum at $\Delta pK = 0$ (the condition for a symmetrical transition state)[87].

(b) The effect on the pre-exponential factor

The statistical thermodynamic formulation of the transition state theory (see Chapter 3, section 4.3) uses partition functions to replace the entropy of activation term used in this chapter. Bell[8] has shown that if the only difference between the partition functions for reactants and activated complex arises from the disappearance of the $X \cdot\cdot H$ stretching vibration and a change in the bending modes, then the calculated difference in A factors assuming simply harmonic vibrations is

$$\frac{A_H}{A_D} = \left(\frac{m_D}{m_H}\right)^{\frac{1}{2}} \times \prod_3 \frac{u_D}{u_H} \left(\frac{1-e^{-u_H}}{1-e^{-u_D}}\right) \times \prod_2 \frac{u_H^{\ddagger}}{u_D^{\ddagger}} \left(\frac{1-e^{-u_D\ddagger}}{1-e^{-u_H\ddagger}}\right) \tag{136}$$

The m_D and m_H refer to the *reduced* masses for motion along the reaction coordinate, but since the masses of X and Y are much larger than those of the deuteron and proton, m_D and m_H are approximately equal to the isotopic masses so that

$$\left(\frac{m_D}{m_H}\right)^{\frac{1}{2}} \simeq \frac{u_H}{u_D} \simeq \frac{u_H^{\ddagger}}{u_D^{\ddagger}} \tag{137}$$

where

$$u = \frac{h\nu}{kT} \tag{138}$$

(ν and $\nu^{\ddagger}$ refer to the vibration frequencies in the initial and transition states respectively).

It can be seen that (136) and (137) predict that the isotope effect on the pre-exponential factor can never be very great. The two extreme cases are

$$u^{\ddagger} \ll 1 \ll u \qquad \frac{A_H}{A_D} \rightarrow \left(\frac{m_D}{m_H}\right)^{\frac{1}{2}} \times \left(\frac{u_D}{u_H}\right)^3 \times (1)^2 = \frac{1}{2} \tag{139}$$

$$u, u^{\ddagger} \ll 1 \qquad \frac{A_H}{A_D} \rightarrow \left(\frac{m_D}{m_H}\right)^{\frac{1}{2}} \times (1)^3 \times (1)^2 = \sqrt{2} \tag{140}$$

The latter extreme (140) is never attained in practice, since it requires a temperature higher than is possible in the liquid phase. The usual situation is that in which both ν and $\nu^{\ddagger}$ are large enough to make e^{-u} negligible compared with unity. Thus for most proton transfer reactions, $A_H/A_D \rightarrow 1$, and the isotope effect on A is very small.

In order to test predictions of a kinetic isotope effect on E or A for a given reaction at least three fairly restrictive conditions must be met. The proton transfer process must be rate determining, the isotopic substitution must be confined to the hydrogen atom to be involved in the transfer (since isotopes in other positions can produce significant secondary isotope effects) and, thirdly, the reactions with both isotopes should be studied in the same medium since the isotopic composition of the medium has also been known to affect reaction rates. One series of reactions which very nearly satisfies these conditions is the base catalyzed bromination of ketones for which values of k_H/k_D have been found to range[8] between 3 and 12.

3.4.2 The tunnelling correction (cf. sections 3.6.4 and 4.5.6, Chapter 3)

The small mass of the proton introduces the possibility of a non-classical proton transfer process in which the system can no longer be considered to pass through a transition state at the top of the potential barrier during its passage from reactants to products. At room temperatures the behaviour of protons may deviate significantly from that predicted by classical mechanics. The de Broglie wavelength $\lambda = h/mv$ for the proton is about one Ångstrom, which is of the same order of magnitude as the width of the potential barriers for classical proton transfer reactions.

Bell[88] has shown that if the barrier is assumed to be parabolic with base width of $2a$, the rate of passage of the system from reactants to products "through" the barrier, compared with the classically expected rate over the barrier, is

$$\frac{k_{tunn}}{k_{class}} = \frac{u_t/2}{\sin u_t/2} = 1 + \frac{u_t^2}{24} + \frac{7u_t^4}{5760} + \ldots \tag{141}$$

where

$$u_t = \frac{h}{a\pi kT}\sqrt{\frac{E}{2m}} < \frac{\pi}{2} \tag{142}$$

It is seen that the deviations from the classically expected rate coefficient would be greatest at low temperatures. Caldin and Kasparian[89] have observed this effect in the proton transfer reaction of the 2, 4, 6-trinitrobenzyl anion with HF. The variation of the tunnelling correction with temperature leads to a curvature of the Arrhenius plot, cf. p. 249, and Fig. 2, Chapter 3, which show the kind of behaviour observed. By fitting the experimental curve to the deviation predicted by eqn. (141) the width of the assumed parabolic barrier was calculated to be 1.46 A and the height to be 12.2 kcal. mole^{-1}. Such work has shown that in most cases the calculated barrier for proton transfer is similar (about 80–90 % of the classical height) to that expected without tunnelling. Consequently it is usually possible to consider proton transfer reactions in terms of the transition state concept with tunnelling processes taken into account by a correction term.

If tunnelling does have a measurable effect on the experimental rate coefficient, the separate effect on the activation energy and the pre-exponential factor can be calculated from (141) by comparing the *experimental* values of E and A defined by

$$E = RT^2 \frac{d \ln k}{dT} \qquad \ln A = \ln k + \frac{E}{RT} \tag{143}$$

with those expected classically from

$$k_{class} = A_{class} e^{-E_{class}/RT} \tag{144}$$

Combining (142), (143), and (144) yields[88]

$$E = E_{class} - RT \left(1 - \frac{u_t}{2} \cot \frac{u_t}{2}\right)$$

$$= E_{class} - RT \left(\frac{u_t^2}{12} + \ldots\right) \tag{145}$$

$$A = A_{class} \frac{u_t/2}{\sin u_t/2} \exp \left(\frac{u_t}{2} \cot \frac{u_t}{2} - 1\right)$$

$$= A_{class} \left(1 - \frac{u_t^2}{24} + \ldots\right)$$

These relations show that if tunnelling is important the experimental values of both E and A will be temperature dependent and less than the values expected classically. The appearance in the expressions in (145) of u_t, and thus of mass, indicates that the isotopic effect of E and A due to tunnelling would be in the same direction as the non-tunnelling isotope effect.

However, the uncertainties in the expected value of the zero point isotope effect on E make the magnitude of the tunnelling isotope effect on E difficult to ascertain. Studies of the isotope effect on A are more fruitful, since it has been shown in section 3.4.1(b) that the classical effects are small and have a limit of $A_H/A_D = 0.5$. Values of A_H/A_D less than 0.5 should therefore suggest the existence of a tunnelling process operating in the reaction. Several such cases have been recorded, and, for example[90], the bromination of 2-carbethoxycyclopentanone in D_2O catalysed by F^- has a measured A_H/A_D of 0.04.

Very large isotope effects have been found in the sterically hindered base catalysis of the iodination of 2-nitropropane[91]. The largest effect was for catalysis by 2, 4, 6-trimethyl pyridine where k_H/k_D was 24.3 corresponding to $A_H/A_D = 0.15$ and $E_D - E_H = 3$ kcal. mole^{-1}. Lewis and Funderbunk[91] invoke a two dimensional potential barrier model to explain the abnormally large isotope effect for these reactions with large steric hindrance. The reaction is considered in terms of a linear three centre system with potential energy contours plotted against X–H and H–Y distance (cf. Fig. 4 of Chapter 3, p. 233). On this model the tunnelling process would occur in a straight line from a point on the reaction co-ordinate well below the saddle point to a point on the reaction co-ordinate on the product side of the energy barrier. Steric repulsion has the effect of raising the energy surface in the region of small interatomic distances, i.e. near the origin. Consequently, steric hindrance tends to reduce the fraction of classical passages over the barrier but has little effect on the tunnelling process. It is therefore expected that the relative importance of the tunnelling process (and hence the isotope effect) would be greatest for the most sterically hindered reactions in a series.

Bell[85] has suggested a further possible contribution to the primary kinetic isotope effect which is not accounted for by zero point or tunnelling considerations. This effect is similar in nature to those caused by the Franck–Condon restrictions in electron transfer reactions (section 2.5.3). The dielectric relaxation rate for water is one to two orders of magnitude slower than is the proton transfer process[92,93] after a suitable hydrogen bridge has been formed. For this reason the surrounding solvent shells must be reorganized in order that the transfer may take place, and the free energy of activation for the transfer process will include a reorganization term, $\Delta G_r^{\ddagger}$. Because of its greater mass a deuteron transfer will be slower than a proton transfer, will require less solvent reorganization, and will therefore have a smaller $\Delta G_r^{\ddagger}$. Unlike the zero point energy and tunnelling contributions, this isotope effect favours deuteron transfer and would tend to reduce k_H/k_D.

4. Linear free energy relationships

4.1 GENERAL CONSIDERATIONS

One of the primary goals of chemistry is to relate the rate and the extent of a reaction to the structure of the reactants. Although prediction of absolute rate and equilibrium constants from a knowledge of structure is not yet possible for even the simplest reactants, the first step toward such a goal is the correlation of the reaction parameters with identifiable properties of reactant structure.

Both rate coefficients and equilibrium constants depend on free energy values for the reaction according to the relations

$$\log k = \log \frac{RT}{N_0 h} - \frac{\Delta G^{\ddagger}}{2.3RT}$$

$$\log K = -\frac{\Delta G^0}{2.3RT}$$

(146)

where $\Delta G^{\ddagger}$ is the free energy of activation and ΔG^0 is the overall standard free energy change. For this reason correlations involving the logarithm of rate coefficients or equilibrium constants of different reactions are known as *free energy relationships* and will constitute the subject matter of this last section. Extensive treatments examining the present role of linear free energy relationships in solution chemistry have been made by Wells[94] and Leffler and Grunwald[95].

Wells[94] has demonstrated that, for a given type of reaction, if two species differ only in a variable x which satisfies certain requirements, then the following ideal generalised linear free energy relationship holds

$$\log \left(\frac{\mathscr{K}_i}{\mathscr{K}_0}\right)_j = X_i R_{xj} \tag{147}$$

$(\mathscr{K}_i/\mathscr{K}_0)_j$ is the ratio of the rate coefficient *or* equilibrium constant for the reactant species i to that for a standard reactant, for a given type of reaction designated by j. For a given variable, x, X_i is a function only of the *reactant species* characterised by i, and R_{xj} is a function only of the nature and conditions of the *reaction type j*. It can be seen from (147) that if either the reaction type or the reactant is the same for a series of reactions then $\log (\mathscr{K}_i/\mathscr{K}_0)_j$ is expected to vary linearly with the variable parameter X_i or R_{xj}.

The derivation of (147) uses the relationships (146) between the rate coefficients and equilibrium constants and the free energies in the following manner. $\mathscr{K}$ will be used to represent *either k or K* and the corresponding free energy $\Delta G^{\ddagger}$ or ΔG^0 will be denoted by ΔG.

If the reactant properties that can be altered by a structural change are denoted $x, y, z \ldots$ the change in ΔG produced by an infinitely small structural change is

$$d\Delta G = \left(\frac{\partial \Delta G}{\partial x}\right)_{T,y,z\ldots} dx + \left(\frac{\partial \Delta G}{\partial y}\right)_{T,x,z\ldots} dy + \ldots \tag{148}$$

where the partial derivative terms indicate the susceptibility of the free energy to a change in a particular reactant variable. If the structural change, i, is finite (*e.g.* the exchange of $-CH_3$ for $-H$ at the *para* position in an aromatic ring) and is accompanied by a change in only one of the properties, x, (*e.g.* the electronic inductive effect of the *para* substituent), and if $(\partial \Delta G/\partial x)_r$ is a constant called b_x then we can write

$$\Delta G_i - \Delta G_0 = b_x(x_i - x_0) \tag{149}$$

where ΔG_0 and x_0 refer to the reference reactant structure (*e.g.* $-H$ in the *para* position). The requirement that $(\partial \Delta G/\partial x)$ be constant is fairly restrictive since ΔG involves ΔH and ΔS which could not, in general, be expected to show the same dependence on the variable x. Constant $(\partial\Delta G/\partial x)$ requires either a special relation between ΔH and ΔS or that one of them is small enough to be neglected.

By substitution of (149) into (146), $\log(\mathcal{K}_i/\mathcal{K}_0)$ can be written for a *reference* reaction type (*e.g.* the ionization of substituted benzoic acids) and again for a second reaction type j (*e.g.* the bromination of substituted acetophenones) to yield

$$\log(\mathcal{K}_i/\mathcal{K}_0)_0 = -(b_x)_0(x_i-x_0)/2.3RT_0$$
$$\log(\mathcal{K}_i/\mathcal{K}_0)_j = -(b_x)_j(x_i-x_0)/2.3RT_j$$
$$(150)$$

These two equations can be combined in order to express $\log(\mathcal{K}_i/\mathcal{K}_0)_j$ in terms of $\log(\mathcal{K}_i/\mathcal{K}_0)$ for the reference reaction

$$\log(\mathcal{K}_i/\mathcal{K}_0)_j = [\log(\mathcal{K}_i/\mathcal{K}_0)_0][(b_x)_j T_0/(b_x)_0 T_j]$$
$$= X_i R_{xj} \qquad (151)$$

If this linear relation holds then the $\mathcal{K}$ for any reaction of any reactant in a series of closely related structures (*e.g.* the rate coefficient for the bromination of p-methylacetophenone) can be calculated if $\mathcal{K}_0$ for the reference reaction (the bromination of acetophenone itself) and only two further parameters are known. The first is X_i, characteristic of the variable (*e.g.* substituent inductive effect) and the reactant structure (*e.g.* the p-CH_3 substituent) and the second is R_{xj} a parameter indicating the susceptibility of $\mathcal{K}$ for the reaction type (bromination of substituted acetophenones at temperature T_j) to a change in the particular variable.

It will be seen in section 4.2.1 that, in the Hammett equation, the $X_i = \log(\mathcal{K}_i/\mathcal{K}_0)_0$ term, which is designed to characterise the total substituent polar effect for structural (substituent) changes, is defined as $\sigma_i = \log K_i/K_0$ for the ionization of the ith substituted benzoic acid relative to unsubstituted benzoic acid. In practice the linear relation between $\log(\mathcal{K}_i/\mathcal{K}_0)_j$ and reactant parameters such as σ_j is less than perfect. The failure is due first to the inability of the chosen definition of the reference $\log(\mathcal{K}_i/\mathcal{K}_0)_0$ to characterise precisely the effect of a structural change on the variable and, secondly, to the variable being such that $(b_x)_j/(b_x)_0$ does not remain exactly constant for all structural changes. In fact in the case of the Hammett equation the polar effect variable, σ, is actually composed of at least two separate variables which can be identified with inductive and resonance polar effect contributions.

In general, if the other independent variables $y, z \ldots$ do not remain constant during the structural change, extra terms are required in (151). For example, if y varied, then a second reference reaction type would be required, and (151) would become a four parameter equation

$$\log(\mathcal{K}_i/\mathcal{K}_0)_j = X_i R_{xj} + Y_i R_{yj} \qquad (152)$$

where Y_i is characteristic of the change in the variable y caused by the structural change i, and R_{yj} is the susceptibility of the jth type of reaction to a change in the y variable.

4.2 SOME COMMON LINEAR FREE ENERGY RELATIONS

4.2.1 The Hammett equation, $\log (\mathcal{K}/\mathcal{K}_0) = \sigma\rho$

The Hammett equation was first proposed in 1937 to correlate the rate coefficients and equilibrium constants for side chain reactions of *meta* and *para* substituted benzene derivatives. The *substituent constant*, σ, depends only on the position and nature of the substituent, and was defined by

$$\sigma_i = \log \frac{K_i}{K_0} \tag{153}$$

for the ionization of the particular substituted benzoic acid in water at 25 °C relative to that of benzoic acid itself. Typical values[96] of σ were found to range from $\sigma_{p\text{-NH}_2} = -0.66$ to $\sigma_{p\text{-NO}_2} = 1.27$. The *reaction constant*, ρ, is independent of the substituent and depends only on the nature of the reaction in question. From the generalized equation (151) it is seen that

$$\rho_j = R_{xj} = \frac{(\partial\Delta G/\partial x)_j}{(\partial\Delta G/\partial x)_0} \cdot \frac{T_0}{T_j} = \frac{(\partial\Delta G_j/\partial\sigma)}{(\partial\Delta G_0^0/\partial\sigma)} \cdot \frac{298}{T_j} \tag{154}$$

where ΔG_0^0 is the standard free energy change for unsubstituted benzoic acid and ΔG_j is either equal to $\Delta G_j^{\ddagger}$ for the jth reaction if rate coefficients are to be correlated, or equal to ΔG_j^0 if the correlation is between equilibrium constants. The original Hammett equation with σ defined by (153) was used to correlate rate coefficients or equilibrium constants for 43 different substituents and 52 different reactions. The resulting σ and ρ values are tabulated by Hammett[96]. Four correlation plots are shown in Fig. 13.

Jaffé[97] has since produced correlations among 204 reactions, but has used σ values which give the best statistical fit for all the available reactions rather than values calculated from the original definition (153). For theoretical interpretation of the variable that σ represents, the more precise original definition of σ is preferred to one based on a statistically best fit for many reactions[98].

The σ parameter has been associated with the ability of the substituent group to withdraw electric charge density from the reaction site. Theoretical calculations of electron densities for benzene derivatives have been made[99] using Hückel Molecular Orbital Theory and although absolute σ values could not be calculated,

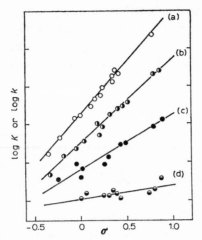

Fig. 13. The relationship between $\log \mathscr{K}$ and σ for four different types of reaction (from Hammett[96]).

(a) Equilibrium constants of substituted anilinium ions in water at 25° C, $\rho = 2.73$.
(b) Ionization constants of substituted phenylboric acids in 25 % ethanol at 25° C, $\rho = 2.14$.
(c) Rate coefficients for solvolysis of substituted benzoyl chlorides in methanol at 25° C, $\rho = 1.47$.
(d) Rate coefficients for the bromination of substituted acetophenones in acetic acid-water solvent at 25° C, $\rho = 0.42$.

good correlation was obtained between σ and the electron density on a *meta* or *para* carbon in a monosubstituted benzene. If σ is identified with a polar effect variable then the substituent constant ρ is a measure of the way in which $\mathscr{K}$ for a given reaction is affected by changes in electron density at the reaction site. An abrupt change of slope in a $\log \mathscr{K}$ *vs.* σ plot may therefore indicate a change of mechanism that takes place when a certain minimum electron density at the reaction site is attained.

There are two distinct ways in which a substituent group can influence the charge density at the reaction site. The first is the *inductive effect* in which charge is electrostatically donated or withdrawn from the aromatic ring. Because of the delocalised aromatic π orbitals the excess ring charge is placed almost evenly on all the aromatic carbon atoms. The second is a *resonance effect* which is due to the ability of some *para* substituent groups to produce a π bond resonance interaction between the substituent and the reaction site. *Meta* substituents can only induce secondary resonance effects which are much smaller. For some substituents such as the methoxy group the inductive and resonance effects act in opposite directions.

To distinguish between the two variables a four parameter equation can be written, *viz.*

$$\log (\mathscr{K}/\mathscr{K}_0) = \sigma_I \rho_I + \sigma_R \rho_R \tag{155}$$

where the subscripts I and R refer to inductive and resonance polar effect variables. However, since the main effect of both variables on the reactivity is to alter the charge on the reactant site, the ρ values for both variables would be expected to be approximately equivalent. Both ρ values are usually equated to the ordinary Hammett ρ value so that (155) becomes

$$\log (\mathcal{K}/\mathcal{K}_0) = (\sigma_I + \sigma_R)\rho \tag{156}$$

Reactions which involve a transition state or intermediate species with a highly electron-deficient centre would be expected to be very sensitive to the stabilising effect of substituents able to donate charge to the centre. In this kind of reaction, *para* substituents able to undergo direct resonance interaction with the electron-deficient centre are found to have a substituent constant greater than the σ which applies to the substituent for other reactions. Brown and Okamoto[100] have defined such an "*electrophilic substituent constant*", σ^+, using the solvolysis of phenyl dimethylcarbinyl chlorides as the reference reaction. Linear correlations have been demonstrated for several electron-deficient reaction centre reactions.

A similar situation exists for reactions with reaction centres having an excess of electronic charge, and enhanced sigma values, σ^-, are required for electron-withdrawing *para* substituents. Reactions of phenols and even non side-chain reactions involving nucleophilic substitution into the aromatic ring[101] have been successfully correlated by the use of σ^- values. Several other substituent constants have been defined for application to particular types of reaction[94]. Some σ constants have been found to correlate well with physical properties of the reactants not related to reactivity. For example, Brown[102] has demonstrated that a correlation exists between intensities of infrared absorption and σ^+ values.

4.2.2 The Taft equation, $\log (\mathcal{K}/\mathcal{K}_0) = \sigma^* \rho^*$

The Hammett equation is only applicable to *meta* and *para* substituted aromatics, where the substituents are far enough from the reaction site for their steric properties to have little effect on the reaction. However, to correlate reactions of aliphatic or *ortho* substituted aromatic compounds the steric as well as the inductive and resonance effects of a structural change must be accounted for. Taft's treatment[103] attempts to isolate a polar (inductive) variable and to use its values to correlate reactions where only the polar variable is significantly affected by the substituent change. The Taft relation is consequently a linear polar energy relationship rather than a total free energy relationship. As with the Hammett relation, σ^* is a substituent constant and ρ^* is the reaction constant corresponding to the susceptibility of the particular reaction to changes in the σ^* variable.

In order to isolate the polar variable in the definition of σ^*, Taft makes the assumption that the relative rate of acid hydrolysis, $\log (k/k_0)_A$, is governed only by steric and resonance effects, while the corresponding term, $\log (k/k_0)_B$, for base hydrolysis, is affected by the polar effect of the R group as well as by the same resonance and steric effects present in acid hydrolysis. The difference between the two terms should then be a measure of the substituent polar (inductive) effect of the R group and the definition of σ^* becomes

$$\sigma^* = \frac{1}{2.48} \left[\log (k/k_0)_B - \log (k/k_0)_A\right] \tag{157}$$

k/k_0 is the ratio of the rate coefficient of hydrolysis of the R substituted ethyl ester $RCO_2C_2H_5$ to the coefficient for the standard $R = CH_3$ ester. The $1/2.48$ factor is arbitrarily chosen[104] to place the σ^* values on about the same scale as the Hammett σ values.

The theoretical interpretation of σ^* is therefore much the same as that of the inductive part, σ_I, of the general Hammett σ. The existence of a qualitative correlation of σ^* with bonding properties and electronegativities has been demonstrated[104]. σ^* values have also been found to exhibit the property of additivity which would be expected for inductive polar effects but not for resonance or steric effects.

If the experimental data obey the relation $\log (\mathscr{K}/\mathscr{K}_0) = \sigma^*\rho^*$ for a reaction series it implies that resonance and steric effects do not change significantly throughout the series. Deviations occur, as expected, for unsaturated substituents or for bulky substituents whose steric hindrance differs greatly from that of the reference $-CH_3$ group. Variable substituent steric effects can be accounted for by adding another term to the original Taft relation to make the four parameter equation (158)

$$\log (\mathscr{K}/\mathscr{K}_0) = \sigma^*\rho^* + E_s\delta \tag{158}$$

E_s is a substituent constant for the steric variable, and typical values are $(E_s)_{C_6H_5} = -0.9$ and $(E_s)_{Cl} = 0.18$. A plot of $\log (\mathscr{K}/\mathscr{K}_0) - \sigma^*\rho^*$ against E_s has been found to be linear for several reactions[105], and can be called a linear steric energy relationship. However, equation (158) is of limited applicability, since other variables influenced by structural change such as hyperconjugation and entropy of solvation will be included in the empirical "steric" term.

As in the case of Hammett σ values, many properties not directly involving reactivity, such as dipole moments and quadrupole coupling constants[106], have been correlated with σ^*. Such correlations have obvious value in helping to yield an exact definition of the σ^* variable in an attempt to provide a better theoretical interpretation of similar variables in the future.

4.2.3 The Brønsted equation, $\log (k/k_0) = [\log (K_A/K_{A0})]\alpha$

The Brønsted equations for acid and base catalysis were the first linear free energy relations to be applied to the rates of reactions in solution. The original equations (125) can be written in the form

$$\log (k/k_0) = [\log (K_A/K_{A0})]\alpha \tag{159}$$

if the measured rate coefficient and the acid dissociation constant K_A are referred to a particular reactant. It can be seen that $\log (K_A/K_{A0})$ corresponds to a reactant constant, X_i, where the structure-dependent variable is related to the ability of the species to release a proton. The parameter α is thus a reaction constant governed by the susceptibility of the acid catalysed rate coefficient to a change in proton donating power of the catalyst. In the Brønsted equation for base catalysis $\log (K_B/K_{B0})$ corresponds in an exactly analogous manner to the ability of the base catalyst to accept a proton. The range of applicability and the interpretation of the Brønsted relations have been discussed in more detail in section 3.3.

4.2.4 The Grunwald–Winstein equation, $\log (k/k_0) = Ym$

This equation is an attempt to correlate the rates of solvolysis reactions with properties of the solvent. Grunwald and Winstein[107] have applied it to first order nucleophilic substitution (S_{N1}) reactions where the rate is determined by the extent of ionization of the substrate. For such reactions the required solvent (reactant) property will be its ionizing power, Y, which is defined by

$$Y = \frac{\log k}{\log k_0} \tag{160}$$

for the rate coefficient, k, of solvolysis of t-butyl chloride in a given solvent relative to that in the reference solvent of 80 % aqueous ethanol. The usual solvent complications, such as ion pair formation, produce departures from linearity when $\log k$ is plotted against Y for a series of different pure solvents. Much better correlations have been obtained by the use of solvent pairs to vary the Y parameter as shown in Fig. 14. The solvent parameter Y appears to have some independent physical significance, since it has been shown to correlate remarkably well with the frequencies of charge transfer optical absorption bands in a series of solvents[110].

4.2.5 The Swain equation, $\log (k/k_0) = ns + es'$

The equation proposed by Swain and Scott[111] is based on the assumption that the specific effects that electrophilic (electron seeking) and nucleophilic (electron

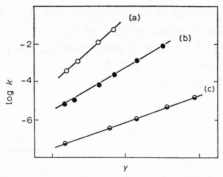

Fig. 14. The relationship between log k and Y of an acetic acid/water solvent pair, for three types of solvolysis reactions (data from Winstein *et al.*[108, 109]).

(a) solvolysis of benzhydryl chloride at 25° C, m = 1.561.
(b) solvolysis of α-phenylethyl chloride at 25° C, m = 1.136.
(c) solvolysis of neophyl bromide at 50° C, m = 0.733.

donating) reagents have on a reaction could be described by independent variables. The parameters n and e in the equation

$$\log (k/k_0) = ns + es' \tag{161}$$

denote the nucleophilicity and electrophilicity of the reagent species (*i.e.* the solvent for solvolysis reactions) relative to $n = e = 1$ for water. s and s' are the corresponding susceptibilities of the substrate to variations in n and e.

For the special cases of reactions in which either s or s' approaches zero, two parameter forms of (161) result. For example, for unimolecular solvolysis reactions, which are relatively insensitive to reagent nucleophilicity, $s \rightarrow 0$, and if the reagent electrophilicity is equated to its ionizing power Y then (161) reduces to the Grunwald–Winstein equation (160).

On the other hand, for nucleophilic displacement reactions $s' \rightarrow 0$ and the following two parameter equation results

$$\log (k/k_0) = ns \tag{162}$$

Swain and Scott[111] defined n for a nucleophile as $\log(k/k_0)$, where k is the rate coefficient for the methyl bromide displacement reaction in the particular nucleophilic solvent and k_0 is the rate coefficient in pure water. Nucleophilic parameters measured in this way were found to give a fairly good correlation for several nucleophilic displacement reactions.

A more general four parameter equation (163)

$$\log (k/k_0) = d_1 c_1 + d_2 c_2 \tag{163}$$

has been proposed for solvolysis reactions[112]. The solvent parameters d_1 and d_2

in (163) are not strictly related to nucleophilicity and electrophilicity, as are n and e. The standard substrate values are defined by $c_1 = c_2 = -1.00$ for t-butyl chloride and $c_2 = 3.00\ c_1$ for triphenylmethyl fluoride. The reference solvent is 80 % ethanol in water where $d_1 = d_2 = 0$. 25 values for c_1 and c_2 and 17 values for d_1 and d_2 are listed by Swain et al.[112].

4.3 CONCLUSION

The main application of linear free energy relationships to solution kinetics has been to store and predict reaction data. The usefulness of such a procedure obviously decreases as the number of required parameters is increased since any body of reactivity data can be "correlated" if enough parameters are used. Refinements in correlative expressions of the form of eqn. (147) are still possible, however, and if the appropriate variables can be identified in a more fundamental manner, linear free energy relationships may provide a valuable insight into the problem of providing a quantitative relation between structure and kinetic behaviour.

GENERAL BIBLIOGRAPHY

1 A. M. NORTH, The Collision Theory of Chemical Reactions in Liquids, Methuen, London, 1964.
2 S. W. BENSON, The Foundations of Chemical Kinetics, McGraw-Hill, New York, 1960.
3 K. J. LAIDLER, Chemical Kinetics, McGraw-Hill, New York, 1965.
4 A. A. FROST AND R. G. PEARSON, Kinetics and Mechanism, Second Edition, John Wiley, New York, 1961.
5 E. S. AMIS, Solvent Effects on Reaction Rates and Mechanism, Academic Press, New York, 1966.
6 S. GLASSTONE, K. J. LAIDLER AND H. EYRING, The Theory of Rate Processes, McGraw-Hill, New York, 1941.
7 W. L. REYNOLDS AND R. W. LUMRY, Mechanisms of Electron Transfer, Ronald Press, New York, 1966.
8 R. P. BELL, The Proton in Chemistry, Methuen, London, 1959.
9 R. P. BELL, Acid–Base Catalysis, Oxford University Press, Oxford, 1941.

REFERENCES

10 E. RABINOWITCH AND W. C. WOOD, Trans. Faraday Soc., 32 (1936) 1381.
11 J. FRENKEL, Kinetic Theory of Liquids, Oxford University Press, Oxford, 1946.
12 T. L. COTTRELL AND J. C. McCOUBREY, Molecular Energy Transfer in Gases, Butterworths, London, 1961.
13 J. CRANK, The Mathematics of Diffusion, Oxford University Press, Oxford, 1956.
14 W. JOST, Diffusion in Solids, Liquids, and Gases, Academic Press, New York, 1952.
15 A. M. NORTH, Quart. Rev., 20 (1966) 421.
16 R. M. NOYES, Progr. Reaction Kinetics, 1 (1961).
17 S. CHANDRASEKHAR, Rev. Mod. Phys., 15 (1943) 1.

18 M. W. SMOLUCHOWSKI, *Z. Physik. Chem.*, 92 (1917) 129.

19 F. C. COLLINS AND S. E. KIMBALL, *J. Colloid Sci.*, 4 (1949) 425.

20 T. R. WAITE, *J. Chem. Phys.*, 28 (1958) 103; 32 (1960) 21.

21 J. H. HILDEBRAND AND R. L. SCOTT, *Regular Solutions*, Prentice Hall, Englewood Cliffs, N.J., 1962.

22 J. H. HILDEBRAND AND S. E. WOOD, *J. Chem. Phys.*, 1 (1933) 817.

23 M. RICHARDSON AND F. S. SOPER, *J. Chem. Soc.*, (1929) 1873.

24 S. D. HAMANN, *Physico-Chemical Effects of Pressure*, Butterworths, London, 1957.

25 E. WHALLEY, *Advances in Physical Organic Chemistry*, ed. V. GOLD, 2 (1964) 93.

26 K. J. LAIDLER AND D. J. Y. CHEN, *Can. J. Chem.*, 37 (1959) 599.

27 M. W. PERRIN, *Trans. Faraday Soc.*, 34 (1938) 144.

28 S. SCATCHARD, *Chem. Rev.*, 10 (1932) 229.

29 E. A. MOELWYN-HUGHES, *Kinetics of Reactions in Solution*, 2nd. Edn., Oxford University Press, Oxford, 1947.

30 K. J. LAIDLER AND H. EYRING, *Ann. N.Y. Acad. Sci.*, 39 (1940) 303.

31 C. V. KING AND J. J. JOSEPHS, *J. Am. Chem. Soc.*, 65 (1944) 767.

32 C. P. SMYTH, *Dielectric Behaviour and Structure*, McGraw-Hill, New York, 1955.

33 K. J. LAIDLER, *Can. J. Chem.*, 37 (1959) 138.

34 R. A. ROBINSON AND R. H. STOKES, *Electrolyte Solutions*, Butterworths, London, 1955.

35 H. S. HARNED AND B. B. OWEN, *The Physical Chemistry of Electrolyte Solutions*, 3rd. Edn., Reinhold, New York, 1958.

36 J. N. BRØNSTED, *J. Am. Chem. Soc.*, 44 (1922) 877; *Z. Physik. Chem.*, 102 (1922) 169; 115 (1925) 337.

37 C. W. DAVIES, *Progr. Reaction Kinetics*, 1 (1961) 161.

38 A. R. OLSON AND T. R. SIMONSON, *J. Chem. Phys.*, 17 (1949) 1167.

39 N. BJERRUM, *Kgl. Danske Videnskab. Selskab, Mat. Fys. Medd.*, 7 (1926) No. 9.

40 C. A. KRAUS AND R. M. FUOSS, *J. Am. Chem. Soc.*, 55 (1933) 1019.

41 C. A. KRAUS, *Ann. N.Y. Acad. Sci.*, 51 (1949) 789.

42 R. P. BELL AND F. J. LINDARS, *J. Chem. Soc.*, (1954) 4601.

43 W. L. SVIRBELY AND J. C. WARNER, *J. Am. Chem. Soc.*, 57 (1935) 1883.

44 E. S. AMIS AND V. K. LAMER, *J. Am. Chem. Soc.*, 61 (1939) 905.

45 W. L. SVIRBELY AND A. SCHRAMM, *J. Am. Chem. Soc.*, 60 (1938) 330.

46 P. A. H. WYATT AND H. L. KORNBERG, *Trans. Faraday Soc.*, 48 (1952) 454.

47 I. WEIL AND J. C. MORRIS, *J. Am. Chem. Soc.*, 71 (1949) 1664.

48 J. C. WARNER AND F. B. STITT, *J. Am. Chem. Soc.*, 55 (1933) 4807.

49 *International Critical Tables*, 1st. Edn., Vol. III, McGraw-Hill, New York, 1926, p. 25.

50 E. S. AMIS, *J. Chem. Educ.*, 30 (1953) 351.

51 K. J. LAIDLER AND P. J. LANDSKROENER, *Trans. Faraday Soc.*, 52 (1956) 200.

52 J. S. KIRKWOOD, *J. Chem. Phys.*, 2 (1934) 351.

53 E. S. AMIS AND G. JAFFÉ, *J. Chem. Phys.*, 10 (1942) 598.

54 E. HÜCKEL, *Physik. Z.*, 26 (1925) 93.

55 P. DEBYE AND J. MCAULAY, *Physik. Z.*, 26 (1925) 22.

56 J. N. BRØNSTED AND W. F. K. WYNNE-JONES, *Trans. Faraday Soc.*, 25 (1929) 59.

57 J. HALPERN, *Quart. Rev.*, 15 (1961) 207.

58 N. SUTIN, *Ann. Rev. Nucl. Sci.*, 13 (1962) 285.

59 R. A. MARCUS, *Ann. Rev. Phys. Chem.*, 15 (1964) 155.

60 H. TAUBE, H. MYERS AND R. L. RICH, *J. Am. Chem. Soc.*, 75 (1953) 4418; H. TAUBE AND H. MYERS, *J. Am. Chem. Soc.*, 76 (1954) 2103.

61 B. J. ZWOLINSKI, R. J. MARCUS AND H. EYRING, *Chem. Rev.*, 55 (1955) 157.

62 E. SACHER AND K. J. LAIDLER, *Trans. Faraday Soc.*, 59 (1963) 396.

63 N. S. HUSH, *J. Chem. Phys.*, 28 (1958) 962; *Trans. Faraday Soc.*, 57 (1961) 557.

64 R. A. MARCUS, (a) *J. Chem. Phys.*, 24 (1956) 979; 26 (1957) 867; 26 (1957) 872; 43 (1965) 679; (b) *Discussions Faraday Soc.*, 29 (1960) 21.

65 F. BASOLO AND R. G. PEARSON, *Mechanisms of Inorganic Reactions*, Wiley, New York, 1958, p. 48.

66 J. SILVERMAN AND R. W. DODSON, *J. Phys. Chem.*, 56 (1952) 846.

67 N. SUTIN AND M. WOLFSBERG, unpublished calculations (see ref. 58, p. 320).
68 R. E. RICHARDS AND J. A. S. SMITH, *Trans. Faraday Soc.*, 47 (1951) 1261.
69 J. N. BRØNSTED, *Rec. Trav. Chim.*, 42 (1923) 718.
70 G. N. LEWIS, *Valency and Structure of Atoms and Molecules*, Reinhold, New York, 1923.
71 M. EIGEN, (a) *Angew. Chem.*, 75 (1963) 489; (b) *Discussions Faraday Soc.*, 39 (1965) 7;
    (c) M. EIGEN, W. KRUSE, G. MAASS AND L. DEMAEYER, *Progr. Reaction Kinetics*, 2 (1964)
    285.
72 W. J. ALBERY, *Progr. Reaction Kinetics*, 4 (1967) 353.
73 P. DEBYE, *Trans. Electrochem. Soc.*, 82 (1942) 265.
74 W. P. JENCKS, *Progr. Physical Organic Chemistry*, 2 (1964) 63.
75 J. N. BRØNSTED AND K. PEDERSON, *Z. Physik. Chem.*, A108 (1923) 185.
76 A. J. KRESGE AND Y. CHIANG, *J. Am. Chem. Soc.*, 83 (1961) 2877.
77 R. P. BELL AND W. C. E. HIGGINSON, *Proc. Roy. Soc. (London), Ser. A*, 197 (1949) 141.
78 R. P. BELL AND G. L. WILSON, *Trans. Faraday Soc.*, 46 (1950) 407.
79 E. J. COHN, F. F. HEYROTH AND M. F. MENKIN, *J. Am. Chem. Soc.*, 50 (1928) 696.
80 M. A. PAUL AND F. A. LONG, *Chem. Rev.*, 57 (1957) 1, 935.
81 L. P. HAMMETT AND A. J. DEYRUP, *J. Am. Chem. Soc.*, 54 (1932) 2721.
82 F. A. LONG AND M. A. PAUL, *Chem. Rev.*, 57 (1957) 935; *J. Am. Chem. Soc.*, 72 (1950) 3267.
83 L. ZUCKER AND L. P. HAMMETT, *J. Am. Chem. Soc.*, 61 (1939) 2791.
84 J. F. BUNNETT AND F. P. OLSEN, *Can. J. Chem.*, 44 (1966) 1899, 1917.
85 R. P. BELL, *Discussions Faraday Soc.*, 39 (1965) 16.
86 F. H. WESTHEIMER, *Chem. Rev.*, 61 (1961) 265.
87 R. P. BELL AND D. M. GOODALL, *Proc. Roy. Soc. (London), Ser. A*, 294 (1966) 273.
88 R. P. BELL, *Trans. Faraday Soc.*, 55 (1959) 1.
89 E. F. CALDIN AND M. KASPARIAN, *Discussions Faraday Soc.*, 39 (1965) 25.
90 R. P. BELL, J. A. FENDLEY AND J. R. HULETT, *Proc. Roy. Soc. (London),
    Ser. A*, 235 (1956) 453.
91 E. S. LEWIS AND L. H. FUNDERBUNK, *J. Am. Chem. Soc.*, 89 (1967) 2322.
92 M. M. KREEVOY AND R. A. KRETCHMER, *J. Am. Chem. Soc.*, 86 (1964) 2435.
93 E. GRUNWALD AND E. PRICE, *J. Am. Chem. Soc.*, 86 (1964) 2965, 2970.
94 P. R. WELLS, *Chem. Rev.*, 63 (1963) 171; see also C. D. RITCHIE AND W. F. SAGER, *Progr.
    Physical Organic Chemistry*, 2 (1964) 323; and *Symposium on Linear Free Energy Correlations*,
    U.S. Army Research Office, Durham, N.C., 1964.
95 J. E. LEFFLER AND E. GRUNWALD, *Rates and Equilibria of Organic Reactions*, Wiley, New
    York, 1963.
96 L. P. HAMMETT, *Physical Organic Chemistry*, McGraw-Hill, New York, 1940.
97 H. H. JAFFÉ, *Chem. Rev.*, 53 (1953) 191.
98 D. H. MCDANIEL AND H. C. BROWN, *J. Org. Chem.*, 23 (1958) 420.
99 H. H. JAFFÉ, *J. Chem. Phys.*, 20 (1952) 279.
100 H. C. BROWN AND Y. OKAMOTO, *J. Am. Chem. Soc.*, 79 (1957) 1913; 80 (1958) 4979.
101 E. BERLINER AND L. C. MONACK, *J. Am. Chem. Soc.*, 74 (1952) 1574.
102 T. L. BROWN, *J. Phys. Chem.*, 64 (1960) 1798.
103 R. W. TAFT, JR., *J. Am. Chem. Soc.*, 74 (1952) 2729.
104 R. W. TAFT, JR., *J. Chem. Phys.*, 26 (1957) 93.
105 R. W. TAFT, JR., *J. Am. Chem. Soc.*, 75 (1953) 4538.
106 R. W. TAFT, JR., *Steric Effects in Organic Chemistry*, Wiley, New York, 1956, p. 614.
107 E. GRUNWALD AND S. WINSTEIN, *J. Am. Chem. Soc.*, 70 (1948) 846.
108 A. H. FAINBERG AND S. WINSTEIN, *J. Am. Chem. Soc.*, 79 (1957) 1597, 1608.
109 S. WINSTEIN, A. H. FAINBERG AND E. GRUNWALD, *J. Am. Chem. Soc.*, 79 (1957) 4146.
110 E. M. KOSOWER, *J. Am. Chem. Soc.*, 80 (1958) 3258, 3261, 3267.
111 C. G. SWAIN AND C. B. SCOTT, *J. Am. Chem. Soc.*, 75 (1953) 141.
112 C. G. SWAIN, R. B. MOSLEY AND D. E. BROWN, *J. Am. Chem. Soc.*, 77 (1955) 3731.

Chapter 5

The Theory of Solid Phase Kinetics

L. G. HARRISON

Introduction

Solid state reactions are taken here to comprise the decompositions of solids, the reactions of two or more solids with each other, and the reaction of a solid with a gaseous or liquid phase. Phenomena such as diffusion and electrical conduction, which are fairly directly related to the transport of atoms or electrons, are mentioned in so far as they throw any light on the steps in reaction mechanisms. Phenomena involving the interplay of two gradients, such as the thermoelectric effect and Soret effect, are excluded as not being of current importance in chemical kinetics. Hence no very detailed exposition of the formalisms of irreversible thermodynamics is required.

A characteristic feature of solid state reactions is the total destruction of the reactant solid phases. These processes are thus sharply distinguished from reactions in gaseous or liquid solution, in which the fluid reacting phase exists continuously throughout the reaction, with a continuous variation in the activities of the components. Consequently, the notion of "order of reaction", as it is understood in the fluid phases, has only very limited applicability to solids. The order of a reaction is defined in terms of reactant concentrations. For the definition to have any meaning, these concentrations must have constant values throughout the reacting material; they may change with time, but not with position. Rate laws involving powers of concentration will normally arise only if the concentrations of intermediates are related directly to those of the reactants, so that the intermediates do not have distinctive time-dependences of their own. As a corollary, any reaction zone must be of fixed extent throughout the reaction.

One or more of these conditions is violated in most solid state reactions. The reaction zone often consists of the surfaces of growing "nuclei" of the products. The number of nuclei and the surface area of each nucleus both increase with time, and either of these dependences may determine the overall rate law. Reaction curves for solid state reactions are often sigmoid or even more complicated, and cannot conveniently be represented by a single function throughout. The main accelerating stage, however, can often be well represented by a power of the time, viz. extent of reaction $\alpha = kt^n$. Laws with $n > 1$ can often be interpreted in terms of the formation or growth of nuclei. Thus a cubic law, $n = 3$, may arise either through the continuous formation of small nuclei each of which needs three steps for its formation or through the linear advance in three dimensions of the surfaces of the nuclei (section 1.2.1).

Even where there is no definite reaction interface, there are usually concentration gradients in a solid phase; *i.e.* there is no such thing as a well-stirred solid. Kinetic expressions are thus usually complicated even if the products are formed in solid solution; and solid reactants in general are rather intolerant of dissolved products. Spinel formation may be envisaged as a solution process of two solid oxides, by migration of cations within a fixed framework of oxide ions; its rate laws are the time-dependent solutions of the diffusion equation for the appropriate particle shape (section 2.4.2).

Isotopic exchange is widely studied because it disturbs the structure of the reacting phases less than any other type of process. By the same token, isotopic exchange might be expected to follow simpler rate laws than most other solid reactions. In the absence of concentration gradients or time-dependent intermediates, the rate law should be first-order. Isotopic exchange is, however, commonly used in situations involving concentration gradients, to study self-diffusion. The rate law then becomes the appropriate solution of the diffusion equation. A common case to which practical situations often approximate well is the semi-infinite solid in contact with a well-stirred fluid phase, which gives the power law $\alpha = kt^n$ with $n = \frac{1}{2}$ (section 2.2.1). Preferential diffusion along grain boundaries may further complicate the kinetic behaviour; one possible limiting case is a power law with $n = 3/4$ (section 2.3.2).

The so-called parabolic law ($n = \frac{1}{2}$) arises not only in self-diffusion, but also in the case of a rate-determining step in which one reactant must traverse the entire thickness of a growing product layer, that thickness being proportional to extent of reaction. This situation is well-known experimentally in the formation of thin oxide layers on metals (section 2.4.3).

There is no single type of elementary process which can be thought of as generally dominant in solid state reaction mechanisms. Rate of reaction may be limited by the transport of matter, electricity, heat or mechanical stress. Examples of the two last-mentioned are propellant burning (section 1.4.1) and detonation (1.4.2).

Transport of matter and electricity nearly always require the presence of defects in the solid phase (except for electronic conduction in metals). The jump of a point defect from one lattice site to another is a simple unimolecular reaction step, analogous in some respects to reaction steps in the fluid phases. Theoretical treatment involves analysis of the vibrational motions of the environment of the reacting site; for a solid, that environment is the whole crystal (section 3.2.1).

A wide variety of point defects may serve as reaction intermediates in solids (section 3.1). The type of defect which is best understood is the lattice vacancy, especially in ionic crystals (section 3.1.3); but electronic defects (sometimes known generically as colour centres, although only a minority of them absorb in the visible) are likely to be important in some instances (section 3.1.6). Point defects are often capable of rapid motion at ordinary temperatures, and their interactions can sometimes give rise to simple first or second order rate laws. Such processes are perhaps the only ones in solids which are at all closely analogous to the majority of fluid phase reactions. A power law $\alpha = kt^n$ with values of n between 0.3 and 0.125 has been explained in terms of a time-dependent concentration of an electronic defect acting as intermediate (section 3.1.6). It will be seen that power laws arise in solid phase reactions from a great variety of different mechanisms. Observed values of n have ranged from 22.8 (section 1.2.1) to 0.125 (section 3.1.6).

The mobility of point defects is not well understood even for those which have been most widely studied. Methods of calculation of activation energies are still far from reliable. Experimental evidence in the last few years has tended generally to indicate that the mobility of point defects at low temperatures is greater than had formerly been supposed. For example, it was long believed that vacancy concentrations in alkali halides are in frozen equilibrium below about 300 °C. Even yet, few studies of diffusion or conductivity have been made at lower temperatures; but the most complete study which has been made shows that the conductivity of sodium chloride can be interpreted very precisely in terms of equilibrium concentrations down to 0°C. At the lowest temperatures, the approach to equilibrium can be followed and is first-order with a rate coefficient of the order of one hour^{-1} (section 3.1.3).

This is not the most remarkable example of unexpected mobility at low temperatures. Structural studies of electronic defects produced by X-irradiation of alkali halides below 20 °K have provided unequivocal evidence for the formation of interstitial halogen structures which are not adjacent to a vacancy. It has been proposed that the formation of these structures involves rapid migration of positively-charged halogen atoms through the lattice at these low temperatures (section 3.1.6). The possibility that a crystal lattice is effectively "transparent" to multiply-ionized atoms at all temperatures may become important in reactions of irradiated solids, or ones involving very strong oxidants. However, no such highly mobile intermediate has yet been persuaded to give account of itself in any structural study.

1. Reactions at a moving interface

Most of the equations relating formation and growth of nuclei to observed rate laws were developed some time ago and have already been comprehensively reviewed, for example by Jacobs, Tompkins and Garner in Chapters 7, 8 and 9 of a compilation edited by Garner[1]. The author has made extensive use of these earlier reviews in writing sections 1.1, 1.2 and 1.3 of this account.

1.1 DEPENDENCE OF RATE LAW ON SHAPE OF INTERFACE

A reaction interface may be a surface of large extent, readily observable, and sometimes, as in the burning of a solid propellant, of a shape which is initially under the experimenter's control—which gives him the opportunity to "preselect" the kinetic rate law. On the other hand, the interface may comprise the surfaces of many minute nuclei. These are often observable microscopically; but in opaque materials or in the early stages of a reaction, the existence of the nuclei often has to be inferred from the rate law and other indirect evidence.

The interface usually advances at a constant linear rate k_ℓ in any given direction; but the rate may be anisotropic, leading to the development of nuclei of complicated shape (Garner[1], Chapter 8). The advance of a large interface may be assisted by the formation of microscopic nuclei in front of the interface, and there are occasional instances of a periodic advance of the interface leading to an appearance of "growth rings" in the reacted material. These phenomena were observed in the work of Garner et al.[1-3] on dehydration of mixed alums. In unpublished work in the present author's laboratory on the oxidation of KBr by Cl_2, the reaction interface apparently overtook successive layers of bromine-filled cavities, leading to concentric brown stains in the reacted material[4].

The following examples illustrate some of the types of rate law which may arise from the constant isotropic motion of a reaction interface at a constant linear rate k_ℓ. α represents the extent of reaction, i.e. fraction of initial volume of reactant destroyed. t is the time from beginning of motion of the interface. Where a macroscopic interface is formed by the coalescing of growing nuclei, the zero of t may be difficult to establish accurately, which sometimes hinders precise application of the rate laws. The laws are of course as numerous as the possible initial shapes of the reaction interface and only a few important cases are given.

(a) Rectangular prism (or cylinder) reacting inwards on prism faces but not on ends, so that the length does not appear in the equations. For a prism of sides a and b

$$\alpha = 2k_\ell t(a+b-2k_\ell t)/ab \tag{1}$$

$$d\alpha/dt = 2k_\ell(a+b-4k_\ell t)/ab \tag{2}$$

For a cylinder the expressions are exactly the same with $a = b =$ diameter. These equations have been applied to the dehydration of potassium hydrogen oxalate hemihydrate[5].

(b) Rectangular parallelepiped, sides a, b, c; and sphere, radius r; in both cases reacting inwards. For the parallelepiped

$$\alpha = [2k_\ell t(ab+bc+ca)-4k_\ell^2 t^2(a+b+c)+8k_\ell^3 t^3]/abc \tag{3}$$

This equation (with $a = c$) has been applied to the dehydration of copper sulphate pentahydrate[6]. The problem of an uncertain time zero was tackled by measuring α at various times t' from an arbitrary zero and solving the cubic equation for $k_\ell t$. A plot of $k_\ell t$ against t' has slope k_ℓ and intercept at the true time zero. The same equations can be applied to a sphere by setting $a/2 = b/2 = c/2 = r$.

$$\alpha = (3k_\ell tr^2 - 3k_\ell^2 t^2 r + k_\ell^3 t^3)/r^3 \tag{4}$$

$$d\alpha/dt = (3k_\ell r^2 - 6k_\ell^2 tr + 3k_\ell^3 t^2)/r^3 \tag{5}$$

A convenient form of equation (4) is

$$(1-\alpha)^{\frac{1}{3}} = 1 - (k_\ell t/r) \tag{6}$$

These equations have been applied to the dehydration of calcium carbonate hexahydrate[7], the decomposition of silver carbonate[8] and the decomposition of magnesium hydroxide[9].

(c) *Shapes giving a constant rate of reaction.* These shapes are important in connection with the design of solid propellant charges, for which a constant rate of production of gas is normally required. An obvious shape is a hollow cylinder, initial radii a (outside) and b, burning on both curved surfaces. The decrease in area of the outer surface is compensated by the increase in area of the inner surface to give constant total burning surface.

$$\alpha = 2k_\ell t/(a-b) \tag{7}$$

$$d\alpha/dt = 2k_\ell/(a-b) \tag{8}$$

Such a shape has the disadvantage that burning must take place on the outer surface, detaching the charge from its casing and exposing the casing to hot gases. A common way out of the difficulty is the use of a "star-centred" charge in which the charge burns from an inner cavity shaped like a star. The evolution of this shape by constant regression of the surface leaves the total burning area roughly constant through the greater part of the reaction. Eventually, there is always some material left as cusps—in technical jargon, "sliver"—which burn at a rapidly decreasing rate. Detailed variations in charge design are directed towards minimizing the "sliver". Fig. 1 shows some designs of solid propellant charge. See also Taylor[10].

(d) *Growing nuclei.* The previous cases are relevant chiefly to the motion of macroscopic reaction interfaces, which may often be established by the coalescence of a number of small nuclei of solid product. Solid state reactions are often studied, however, in circumstances in which the nuclei are being formed and are growing separately without coalescing. The volume v of a single nucleus is then

$$v = K_g k_\ell^m t^m \tag{9}$$

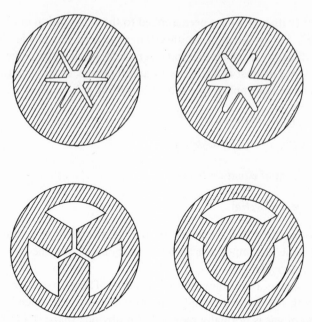

Fig. 1. Cross-sections of some shapes which maintain essentially constant area of reaction interface (used for solid propellant charges). In all cases, the outer cylindrical surface is not reacting.

where K_g is a shape factor (*e.g.* $4\pi/3$ for spheres, 8 for cubes, etc.) and m is the number of linear dimensions in which the nucleus is growing. If the growth rate is anisotropic, k_l is the geometric mean of the k_l values for the principal directions of growth. The cubic law indicated by equation (9) for three-dimensional growth of nuclei which all start to grow at the same time has been observed in the decompositions of lithium aluminum hydride[11], calcium azide[12] and silver oxide[13]. Simultaneous formation of all nuclei seems, however, to be relatively unusual, and in general it is necessary to consider both the rate of formation and the rate of growth of nuclei. This topic is taken up in the next section.

1.2 NUCLEATION AND DEVELOPMENT OF REACTION INTERFACE

Fig. 2 is a hypothetical reaction curve showing some features commonly observed in solid state reactions which involve nucleation. The curve OABCD shows three successive stages which might correspond to the following steps in the mechanism: (*i*) a pre-nucleation process leading to the formation of a large number of nuclei roughly simultaneously at point B; (*ii*) an accelerating region BC, which might follow, for example, a cubic law with time-zero close to B; (*iii*) the portion CD in which the nuclei are coalescing. The previously-mentioned example of lithium aluminum hydride[11] shows the two distinct regions OAB and BC; the separate

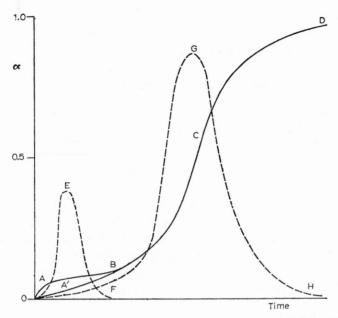

Fig. 2. Typical curves for extent of reaction α (solid lines) and concentrations of intermediates (dotted lines) for a reaction with nucleation of a new solid phase.

growth and decay of intermediates in those two regions, as suggested by curves OEF and OGH, has been followed by measuring electrical conductivity changes. By the same method, the curve OGH has been determined for the reaction of potassium iodide with chlorine or fluorine[14, 15]. In that case, the accelerating portion of OG obeyed a t^2 law, which is to be expected for an intermediate if the extent of reaction follows a t^3 law.

The curve OA'BC, showing no clear separation of pre-nucleation and growth periods, suggests that nucleation and growth may take place together over extended periods of time. If a nucleus starts to grow at a constant linear rate k_ℓ at time y, then at time t its linear size parameter ℓ (radius of circle or sphere, half-side of square or cube, etc.) is

$$\ell = k_\ell(t - y) \tag{10}$$

and from equation (9) its volume is

$$v = K_g k_\ell^m (t - y)^m \tag{11}$$

Now if there are N nuclei at time t, and of these dN were formed in the time interval from y to $(y + dy)$, the part dV of the total volume of nuclei represented by these dN is

$$dV = v(t, y)dN = v(t, y)(dN/dt)_{t=y}dy \tag{12}$$

and the total volume of reacted material at time t is

$$V(t) = \int_0^t K_g k_l^m(t-y)^m(dN/dt)_{t=y}dy \tag{13}$$

1.2.1 Simple nucleation; power laws, $\alpha = C_\alpha t^n$

A nucleus may be formed in a single act from any one of N_0 sites capable of reacting to form nuclei; or its formation may involve the conversion of a number of the sites into intermediates (concentration C_I), of which a number $(\beta-1)$ must combine to yield a nucleus. In the latter case, if we assume that N_0 is large compared to the number of intermediates formed, and that the rate of formation of intermediates is large compared to their rate of reaction to form nuclei, we have

$$dC_I/dt = k_b \tag{14}$$

$$dN/dt = k(k_b t)^{\beta-1} \tag{15}$$

Thus

$$N = (kk_b^{\beta-1}/\beta)t^\beta = Dt^\beta \tag{16}$$

where

$$D = (kk_b^{\beta-1}/\beta) \tag{17}$$

This law includes the first-mentioned case of formation of a nucleus in a single step, for which $\beta = 1$, $dN/dt = k$ and $N = kt$.

A power law of the same form as equation (16) was derived by Bagdassarian[16] (in relation to the problem of formation of a silver nucleus in a photographic film) for a different model in which β successive events, each with a rate coefficient k_1, are required to form a nucleus. Then the constant D in equation (16) represents

$$D = N_0 k_1^\beta/\beta! \tag{18}$$

If a rate law for nucleation of the form of equation (16) is found, it is not always easy to distinguish whether the mechanism is that represented by equation (17) or equation (18). In the case of barium azide decomposition, for which $\beta = 3$, Thomas and Tompkins[17] noted the presence of a few large reaction zones rather than a large number of small ones, and concluded that the activation energy for growth must be markedly less than that for any step in the nucleation. Observed values excluded the mechanism leading to equation (18) on this basis, and they proposed nucleation by combination of two F-centres as the step represented kinetically by equation (15).

The rate law for extent of reaction α as a function of t may be derived as follows. From equation (16), $(dN/dt)_{t=y} = D\beta y^{\beta-1}$. Substituting in equation (13)

$$V(t) = K_g k_\ell^m D \int_0^t (t-y)^m \beta y^{\beta-1} dy \tag{19}$$

whence

$$V(t) = K_g k_\ell^m D K_{m\beta} t^{m+\beta} \tag{20}$$

where

$$K_{m\beta} = 1 - \frac{m\beta}{\beta+1} + \frac{m(m-1)}{2!} \frac{\beta}{\beta+2} \quad \text{etc.} \tag{21}$$

If the volumes of nuclei in the above expressions are taken to be measured in such a way that they represent volume of reactant consumed rather than volume of product formed, then for an initial reactant volume V_0, the extent of reaction α is

$$\alpha = V(t)/V_0 \tag{22}$$

From equation (20)

$$\alpha = C_\alpha t^n \tag{23}$$

or for the pressure of a gaseous product

$$P = C_p t^n \tag{24}$$

where

$$n = m + \beta \tag{25}$$

and the constants C_α, C_p are obtainable from equations (20), (21) and (22) with $\alpha = P/P_{final}$.

The constancy of the linear rate of growth of nuclei assumed in the above treatment may seem a doubtful assumption, particularly in the early stages in which the nucleus is developing from essentially molecular dimensions to a macroscopic object. There is evidence to suggest that nuclei sometimes grow at an accelerating rate (Fig. 3, dotted curve AB) for a short period t' before reaching the condition for linear growth. In this case, of all nuclei existing at time t, those formed later than $(t-t')$ are still in their "slow growth" period AB. Thomas and Tompkins[17] (see also ref. 1) treated this situation approximately by replacing the curve for the slow growth period by a straight line AB (shown solid in Fig. 3). The integral in equation (13) or (19) must then be resolved into two parts with different values of k_ℓ; the part with normal k_ℓ has limits 0 to $(t-t')$ and the "slow growth" part has limits $(t-t')$ to t. The simple result of this treatment is that the second integral is negligible for $t \gg t'$, and equation (20) can be replaced by

$$V(t) = K_g k_\ell^m D(t-t')^{m+\beta} K_{m\beta} \tag{26}$$

The power law of equation (23), (24) or (26) has been found to hold for many reactions. Its interpretation in terms of the generation and growth of nuclei is of course assisted if the nuclei can be observed directly, so that either their number or their size or both can be determined as a function of time. Even a static picture of the size distribution at one time can be very informative, particularly if it in-

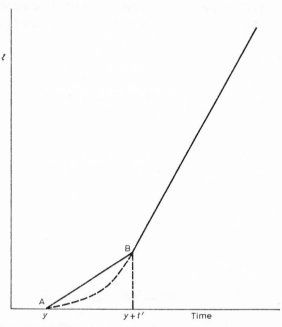

Fig. 3. Variation of size parameter ℓ of a nucleus with time in a case of initial slow growth (curved region diagrammatic only).

dicates uniform size and hence simultaneous formation of all nuclei. This has been seen, for example, in the photolytic decomposition of silver bromide[18]. Formation of nuclei linearly with time ($\beta = 1$) has been observed in the dehydration of chrome alum[2] and of copper sulphate pentahydrate[2], while number of nuclei varying as the square of the time ($\beta = 2$) has been seen in the dehydration of nickel sulphate heptahydrate[19]. Nucleation according to a cubic law ($\beta = 3$) followed by three-dimensional growth ($m = 3$) to give an overall sixth-power law ($n = 6$) has been observed in the decomposition of barium azide[20]. This provides one of the clearest cases of a power law arising from the combined effects of nucleation rate and growth rate. There are many other cases in which the mechanism seems fairly clear from kinetic data, but is not so clearly established by direct observation of all factors contributing to the power law. There appear to be numerous known instances of simultaneous nucleation ($\beta = 0$, so that $m = n$). Among these, the decompositions of lead and barium styphnates[21] show two-dimensional growth of nuclei leading to a quadratic rate law, $n = m = 2$. Three-dimensional growth leading to the cubic law $n = m = 3$ seems rather commoner, being observed for example in the decompositions of lithium aluminum hydride[11], calcium azide[12], silver oxide[13] and mercury fulminate[22]. In most of these cases, there is an induction period, or "slow growth" period, so that, as in equation (26), t must usually be replaced by $(t-t')$. The introduction of the parameter t' resulted in some of these reactions being recognized as obeying power laws, whereas they had formerly been assigned other rate expressions, but it also means that a clear decision between possible alternative rate expressions cannot always be made from overall rate of reaction data alone. The case of silver oxide is unusual in showing a negative value of t', corresponding to initial "fast growth" of nuclei. For silver oxalate decomposition, $n = 4$, the exponent apparently arising from first-order nucleation, $\beta = 1$, with three-dimensional growth, $m = 3$.

Much higher powers have been observed in some cases, but there is often some doubt as to whether the power law is the appropriate expression in such instances. For the decomposition of large crystals of mercury fulminate, Garner and Hailes[22] found values of n between 11.2 and 22.8, and were led to suggest that the formation of nuclei was a "branched chain" process. This mechanism, however, would generally be expected to lead to a different type of rate law (see section 1.2.2). More recently, Morrison and Nakayama[23] reported very variable values of n, between 3.3 and 11.3, for the reaction of potassium bromide with chlorine gas. The rate of nucleation was also observed directly and found to obey a power law with variable β between 3.8 and 8.4. Nuclei were three-dimensional, so that $(n-\beta) = m$ should be 3. In one run both n and β were observed and gave $m = 8.7 - 5.4 = 3.3$. This is a puzzling case which is certainly not yet fully resolved. Later work in the author's laboratory[24] suggests that the rate curves can sometimes be fitted by the type of exponential law discussed in the next section.

1.2.2 Self-catalyzed nucleation; exponential laws, $\alpha = Ce^{kt}$

Experimentally, it is often not possible to make an unequivocal decision between empirical rate laws of the form $\alpha = C_{\alpha}t^n$, with high n, and those of the form

$$\alpha = Ce^{kt} \tag{27}$$

since reasonably linear plots can often be obtained over fairly wide ranges of t and α according to both expressions. The t^n law seems physically reasonable in terms of reaction mechanism only when the exponent n is small ($n \leqslant 6$) or, on an experimental basis, when n has been resolved clearly into its components m and β by direct observation of the generation and growth of the nuclei. Otherwise, it is reasonable to suspect a mistaken identification of the form of the rate law. Equation (27) then supplies a good alternative on an empirical basis, and, mechanistically, it corresponds to a type of nucleation process which is, *a priori*, quite likely. This process is one in which the rate of nucleation is affected by the amount of reaction α which has already occurred, so that nucleation is self-catalyzed.

One model for self-catalytic nucleation (Garner and Hailes[22], Jacobs and Tompkins[1]) envisages the formation of nuclei in two ways: (*i*) from N_0 potential nucleus-forming sites present at $t = 0$; (*ii*) by a mechanism (referred to as "branching", *cf.* p. 88) which gives a contribution to dN/dt proportional to the N nuclei already formed. Hence

$$dN/dt = k_1 N_0 + k_2 N \tag{28}$$

which gives

$$N = (k_1 N_0/k_2)(e^{k_2 t} - 1) \tag{29}$$

and

$$(dN/dt)_{t=y} = k_1 N_0 e^{k_2 y} = C_N e^{k_2 y} \tag{30}$$

Substitution in equation (13) yields, for the total volume of reactant consumed at time t

$$V(t) = \int_0^t K_g k_{\ell}^m (t-y)^m C_N e^{k_2 y} dy \tag{31}$$

Integration gives

$$V(t) = K_g k_{\ell}^m C_N k_2^{-(m+1)} m![e^{k_2 t} - P_m(k_2 t)] \tag{32}$$

where $P_m(x)$ is the polynomial

$$P_m(x) = 1+x+(x^2/2!)+ \ldots +(x^m/m!) \tag{33}$$

The expression in the square bracket in equation (32) is thus $e^{k_2 t}$ minus the first $(m+1)$ terms of the series expansion of the same quantity; for $k_2 t \gg 1$, the polynomial P_m is a negligible part of the complete value of the exponential, and equation (32) then reduces approximately to the form of equation (27).

The above treatment supposed that $N_0 \gg N$ at all times to which the equations are to be applied. If this is not so, the form of the rate law is unchanged, but either C_N or k_2 must be replaced by a different coefficient. If the first-order decay of the nucleus-forming sites becomes important, k_2 must be replaced in all equations from (28) onwards by

$$k_2' = k_2 - k_1 \tag{34}$$

If the N_0 nucleus-forming sites are used up almost instantaneously, so that the term $k_1 N_0$ vanishes in equation (28) for $t > 0$, but there remains a boundary condition $N = N_0$ at $t = 0$, then C_N, which had the value $k_1 N_0$, must be replaced by

$$C_N' = k_2 N_0 \tag{35}$$

The significance of an exponential law in mechanistic terms remains somewhat obscure at the present time. Garner and Hailes[22] envisaged a tree-like growth of linear branching chains, and it has more recently been suggested[25] that such nucleation might occur along dislocation lines. An alternative which has been considered in the present author's laboratory[24] would allow the exponential law to arise even for the growth of ordinary separated "compact" nuclei in three dimensions. Electrical conductivity studies on the KI/Cl_2 reaction showed that the number of charge carriers was proportional to t^2, which might represent the derivative of a t^3 rate law, and which thus suggested that the number of charge carriers at time t was proportional to the total area A_t of the growing nuclei at that time. For KBr/Cl_2, the kinetics of which involve either an exponential law or a power law with very high n, it is suggested that the number of charge carriers is again proportional to A_t, but that they can create new nuclei at a rate proportional to their number. Thus

$$dN/dt = k_3 A_t + k_1 N_0 \tag{36}$$

Now the area A_i of each nucleus will be given by an expression of the form

$$A_i = K_g' k_\ell^2 (t-y)^2 \tag{37}$$

in which K_g' is a geometrical factor for area analogous to K_g for volume. Double differentiation of equations (36) and (37) eliminates y and gives the following equation for N

$$d^3N/dt^3 = k_3(d^2A_t/dt^2) = k_3 \sum_{i=1}^{N} (d^2A_i/dt^2) = 2k_3 K_g' k_\ell^2 N \tag{38}$$

The solution for dN/dt, with the condition $A_t = 0$ at $t = 0$, is

$$dN/dt = (k_1 N_0/k_4)e^{k_4 t} \tag{39}$$

where

$$k_4 = (2k_3 K_g' k_\ell^2)^{\frac{1}{3}} \tag{40}$$

Equation (39) is of the same form as (30) and leads to the exponential law of equation (27) on substitution in the general expression [equation (13)] for $V(t)$.

1.3 MUTUAL INTERFERENCE OF NUCLEI

1.3.1 Reaction ending at different times in different crystals

Geometrically the simplest type of situation in which the advancing reaction interface eventually destroys itself is exemplified by the meeting of two plane interfaces growing from opposite sides of a rectangular slab, or the shrinking of a spherical reaction interface, to vanish finally at the centre of the sphere. In such situations, reaction in any one crystal will continue only for a period t_1, after which the entire crystal has reacted, and in general

$$t_1 = \ell_c/k_\ell \tag{41}$$

where ℓ_c is some critical dimension (radius of sphere, half-thickness of slab, etc.).

There are two obvious limiting situations in which all particles may not finish reacting simultaneously:

(i) Reaction interface formed at $t = 0$ in all crystals; particle size distribution $N(t_1)$, such that the number of crystals with sizes corresponding to lifetimes between t_1 and $t_1 + dt_1$ is $dN = N(t_1)dt_1$. (N has previously been used for number of nuclei; but since the entire reaction zone is here envisaged as starting to grow simultaneously at all points, there is in effect only one nucleus per crystal and number of nuclei is equal to number of reacting crystals.)

If the volume of any one crystal reacted at time t is $v(t_1, t)$, so that the total volume of a crystal of dimension ℓ_c is $v(t_1, t_1)$, then the total volume reacted at time t is

$$V(t) = \int_0^t v(t_1, t)N(t_1)dt_1 + \int_0^{t_1} v(t_1, t_1)N(t_1)dt_1 \tag{42}$$

$$\text{for } t_1 > t \qquad\qquad \text{for } t_1 < t$$

Equation (42) can be solved for $V(t)$ as a function of t if the particle size distribution is known.

(ii) All crystals of uniform size, but nucleation and movement of reaction interface starting randomly at time y in each crystal. N again represents number of nuclei or number of reacting crystals at any time, the two being identical. Let $v(t, y)$ be the volume of a crystal reacted by time t and v_{total} be the volume of one crystal. Then the total volume of the sample reacted by time t is

$$V(t) = \int_0^t v(t, y)(dN/dt)_{t=y}dy$$

$$= \int_{t-t_1}^t v(t, y)(dN/dt)_{t=y}dy + \int_0^{t_1} v_{\text{total}}(dN/dt)_{t=y}dy \tag{43}$$

Expressions derived from equation (43) for rectangular slabs and for spheres have been applied[26] to the dehydration of calcium carbonate hexahydrate. For a rectangular slab with reaction fronts of area ab approaching each other across an initial thickness c

$$v(t, y) = 2abk_\ell(t-y) \tag{44}$$

and for random nucleation

$$dN/dt = k_1 N_0 e^{-k_1 t} \tag{45}$$

whence for $t < t_1 = c/2k_\ell$

$$\alpha = (2k_\ell/k_1 c)(e^{-k_1 t} + k_1 t - 1) \tag{46}$$

and for $t > t_1$

$$\alpha = 1 + (2k_\ell/k_1 c)(1 - e^{k_1 c/2k_\ell})e^{-k_1 t} \tag{47}$$

Equation (47) indicates a linear relationship between $\ln(1-\alpha)$ and t, which was observed in the work cited[26]. If nucleation is very slow, so that each crystal reacts completely in a time small compared to the time required for a majority of the crystals to have started reacting, *i.e.* $k_1 < (2k_\ell/c)$, equation (47) reduces to a simple first-order law, $\alpha = 1 - e^{-k_1 t}$.

For spherical particles of radius r, dN/dt is given by equation (45) but $v(t, y)$ now becomes

$$v(t, y) = (4\pi/3)[r^3 - \{r - k_\ell(t - y)\}^3] \tag{48}$$

To simplify the formidable array of constants in the final result, let

$$B(x) = x^3 + 3x^2(k_\ell/k_1 r) + 6x(k_\ell/k_1 r)^2 + 6(k_\ell/k_1 r)^3 \tag{49}$$

The solutions then become
for $t < t_1 = r/k_\ell$

$$\alpha = 1 + e^{-k_1 t}[B(1) - 1] - [B\{(r - k_\ell t)/r\}] \tag{50}$$

and for $t > t_1$

$$\alpha = 1 + e^{-k_1 t}[B(1)(1 - e^{k_1 t_1}) - 1] \tag{51}$$

which again indicate a linear relation between $\ln (1 - \alpha)$ and t.

1.3.2 Overlap of nuclei; the Avrami–Erofeev equation

In the later stages of the growth of reaction zones in a single crystal, some of the nuclei will begin to touch each other, and growth will cease at the areas of contact. It must also be recognized that the number of potential nucleus-forming sites (initially N_0) decreases not only by formation of nuclei, but also by the "ingestion" of some of these sites by the advancing reaction zones. If no correction is made for this ingestion, the calculated number of nuclei at time t includes a number of "phantom" nuclei which never actually appeared, because of previous ingestion of the nucleus-forming site.

Let α_{ex} be the "extended" value of α in which the number of nuclei includes all phantoms and each nucleus is considered to grow regardless of overlap; and let α represent the true extent of reaction. For random nucleation, if any point in the crystal is indicated as a site at which reaction is about to take place by further growth of an existing nucleus or formation of a new one, there is a probability α that the site has already reacted and a probability $(1 - \alpha)$ that it is still available to react. Thus an increase $d\alpha$ in the true extent of reaction is related to the corresponding "extended" quantity $d\alpha_{ex}$ by

$$d\alpha = d\alpha_{ex}(1-\alpha) \tag{52}$$

whence

$$\alpha_{ex} = \int_0^{\alpha_{ex}} d\alpha_{ex} = \int_0^{\alpha}(1-\alpha)^{-1}d\alpha = -\ln(1-\alpha) \tag{53}$$

For first-order formation of nuclei, the total number, including phantoms, N_{ex} at time t, is

$$N_{ex} = N_0(1-e^{-k_1t}) \tag{54}$$

so that

$$dN_{ex}/dt = k_1 e^{-k_1t} \tag{55}$$

For growth of nuclei in three dimensions, V_{ex} and $\alpha_{ex}(= V_{ex}/V_0)$ may be found from equation (13) with $m = 3$ and dN_{ex}/dt used in place of dN/dt. Thus

$$\alpha_{ex} = (K_g k_\ell^2 k_1/V_0)\int_0^t (t-y)^3 e^{-k_1y}dy \tag{56}$$

From (53) and (56), the integral being the same as equation (32) with $m = 3$ and k_2 replaced by $-k_1$

$$-\ln(1-\alpha) = (6K_g k_\ell^3 N_0/V_0 k_1^3)[e^{-k_1t}-1+k_1t-(k_1^2t^2/2!)+(k_1^3t^3/3!)] \tag{57}$$

For large values of t, which is the situation of interest in which ingestion and overlap play a significant role, the square bracket can be approximated by its last term, and we have

$$\alpha = 1-\exp(-K_g k_\ell^3 N_0 t^3/V_0) \tag{58}$$

The general form

$$\alpha = 1-e^{-kt^3} \tag{59}$$

is commonly referred to as the Avrami–Erofeev equation. It has been derived in various ways by Avrami[27], Erofeev[28] and Mampel[29], the last-named having given the most comprehensive treatment of the combined effects of ingestion, overlap and particle size distribution (see also Young[29a]). The approach in terms of α_{ex} used above is that of Avrami, and (57) is sometimes called the Avrami equation. Erofeev considered also the possibility that a nucleus may require β successive events to form it, before it starts to grow at a constant rate k_ℓ (as discussed by Bagdassarian[16]—see section 1.2.1 above, discussion of equations 16 to 18). This possibility leads to a more general form of equation (59), sometimes called the Erofeev equation, of the form

$$\alpha = 1 - e^{-kt^{\beta+3}} \tag{60}$$

Allnatt and Jacobs[30] have recently given a generalized form of the Bagdassarian – Erofeev approach, in which the β successive steps are allowed to have different rate coefficients, $k_0 \ldots k_i \ldots k_{\beta-1}$. Their principal conclusions are that, provided $k_i t \ll 1$ for all i, the power law for the extended number of nuclei [equation (16)] is unaltered except that in the expression for the constant D [equation (18)], k_1^β is replaced by $k_0 k_1 k_2 \ldots k_{\beta-1}$. With the same condition of short time, equation (60) remains valid except that the expression for k contains a rather complicated function of the k_i's as well as the other factors indicated in equation (58).

Confusion is possible over the question of whether equation (60) is a short time or a long time approximation. In this connection, it is necessary to distinguish the time-scales for formation and growth of nuclei. If the growth of nuclei is not very far advanced, there is of course little practical significance in considering equation (60), since problems of ingestion and overlap are not yet important in the system. Thus for $\alpha \ll 1$, equation (60) reduces to the familiar power law, $\alpha = kt^{\beta+3}$. The validity of equation (60) in circumstances in which it does not reduce to a power law apparently requires that the process should be simultaneously at an advanced stage in the growth of such nuclei as have already been formed, but, as indicated by Allnatt and Jacobs' "short time" condition, at an early stage in the overall formation process. This would seem to imply a system in which a relative small number of nuclei have grown to large dimensions.

Kahlweit[30a] has recently suggested that the variations in nucleus size and concentration in the very early stages of a reaction may be complicated. As in the maturing of a precipitate thrown down from a liquid solution, large nuclei may grow at the expense of small ones, many of which may disappear without having reached observable size.

1.3.3 Termination of self-catalyzed nucleation; the Prout–Tompkins equation

Suppose that self-catalyzed nucleation is proceeding according to equation (28), but that the N_0 initial sites are exhausted, so that the equation reduces to

$$dN/dt = k_2 N \tag{61}$$

Suppose now that the carrier of the chain-branching process represented by this equation (*e.g.* the lattice defect produced along with each nucleus) can react in two ways:

(*i*) It can migrate to an unreacted site and there give rise to a new nucleus; if this is the only possibility, equation (61) correctly represents the development of the system.

(*ii*) It can interact with a site already reacted and so be destroyed without production of a new nucleus.

If both processes occur, and N_c is the total number of reaction sites initially present in the crystal, then the probability of a chain-carrier propagating nucleation according to process (*i*) at any stage in the reaction is $(N_c - N)/N_c$. Equation (61) must then be modified to

$$dN/dt = k_2 N(N_c - N)/N_c \tag{62}$$

and since $\alpha = N/N_c$, this may be rewritten

$$d\alpha/dt = k_2 \alpha(1 - \alpha) \tag{63}$$

whence

$$\ln [\alpha/(1 - \alpha)] = k_2 t + c \tag{64}$$

This equation was first proposed by Prout and Tompkins[31] and applied successfully to the decomposition of potassium permanganate. Their derivation was somewhat different from that given above, and depended not on an assumed mechanism but on an assumed linear relationship between extent of reaction α and a termination constant k_3 for chain-branching. They wrote, in place of equation (62), the expression

$$dN/dt = (k_2 - k_3)N \tag{65}$$

and for the rate of decomposition

$$d\alpha/dt = k'N. \tag{66}$$

They then assumed that $k_3 = k_2 \alpha/2$, which embodies both the linear relation between α and k_3 and the observed fact that, for the potassium permanganate decomposition, the curves of α against time were symmetrical with a point of inflection at $\alpha = \frac{1}{2}$. With this assumption, equations (65) and (66) can be solved to yield equations (63) and (64). The condition of maximum rate at $\alpha = \frac{1}{2}$ is not special to any one method of deriving the Prout–Tompkins equation. It may be obtained from equation (63) and should be observed with any reaction which obeys the Prout–Tompkins equation over most of the range of α.

1.4 THE RATE OF ADVANCE k_ℓ OF THE REACTION INTERFACE

1.4.1 Exothermic reaction propagated by thermal conduction

The propagation of reaction in the burning of a solid propellant, or in some exothermic processes in which all the products are solid, such as oxidation–reduction in metal–metal oxide mixtures, requires the transport of heat from the reaction zone into unreacted parts of the solid. For very high temperature reactions, in which the final stages may take place in the gas phase, it is necessary to consider radiation of heat back to the solid surface, and conduction through the gas phase. It is probably for this reason that the burning rate of a solid propellant is usually pressure-dependent [see equation (73) and related discussion below]. The simplest treatment of this type of reaction, however, is that for situations in which heat transport can be assumed to take place only by conduction within the solid phase. The derivation below of a differential equation for this situation follows Booth[32].

Consider a cylindrical block of reactant, of unit cross-section. Distances along its axis are designated x. Suppose that reaction takes place adiabatically, complete reaction raising the temperature of the material from T_1 to T_2, the reaction zone being perpendicular to the axis of the cylinder and moving in the direction of x negative at velocity k_ℓ. Let the extent of reaction at any point be α_u, and let the specific heat of the reacting material be c, assumed independent of both α_u and T. This is a reasonable approximation for reactions entirely in the solid phase, with only solid products, since every atom, regardless of its state of combination, contributes roughly an equal amount (three times Boltzmann's constant) to the heat capacity.

If the heat liberated by complete reaction of unit mass of the reaction mixture is Q, and the thermal conductivity of the mixture is λ, which, like c, is assumed independent of α_u and T, then the rate of addition of heat to the region between x and $(x+\mathrm{d}x)$ is $Q(\partial\alpha_u/\partial t)-\lambda(\partial T/\partial x)_x+\lambda(\partial T/\partial x)_{x+\mathrm{d}x}$. If the density of the solid is ρ, the rate of addition of heat to this volume element is also $\rho c(\partial T/\mathrm{d}t)\mathrm{d}x$. Equating the two expressions, we have

$$\lambda(\partial^2 T/\partial x^2) = \rho[c(\partial T/\partial t)-Q(\partial\alpha_u/\partial t)] \tag{67}$$

In the steady state of propagation of reaction at rate k_ℓ, T and α_u are functions of u only, where

$$u = x+k_\ell t \tag{68}$$

Transformation to this variable yields

$$\lambda(\mathrm{d}^2 T/\mathrm{d}u^2) = \rho k_\ell[c(\mathrm{d}T/\mathrm{d}u)-Q(\mathrm{d}\alpha_u/\mathrm{d}u)] \tag{69}$$

which, with the boundary condition $\alpha_u = 1$ at $T = T_2$, can be integrated to

$$\lambda(dT/du) = \rho k_\ell [Q(1-\alpha_u) - c(T_2 - T)] \tag{70}$$

At any point, the rate of reaction $(d\alpha_u/dt) = k_\ell(d\alpha_u/du)$ is a function θ of α_u, T and a number of factors such as particle size in the reaction mixture, which may be designated collectively by a parameter, P, i.e.

$$k_\ell(d\alpha_u/du) = \theta(P, \alpha_u, T) \tag{71}$$

From equations (70) and (71), we obtain

$$d\alpha_u/dT = \frac{\lambda\theta(P, \alpha_u, T)}{\rho k_\ell^2 [Q(1-\alpha_u) - c(T_2 - T)]} \tag{72}$$

The solution of this equation for any case in which the form of θ is known involves the determination (usually by numerical methods) of the value of k_ℓ for which the solution α_u satisfies the boundary condition $\alpha_u = 1$ at $T = T_2$. Booth[32], however, has pointed out some general features which are worthy of note in comparing burning rates of closely similar compositions. For a range of compositions in which the specific heat c is the same for all, and the rate law and particle size factors embodied in θ are the same for all, and Q is also constant, it follows that $(\lambda/\rho k_\ell^2)$ must be constant. Thus $k_\ell \propto (\lambda/\rho)^{\frac{1}{2}}$. This is qualitatively consistent with the known facts that loosely-packed powders commonly burn more rapidly than tightly-packed ones, and that addition of highly-conducting materials such as metal gauze also enhances the burning rate. Booth points out also that if the particle size factors P can be separated out of θ in the form $\theta(P, \alpha_u, T) = \theta_1(P)\theta_2(\alpha_u, T)$ then, other things being equal, $k_\ell \propto \theta_1^{\frac{1}{2}}$. If, for example, for particles of radius R, θ_1 happens to have the form $\theta_1 \propto R^n$, then burning rate $k_\ell \propto R^{n/2}$. Booth indicates that for compositions consisting of a finely-divided oxidant and a reducing agent formed of larger spheres (radius R), it has often been observed that $k_\ell \propto R^{-1}$, indicating according to the above treatment that $\theta_1 \propto R^{-2}$, and he demonstrates that this law is to be expected if diffusion through the oxidized layer around each sphere is rate-determining. He gives a full discussion of the application of his treatment to the results of Spice and Staveley[33] for Fe–BaO$_2$ mixtures.

The above treatment assumed, for generality, that the chemical reaction took place over a wide range of distance from the solid surface (u) and of temperature, but in consequence of the high temperature coefficient of most reaction rates, a more realistic model for many practical situations is obtained by considering the reaction to take place in a very narrow zone at a fixed temperature T_2. For reactions with gaseous products, this zone may be at the gas–solid interface, and in the combustion of solid propellants, it is well established that a substantial part of the exothermic reaction takes place in the gas phase, and that conduction of heat through the gas to the solid surface is important in determining the burning rate k_ℓ. For example, the burning rate is often strongly dependent on gas pressure P. The dependence is often represented empirically by Vieille's Law

$$k_\ell = bP^n \tag{73}$$

in which the "pressure exponent" n may have values ranging from much less than unity to much greater than unity. In a simple rocket motor with discharge of gas through a nozzle of fixed size, stable burning at a finite pressure is possible only if $(\mathrm{d}\ln k_\ell/\mathrm{d}P) < 1$. Consequently, rocket motor technologists prefer to quote values of n, which represents this derivative, even if a different equation, such as $k_\ell = a+bP$, or $k_\ell = a+bP^m$, gives a better representation of the burning rate over a wider range of P.

No theory exists to account for Vieille's Law or any of the other empirical expressions used, but qualitatively the increase in k_ℓ with P is explained by the known fact that the flame region in the gas phase moves nearer to the solid surface with increasing pressure[34]. In such cases, the physical properties of the solid, such as the thermal conductivity λ, may be relatively unimportant in determining k_ℓ. In general, if heat from the solid surface at temperature T_2 flows into the solid at a rate q, then

$$k_\ell = q/(T_2-T_1)\rho c \tag{74}$$

q may be determined largely by heat conduction from the flame region in the gas phase. T_2 may involve λ if it is determined by a balance between conduction in gas and solid phases. On the other hand, T_2 may be a melting-point or boiling-point of one of the reactants.

For this case of no reaction within the solid, the temperature profile in front of the burning surface has a particularly simple form. Equation (69) is amended by omission of the term in Q to yield

$$\lambda(\mathrm{d}^2T/\mathrm{d}u^2) = \rho k_\ell c(\mathrm{d}T/\mathrm{d}u) \tag{75}$$

and the solution of this equation with the boundary conditions $u = 0$, $T = T_2$, and $u = -\infty$, $T = T_1$ gives for the temperature profile in the solid phase

$$\ln\,[(T-T_1)/(T_2-T_1)] = (\rho k_\ell c/\lambda)u \tag{76}$$

Suppose that a small thermocouple is imbedded in a composition of known ρ and c at a fixed position x, and that temperature is recorded against time as the solid burns. k_ℓ can be measured accurately quite easily (*e.g.* by using two thermo-couples at different x). In favourable cases, T_2 may be estimated—although this is often difficult in practice because the point corresponding to T_2 may be difficult to identify precisely. Since the time variable can be converted to u by equation (68), it is also possible in principle to obtain λ from a plot of ln $(T-T_1)$ *versus* u; the linearity of this plot is of course diagnostic of the absence of reaction in the solid phase.

1.4.2 Detonation; propagation by a shock wave

In the preceding discussion of combustion processes, the linear rate k_ℓ depended in general both on physical properties and on the kinetic characteristics of the chemical reaction. By contrast, the rate-determining step in a detonation is the motion of a shock wave in a manner similar to the propagation of sound. The velocity of detonation k_ℓ depends only on physical properties of the reactants and products. Detonation is maintained by the occurrence of an exothermic reaction, but the only characteristics of that reaction which need be considered are those relating to the amount of energy released and the nature of the reaction products, *i.e.* equilibrium thermodynamic properties. Hence the usual development of the theory of detonation is referred to as the "hydrodynamic" or "thermohydro-dynamic" theory. The basic equations of this theory have been given in many books and articles[35–38]. Perhaps the clearest statement of the distinction between detonation and combustion as represented in formal thermodynamic terms on the Rankine–Hugoniot curve is in a very brief article by Paterson[38].

Consider a disturbance moving at a rate k_ℓ into a solid of specific volume v_1 at pressure p_1. The disturbance converts the solid to gaseous products which, *in the immediate vicinity of the disturbance*, are at pressure p_2 with specific volume v_2 and are moving at a streaming velocity w. *A priori*, w may be in the same direction as k_ℓ (positive w) or in the opposite direction. Both situations in fact occur: the first (products moving in the same direction as reaction front) is detonation; the other is combustion.

If the specific internal energies of the reactants and products at the reaction front are designated E_1 and E_2 respectively, the application of the laws of con-servation of mass, momentum and energy to the process leads to the following equations

mass
$$k_\ell/v_1 = (k_\ell-w)/v_2 \tag{77}$$

momentum

$$(k_\ell^2/v_1)+p_1 = [(k_\ell-w)^2/v_2]+p_2 \tag{78}$$

energy

$$E_1 + (\tfrac{1}{2})k_\ell^2 + p_1 v_1 = E_2 + (\tfrac{1}{2})(k_\ell - w)^2 + p_2 v_2 \tag{79}$$

From these conservation conditions, the velocity of detonation k_ℓ, stream velocity w and gain in internal energy $\Delta E = E_2 - E_1$ may be expressed in terms of the pressures and specific volumes as follows

$$k_\ell = v_1[(p_2 - p_1)/(v_1 - v_2)]^{\tfrac{1}{2}} \tag{80}$$

$$w = (v_1 - v_2)[(p_2 - p_1)/(v_1 - v_2)]^{\tfrac{1}{2}} \tag{81}$$

$$\Delta E = (\tfrac{1}{2})(p_1 + p_2)(v_1 - v_2) \tag{82}$$

If the conditions of chemical equilibrium are known, together with the equations of state of the products, equation (82) can be expressed in terms of p_2 and v_2 only, to yield a curve in the (v_2, p_2) plane known as the Rankine–Hugoniot curve or "dynamic adiabatic" (Fig. 4). It corresponds to (and is similar in general form to) the adiabatic, or isentrope, derived from $dE = -pdv$ in a static system, i.e. one with no streaming velocity w. The slope of the Rankine–Hugoniot curve is $(\partial p/\partial v)_S$.

In Fig. 4, A represents the initial state (v_1, p_1) of the reactant mixture. The solid curve BCDE is the dynamic adiabatic curve for the products at chemical equilibrium. Now in equation (80), only real positive values of k_ℓ can have physical significance. Thus the region CD, with $p_2 > p_1$ and $v_2 > v_1$, must be excluded from possible representations of the reaction. This leaves the two regions BC and DE. The latter, with pressure in the products always less than that at the reaction front and, by equation (81), with w negative (products flowing away from reactant), corresponds to propellant burning. The region BC represents the effect of a compression wave, with the products following the reaction front; this is detonation.

There is, however, only one point on BC which represents stable propagation of detonation at a fixed velocity. This is the point F, determined by the Chapman–Jouguet condition, which can be expressed in various ways. In terms of the propagation of disturbances through the products, it is evident that the region of compression, with w positive, will be quite small. The explosion which results from detonation consists of a rapid motion of products away from the detonation front at decreasing pressure. Thus rarefaction begins not very far from the compression region. If such rarefaction can overtake the wavefront it will weaken the front and ultimately quench the detonation. Suppose that the velocity of sound in the products near to the shock wave is a. Then, in the frame of reference fixed in the unreacted material, the shock wave is moving at velocity k_ℓ, and rarefactions in the products are attempting to overtake the shock wave at velocity $(w + a)$. For the

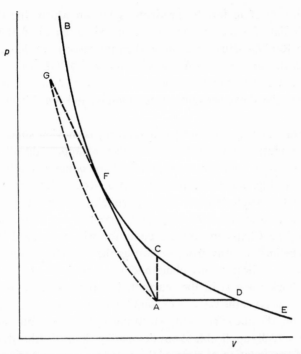

Fig. 4. Pressure–volume curves for detonation. A, initial state of reactant; AG, dynamic adiabatic (Rankine–Hugoniot curve) for reactant (dotted curve); BE, dynamic adiabatic for products; AG, (straight line) the Rayleigh line; F, Chapman–Jouguet point; BC, region representing detonation products; CD, region without physical meaning; DE, region representing burning without detonation.

detonation to be maintained, it is necessary that $k_\ell \geqslant w + a$. The situation $k_\ell > w + a$ implies, however, that the shock wave is moving away from the region of rarefaction. This could only happen if the region of release of the energy of reaction became spread out; and since the energy of reaction is needed to sustain the shock wave, the inequality is not acceptable as a condition for stable propagation of detonation. We are thus left with the condition

$$k_\ell = w + a \tag{83}$$

which is one way of stating the Chapman–Jouguet condition.

In relation to the Rankine–Hugoniot diagram, the Chapman–Jouguet postulate is that the conditions (v_2, p_2) for stable detonation are found at the point F at which a line from A is tangential to BC; or in algebraic terms, where

$$-(p_2 - p_1)/(v_1 - v_2) = (\partial p_2 / \partial v_2)_s \tag{84}$$

For discussion of the postulate, see for example Cook[35], or the earlier references on which his account is based[39–42].

If an equation of state for the products is known which, together with the thermodynamic data for the chemical reaction enables equation (82) to be converted into the Rankine–Hugoniot curve, then the velocity of detonation can be calculated from the known point A, the known curve BCDE, and equations (80) and (84). The actual structure of the reaction zone may, however, be much more complicated than the above account seems to imply, and is generally supposed to be as follows.

The shock wave triggers chemical reaction, but is moving sufficiently rapidly that the reaction must be envisaged as taking place an appreciable distance behind it. Thus the wavefront itself represents an adiabatic compression of the *reactant* mixture, shown in Fig. 4 as the dotted curve AG. As reaction occurs, the system approaches the Chapman–Jouguet point F roughly along the upper part GF of the tangent at A to the Rankine–Hugoniot curve (GFA is sometimes called the Rayleigh line). The Chapman–Jouguet condition will be realized in practice at some position behind the wavefront known as the Chapman–Jouguet plane. As the above account indicates, this plane and the wavefront are so related that, although the shock wave is in the reactant, it moves (relative to the stream velocity w of products) at the velocity of sound in the products.

For this reason, the above account, which purports to be about the detonation rate of solids, is actually applicable to any detonating phase. Further development of the question of the equation of state of the products is an extensive topic, is not primarily related to properties of the solid phase, and is beyond the scope of this account. The most widely-discussed equations of state for this purpose are those of Kistiakowsky and Wilson[43], and Paterson[44]. Cook[35] gives an extensive account of this topic and also discusses the possibility of getting direct information on the true equation of state from measurements of detonation temperature and related properties. Another approach is the direct determination of the Rankine–Hugoniot curve by measurement of both k_ℓ and w. The latter can be obtained by measuring the velocity imparted to the free surface of the solid by reflection of the shock (see, for example, Lawton and Skidmore[45]).

1.4.3 Processes requiring matter transport in the solid phase

As already indicated, a remarkable feature of solid state kinetics is the widespread occurrence of a constant linear rate of motion k_ℓ of the reaction interface. Except where physical properties determine k_ℓ (as in the case of detonation discussed above), it is to be expected that k_ℓ will be governed by either (*a*) the rate of reaction in the interface or (*b*) the rate of transport of material to or from the interface through reactant or product phases. These two possibilities will be discussed separately.

(*a*) Reaction at interface rate-controlling. In this case, a constant k_ℓ is obviously to be expected, provided that reaction is possible everywhere in the interface and does not require some special site which might not always be formed in a number proportional to area. This situation has already been exemplified in the discussion of propellant burning (section 1.4.1). A very different type of reaction which frequently appears to fall in this mechanistic class is the dehydration of hydrated salts (see Chapter 8 of ref. 1 for an extensive discussion). Transport of water is commonly rapid enough that it is not rate-controlling, especially if, as is often the case, the dehydrated product is severely cracked. Calculations of dehydration rate have been made according to an expression $Cv \exp(-E/RT)$, where C is related to the concentration of water molecules in the interface, and v is a vibration frequency for the decomposition step. In some cases, the dehydration rate correlates well with reasonable values of v, of order 10^{12} sec^{-1}, and in other cases the observed rates are much greater[1–3, 7, 46, 47].

(b) Transport processes rate-controlling. Reactions such as decomposition of ionic compounds, or oxidation–reduction processes in general with ionic reactants or products, involve migration of electrically-charged species, so that the reacting system may commonly be thought of as a solid-state voltaic cell. This type of mechanism was developed extensively in the 1930's by Wagner[48, 49], generally with reference to large-scale migration between separated reactants and products. The migration processes usually include at least one in which ions have to move, and this step will commonly have an activation energy at least as high as that of the interface reaction, so that migration is quite likely to be rate-determining. In such cases, a uniform k_ℓ seems, *a priori*, unlikely. Such processes would be expected to follow rate laws which are solutions of the diffusion equation, to which the whole of section 2 of this account is devoted. A well-known example from the work of Wagner[50] is the tarnishing of metals, which most often follows the so-called parabolic law (see section 2.4.3 below).

For processes in an ionic crystal in which many small nuclei develop in the bulk, it is quite likely that many small voltaic cells should be considered to exist with the nucleus in the bulk acting as one half-cell and a nucleus of another product on the surface acting as the other half-cell (Fig. 5). Electrical conduction between bulk and surface requires both electronic conduction (the term being intended to include positive-hole conduction) and ionic conduction, and the charge carriers for both must be produced at the reaction interfaces. If the production of carriers and the trapping sites for them is related to the area-increasing steps at the reaction interfaces, then the conductance between bulk and surface will be proportional to area for each nucleus. This should lead to a constant linear rate of advance k_ℓ of the interfaces. This suggestion was made in connection with work on the KI/Cl$_2$ reaction in the author's laboratory[14], and it was pointed out that the growth rate k_ℓ, while constant for any one nucleus, should not on this basis have the same value for all nuclei, but should decrease sharply with distance below the solid surface.

Extensive further work is necessary to determine whether such a mechanism

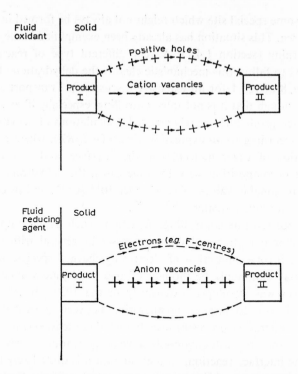

Fig. 5. Possible conduction processes in an electron transfer reaction with nuclei of products as the electrodes. Curved paths imply transport through the bulk. The straight paths for vacancies imply transport in regions of accelerated diffusion, such as grain boundaries.

has any general validity, and what may be the details of the mechanism in any particular case.

2. Diffusion-controlled reactions

2.1 THE DIFFERENTIAL EQUATIONS OF DIFFUSION
(*cf*. Section 1.2, Chap. 4)

The law of diffusion proposed by Fick in 1855 arose from the recognition that there is a close analogy, in terms of mathematical treatment, between diffusion and heat conduction. Fick's hypothesis was that the rate of transfer F of diffusing material across unit area of cross-section is proportional to the concentration gradient in the direction of diffusion, *viz*.

$$F = -D(\partial C/\partial x) \qquad (85)$$

The analogous equation for rate of flow of heat H is

$$H = -\lambda(\partial T/\partial x) \tag{86}$$

Converted into partial differential equations for three-dimensional time-dependent flow, to be solved for $C(x, y, z, t)$ and $T(x, y, z, t)$, these expressions become respectively

$$\partial C/\partial t = D\nabla^2 C = D \text{ div grad } C \tag{87}$$

$$\partial T/\partial t = (\lambda/\rho c)\nabla^2 T \tag{88}$$

where ρ and c are the density and specific heat of the material concerned. Consequently, solutions of the heat conduction equation can usually be converted to solutions of the diffusion equation simply by writing C for T and D for $(\lambda/\rho c)$. Books on heat conduction, such as Carslaw and Jaeger[51], have thus been extensively consulted in connection with diffusion problems. Several texts on diffusion have been written[52-54], of which the most comprehensive in discussing the mathematics of the diffusion equation is that of Crank[54].

Equation (87) might be expected to be a very accurate representation of self-diffusion, as commonly followed by isotopic tracer techniques, in a material at constant temperature throughout. For interdiffusion, in which the process alters the chemical composition of the material through which the transport is taking place, D may well be a function of C. The diffusion equation must then be written

$$\partial C/\partial t = \text{div } D \text{ grad } C \tag{89}$$

For the one-dimensional case

$$\partial C/\partial t = \frac{\partial}{\partial x}\left(D\frac{\partial C}{\partial x}\right) \tag{90}$$

Boltzmann showed that a convenient method of starting the solution in many cases is to use the transformation

$$\eta = (\tfrac{1}{2})x/t^{\frac{1}{2}} \tag{91}$$

which yields the ordinary differential equation

$$-2\eta(dC/d\eta) = \frac{d}{d\eta}\left(D\frac{dC}{d\eta}\right) \tag{92}$$

This transformation is legitimate if the initial and boundary conditions can be expressed in terms of η only, without explicit use of x or t; *e.g.* the conditions at $x = 0$ (all times) and $t = \infty$ (all x) must be the same. For a more detailed discussion of the conditions, see Jost[53].

If, for example, concentration gradients and temperature gradients are present simultaneously, so that D, T and t are all inter-related, it is necessary to apply the complete Laplacian operator to D as well as C

$$\partial C / \partial t = \nabla^2 (DC) \tag{93}$$

This is the most general form of the differential equation for diffusion.

2.2 SOME SOLUTIONS OF THE DIFFUSION EQUATION FOR SELF-DIFFUSION

Solutions of the diffusion equation are of course as numerous as the possible initial conditions and boundary conditions, but in experimental practice a few situations are very much more commonly encountered than any others. These are discussed in this section.

2.2.1 *"Sectioning" and "exchange" experiments: semi-infinite solid*

There are two common methods of studying self-diffusion in solids, the "sectioning" method and the "exchange" method.

(*a*) In a sectioning experiment, a thin layer of isotopically-labelled material is deposited (often by sublimation) on to one end-face of a prismatic solid sample, and diffusion takes place in one dimension (x) perpendicular to the plane of deposition. After diffusion has taken place at an elevated temperature for time t, the sample is cooled and the concentration distribution is determined by cutting slices off the crystal and analyzing them by any suitable method (often radioactive counting). If the region of varying concentration is much wider than the original deposited layer, the initial condition can be regarded as the "instantaneous plane source" with all diffusing material (amount M) concentrated in the plane $x = 0$; *i.e.* at $t = 0$, $C = M\delta(x)$, where δ is the Dirac delta-function. With diffusion in the positive x-direction only, the appropriate solution of

$$(\partial C / \partial t) = D (\partial^2 C / \partial x^2) \tag{94}$$

is

$$C = (M / \pi D t)^{\frac{1}{2}} e^{-x^2 / 4 D t} \tag{95}$$

Since t is a constant for any one sample, a plot of $\ln C$ against x^2 should be linear, and D can be determined from its slope.

(b) In the exchange method, a solid sample is in contact with a well-stirred fluid phase (gas or liquid) and the labelled species diffuses either into or out of the solid. There are several solutions of the diffusion equation which are commonly needed in such cases, depending on the geometry of the solid, the extent to which the diffusion process penetrates into the solid, and the relative amounts of solid and fluid phase.

If the region of varying concentration is narrow compared to the overall dimensions of the solid, *i.e.* diffusion is followed for only a small fraction of its total possible course, the process may be considered as a good approximation to one-dimensional diffusion in a semi-infinite solid. Suppose that the initial concentration of diffusing material in the solid is C_0 throughout, and the concentration during diffusion is $C(x, t)$. If the surrounding fluid is of sufficient volume that the concentration therein effectively remains constant at C_1, then the solution of equation (94) is required for the conditions $C = C_0$, $x > 0$, $t = 0$ and $C = C_1$, $x = 0$, $t \geqslant 0$. This solution is

$$(C - C_1)/(C_0 - C_1) = \text{erf}\left[x/2(Dt)^{\frac{1}{2}}\right] \tag{96}$$

where erf is the error function

$$\text{erf}(x) = (2/\pi^{\frac{1}{2}}) \int_0^x e^{-\beta^2} d\beta \tag{97}$$

A convenient quantity to follow experimentally, and one which obeys a particularly simple law, is the total amount of material M_t which has diffused out of or into unit area of the solid by time t. M_t is of course proportional to the extent of reaction α. For the case $C_1 = 0$ this quantity is

$$M_t = \int_0^t D(\partial C/\partial x)_{x=0} dt = 2C_0(Dt/\pi)^{\frac{1}{2}} \tag{98}$$

This is one instance of the $t^{\frac{1}{2}}$ law, or "parabolic law", which can also arise in a different way in the case of diffusion across a growing layer of product (see section 2.4.3). In either case, however, a $t^{\frac{1}{2}}$ law may be taken as a clear indication of a diffusion-controlled process.

In all solutions of the diffusion equation for a constant diffusivity, D and t always appear multiplied together, *i.e.* the time variable could be replaced by $\tau = Dt$. This is obvious since the diffusion equation can clearly be transformed to

$$\partial C/\partial \tau = \nabla^2 C \tag{99}$$

The fact that the concentration distribution in equation (96) contains D and t only as the product τ is important in connection with possible experimental procedure. Suppose diffusion is started at temperature T_1 (diffusivity D_1), and that at time t_1 the temperature is raised to T_2 (diffusivity D_2). If the change can be effected rapidly in comparison with the time-scale of diffusion, so that the concentration profile does not alter significantly during the temperature change, then the profile at the beginning of diffusion at the new temperature T_2 is

$$(C-C_1)/(C_0-C_1) = \text{erf}\left[x/2(D_1 t_1)^{\frac{1}{2}}\right]$$

This is identical to the profile which would have been obtained if diffusion had been started at the temperature T_2, but had proceeded only for a shorter time

$$t_2 = D_1 t_1/D_2. \tag{100}$$

The experimental consequence of this is that, in an exchange experiment in which the diffusion process is slow compared to attainable rates of temperature change, diffusion can be studied successively at several different temperatures in the same experiment. A plot of M_t^2 against t will then consist of a number of linear portions, each corresponding to diffusion at one temperature, and each extrapolating to a different intercept on the time axis. At each temperature, D can be evaluated from the slope of the appropriate portion of the curve.

2.2.2 Exchange experiments with spherical particles

When the extent of penetration of the diffusion process into the solid is sufficiently great that the solid can no longer be assumed semi-infinite, the solution of the diffusion equation depends on the shape of the solid. The most widely used solutions are those for spherical particles (radius a, distance of any point from centre r). There are three cases of particular importance:

(a) Concentration inside sphere initially uniform at C_1, surface concentration maintained at C_0 for all $t > 0$. The concentration profile in the sphere is then

$$(C-C_1)/(C_0-C_1) = 1+(2a/\pi r)\sum_{n=1}^{\infty}[(-1)^n/n]\sin(n\pi r/a)e^{-Dn^2\pi^2t/a^2} \tag{101}$$

The extent of reaction (total amount of material entering or leaving sphere as a fraction of final amount, $\alpha = M_t/M_\infty$) is

$$\alpha = 1-(6/\pi^2)\sum_{n=1}^{\infty}(1/n^2)e^{-Dn^2\pi^2t/a^2} \tag{102}$$

Plots of $(C-C_1)/(C_0-C_1)$ against r/a at various values of Dt/a^2 are given by Carslaw and Jaeger[51] (p. 201) and by Crank[54] (p. 86). The determination of D from data of α against t may be tackled as follows. A curve of α against the dimensionless parameter Dt/a^2 may be computed (Crank[54], p. 90, curve marked zero). This may be used to convert values of α to values of Dt/a^2. A plot of these values against t should be a straight line with slope D/a^2.

(b) If the fluid phase surrounding the solid is of limited volume, the condition of constant surface concentration will not be realized. Suppose that the ratio of volume of fluid to that of solid is β. (If there is a distribution coefficient K between the two phases, the volume of the favoured phase must be increased by the factor K in computing β.) The rate law then becomes

$$\alpha = 1 - \sum_{n=1}^{\infty} \frac{6\beta(\beta+1)e^{-Dq_n^2 t/a^2}}{9+9\beta+q_n^2\beta^2} \tag{103}$$

where the q_n's are the non-zero roots of

$$\tan q_n = \frac{3q_n}{3+\beta q_n^2} \tag{104}$$

These roots are tabulated by Crank[54] (p. 330), who also gives curves of α against Dt/a^2 (p. 90).

(c) Sometimes the surface concentration may be known explicitly as a function of time, $C(a) = \phi(t)$. In that case, for initial concentration zero, the concentration profile inside the sphere is given by

$$C = -(2D/ra) \sum_{n=1}^{\infty} (-1)^n e^{-Dn^2\pi^2 t/a^2} n\pi \sin(n\pi r/a) \int_0^t e^{Dn^2\pi^2\lambda/a^2}\phi(\lambda)d\lambda \tag{105}$$

For non-zero initial concentration, the solution is the sum of two terms of the forms indicated in equations (105) and (101).

2.2.3 Exchange experiments with plate-like crystals

For finite solids of any shape less symmetric than a sphere or an infinite cylinder it is in general very difficult to obtain solutions which are readily usable in connection with experimental data. Only one other situation will be discussed here. That is the infinite plane sheet of finite thickness $2l$ (extending from $x = +l$ to $-l$). These solutions should be a good representation of diffusion in thin plate-like crystals. The solutions have a formal resemblance to those for a sphere.

(a) Infinite plane sheet, uniform initial concentration C_0, surface concentration C_1 at $t > 0$.

$$(C - C_0)/(C_1 - C_0) = 1 - (4/\pi) \sum_{n=0}^{\infty} [(-1)^n/(2n+1)] e^{-D(2n+1)^2 \pi^2 t/4l^2}$$
$$\times \cos [(2n+1)\pi x/2l] \quad (106)$$

$$\alpha = 1 - \sum_{n=0}^{\infty} [8/(2n+1)^2 \pi^2] e^{-D(2n+1)^2 \pi^2 t/4l^2} \quad (107)$$

Computed data for the concentration profile and for α have been given by Crank[54], Henry[55] and McKay[56].

(b) Infinite plane sheet in contact with stirred solution of limited volume. Let the ratio of the volume of fluid to volume of solid be β (corrected for any distribution coefficient as in the case of a sphere). Then

$$\alpha = 1 - \sum_{n=1}^{\infty} [2\beta(\beta+1)/(1+\beta+\beta^2 q_n^2)] e^{-Dq_n^2 t/l^2} \quad (108)$$

where the q_n's are the positive non-zero roots of

$$\tan q_n = -\beta q_n \quad (109)$$

Crank[54], Carslaw and Jaeger[51] and Carman and Haul[57] have given tabulations of q_n.

2.3 GRAIN BOUNDARY DIFFUSION

Both in metals and in the alkali halides, the direct study of concentration distributions (e.g. by autoradiography) has shown that such defects as dislocations and grain boundaries may provide preferential paths for diffusion[58-60]. The problem of the overall rate laws which may be observed for transport of material in such a system is a very complex one. The present author has compared the various approximate expressions which have been proposed, and has attempted to establish the ranges of their validity[61]. The discussion was based on a classification into three general types of behaviour.

2.3.1 Type A

This is the limiting form at long times. The average diffusing particle has wandered sufficiently far to have spent many separate portions of its time both in the bulk and in the "grain boundary" region of rapid diffusion. Its overall motion arises from the combined effects of its mobilities in both regions. In this situation, the overall kinetics of diffusion should be in accordance with Fick's Law, i.e. solutions of equation (87), with an apparent diffusivity D_c which is a suitably weighted average of the bulk diffusivity D_g and the diffusivity in grain boundaries D_s. If the fraction of the total volume having the enhanced diffusivity D_s is f, then

$$D_c = fD_s + (1-f)D_g \tag{110}$$

This type of behaviour was first discussed by Hart[62], who used a very concise random-walk approach, and indicated that, for this approximation to be valid, t must be much greater than the migration time of particles through the bulk between encounters with grain boundaries. For grain radius a_g this condition becomes:

$$a_g^2/D_g \ll t \tag{111}$$

A later discussion by Lidiard and Tharmalingham[63] gave the condition that the scale of the dislocation network (*i.e.* a_g) should be small compared to the diffusion distance [of the order of $(D_c t)^{\frac{1}{2}}$, see section 3.2.2]. This replaces D_g by D_c in the inequality (111) and is a less restrictive condition. The present author[61] studied the same situation, using solutions of the diffusion equation in place of Hart's random-walk approach. This procedure is mathematically more cumbersome than Hart's, but appears to give a more precise indication of the nature of the restrictive condition.

The solid was assumed to consist of a spherical assembly of spherical grains. The concentration ϕ at the surface of each grain was assumed to be given by an expression of the form of equation (101), with the distances r and a referring to the large sphere, and D being the overall diffusivity D_c. Within each grain, the concentration C was expressed as a sum of two terms as given by equations (101) and (105). ϕ was inserted as the time-dependent surface concentration in (105), and the mathematical model was examined for the conditions which would give $C = \phi$ to a good approximation, *i.e.* the conditions in which there would be no large concentration gradients between a grain and its boundary. This is the essential character of type A diffusion, and implies that the paths of rapid diffusion should not be detectable by autoradiography. The condition derived was identical to (111), with a rough indication that the inequality involved is a very large one, of the order of a factor of 10^5.

2.3.2 Type B

This is the most general situation, in which diffusion distance is comparable to the scale of the dislocation network. The concentration distribution does not approximate to any simple form. For diffusion out of the solid, a concentration contour diagram will have the appearance of "river-valleys" cutting back into the crystal along the dislocations or grain boundaries. For a semi-infinite solid with a plane surface, and a single plane grain boundary at right-angles to that surface, Whipple[64] has given an exact solution. For the concentration gradient in the grain boundary at its point of emergence on the external surface, he obtained

$$(\partial\phi/\partial x)_{x=0} = (\pi D_g t)^{-\frac{1}{2}} \left[1 - (\tfrac{1}{2}) \int_0^{\Delta} \sigma^{-\frac{3}{2}} \operatorname{erfc} \left(\frac{\Delta-1}{\Delta-\sigma} \frac{\sigma-1}{2\beta} \right) d\sigma \right] \tag{112}$$

where $\Delta = D_s/D_g$ and $\beta = (\Delta-1)\ell/2(D_g t)^{\frac{1}{2}}$. ℓ is the thickness of the grain boundary (region to which D_s applies) and erfc is the "error function complement", $\operatorname{erfc} x = 1 - \operatorname{erf} x$, where $\operatorname{erf} x$ is defined by equation (97).

The possibility of making any simplifying approximations in equation (112) depends on the comparative values of the width ℓ of a grain boundary and the distance [roughly $(D_g t)^{\frac{1}{2}}$] which a particle would have travelled at the bulk diffusivity D_g. If these values are comparable, the concentration contour diagram shows maximum deviation from any simple form and no approximation can be made. If $D_g t \gg \ell^2$, an approximation is possible resulting in $(\partial\phi/\partial x)_{x=0}$ proportional to $t^{\frac{1}{4}}$, so that the amount of material which has diffused out of the solid by time t (or the extent of reaction α) is given by

$$\alpha = (\text{const})t^{\frac{3}{4}} \tag{113}$$

This rate law was indicated by Lidiard and Tharmalingham[63]. The method of approximation by which it is derived was indicated by the present author[61], who also pointed out the required condition, *i.e.* $D_g t \gg \ell^2$.

2.3.3 Type C

This is the limiting form at short times, or when D_s is enormously greater than D_g, so that diffusion may be considered to take place in the dislocation network only, while material in the bulk is immobile. In terms of a concentration contour-diagram, the river-valleys have become vertical-walled gorges or canyons, and in terms of equation (112), this is the opposite extreme from the approximation last considered. For $D_g t \ll \ell^2$, equation (112) approximates to

$$(\partial\phi/\partial x)_{x=0} = (\pi D_s t)^{-(\frac{1}{2})} \tag{114}$$

This corresponds to equation (98), for simple diffusion out of a semi-infinite solid; but an important question arises regarding the relation of the true grain boundary diffusivity D_s and the apparent diffusivity D_a, which an observer will calculate if he carries out an experiment and performs his calculations unaware that diffusion is not a true bulk process. The present author pointed out[61] that the relation of D_s to D_a is not the same for all experimental procedures. In a sectioning experiment, in which diffusion into a slab from an instantaneous plane source on the surface is described by equation (95), it is immaterial whether the experimenter realizes that the average C for a section is not the true C in the grain boundaries. D_a is found from a plot of x^2 *versus* ln C, and the scale factor is destroyed in taking the logarithm. The experimenter will thus calculate D_c, the true diffusivity in the grain boundaries. His experiment is a direct determination of the distribution of diffusion distances in the x direction from the source, and, unless he determines the concentration distribution across one of his slices of the crystal, he has no way of knowing whether diffusion in the x-direction can take place anywhere or only along certain paths.

In an exchange experiment, the experimentally measured quantity is the total amount of material which has diffused out of the crystal, as a function of time. The experimenter obtains a very different impression of diffusion distances in the crystal, according to whether he believes this material to have been furnished at random from all points in the bulk or only from a small fraction of the solid of unusual mobility. If a fraction f of the cross-section of the crystal (perpendicular to diffusion direction) is occupied by grain boundaries, the amount of material which has diffused out is, from equation (98)

$$M_t = f(2D_s t/\pi)^{\frac{1}{2}} \tag{115}$$

If the observer believes that he is measuring ordinary diffusion from the whole bulk, he will calculate an apparent diffusivity D_a according to

$$M_t = (2D_a t/\pi)^{\frac{1}{2}} \tag{116}$$

From equations (115) and (116)

$$D_a = f^2 D_s \tag{117}$$

For a system in which the grain size, or concentration of dislocations, can be varied in a manner permitting measurement of f, equations (110) and (117) indicate a way of distinguishing between the type A and type C situations. The dependence on f is linear in the former and quadratic in the latter.

Correlation of these theoretical indications with experiment is not satisfactory at the present time. On the basis of the last-mentioned test, diffusion of anions in the alkali halides is apparently type A[60], but autoradiographs indicate a type B concentration distribution, in contradiction to the simple Fick's Law kinetics observed in the same work[60]. In this and other work on alkali halides[65], it is doubtful whether the inequality (111) is satisfied to a sufficient order of magnitude for type A diffusion.

2.4 INTERDIFFUSION AND DIFFUSION ACCOMPANIED BY REACTION

2.4.1 Some general forms of the diffusion equation and its solutions

Mathematically the simplest situation in which diffusion and chemical reaction occur together is that in which the moving species, concentration $C(x, y, z, t)$ reacts reversibly to form an immobile species, concentration $S(x, y, z, t)$, and in which the reaction is sufficiently rapid that equilibrium can be assumed at all times and positions according to

$$S = KC \tag{118}$$

The diffusion equation for this situation is

$$(\partial C/\partial t) = D\nabla^2 C - (\partial S/\partial t) \tag{119}$$

and from equations (118) and (119)

$$(\partial C/\partial t) = [D/(K+1)]\nabla^2 C \tag{120}$$

Thus the situation is kinetically identical to simple diffusion with the diffusivity decreased by the factor $(K+1)$.

If the functional dependence of S on C is more complicated than that of equation (118), the resulting equation assumes the form of the equation for diffusion with a concentration-dependent diffusivity, provided that the equilibrium constant is large (in favour of the immobile form). This may be shown as follows. Equation (118) is replaced by

$$S = Kf(C) \tag{121}$$

whence

$$(\partial S/\partial x) = K[\partial f(C)/\partial x] = Kf'(C)(\partial C/\partial x) \tag{122}$$

If $(\partial S/\partial t)$ is sufficiently large compared to $(\partial C/\partial t)$ that the latter can be neglected, equation (119) can be replaced by

$$(\partial S/\partial t) = \operatorname{div}(D \operatorname{grad} C) = \operatorname{div}\left[\frac{D}{Kf'(C)} \operatorname{grad} S\right] \tag{123}$$

The expression $D/Kf'(C)$ is then mathematically equivalent to a concentration-dependent diffusivity. In practice, the concentration dependence of diffusivity can be found directly from a concentration–distance curve provided that the conditions for the Boltzmann transformation leading to equation (92) are satisfied. Suitable conditions are those of a sectioning experiment, or two large slabs of material placed in contact, the diffusing material being at first uniformly distributed through one slab only: $C = C_0, x < 0, t = 0$; $C = 0, x > 0, t = 0$. On integrating equation (92) with respect to the Boltzmann variable η [equation (91)], we have

$$-2\int_0^C \eta \, dC = D(dC/d\eta) \tag{124}$$

and converting from η back to x and t

$$D = -(1/2t)(dx/dC)\int_0^C x \, dC \tag{125}$$

This expression enables a concentration-dependent D, or the quantity $D/Kf'(C)$ which replaces D for diffusion accompanied by a non-ideal equilibrium, to be evaluated as a function of C from known C, x data at a fixed time, *i.e.* from the kind of data usually obtained in a sectioning experiment.

For the case of an irreversible reaction proceeding simultaneously with diffusion, if the diffusing species is destroyed at a rate which is proportional to its concentration C, the diffusion equation becomes

$$(\partial C/\partial t) = D\nabla^2 C - kC \tag{126}$$

Danckwerts[66] has indicated a general method of solution, applicable to the boundary conditions $C = 0$, $t = 0$ in the solid medium and $C = C_0$, $t > 0$ at the surface. If C_1 is the appropriate solution of the simple diffusion equation, *i.e.* equation (126) without the term $-kC$, Danckwerts showed that the corresponding solution of (126) is

$$C = k\int_0^t C_1 e^{-kt'} dt' + C_1 e^{-kt} \tag{127}$$

The most complicated cases of simultaneous diffusion and reaction which have been solved exactly[67, 68] involve a reversible reaction proceeding at a finite rate in both directions, when

$$(\partial S/\partial t) = \lambda C - \mu S \tag{128}$$

The formidable solution of the diffusion equation for this case is given, together with a thorough discussion of the topic summarized in this section, in Chapter 8 of Crank's book[54].

2.4.2 Interdiffusion and the Kirkendall effect

Interdiffusion causes large changes in the physical and chemical characteristics of the region affected, and might be expected in general to lead to concentration-dependent diffusion coefficients. There are, however, certain cases in which Fick's Law kinetics are observed, and which have important features in relation to the mechanisms of diffusion. Among these are the interdiffusion of copper and zinc, as studied by Kirkendall et al.[69], and the interdiffusion of various metallic oxides, especially the processes resulting in the formation of spinels. These processes were first studied in detail by Wagner[48], and extensive subsequent work has in general tended to confirm the Wagner mechanism[70-72].

Up to about 1940, it was generally supposed that interdiffusion of similar metals such as copper and zinc took place with conservation of lattice sites, by a simple place-exchange of Cu and Zn atoms. On this basis, markers in the form of, for example, non-diffusing metal wires placed at the original interface between diffusing phases should not move as the diffusion proceeds. Kirkendall et al., using a brass core with molybdenum wire markers electroplated with copper (Fig. 6) showed that the distance ℓ between the markers decreased as diffusion proceeded. Movement of the markers constitutes what is generally called "the Kirkendall effect". It indicates loss of lattice sites in the region towards which the markers move. In the original experiments of Kirkendall, it indicated that Zn atoms were diffusing out of the brass more rapidly than Cu atoms were entering it, so that the mechanism of diffusion is not a simple place exchange, but requires the participation of point defects.

The decrease in ℓ was proportional to $t^{\frac{1}{2}}$. This may be seen to be a consequence

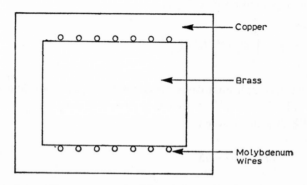

Fig. 6. Type of sample in which the Kirkendall effect was first observed.

of Fick's Law diffusion of both components at different rates, as follows. Consider two semi-infinite bars of material placed in contact at the plane $x = 0$, such that, at zero time, one bar contains only the mobile species A (concentration C_A) and the other only the mobile species B (concentration C_B), *viz.*

$$t = 0, \; x < 0, \; C_A = C_A^0; \qquad t = 0, \; x > 0, \; C_B = C_B^0.$$

Consider the diffusion of species A. At any $t > 0$, its concentration distribution has the form of Fig. 7, with a constant concentration $(\tfrac{1}{2})C_A^0$ at the plane $x = 0$. Thus the concentration of A at $x > 0$ is given by the solution of the diffusion equation for the right-hand slab with a constant surface concentration $(\tfrac{1}{2})C_A^0$. This is equation (96), with $C_1 = (\tfrac{1}{2})C_A^0$, $C_0 = 0$, so that

$$C = (\tfrac{1}{2})C_A^0 \, \text{erfc} \left[x/2(D_A t)^{\frac{1}{2}} \right] \tag{129}$$

where erfc $x = 1 - \text{erf}\, x$. The amount of A which has diffused into the right-hand slab by time t is then, by the same procedure indicated in equation (98)

$$M_A = C_A^0 (D_A t/\pi)^{\frac{1}{2}} \tag{130}$$

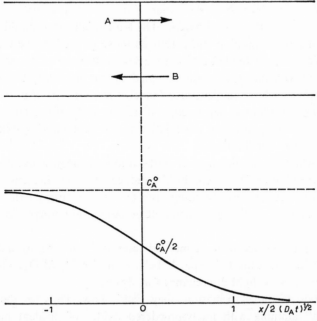

Fig. 7. Interdiffusion in two bars with mobile species A and B, initially confined to left and right hand bar respectively. Concentration against distance for A. Distance given as dimensionless parameter $x/2(D_A t)^{\frac{1}{2}}$.

Similarly the amount of species B which has diffused out of the right-hand slab is

$$M_B = C_B^0 (D_B t/\pi)^{\frac{1}{2}} \tag{131}$$

If the concentrations are both in moles per unit volume of a species which occupies one lattice site in the same crystal lattice, the net loss in moles of lattice sites in the region $x > 0$ is

$$M_B - M_A = (C_B^0 D_B^{\frac{1}{2}} - C_A^0 D_A^{\frac{1}{2}})(t/\pi)^{\frac{1}{2}} \tag{132}$$

In an experimental arrangement in which the (nominally semi-infinite) bars are in fact fixed at their free ends, markers originally placed at $x = 0$ will move to the right by a distance proportional to $M_B - M_A$, and therefore to $t^{\frac{1}{2}}$.

For the formation of spinels, Kooy[72] has given a detailed analysis of possible mechanisms from the point of view of whether they should give rise to a Kirkendall effect. He classifies the possible mechanisms as follows:

Case A. Counter-diffusion of cations (three doubly-charged ions in one direction, two triply-charged ions in the reverse direction) in a static lattice of oxide ions. This is the Wagner mechanism. No Kirkendall effect should be observed, because the oxide ions define the framework of lattice sites.

Case B. One cation immobile, the other diffusing together with an equivalent diffusion of anions in the same direction. The Kirkendall effect should be observed. If the diffusing species (anion and cation B) move to the left, there will be a net gain in lattice sites on the left and the markers will move to the right. It should be noted, however, that the region occupied by the immobile species A extends right up to the markers, so that, in a frame of reference fixed in the laboratory, some A atoms have moved macroscopic distances to the right. They have moved only relative to the laboratory, not relative to the neighbouring oxide ions, and their movement is not diffusion.

Case C. A Wagner-type diffusion of both cations through a fixed oxide lattice, but with the trivalent ions (e.g. Fe^{3+}) reduced to the divalent state during transport, and with a corresponding transfer of discharged oxide ions as O_2 through the gas phase. The Kirkendall effect should be observed, because of the oxygen transfer.

Experimental evidence using a marked interface[71] apparently indicated close agreement with the expected type A behaviour for $MgAl_2O_4$, and a similar result with possibly a slight discrepancy for $MgFe_2O_4$.

The kinetics of spinel formation obey Fick's Law. Thus for the reaction of "spherulized" alumina with finely-powdered oxides of divalent metals, it has been found[70] that equation (102), for diffusion into a sphere, is obeyed. It seems that each diffusing cation takes note principally of its immediate environment of anions, and is hardly affected by more distant neighbours. The diffusion paths probably involve an alternation of octahedral and tetrahedral positions, and the ease of formation of spinels can be correlated with site-preference energies[70].

2.4.3 Diffusion across a growing layer of product

(a) The parabolic law

Consider a film of metal oxide on the surface of the metal, extending from its outer surface at $x = 0$ to contact with the metal at $x = X$. Suppose that the rate of advance of the interface at X is determined by diffusion of oxygen (surface concentration C_0) through the oxide layer to the reaction interface, where its concentration drops to zero as a result of the chemical reaction. (It will later be shown that the same rate law can be derived when the diffusion process involves species other than gaseous oxygen.)

If we assume a quasi-stationary state in which the concentration gradient across the oxide layer is uniform

$$\partial C/\partial x = C_0/X \tag{133}$$

The rate of transport dM/dt of oxygen across the oxide layer is then

$$dM/dt = DC_0/X \tag{134}$$

If C is defined in terms of moles of oxygen atoms per unit volume, and V_M is the volume of oxide containing one mole of oxygen atoms, then

$$dM/dt = (1/V_M)(dX/dt) \tag{135}$$

Eliminating dM/dt from (134) and (135) and integrating

$$X = (2DC_0 V_M t)^{\frac{1}{2}} \tag{136}$$

This is the well-known parabolic law for tarnish reactions, which were first studied extensively by Wagner[50].

The above quasi-stationary state treatment involves an approximation, the nature of which is rather obscure without a more detailed treatment. The general problem of diffusion with a moving boundary has been treated by Danckwerts[73], and the parabolic law has also been derived by Booth[74]. An account of Danckwerts' treatment is given in Chapter 7 of Crank's book[54]. In practical circumstances, the parabolic law is a good approximation, and the condition for its validity is

$$C_0 \ll (1/V_M) \tag{137}$$

i.e. that the concentration of oxygen diffusing is much less than the concentration of immobilized oxygen in the solid product. This implies that large fractional readjustments of concentration in the diffusion region can take place while the interface at X is moving very little, which is a common-sense condition for establishment of a quasi-stationary state concentration distribution.

The above derivation is in terms of the diffusion of electrically neutral oxygen atoms, and does not correspond at all closely to the actual mechanisms which appear to operate in cases of oxidation of metals which have been studied in practice. It is beyond the scope of the present account to review the experimental evidence in detail, but the general features of one widely-investigated example will be indicated. A very detailed account has been given by Grimley[75].

In the oxidation of Cu to Cu_2O by gaseous oxygen, the main transport processes appear to be the motion of cation vacancies and positive holes inwards from the gas/oxide interface to the oxide/metal interface. Alternatively, these processes may be thought of as motion of cations and electrons outwards. Thus the expansion of the oxide lattice takes place at the gas/solid interface, where the chemical process may be written

$$(\tfrac{1}{2})O_2(g)' + 2Cu^+(\text{lattice}) \rightarrow O^{2-}\begin{pmatrix}\text{new lattice}\\\text{site}\end{pmatrix} + 2Cu^+\begin{pmatrix}\text{new lattice}\\\text{sites}\end{pmatrix}$$

$$+ 2\,(\text{cation vacancy}) + 2\,(\text{positive hole})$$

The migration of the defects is an electrical conduction process driven by a potential difference E across the oxide layer, which may be calculated from the free energy change of the overall reaction. E corresponds, in the diffusion analogy, to the drop in concentration from C_0 to zero. The quasi-stationary state assumption is represented by assuming a current I which is the same at all depths (x) in the oxide layer. The situation is then kinetically identical to the previously-discussed diffusion situation, and the parabolic law can be derived in a similar way.

One restriction on the use of the parabolic law is important when transport is by charged defects through extremely thin layers. At the surface of an ionic crystal, and also at dislocations (which act in many ways effectively as "internal surfaces"), there will usually exist, in thermal equilibrium, an electric double layer in the form of a charged surface layer counterbalanced by a discrepancy in concentration between positive and negative defects extending some distance into the bulk (the "space charge" layer). This has been discussed by Lehovec[76] for the surface of pure crystals in equilibrium, by Grimley[75] for both interfaces in the $Cu/Cu_2O/O_2$ system, and by Eshelby *et al.*[77] for dislocations, including the effect of cation vacancies generated by impurities. The space-charge has no sharp boundary, but all authors give similar formulae for the effective thickness λ of the layer for a gas/solid interface in stoichiometric crystals, *viz.*

$$\lambda \sim (\varepsilon kT/8\pi e^2 n_0)^{\frac{1}{2}} \tag{138}$$

where ε is the static dielectric constant of the ionic crystal, and n_0 is the number of defects of each sign per unit volume of the crystal. At a metal/oxide interface, the space charge is not counterbalanced by a charge in the surface layer, and the factor 8 in the denominator of (138) is replaced by 24.

λ may vary widely in different circumstances, but typically n_0 may be of the order 10^{16} defects·cm^{-3} and λ is then about 100 A. For the parabolic law to be obeyed in oxidations, it is thus necessary that the thickness of the oxide layer throughout the range of time concerned should be substantially above 200 A.

(b) Reaction of spherical particles

There is some obscurity in the literature on this topic. Three equations, sometimes referred to[78,79] as the Jander[80], Ginstling–Brounshtein[81] and Dünwald–Wagner[82] equations have been used; the names of Serin and Ellickson[83] are also commonly referred to in relation to the last-mentioned. In references to these equations, there is not always a clear indication of the nature of the approximations involved in them, or of the fact that they do not all refer to the same situation. The object of this section is to attempt to clarify these points.

(i) The Dünwald–Wagner equation is equation (102) in section 2.2.2 above. It is applicable to diffusion with constant D into or out of a system of spheres of uniform radius with constant surface concentration of the diffusing species. It is *not* appropriate to the situation of a sharp reaction interface advancing into the spherical particles. It is obeyed in certain cases of interdiffusion, such as spinel formation[70], in which the original interface will gradually become "blurred" and ultimately disappear altogether, but this situation is quite strictly one of constant surface concentration at the original radius. It corresponds, in spherical geometry, to the situation of two semi-infinite bars in contact, discussed in section 2.4.2 and illustrated in Fig. 7, in which the concentrations of the two diffusing species remain constant at all $t > 0$ at the original interface $x = 0$, but the interface is sharp only at $t = 0$.

(ii) The Jander equation and the Ginstling–Brounshtein equation are both attempts to treat, by the "quasi-stationary state" approach, the case of an advancing reaction interface, *i.e.* the situation which in semi-infinite solids gives the "parabolic law", but now in spherical geometry. The Jander equation is a very rough approximation, which should be used *only* for small extents of reaction α. The Ginstling–Brounshtein equation is the proper analogue, in spherical geometry, of the parabolic law.

Consider a sphere of initial radius b, at which the concentration C of the diffusing species is C_1 at all times. At time t, let the reaction interface have penetrated to radius a, at which the reaction removes the diffusing species so that $C = 0$. The extent of reaction is

$$\alpha = 1 - (a^3/b^3) \tag{139}$$

Jander[80] assumed that the thickness $X = (b-a)$ of the reacted layer is given by the parabolic law applicable to semi-infinite geometry, *i.e.* equation (136). With this assumption

$$(b-a)^2 = 2DC_1 V_M t \tag{140}$$

where, as in (136), V_M represents the volume of the immobile reaction product containing one mole of the diffusing species. Equations (139) and (140) yield

$$[1-(1-\alpha)^{\frac{1}{3}}]^2 = 2DC_1 V_M t/b^2 \tag{141}$$

This, with the collection of constants on the right-hand side sometimes represented by K_J, is the Jander equation. It clearly takes no account of the convergence of the diffusion paths as the centre of the sphere is approached. It is just a way of writing the parabolic law, which may readily be used without casting it in the form of equation (141), and which is commonly applied to many shapes of solid sample if α is sufficiently small.

The steady-state solution of the diffusion equation for a spherical shell of radii b and a, with concentrations respectively C_1 and 0, is[54]

$$C = bC_1(r-a)/r(b-a) \tag{142}$$

from which the rate of flow of material through the shell may be calculated as

$$dM_t/dt = -4\pi a^2 D(\partial C/\partial r)_{r=a} = 4\pi DC_1 ab/(b-a) \tag{143}$$

In the usual manner of the quasi-stationary state approximation, having derived the above expression on the basis of constant a, we now allow a to vary with t. Writing

$$dM_t/da = -4\pi a^2/V_M \tag{144}$$

we have

$$da/dt = -DC_1 V_M b/a(b-a) \tag{145}$$

whence

$$DC_1 V_M t = \int_b^a [(a^2/b)-a]da \tag{146}$$

Integrating, and using equation (139) to convert from a to α, gives

$$1-(\tfrac{2}{3})\alpha-(1-\alpha)^{\frac{2}{3}} = 2DC_1 V_M t/b^2 \tag{147}$$

which is the Ginstling–Brounshtein equation. It should be applicable up to high values of α in practical circumstances. Giess[79] has tabulated the functions of α in both equations (141) and (147), and comparison of the values indicates that (141), the Jander equation, should not be trusted beyond about $\alpha = 0.15$.

The validity of (147) depends of course on the constancy of D and the degree to which the system approaches exact spherical geometry. The linearity of a plot of the Ginstling–Brounshtein function, as we may call the left-hand side of (147), against t does not seem to be very sensitive to either of these. In the author's laboratory, linear Ginstling–Brounshtein plots have been obtained for a strongly exothermic reaction ($SrCl_2$ with F_2) in which there was undoubtedly a large temperature gradient across the reacted material, and in which the sample was a large crystal often of rather irregular initial shape[84]. In this connection, it should be noted that the Ginstling–Brounshtein plot expands the region occupied by the second half of the reaction $[\alpha > (\frac{1}{2})]$ greatly at the expense of the first half. In the second half of a reaction, the interface is quite likely to have assumed a more regular form than it had at the beginning, and temperature gradients in an exothermic process will also have diminished concordant with the diminishing rate of reaction.

3. Point defects in reaction mechanisms

3.1 TYPES OF DEFECT INVOLVED IN REACTION MECHANISMS

It is beyond the scope of this account to attempt a detailed discussion of the mechanisms of a wide variety of specific reactions. But if the possibilities of making theoretical calculations on concentrations and rates of migration of reaction intermediates are to be discussed, it is necessary to establish, at least in general terms, what types of defect may be important as intermediates. There are of course some processes in which defects are not important, and the reaction interface can be treated as a phase in which the rate-controlling processes take place. These include reactions in which a gaseous reactant or product is involved and the molar volume of the solid product is small compared to that of the reactant, so that large cracks develop to allow access or escape of gas. Among such processes, dehydration of salt hydrates has been mentioned above (section 1.4.3). For certain oxidations of metals, which obey a linear rather than a parabolic law[85], the reason appears to be a rate-controlling step at the reaction interface. In general, however, reactions with an ionic reactant or product usually require the presence of one or more of the following types of defect.

3.1.1 Conduction electrons and positive holes

Electronic conduction by means of either conduction-band electrons or valence-band positive holes has been indicated as probably occurring in the mechanisms of many reactions involving ionic crystals, including for example Wagner's work on tarnishing reactions[50, 75], the work of Mitchell on the photographic process in silver halides[86], and the present author's work[14, 15] on oxidation of KI by Cl_2 and F_2. From a chemist's viewpoint, a conduction electron and a hole may often be thought of as roughly equivalent to a discharged cation and anion respectively. The activation energy of formation of electrons or holes will be determined by oxidation–reduction processes at the reaction interface, and not by the band gap, which is much too large in most ionic crystals for thermal excitation across it to be of any significance (the band gap in alkali halides ranges from about 6 eV upwards[87]).

The mobility of electrons and holes in semiconductors is commonly very high, of the order of 1000 $cm^2 . volt^{-1} . sec^{-1}$. This compares, for example, with a range of about 10–100 $cm^2 . volt^{-1} . sec^{-1}$ for conduction electrons in metals (see tabulations of McKelvey[88], p. 312, and Wert and Thomson[87], pp. 216, 257). A reaction mechanism commonly involves simultaneous transport of vacancies and electrons or holes, and since the mobility of a vacancy is often 10^{-6} $cm^2 . volt^{-1} . sec^{-1}$ or even lower (see formulae relating to motion of vacancies in section 3.2.3), it is evident that mobility of electrons or holes will rarely be rate-determining, provided that all these defects are available in adequate concentration. In ionic systems, however, conduction electrons are absent, and reaction rates may be greatly enhanced by short-circuiting[88a], just as in the operation of voltaic cells with liquid electrolytes.

3.1.2 Surface defects

One of the most obvious examples of the advance of a reaction interface with a solid phase is the growth of a crystal from vapour or liquid. The theory of crystal growth now generally accepted was proposed by Frank[89] in 1949, and introduced into solid state chemistry the theory of dislocations which had already been developed extensively by metallurgists since 1934 in connection with the theory of plastic flow in metals[90, 91].

This modern theory of crystal growth envisages the addition of successive atoms to the crystal in a manner which, for each addition, releases the lattice energy per atom, and which leaves the surface ready to accept another atom in the same manner. Consider first a monatomic crystal, in the rigid lattice approximation. The potential energy of the whole crystal is

$$V_{cryst} = (\tfrac{1}{2}) \sum_{i, j} V(r_{ij}) \tag{148}$$

where i and j represent atoms and r_{ij} is the distance between them. The factor ($\frac{1}{2}$) compensates for counting each pair interaction twice, as ij and ji. The potential energy per atom is then

$$V_L = (\tfrac{1}{2}) \sum_j V(r_{ij}) \tag{149}$$

If one atom is removed from a site in the bulk of the crystal to infinity, the energy absorbed is, however, the doubled amount

$$\sum_j V(r_j) = 2V_L \tag{150}$$

The required repeating step for successive addition or removal of atoms must be associated with the lattice energy V_L, not $2V_L$. It must therefore involve an atom with one-half of the interactions of a bulk atom. Such an atom is one occupying a kink site in a step on the surface (Fig. 8, position K). It is in contact with a semi-infinite block B below it, a semi-infinite sheet S to the left and a semi-infinite row R behind it. In the infinite crystal each of these structures has its counterpart, above, to the right of and in front of atom K.

The important energetic property of a kink site is not restricted to a monatomic crystal, but remains true for an ionic crystal. The lattice energy is commonly expressed as a sum of electrostatic terms between point ions on lattice sites and

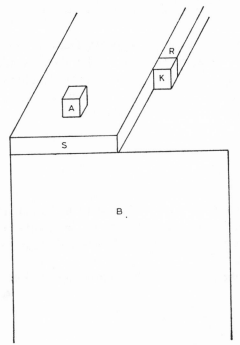

Fig. 8. An adsorbed atom (A) and a kink site (K), with the semi-infinite row (R), sheet (S) and block (B) in contact with K.

overlap repulsions (V^{rep}) for the neighbours j of any ion i, *viz.*

$$V_L = -\alpha_M(e^2/a) + V^{\text{rep}} \tag{151}$$

where the "Madelung constant" α_M is

$$\alpha_M = \sum_j \pm(1/r_j) \tag{152}$$

the sign being positive or negative according as the ion j is the same sign as or of opposite sign to the ion i. The repulsion term, if nearest neighbour repulsions only are important, can be written

$$V^{\text{rep}} = \sum_j V_j^{\text{rep}} \tag{153}$$

in which the summation is taken over nearest neighbours only. If more distant neighbours are important, the terms for anion and cation are different, and (153) must be replaced by the mean value for anion and cation

$$V^{\text{rep}} = (\tfrac{1}{2})\sum_j V_j^+ + (\tfrac{1}{2})\sum_{j'} V_{j'}^- \tag{154}$$

Equations (152) to (154) do not contain the factor $(\tfrac{1}{2})$ in (149) and V_L thus represents the lattice energy per ion pair, which is the conventional unit for which energies are quoted in an ionic crystal. In terms of the subdivision of the crystal according to Fig. 8, each of the energy terms can be resolved into contributions from the blocks, sheets and rows indicated in the diagram

$$\alpha_M = 2\alpha_B + 2\alpha_S + 2\alpha_R \tag{155}$$

$$V^{\text{rep}} = 2V_B^{\text{rep}} + 2V_S^{\text{rep}} + 2V_R^{\text{rep}} \tag{156}$$

For a kink site, each of these expressions must be halved, and for many other positions, such as an ion in a flat surface or an ion adsorbed above a flat surface (Fig. 8, position A), rough estimates of the energetics may be made by selecting the appropriate terms.

For example, in the NaCl lattice the contributions to the Madelung constant are[92, 93]

$$\alpha_B = 0.067$$
$$\alpha_S = 0.114$$
$$\alpha_R = 0.693 \quad (\text{\textit{i.e.} ln 2, representing the summation } 1 - (\tfrac{1}{2}) + (\tfrac{1}{3}) \text{ etc.})$$

Whence

$$\alpha_{bulk} = 2\alpha_B + 2\alpha_S + 2\alpha_R = 1.748$$
$$\alpha_{kink} = \alpha_B + \alpha_S + \alpha_R \quad = 0.874$$
$$\alpha_{surf} = \alpha_{bulk} - \alpha_B \quad\quad = 1.681$$
$$\alpha_{ads} = \alpha_B \quad\quad\quad\quad\quad = 0.067$$

An important feature, in relation to the mechanism of addition of ions at a surface is the unfavourable energetic situation of an adsorbed ion, which has less than 8 % of the binding energy of an ion at a kink site.

The actual structure of a real crystal surface will of course involve distortion of the lattice with consequent changes in the energetics, and the accurate calculation of surface energies requires the determination of this distortion. This is very difficult to do precisely, since there are many possible ways in which the surface might deform, and the potential energy surface for the system apparently has many shallow minima. The calculations of Benson et al.[94] for sodium chloride provide a good exposition of the nature of the problem and the types of approach used in its solution.

Such distortions, however, do not in any way affect the principle that addition of an atom at a kink site will be accompanied by release of energy exactly equal to the lattice energy. Such addition of an atom will produce, so to speak, an all-move-down-one-place effect in which the sum of all distortion energies is unaffected, provided that the crystal is large enough to be thought of in terms of a crystal of infinite extent. No position other than a kink site has this special property of replication in the addition of atoms.

If steps are present on a surface, they will at thermal equilibrium contain a high concentration of kinks[95]. Thus, once the steps are present, the preservation of kinks in the steps as reaction proceeds presents no difficulty. If, however, the step extends right across the surface, as Fig. 8 suggests, the disappearance of the step after the addition of a complete layer is a serious discontinuity in the growth mechanism. The contribution of Frank in this regard was to point out that the disappearance of the step would be avoided if one end of the step was anchored somewhere in the crystal surface, and that the point of emergence of a dislocation with any "screw" character (i.e. with its Burgers vector not perpendicular to the dislocation line, see section 3.1.5) would act as such an anchor point. The step may then be expected to wind up into a spiral about the anchor point, since addition of atoms giving a constant linear rate of advance of all parts of the step causes the more distant parts of the step to move more slowly in terms of angular velocity about the anchor point. The spiral should eventually reach a steady state configuration in which the crystal can grow, in principle indefinitely, by rotation of the spiral without change of its structure.

For an extensive discussion of surface disorder in crystal growth, see Frank *et al.*[95]. For an indication of the possible effect of adsorbed gases in increasing surface disorder, see Cabrera[96].

3.1.3 Schottky defects in thermal equilibrium

(a) The equilibrium constant expression

Schottky defects[97] are essentially lattice vacancies, but in considering the thermodynamics of their formation it is necessary to specify that a vacancy is formed by expansion of the lattice by one lattice site of the appropriate type. This may be envisaged as occurring by removal of an atom (or ion) followed by its replacement *at a kink site on a surface step*, as discussed in section 3.1.2 above. The expansion of the lattice has sometimes been misrepresented in textbook diagrams by indicating addition of atoms in "adsorbed" positions above a flat surface, which is not a true expansion by one lattice site. The correct and incorrect two-dimensional analogues are shown in Fig. 9.

For a monatomic crystal with N atoms and n vacancies, provided that $n \ll N$, the equilibrium number of vacancies can be considered in practice to be given by

$$n/N = A e^{-W/kT} \tag{157}$$

where W is the energy of formation of a Schottky defect. A represents the entropy effect of changes in lattice vibrations arising from the formation of the vacancy. If there are no such changes, $A = 1$, but it is generally to be expected that $A > 1$.

There is some confusion in the literature regarding the rigorously correct form of equation (157). Derivations have been given[98-100] in which the left-hand side appears as $n/(N-n)$, n/N and $n/(N+n)$. The last-mentioned, as given by Howard and Lidiard[100], appears to be the correct one. Confusion on the point can arise for two reasons: firstly, that derivations from "dynamic equilibrium" considerations between surface and bulk sites do not give the precisely correct expression; secondly, that Frenkel defects (see section 3.1.4) and Schottky defects are often discussed together without a clear indication that lattice sites are conserved in formation of Frenkel defects, but not in the formation of Schottky defects, for which the number of lattice sites is $(N+n)$.

The left-hand side of equation (157) arises from the configurational entropy of a random arrangement of the N atoms and n vacancies on $(N+n)$ sites, and is

$$\exp\left[-(1/k)(\partial S_{\text{config}}/\partial n)_N\right] = \exp\left[-\frac{\partial}{\partial n} \ln\{(N+n)!/N!\,n!\}\right] = n/(N+n) \tag{158}$$

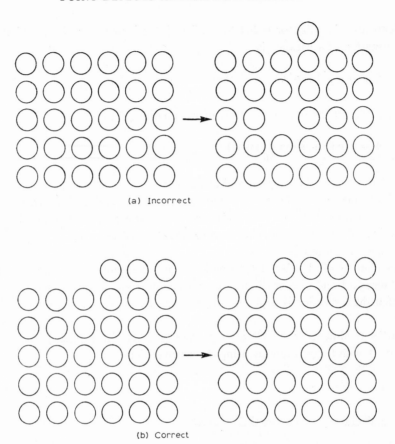

Fig. 9. Ways of depicting the formation of a Schottky defect (two-dimensional analogue): (a) incorrect — atom replaced on surface in an adsorbed state; (b) correct — atom replaced at a step — in three dimensions, at a kink.

(b) The calculation of A

The various expressions which have been proposed for the pre-exponential term A have been reviewed by Kröger[101]. Some difficulties of definition arise for this quantity, chiefly because equation (157) gives only an approximate form for the temperature dependence of n/N. In any real crystal, we may expect both A and W to be volume-dependent, since both are dependent on interactions between atoms which change with the interatomic spacing; and hence A and W should both be temperature-dependent, as a consequence of thermal expansion. The various theoretical treatments of A differ in the way this dependence is handled.

All treatments agree in including a vibrational entropy term in A. If the formation of a vacancy changes the frequency of the ith normal mode of the crystal from v_i to v_i', then, in the high-temperature harmonic approximation in which the entropy of a vibrator is $k[1+\ln(kT/hv)]$, the increase in vibrational entropy on formation of a vacancy is

$$S_{\text{vib}} = k \sum_i \ln(v_i/v_i') \tag{159}$$

If this is the only effect causing A to differ from unity, then

$$A = \exp(S_{\text{vib}}/k) = \prod_i (v_i/v_i') \tag{160}$$

The view which is the simplest and at the same time probably the most rigorous is that of Vineyard and Dienes[102], who give A the form (160), with no other terms. The v_i and v_i' are then the frequencies at the actual experimental conditions, and the value of W used in association with this A must also refer to the actual experimental conditions.

Calculations of W often determine in the first instance an absolute zero value W_0. Vineyard and Dienes discuss the correction from W_0 to W, but they point out that the correction is probably smaller than the uncertainty in the calculation of W_0.

Earlier treatments[98, 103, 104] attempted to take account explicitly of the variation of W with temperature by using the assumed form

$$W = W_0 - aT \tag{161}$$

Mott and Gurney[98] and Frenkel[103] wrote

$$n/N = Ae^{-W/kT} = Ae^{a/k}e^{-W_0/kT} = A'e^{-W_0/kT} \tag{162}$$

in which A' exceeds A by the factor $e^{a/k}$, which could be quite large. Haven and van Santen[104] suggested that (162) does not account completely for the effect of the $-aT$ term. By the thermodynamic identity

$$(\partial H/\partial T)_P = T(\partial S/\partial T)_P \tag{163}$$

equation (161) implies an actual entropy contribution from the $-aT$ term, as well as the term $e^{a/k}$ which is dimensionally analogous to an entropy term. The new term would represent a decrease in entropy and might even reduce the pre-exponential term more than $e^{a/k}$ increases it. This term is presumably necessary in some form if A has not been calculated for actual experimental conditions, but to the present author the best representation seems to be that of Vineyard and Dienes[102] in which A and W are both calculated for experimental conditions, and A is given completely by (160).

Vacancy equilibria have been studied extensively with ionic crystals, particularly

NaCl and other crystals with the same lattice. Equation (157) must then be replaced by

$$(n_+ n_-/N^2) = A_+ A_- e^{-W/kT} \tag{164}$$

in which n_+ and n_- are numbers of cation and anion vacancies, N is the number of ions of one sign in the crystal, A_+ and A_- are expressions of the type of (160), and W is the energy of formation of a separated pair of cation and anion vacancies. For a pure crystal, $n_+ = n_- = n$, and hence

$$n/N = A e^{-W/2kT} \tag{165}$$

where $A = (A_+ A_-)^{\frac{1}{2}}$ and should have a similar magnitude to the A in (160).

Comparatively little work has been done on the theoretical estimation of A, which is a very difficult problem, especially if the v_i are taken, as specified in connection with equation (160), to be normal modes of the lattice. The usual approach to an approximate calculation is to replace the normal modes by vibrations in an Einstein model, and to assume that only the nearest neighbours of the vacancy are affected, and those only in one vibration, along the co-ordinate in the direction of the vacancy. For the NaCl lattice, each ion having six nearest neighbours

$$A_+ \text{ or } A_- = (v/v')^6 \tag{166}$$

where the v' may be different for the neighbours of an anion vacancy and a cation vacancy. Mott and Gurney[98] suggested, without detailed discussion, that v/v' might be as large as 2. Vineyard and Dienes[102] pointed out that, on the simplest picture, it should be the force constant rather than the frequency which is halved, which would lead to $v/v' = 2^{\frac{1}{2}}$. They took the results of Etzel and Maurer[105] for NaCl, viz.

$$n/N = 5.4 e^{-1.01 eV/kT} \tag{167}$$

and, on the assumption that v/v' is the same for cations and anions, calculated $v/v' = 1.32$, remarkably close to the expected value. Subsequently, two other values of A have been determined: Biermann[106] gives $A = 54$ and Dreyfus and Nowick[107] give $A = 13.4$ (see Kröger's extensive tabulation[101]). These values yield $v/v' = 1.95$ and 1.54. The Dreyfus and Nowick results probably represent the greatest accuracy to which data are currently known for NaCl, and give

$$n/N = 13.4 e^{-1.06 eV/kT} \tag{168}$$

Theimer[108, 109] suggested that the distortion around a vacancy, which is effectively a compression of a small region of the lattice, would compensate for the diminished vibration frequencies in the six vibrations most obviously affected by increases in the frequencies of a number of other vibrations in the distorted region. His detailed calculations[109], based on the distortions calculated by Fumi and Tosi[110] [see subsection (c) below], indicated almost exact compensation, the fairly large entropy term discussed above being replaced by a very small *negative* one. As Theimer points out, this result is not in accord with experimental facts.

Similarly detailed calculations have been made for vacancies in face-centred cubic metals by Huntington *et al.*[111], and have indicated fairly large positive entropy terms.

(c) The calculation of W

For a rigid 1 : 1 ionic lattice which did not distort or polarize in any way around a vacancy, the energy W_+ or W_- needed to remove a single ion to infinity would be equal to the lattice energy V_L *per ion pair*, as discussed in section 3.1.2 above. In a real crystal, W_+ and W_- are greatly diminished by polarization of the lattice around the vacancy. The environment of a cation being different from that of an anion, the potentials produced by polarization V_{pol}^+ and V_{pol}^- will   be different but the form of the mathematical treatment is the same in either case, and in the following account $V_{pol}^\pm$ is used to signify either of these quantities; similarly $W_\pm$.

Consider the hypothetical process of distorting the lattice to its final equilibrium configuration around the site to be vacated, and afterwards removing the ion from that site to infinity. The energy required for the latter step is $(V_L - V_{pol}^\pm)$. Then for the actual removal of an ion accompanied by distortion of the surrounding lattice, the energy required is[98] the average of this value and the rigid lattice value V_L, viz.

$$W_\pm = V_L - (\tfrac{1}{2})V_{pol}^\pm \tag{169}$$

The energy of formation W of a pair of vacancies by removal of the ions to infinity followed by their replacement at kink sites on the surface (energy released V_L) is

$$W = W_+ + W_- - V_L = V_L - (\tfrac{1}{2})(V_{pol}^+ + V_{pol}^-) \tag{170}$$

The problem is then to estimate V_{pol}^+ and V_{pol}^-. The first rough estimate, by Jost[112], used a model of the vacancy as a charged sphere of radius approximately equal to the cation or anion radius r_c or r_a, in a continuous dielectric having the known static dielectric constant k of the substance concerned. Then

$$V_{pol}^\pm = (e^2/r_{c,a})[1 - (1/k)] \tag{171}$$

If a Born–Mayer model is used, with a repulsive potential of the form r^{-n}, so that

$$V_L = (\alpha_M e^2/r_0)[1-(1/n)] \tag{172}$$

where $r_0 = r_a + r_c$, then, from (170), (171) and (172)

$$W = (e^2/r_0)\{\alpha_M[1-(1/n)]-(r_0^2/2r_a r_c)[1-(1/k)]\} \tag{173}$$

This equation, with $r_0 \sim 3$ A, $\alpha_M = 1.748$ (NaCl), $n \sim 9$, k ~ 5, and $(r_0^2/r_a r_c) \sim 2$, gives $W \sim 0$ (within ~ 0.25 eV), although the lattice energy $V_L \sim 8$ eV. This simple treatment indicates correctly that $W \ll V_L$, but underestimates W, which is of the order of 2 eV experimentally for several of the alkali halides (see Kröger's tabulation[101]).

More precise calculations have followed the method of Mott and Littleton[113], in which, for ions close to the vacancy, the displacement of each ion is calculated on the basis of a Born model for interactions between ions. Each displaced ion represents a displacement dipole of strength equal to (charge × displacement from lattice site) which contributes to V_{pol}. In addition, the electron system of each ion is polarizable, and dipoles produced in this way also contribute to V_{pol}. The calculations are complicated, since all the dipoles affect each other. Ions are generally treated individually as far as about fourth neighbours of the vacancy; beyond this distance, the crystal behaves essentially as a continuous dielectric, characterized completely by its static dielectric constant.

Later refinements of the Mott and Littleton type of calculation, particularly by Fumi and Tosi[110], have differed chiefly in the precise values of the parameters used and in the model for the repulsive potential. For NaCl, experimental values of W [found from conductivity studies—see subsection (f) below] range from 2.02 to 2.19 eV. The most reliable is probably that of Dreyfus and Nowick[107, 114], which is 2.12 eV (but the stated limits of ± 0.07 eV on that figure cover most of the reported range of values). In comparison, Mott and Littleton's calculation gave 1.93 eV. Fumi and Tosi obtained 1.91 eV using a Born–Mayer potential, but a Huggins–Mayer exponential form for the repulsive potential gave exact agreement with the experimental value, 2.12 eV.

(d) Origin of Schottky defects

Equilibria involving Schottky defects might be expected to have a close formal analogy to ionic equilibria in aqueous solutions, with equation (164), for the product $n_+ n_-$, taking the place of $[H^+][OH^-] = K_w$. Schottky defects in the alkali halides have been widely studied, especially by electrical conductivity, in which, for most alkali halides at most temperatures, the predominant mobile species is the cation vacancy (as indicated by measurements of transport numbers[99,115]). At temperatures above about 600 °C (higher for the fluorides, lower for iodides), in crystals of ordinary purity, Schottky defects can be considered to be present in equilibrium, with $n_+ = n_-$, i.e. according to equation (165). At lower temperatures, the supply of vacancies arising from the presence of various impurities predominates, and the vacancy concentration is often essentially independent of temperature[99] from about 300 °C to 600 °C. The most widely studied impurities are divalent cations, each of which requires for electrical neutrality the presence of a cation vacancy. The dissolution of a divalent metal halide in an alkali halide may be exemplified by

$$CaCl_2(s) \rightarrow Ca^{2+}(\text{cation site}) + (\text{cation vacancy}) + 2\,Cl^- \begin{pmatrix} \text{anion} \\ \text{sites} \end{pmatrix}$$

If n_+ is increased above the pure crystal value by the presence of impurities, n_- should decrease by the same factor, according to equation (165). This, as Stone[99] has pointed out, is analogous to the "common ion effect" in aqueous solutions. While n_+ has been extensively studied experimentally, n_- is less easily accessible to measurement, since anion vacancies usually make only a minor contribution to conductivity and must therefore be investigated by the more laborious methods of diffusion experiments. Anion migration is apparently sensitive to the presence of dislocations and grain boundaries[60,116], probably through a space-charge effect[77] [as discussed in connection with parabolic oxidations in section 2.4.3 (a) above]. The interrelationship of n_+ and n_- is then more complicated than that of equation (165). This seems to be yet another example of the general principle that the concentrations of kinetically significant species in solids are usually not spatially uniform.

(e) Interactions between imperfections

Point imperfections such as Schottky defects or divalent impurities have effective electric charges equal to the difference between the charge actually present on the lattice site and that which would be present in the perfect crystal. Consequently these defects interact with each other electrostatically. If the crystal is assumed to act as a continuous dielectric, with $k \sim 5$, even for distances of approach down to that of adjacent sites, then, with $(e^2/r_0) \sim 5$ eV, a vacancy pair on adjacent sites should have a binding energy of about 1 eV, and a cation vacancy on a second neighbour site (nearest cation site) to a divalent cation should have a binding energy of about 0.7 eV. Detailed calculations of the binding energies[101,110,117–119] have generally given results of a similar order. Such interactions can therefore be important, especially from about 500 °C downwards.

(i) The vacancy pair, i.e. anion and cation vacancies on adjacent sites. This species, which is electrically neutral and does not contribute to conduction, has been suggested[120] as an intermediate in processes occurring at or below room temperature at rates too great to be attributable to motion of charged particles. The first quantitative estimate of the energy of migration of this species by Dienes[121] supported this suggestion by indicating a low activation energy for migration. Later calculations by Tharmalingham and Lidiard[117], however, indicate activation energies comparable to those for migration of anion vacancies, which are virtually immobile at room temperature. (For discussion of the calculation of activation energies of migration, see section 3.2.1). The status of the vacancy pair as a reaction intermediate is thus doubtful.

(*ii*) Impurity-vacancy complexes, in which a cation vacancy is restricted in its motion to the twelve second nearest neighbours of a divalent cationic impurity, and hence cannot contribute to conduction, are certainly formed and have an influence on the energetics of the electrical conduction process. Suppose that a fraction f_2 of the cation sites is occupied by divalent impurities, including those associated with vacancies. Let a fraction f_1 be occupied by mobile vacancies. Provided that the temperature is sufficiently low for the pure crystal to be neglected as a source of vacancies, the total vacancies are a fraction f_2, and undissociated impurity–vacancy complexes are $(f_2 - f_1)$. The equilibrium constant expression for the dissociation is

$$f_1^2/(f_2 - f_1) = (\tfrac{1}{12})e^{-W_a/kT} \tag{174}$$

in which W_a is the binding energy of the complex (energy of association), and the factor $\tfrac{1}{12}$ arises from the configurational entropy $k \ln 12$ of a complex. Equation (174) is the analogue of a weak acid dissociation equilibrium in aqueous solution. If the degree of dissociation is low

$$f_1 = (f_2/12)^{\frac{1}{2}}e^{-W_a/2kT} \tag{175}$$

and the concentration of free vacancies is governed by an energy $W_a/2$, which appears in the activation energy of electrical conduction.

(*iii*) Debye–Hückel effects. While the short-range effects discussed under (*i*) and (*ii*) above are the most important interactions, longer-range electrostatic effects also occur and can be treated just like those in ionic solutions, *e.g.* by the Debye–Hückel method. Lidiard[122] has developed the necessary expressions. The correction to the complex-formation theory of (*ii*) above, for example, replaces W_a in equation (175) by W_a', where

$$W_a' = W_a(1 - C) \tag{176}$$

and

$$C = \frac{2^{7/4}\pi^{\frac{1}{2}}(W_a/kT)^{\frac{1}{2}}f_1^{\frac{1}{2}}}{1 + 2^{9/4}\pi^{\frac{1}{2}}(W_a/kT)^{\frac{1}{2}}f_1^{\frac{1}{2}}} \tag{177}$$

(f) Activation energies from electrical conductivity

Fig. 10 shows diagrammatically the various regions which give linear plots for ln (σT) against $1/T$, as they are now known for NaCl doped with divalent cationic impurities from the melting-point to somewhat below room temperature[107, 114]. [Equation (219) in section 3.2.3 shows that $\sigma = (\text{const}/T)e^{-E/RT}$, so that it is usual to plot ln(σT) rather than ln σ.] For other alkali halides, very little is known about the low-temperature regions in which lines III, IV and III′ are observed for NaCl, but the regions I and II have been well-known and widely-studied with many compounds for a long time[99, 123]. The explanation of region II as arising from a constant cation vacancy concentration produced by divalent impurities was first given by Koch and Wagner[124] and has been confirmed experimentally, for example, by the work of Etzel and Maurer[105] on cadmium-doped NaCl.

The explanations of the five regions shown in Fig. 10 for NaCl are fairly clearly established, as follows. (E_I, etc., are the activation energies for the appropriate regions. U is the activation energy for migration of a cation vacancy. The theoretical estimation of U is discussed in section 3.2.1 below.)

I. "Intrinsic" region, with vacancy concentration given by equation (165).

$$E_I = U + (W/2) \tag{178}$$

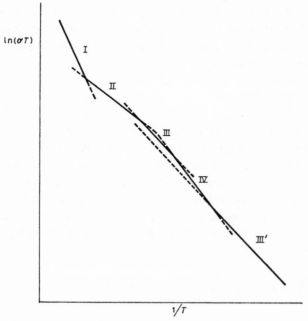

Fig. 10. Electrical conductivity of an alkali halide (diagrammatic; relative slopes are roughly correct for NaCl with common divalent impurities). The lines are extended as broken lines merely to make the linear regions more clearly visible.

II. "Impurity" region. Constant number of vacancies, completely dissociated from divalent impurities.

$$E_{II} = U \tag{179}$$

Hence the energy of formation of a Schottky defect is found experimentally as

$$W = 2(E_I - E_{II}) \tag{180}$$

III. "Association" region, in which formation of impurity–vacancy complexes takes place according to equation (174), or at low enough temperatures (175), so that

$$E_{III} = U + (W_a/2) \tag{181}$$

IV. "Precipitation" region, in which the impurity atoms precipitate out at surfaces, grain boundaries or dislocations, and their concentration in the bulk depends on the heat of solution W_s.

$$E_{IV} = U + (W_a/2) + W_s \tag{182}$$

III'. "Arrested precipitation" region (to coin a phrase), in which the precipitation reaction has become very slow, or perhaps has been arrested by the saturation of all internal sites available for precipitation, such as grain boundaries, so that the activation energy reverts to that of region III.

$$E_{III'} = U + (W_a/2). \tag{183}$$

The above explanations assume thermal equilibrium right down to room temperature, except for the precipitation of impurities. This is in accord with the studies of Dreyfus and Nowick[114], who were able to study kinetically the relaxation of the conductivity to its equilibrium value at temperatures down to $-40\,°C$ after sudden quenching from $100\,°C$. They found a first-order process, with a half-life of a few hours. This evidence appears to rule out earlier explanations[53] of region II in terms of a frozen equilibrium, which similarly gave $E_{II} = U$, but attributed the constancy of vacancy concentration to almost complete arrest of the processes destroying them.

Anion diffusion has been studied much less than electrical conductivity, but regions I and II have been found also in anion diffusion, and the association phenomenon and Debye–Hückel correction have been considered in relation to these results[65].

3.1.4 Interstitials

(a) Frenkel defects

A Frenkel defect[97, 125] is formed by the migration of an atom or ion from a lattice site to an interstitial position. The defect consists of a separated vacancy and interstitial (Fig. 11). Consider a monatomic crystal, or the sublattice of ions of one sign in an ionic crystal. Let there be N lattice sites, and N_i equivalent interstitial positions. Let there be n_v vacancies and n_i interstitials, and let W be the energy of formation of a Frenkel defect. Then the equilibrium equation is

$$[n_v/(N-n_v)][n_i/(N_i-n_i)] = A_v A_i e^{-W/kT} \tag{184}$$

in which the A_v and A_i depend on the same factors discussed for Schottky defects in section 3.1.3. If Frenkel disorder is the only type present, $n_v = n_i = n$, and if N and $N_i \gg n$, then

$$n = (N N_i A_v A_i)^{\frac{1}{2}} e^{-W/2kT} \tag{185}$$

Unlike Schottky defects, Frenkel defects in the cation and anion lattices are independent of each other so far as the electrical neutrality of the crystal is concerned. Some writers refer to the cationic variety only as the Frenkel defect, and its anionic analogue as the "anti-Frenkel" defect. In the present account, the term Frenkel defect will be used to cover both.

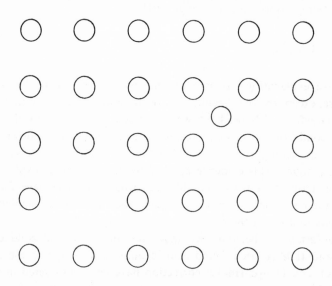

Fig. 11. A Frenkel defect in a simple square lattice.

Cationic Frenkel defects are important chiefly in the silver halides[101,126] in which the energy of formation of such a defect is of the order of 1 eV, while the energy of Schottky disorder is probably no less than in the alkali halides, *i.e.* about 2 eV. For β-AgI, the experimental W, 0.69 eV, compares with calculated values in the range 0.6 to 0.8 eV[101,127].

Anionic Frenkel defects appear to be important chiefly in a variety of compounds crystallizing in the fluorite lattice[128-131]. Studies of these crystals are, however, much less extensive than those of the alkali halides, and, although it is quite clear that, in strong contrast to the alkali halides, the anions are the predominant mobile species in fluorite lattices, very little has been definitely established in detail about the defects responsible for this motion. On the question of activation energy, for example, theoretical and experimental values currently fail to correspond[128], the discrepancy being as much as 2.8 eV. One of the main sources of difficulty seems to be that an interstitial anion might occupy many different positions in these structures. This is a rather similar difficulty, in general terms, to that which is encountered in discussing diffusion in many metals (see section 3.2.1). The present author has attempted to make confusion more confounded[84] by introducing the possibility of a Paneth "crowdion" structure [see subsection (c) below]; this possibility for metals has earlier proved an excellent source of endless controversy.

(b) Interstitial and "interstitialcy" mechanisms in ionic crystals

Interstitials are important in reaction mechanisms in the silver halides[132], zinc oxide and sulphide[133], and in the formation of spinels from two metallic oxides[70]. In all of these (except α-AgI) the anions are arranged in one of the "close-packed" geometries. The interstices are therefore of two types, octahedral sites equal in number to the anions and tetrahedral sites amounting to twice that number. Fig. 12 shows a (110) plane of a face-centred cubic lattice, with the octahedral and tetrahedral sites indicated by the numbers 6 and 4.

In all these substances the important interstitial sites are tetrahedral, and the most likely directions of migration of the interstitial ion lie in a (110) plane. There are three possible migration steps, which will be discussed in relation to Fig. 12. (*i*) The path IC, direct from one tetrahedral site to another, across a shared edge of the tetrahedra. This is likely to be a path of high activation energy, since the migrating ion passes between two anions at their closest approach in the crystal; but in crystals with the NaCl structure, in which all octahedral sites are blocked by cations, this is the most obvious path for diffusion of interstitials. (*ii*) The path IAB, in which migration from one tetrahedral site to another takes place through an intervening octahedral site A. In this path, the migrating ion passes through the centre of the shared triangular faces of the octahedra and tetrahedra, a more favourable path than (*i*) from the viewpoint of overlap repulsion with the anions. In the spinels, $A^{2+}B_2^{3+}O_4$, the cations occupy one-half of the octahedral sites (used by B^{3+} in a normal spinel, such as $MgAl_2O_4$

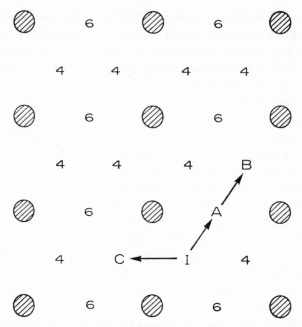

Fig. 12. A (110) plane of an FCC lattice, showing tetrahedral interstices (4) and octahedral interstices (6). IC and IAB are possible paths for migration of interstitials (see text).

itself), and one-eighth of the tetrahedral sites. This structure leaves so many interstices unoccupied that, for example, any occupied tetrahedral site in the perfect lattice is surrounded by six vacant tetrahedral sites and four vacant octahedral sites. The diffusion path in spinel formation is probably IAB, and Stone and Tilley[70] have found appropriate correlations of ease of spinel formation with the site preference energies of the ions concerned.

(*iii*) The so-called "interstitialcy" mechanism involves the use of the migration path IAB even though the octahedral site A is occupied, and thus provides a mechanism which avoids the path IC in the NaCl structure. The mechanism was first proposed in relation to tracer work on diffusion of cations in AgCl and AgBr[132]. It involves the simultaneous motion of the ions at I and A, so that the original interstitial enters the lattice site A, and the ion from that site is displaced to the interstitial position B.

(c) *The Paneth "crowdion"*

The elemental alkali metals crystallize in the body-centred cubic lattice, in which the line between nearest neighbours is a $\langle 111 \rangle$ direction. Paneth[134] proposed a mechanism of self-diffusion in these metals involving what he described as "the rapid transmission of short linear regions of compression" along the $\langle 111 \rangle$ directions. Such a region, for which he coined the term "crowdion", contains an extra atom, with the misfit relieved vernier-fashion over several lattice sites in the

⟨111⟩ line. (There were 9 atoms to 8 sites in Paneth's example, as illustrated for a simple square lattice in Fig. 13a.) Motion of the crowdion along the line can be effected by shifts in atomic positions of a small fraction of a lattice spacing, to advance the whole crowdion structure by one lattice spacing. If the crowdion travels along a complete row of atoms, then that row is moved one atomic spacing, and that motion constitutes the diffusion process. The crowdion mechanism is closely related to three other matters discussed in this account:

(*i*) The "interstitialcy" mechanism [*b*(*iii*) above] is essentially a localized crowdion mechanism which has lost its vernier character by the localization.

(*ii*) The crowdion is a one-dimensional analogue of the three-dimensional defect known as an edge dislocation (compare Fig. 13a and b, and see section 3.1.5). Its motion corresponds to glide of a dislocation. ⟨111⟩ is in fact the glide direction of a BCC lattice, and the theories of plastic deformation of metals and of migration of point defects in chemical kinetics thus meet in the crowdion mechanism.

(*iii*) The defect in alkali halides known as an H-centre (see section 3.1.6) contains an extra halogen atom (neutral) in a line of halide ions in their direction of closest approach, ⟨110⟩, and the term "crowdion" has been extended by Känzig and Woodruff[135] to this situation, although the vernier character, which is essential to the Paneth mechanism, is missing in this structure as envisaged by Känzig and Woodruff. (Their electron spin resonance experiments, which do not provide any precise estimate of interatomic spacing, do not necessarily exclude the vernier structure, but it has not been proposed.) In the present author's laboratory, a defect which is probably the H-centre has been found to be important kinetically as a reaction intermediate in NaCl and KCl[136], and a much more tentative suggestion of an actual vernier-displacement crowdion mechanism in $SrCl_2$ has been made[84].

3.1.5 Dislocations

Although it is often necessary, in discussing dislocations, to refer to a "dislocation line", a dislocation should not be regarded as a one-dimensional defect. The geometrical description of a dislocation requires the specification both of the dislocation line and of its "Burgers vector". Since these are not in general parallel, this is a specification of a plane, and a dislocation should be regarded as a two-dimensional defect in which a three-dimensional distortion of the whole crystal can be described with reference to a plane. (The only difference in this respect in the case of a pure screw dislocation, for which dislocation line and Burgers vector are parallel, is that the plane cannot be specified uniquely.)

The character of a dislocation is displayed most clearly in the description of its hypothetical formation by cutting the crystal part-way through and displacing the cut surfaces relative to each other. The "Volterra cut" is the plane of the defect.

The dislocation line specifies where the cut ends inside the crystal, and the Burgers vector specifies the displacement of the faces of the cut. A general example of a dislocation is usually regarded as a combination of two limiting types:

(*i*) The "edge" dislocation, in which the cut faces are envisaged as being displaced away from each other, perpendicular to their planes, to allow insertion of one or more planes to fill the gap. Clearly the Burgers vector is then perpendicular to the dislocation line, which lies in the plane of the cut. Fig. 13b shows a cross-section of a simple cubic lattice containing an edge dislocation. In the conventional symbol ⊥ marking an edge dislocation, the stem of the inverted T indicates the inserted half-plane in the Volterra cut, and the crossbar indicates the direction of the Burgers vector.

Fig. 13b could also be thought of as representing a crystal of which the upper part ABCD has been compressed. If the dislocation E⊥ moves to the right, it will finally cause the face BC to move out one atomic spacing and create a step on the right-hand surface, relieving the compression. This is a possible elementary step in the plastic deformation of a metal. In the diagram, the compression has been shown as forcing ten atomic spacings into the space normally occupied by nine, to stress the analogy of the misfit to a vernier situation.

Despite the fact that both the Paneth crowdion and the edge dislocation can be thought of as analogous to vernier scales, they are quite distinct defects. In Fig. 13b, the misfit must be relieved differently above and below the plane CD, and the difference is the Burgers vector. In the crowdion of Fig. 13a, there is no difference in the methods of relief of the misfit on all sides of the line.

(*ii*) The "screw" dislocation results from movement of the faces of the Volterra cut parallel to each other and to the dislocation line (Fig. 13c). Thus no gap is produced which needs an extra plane of atoms to fill it. In mechanical terms, a screw dislocation has shearing stresses only associated with it, while an edge dislocation has also tensile stresses. Since the latter involve a slight change in lattice parameter, edge dislocations produce broadening of X-ray diffraction lines while screw dislocations do not.

At the surfaces of the crystal at which a screw dislocation emerges, steps will be produced terminating at the point of emergence of the dislocation line. This is the type of step which is self-perpetuating in crystal growth, as described above in section 3.1.2. If a crystal contains a single screw dislocation, with a Burgers vector equal to one atomic spacing, the whole crystal can be thought of as a single atomic layer wound into a spiral about the dislocation line.

Dislocations have been discussed most widely in relation to the mechanisms of plastic deformation of metals. From the viewpoint of the chemical kineticist, the precise range of their significance in the reactivity of solids is yet to be established, but they are certainly known to have some important functions, as follows:

(*a*) They are known to provide paths for accelerated diffusion both in metals and in ionic crystals (see section 2.3 above).

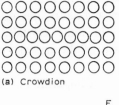

(a) Crowdion

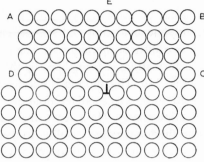

(b) Edge dislocation

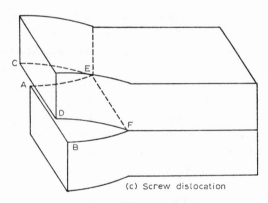

(c) Screw dislocation

Fig. 13. Dislocations and related defects.

(a) The Paneth crowdion as it would be in a simple square lattice. If the lattice is made simple cubic, the extra layers added contain no extra atoms; the defect is in one row only.

(b) Cross-section of an edge dislocation in a simple cubic lattice. The extra row E ⊥ is to be found in all planes parallel to that of the diagram. The upper half of the crystal is supposed to have been compressed by an instrument which keeps all faces vertical (for simplicity in showing the distortion).

(c) An ambiguous perspective, intended to represent a screw dislocation (AB and CD in contact, Burgers vector BD parallel to dislocation line FE). If the lines AB and CD are visualized as being apart, this may also be used as a picture of the Volterra cut for an edge dislocation, opened ready for insertion of the extra half-plane.

(b) They act as sources and sinks for vacancies in single crystals[137], and thus allow establishment of thermal equilibrium concentrations of vacancies at low temperatures (around room temperature) in ionic crystals, when transport between the external surface and inner parts of the crystal would otherwise be prohibitively slow.

(c) They may act as nuclei for precipitation of impurities, or for the commencement of growth of any solid product.

(d) Dislocations have been found to affect the rates of reactions as various as acrylic acid polymerization, graphite oxidation and sucrose caramelization[137a].

Dislocations may move by two distinct mechanisms. The process of "glide", which is important in plastic deformation, is exemplified by the movement of the dislocation ⊥ to the right in Fig. 13b. This motion is effected by the successive adjustment of many atomic positions by only small fractions of a lattice parameter, and requires virtually no activation energy. (In a sense, the whole dislocation is an activated complex. It has no thermodynamic stability. In thermal equilibrium, there would not be even one dislocation in a crystal of any practical size[138].) Glide can take place only in a plane containing both the dislocation line and its Burgers vector.

The process of "climb" is activated, and corresponds to withdrawal of the extra half-plane E⊥ upwards, atom by atom, without motion to left or right. To this end the dislocation line must contain "jogs", which are the analogue of kinks in a surface step[137]. If a vacancy migrates to a jog, the jog is thereby moved one atomic spacing along the line. By repetition of this process, the line may ultimately be annihilated on rising to the surface at E. Climb of dislocations is obviously analogous in mechanism to crystal growth, and is important to the kineticist in connection with the function of dislocations as sources and sinks of vacancies and as sites for precipitation of impurities, i.e. points (b) and (c) above; but note that the crowdion also brings glide within the kineticist's field.

3.1.6 Electronic defects

The literature on electronic defects, especially those produced by various kinds of radiation damage, is very extensive. No attempt is made here to do more than indicate a few features which are likely to be relevant to the establishment of chemical reaction mechanisms. The earlier work on electronic defects was reviewed in two comprehensive articles by Seitz[139,140]. Schulman and Compton's book[141] is of sufficiently recent date to cover the initial development of the concepts of covalent bonding and interstitials in trapped-hole centres. The introduction of these concepts from 1954 onwards has extended considerably the possible range of structures for trapped-hole defects, and has provided evidence for a new type of anion migration in alkali halide lattices which may be of chemical significance.

Electronic defects may be classified into two main types:

(a) Trapped-electron defects

The trapping sites are usually anion vacancies. The most widely studied of all electronic defects since the work of Pohl[142] in the 1930's is the F-centre, which is a single electron in an anion vacancy, and hence has no effective charge. An F'-centre contains two electrons in an anion vacancy, and it can hence contribute to electrical conduction, while an F-centre cannot.

F and F'-centres have been postulated as reaction intermediates in the thermal decomposition of, for example, barium azide[17], lithium aluminum hydride[11] and silver bromide[143]. The proposed mechanisms involve aggregation of F or F' centres as the nucleation process (see section 1.2.1 above). In later work on azides, a simple aggregation product known as the "F_2^+-centre", a single electron occupying two anion vacancies, has been identified by electron spin resonance[144, 145].

It appears from such studies that the F-centre can migrate readily in the region of room temperature. This is surprising, since its rate of migration might be expected to be comparable to that of an anion vacancy, which usually has a sufficiently high activation energy to immobilize the defect at room temperature. By contrast, activation energies of the order of 0.4 eV have been indicated for the migration of F-centres[1, 146]. The conditions in which aggregation of F-centres occurs, and the mechanism for their diffusion are certainly not yet fully understood. The diffusion process may involve transfer of an electron from an F-centre to a cation or to a vacancy (Kröger[101], pp. 799–800). Thermal bleaching of F-centres in KCl between room temperature and 160 °C has been found[140, 147, 148] to proceed with a very low activation energy (two processes, 0.16 eV and 0.31 eV) but also with remarkably low pre-exponential factors of the order of 1 sec^{-1}. This may indicate escape of electrons from F-centres by tunnelling; but it may also indicate some interaction, as yet uncharacterized, with slightly mobile trapped holes.

(b) Trapped-hole centres and Varley mechanisms

Alkali halides X-irradiated at any temperature between 4 °K and room temperature show a number of UV absorption bands. Most of these have been designated V bands, with numerical subscripts, V_1, V_2, V_3 and V_4, although the one which appears at the lowest temperatures is designated the H-band. KBr and KI also develop UV absorption bands, known as V_2 and V_3, on exposure to Br_2 or I_2 vapour at high temperatures. Cl_2 will not similarly form V-centres in alkali chlorides, although V-centres produced by X-irradiation in chlorides have been studied extensively. See Schulman and Compton[141] for a fuller account of this topic.

The work of Mollwo[149] in the 1930's established that centres produced by exposure of KBr and KI to halogen gas are in an equilibrium concentration proportional to gas pressure. These centres are thus probably diatomic. Although some of the centres produced by X-irradiation have been postulated as monatomic, no electron spin resonance signal has been found for any of the V-centres designated by numerical subscripts. Thus they are inaccessible to one of the most powerful methods of structure determination in solids – which has indeed been used in every case of a trapped-hole centre structure currently regarded as definitely confirmed. Although optical studies of the F-centre and V-centres started simultaneously at Göttingen in the 1930's, and the correct structure of the F-centre was postulated by de Boer[150] in 1937, the structures of the V_1, V_2, V_3 and V_4-centres remain uncertain.

The reader of colour centre literature should be on his guard against confused terminology. The terms V_1, etc., are used sometimes to designate optical absorption bands, and sometimes as a convenient way to refer to some structural models proposed in 1954 by Seitz[140]; the correspondence between the bands and the models now looks very uncertain.

The Seitz models envisaged the trapping of a hole as an electrostatic effect, involving a trapping centre with an effective charge. The simplest such structure, originally postulated as the structure of the "V_1-centre", is a positive hole trapped in the vicinity of a cation vacancy. This is the so-called "antimorph of the F-centre". The positive hole might be envisaged as localized on one of the halide ions neighbouring the vacancy, to form a halogen atom, or in various other configurations, e.g. delocalized over all six halide neighbours.

Since 1955, a number of structures of trapped-hole defects have been established definitely by electron spin resonance, principally by Känzig et al. [135, 151–153]. Among these are three variants of the halogen molecule-ion, X_2^-. The most recently-discovered of these[153] occupies two anion sites adjacent to a cation vacancy, and was claimed by Känzig to be the "antimorph of the F-centre", but is not the centre responsible for the V_1 absorption band.

The other two forms of X_2^- are not associated with vacancies, and are apparently stabilized by covalent bonding, a possibility not envisaged in Seitz's 1954 proposals. The bond energies of isolated X_2^- ions are not known, but are likely to be of the order of 1 eV, for the bond order $\frac{1}{2}$. Cl_2^-, for example, is isoelectronic with Ar_2^+, which has a bond energy of 1.04 eV. This is about 40 % of the bond energy of neutral Cl_2.

The two forms of the X_2^- ion which have been studied most extensively are:
(i) The V_k-centre. (The subscript, being Känzig's initial, is omitted in his own papers.) This is an X_2^- ion occupying two anion sites (Fig. 14) and was the first trapped-hole centre for which a structure was definitely established. The centre was at first believed to be the V_1-centre, because its optical absorption has its peak at about the same wavelength as the V_1 band, but the centre was later shown to be distinct from V_1.
(ii) The H-centre. This is an X_2^- ion occupying a single anion site (Fig. 14), aligned along the direction of nearest approach of anions, i.e. $\langle 110 \rangle$. Känzig and Woodruff[135] extended the Paneth term "crowdion" to cover this structure

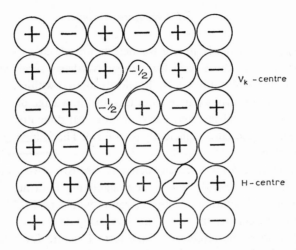

Fig. 14. A (100) face of the NaCl lattice, showing a V_k-centre (X_2^- in two anion sites) and an H-centre (X_2^- in one anion site).

(see section 3.1.4). The H-centre and the F-centre are formed together in X-irradiation of alkali halides at 4 °K. This implies that discharged anions can move in these conditions to leave the anion vacancy and electron which constitute the F-centre, and to form the H-centre which effectively contains an interstitial halogen atom. It is considered probable that the halogen atom moves in a positively charged, multiply-ionized form, a number of electrons being elevated to the conduction band and eventually finding their way to the vacancy and the interstitial to yield centres with no effective charge, the F-centre and the H-centre.

The work of Känzig and Woodruff thus provides spectacular evidence for the occurrence of a transport process at low temperatures not envisaged in the Frenkel–Schottky–Wagner concepts of reaction mechanism in ionic solids which have dominated theoretical discussions from the 1930's. The first suggestion that trapped-hole defects might contain interstitials which arrive by migration in a positively-charged state was made by Varley[154], and it would be appropriate to call a transport mechanism of this type a Varley mechanism. Much work remains to be done to establish whether such mechanisms are of any general importance in chemical kinetics; for their role in radiation damage, see Royce[154a].

Some rate laws for processes in halides with electronic defects as intermediates have been studied in the present author's laboratory[14,15,155]. In the KI/Cl_2 reaction, charged intermediates have been found to increase according to a t^2 law and later to decay, in different circumstances, according to a first- or second-order law. For the NaCl/Cl_2 exchange, a t^n ($n < 0.5$) law first observed in Morrison's laboratory[156] could be changed to a second-order law by deliberate introduction of electronic defects[155].

3.1.7 Notation for defects

The notations currently in use for defects are about as complicated and un-satisfactory as were those used for radioactive decay series at the beginning of this century, and will doubtless be drastically altered within the next two or three decades. Confusion arises partly from lack of knowledge of the precise structure of many electronic defects, and partly from a real difficulty in presenting all the necessary information in a sufficiently brief form. The following types of confusion should be noted:

(*i*) F_2^+ is a relative of the F-centre, being an electron in two vacancies; but F_2^- is the V_k-centre in a fluoride, the F now signifying fluorine, and in the present author's proposed notation[157], this V_k-centre would be $F(a)_2^+$, because it has an effective positive charge. In view of the increasing amount of work which is being done on fluorides, the retention of the term F-centre for a trapped electron is unfortunate. The author would prefer to see it replaced by some such term as "lattice electron" (by analogy, for example, with "hydrated electron", which is a similar species in many ways). The name is one matter, and the abbreviated notation another; alternative notations in current use are discussed below.

(*ii*) "V-centre" is sometimes used as a generic term synonymous with "trapped-hole centre", especially by kineticists wishing to postulate an intermediate of unknown structure; (the author has been guilty of this sort of over-general use of the term[157]). V_{1-4} are sometimes used to specify uv absorption bands, and sometimes for the Seitz models. But Kröger[101] uses V for a vacancy. Confusion with the V-centre terminology can readily arise, especially since Kröger's notation for a cation vacancy in KCl is V_K.

The problem of notation is that it is desirable, ideally, to indicate for any defect in a solid the types of lattice sites (or interstitial sites) used, the contents of each site (vacant, occupied by electrons, atoms or ions) and both the true charge and the effective charge of the defect.

The convention of representing a vacant lattice site by a square, introduced by Wagner and Schottky, has been used by many authors; see for example Seitz[140] and Rees[158]. In the form $\square^+$ and $\square^-$, this notation has the disadvantage of introducing a third meaning for + and − superscripts, in addition to true charge and effective charge. Kröger[101] uses V_A for a site normally filled by atom A, and A_i for atom A at an interstitial site. His system is the only one which has symbols for both effective charge (superscripts $\cdot$ $'$ x for positive, negative and neutral) and true charge (+ −). The present author[157] has proposed the use of a and c for anion and cation sites, with + and − for effective charge. No notation seems to have gained general acceptance at the present time; each author who has proposed a notation uses it extensively in his own writings, and is apparently unable to persuade any significant number of other authors to adopt it. The various existing notations are best clarified by a few examples.

	Seitz	Rees	Kröger	Harrison
Anion vacancy (NaCl)	⊟	□⁻	V_{Cl}°	a^+
F-centre (NaCl)	⊟	(e | □⁻)	V_{Cl}^{x}	a
F'-centre (NaCl)	⊟	(e₂ | □⁻)	V_{Cl}'	a^-
Interstitial cation (AgBr) (*see Note 1*)		(Ag⁺ | Δ)	$Ag_i^{\cdot}$	Ag⁺(i)
V_k-centre (NaCl) (*see Note 2*)	$\begin{smallmatrix}+ & \overline{}\\ & o\\ \overline{} & +\end{smallmatrix}$	(Cl₂⁻ | □₂⁻)	$h_{\text{self trapped}}^{\cdot}$ or $(Cl_{Cl})_2^{\cdot}$	Cl(a)₂⁺
H-centre (NaCl) (*see Note 3*)		(Cl₂⁻ | □⁻)	Cl_i^{x} or $(Cl_2)_{Cl}^{x}$	Cl₂(a)

Note 1 Seitz's 1954 review contained no discussion of interstitials; Harrison's suggestion did not mention interstitials, but the above extension of the notation is obvious.

Note 2 The author has taken the liberty of devising what he believes should be written for the V_k-centre in the Seitz and Rees notations. Seitz used an open circle for a positive hole. Kröger (p. 524, Table 15.3) gives only the peculiar notation $h_{\text{self trapped}}^{\cdot}$, although the alternative $(Cl_{Cl})_2^{\cdot}$ appears consistent with the rest of his notation.

Note 3 No Seitz notation. Kröger gives both alternatives shown.

3.2 RATES OF MIGRATION OF DEFECTS

3.2.1 Calculation of jump frequencies of vacancies

Consider an atom or ion moving from a lattice site to another equivalent site which is initially vacant. Somewhere along the path, the moving particle will cross an energy barrier, height U above the initial or final state. It is to be expected that the probability w of the jump taking place in unit time will be given at least approximately by an equation of the Arrhenius form, *viz.*

$$w = Ae^{-U/kT} \tag{186}$$

In a simple rigid lattice model, the top of the energy barrier will be exactly half-way between the initial and final positions of the moving particle, and A may be thought of as a vibration frequency along the reaction co-ordinate.

If vibrations of the rest of the lattice are taken into account, the discussion becomes much more complicated. The whole crystal may then be thought of as a molecule, which may in principle be treated according to all the methods which are used in the discussion of unimolecular decompositions (see section 5, Chap. 3). There are two general lines of approach to that problem[159] and both have been applied to vacancies in solids.

(*a*) The interaction of the vibrations of the system may be studied to determine the probability of one particular motion, along the reaction co-ordinate q_r, exceeding a critical amplitude q_c which leads to reaction. This is the approach of Rice *et al.*[160-162]. In the fluid phases, the vibrational analysis is carried out for one molecule, the rest of the system acting as source and sink of energy in collisions. In the solid phase, a similar division may be made into "molecule" and "heat bath", but the size of the "molecule" is arbitrary.

In this approach, with a classical description of the vibrations, if the ith normal mode is represented by the variation of a normal co-ordinate Q according to

$$Q_i = E_i^{\frac{1}{2}} \cos \left[2\pi(v_i t + \phi_i) \right] \tag{187}$$

where E is the energy of the vibrator and $E^{\frac{1}{2}}$ is proportional to amplitude, then the reaction co-ordinate q_r is a linear combination of the Q_i, *viz.*

$$q_r = \sum_i \alpha_i Q_i \tag{188}$$

The pre-exponential factor for reaction along q_r is found to be a root-mean-square frequency of the v_i, with the α_i as weighting factors

$$A = (\sum_i \alpha_i^2 v_i^2 / \sum_i \alpha_i^2)^{\frac{1}{2}} \tag{189}$$

The minimum energy U of the whole system which enables q_r to exceed q_c is found to be

$$U = q_c^2 / \sum_i \alpha_i^2 \tag{190}$$

(*b*) The system may be treated according to "transition-state" theory. In this, according to the familiar Glasstone, Laidler and Eyring formulation[163]

$$w = (kT/h)(Z^*/Z)e^{-U/kT} \tag{191}$$

where U is the activation energy in a static lattice, and Z and Z^* are the partition functions for the initial state and the transition state. These are

$$Z = \prod_{i=1}^{N} (kT/hv_i) \tag{192}$$

where the v_i are the normal modes of a crystal containing a vacancy, and

$$Z^* = \prod_{i=1}^{N-1} (kT/hv_i') \tag{193}$$

where the v_i' are the normal modes of the transition state, if it is constrained to

avoid the Nth vibration, which is the one along the reaction co-ordinate.
Then

$$A = \prod_{i=1}^{N} v_i \Big/ \prod_{i=1}^{N-1} v_i'. \tag{194}$$

This approach to the problem was taken by Vineyard[164], and does not depend upon the assumption that the saddle point is half-way between initial and final positions. In an earlier treatment by Wert[165], both the v_i and the v_i' refer to "constrained" systems lacking the vibration along the reaction co-ordinate. The transition state must then lie in the plane half-way between initial and final position. If v represents the vibration along the reaction co-ordinate which is completely separated from the v_i and v_i', A becomes

$$A = v \prod_{i=1}^{N-1} (v_i/v_i') \tag{195}$$

A difference in principle, which does not yet seem to have found any important application in practice, between (194) and (195) is that (195) predicts an isotope effect $w \propto M^{-\frac{1}{2}}$ exactly, while (194) predicts this only as an approximation, M being replaced by an effective mass M^* which contains linear contributions from the masses of neighbouring ions. In practice, it is usually found that $w \propto M^{-\frac{1}{2}}$.

There appears to be some difference of opinion[100,161,162] on the question of whether the Vineyard or the Rice formulation has the greater generality, especially in cases in which there is, for example, a temperature gradient in the crystal. This seems to be a subtle point, depending on the precise nature of the approximations made in any particular treatment. Both approaches are yet capable of further development. Since the products of vibration frequencies needed cannot usually be calculated to anything better than an order-of-magnitude estimate, the distinction between the formalisms has not yet assumed any quantitative importance.

The energy barrier U to migration is much more difficult to calculate than the energy W of formation of a vacancy, because of the difficulty of formulating a precise description of the transition state, and because the results are much more sensitive to the assumed repulsive potential than are the results for W. This sensitivity arises because ions or atoms in migration approach each other appreciably more closely than their equilibrium separations in the crystal. The Born–Mayer potential $V_{rep} \propto r^{-n}$ is often adequate for calculations of equilibrium energies, but for energies of migration of vacancies in NaCl and KCl, Guccione et al.[166] obtained better agreement by using a potential which is "harder" at short distances, the Born–Mayer–Verwey potential

$$V_{rep} = B + Cr^{-12} \tag{196}$$

This potential was used whenever $r < r_0$, the distance of approach of nearest neighbours at equilibrium. B and C were calculated by fitting the potential and its derivative to the Born–Mayer potential at r_0. For anions and cations in NaCl and KCl, this procedure gave values of U in the range 0.87 to 1.18 eV, which agrees with the range of experimental values. By contrast, an unmodified Born–Mayer potential gave a range of 0.08 to 0.64 eV. The sensitivity of the result to the form of the repulsive potential will be clear from these figures.

Calculations have also been made on the migration energies of vacancies and interstitials in metals. These have been reviewed by Lomer[167] and a number of results have been summarized by Howard and Lidiard[100]. There is, in general, much more uncertainty about the configurations of defects in metals than for ionic crystals, and a full discussion of current controversies on mechanisms of diffusion is beyond the scope of this account. There are two main reasons for the uncertainty:

(*i*) The site occupied by an interstitial is difficult to determine unequivocally, either by theory or by experiment, and possibilities of anisotropic distortions such as the "crowdion" [see section 3.1.4(c) above] arise.

(*ii*) Different metals vary greatly in the relative importance of ion-core interactions and conduction electrons in determining configuration. Where ion-core interactions are predominant, as in copper, there is a somewhat better chance of accurate theoretical treatment than in cases, like the alkali metals, in which conduction electrons are very significant.

3.2.2 Relation of jump frequencies to transport parameters

For any moving particle in a solid, each jump will usually have the same length ℓ, but may be in a number of different directions. If, for every jump in the sequence, each crystallographically equivalent direction is equally probable, the process is one of "random walk". Frequently, however, the probabilities will not be equal, and their distribution for one jump will depend on the pattern of the preceding steps. This is called a "correlation effect", and causes the mean square displacement of the particle over a large number of jumps to deviate from the random walk value by a correlation factor f, which is usually less than unity. Extensive work has been done on the calculation of correlation factors[100, 168–170].

The types of situation in which a correlation factor $f \neq 1$ will and will not arise may be seen from the following example. Consider an ionic crystal in which cations only are mobile, with a vacancy mechanism. There are three transport parameters to consider: the diffusivity D_v of vacancies, the diffusivity D_c of cations as it would be determined in a self-diffusion experiment with an isotopic tracer, and the electrical conductivity σ. The unit step in all these processes is the interchange of a cation and a vacancy, yet these are three quite different parameters.

In general, the vacancies will be in such low concentration that every nearest neighbour site to a vacancy in the cation sublattice is occupied. The successive jumps of a vacancy are therefore randomly oriented, and D_v can be calculated from random walk with no need for a correlation factor f. The jump of a vacancy may also be thought of as the process which transports charge. The electrical conductivity σ will of course depend on the concentration of charge carriers, and the activation energy of σ may therefore include an energy term for thermal creation of vacancies, which D_v does not; but there are no directional influences other than that of the electric field itself (see section 3.2.3 below), and no correlation factor f in the relation between D_v and σ.

D_c, however, refers to the motion of particular marked ions, which can be distinguished from the rest of the cations in the lattice. For any one of these marked ions, the first jump is random, because a vacancy may appear on any neighbouring site; but immediately after that jump, the vacancy lies behind the ion which has moved, and the reverse jump becomes more probable than any other possibility. Over many jumps, the displacement of the ion is therefore less than in the "random walk" process. Thus a correlation factor $f < 1$ is required in the calculation of D_c from D_v and in the relation between diffusivity D_c and conductivity σ.

Consider a random walk in one, two or three dimensions, consisting of steps of length ℓ in random directions. Let the displacement of a particle from its initial position be x, r or R in the three cases. If the medium in which the diffusion takes place extends to $\pm\infty$ in all the dimensions concerned, the mean displacement $\bar{x}$, $\bar{r}$ or $\overline{R}$ of a large number of particles (or one after many jumps) is zero; but the mean square displacement is

$$\overline{x^2} \text{ or } \overline{r^2} \text{ or } \overline{R^2} = n\ell^2 \tag{197}$$

if n is the number of jumps in time t. The relation between mean square displacement and diffusivity may be calculated in many ways. One method, given by Jost[53], starts from the solution of the diffusion equation for an instantaneous plane source, which in fact represents the distribution of x values at time t in the one-dimensional case with motion restricted to positive x. This is equation (95) in section 2.2.1 of the present account. Let q represent the co-ordinate x, r or R. Then the probability $P(q)$ dq of a displacement between q and $q + dq$ corresponding to time t is given by multiplying together one, two or three expressions (according to the number of dimensions of motion) of the form $t^{-\frac{1}{2}} \exp(-x^2/4Dt)$, which, with $x^2 + y^2 = r^2$ and $x^2 + y^2 + z^2 = R^2$, yields in general

$$P(q)dq = Kt^{-m/2}e^{-q^2/4Dt} \tag{198}$$

where K is a constant which need not be specified since it cancels out of the final expressions for mean square displacement, and m is the number of dimensions. Then

$$\overline{q^2} = \int_{0,-\infty}^{+\infty} q^2 P(q)\mathrm{d}q \Big/ \int_{0,-\infty}^{+\infty} P(q)\mathrm{d}q \tag{199}$$

where the lower limit $-\infty$ refers to x, and 0 to r or R. From (199)

one-dimensional case $\overline{x^2} = 2Dt$ $\tag{200}$

two-dimensional case $\overline{r^2} = 4Dt$ $\tag{201}$

three-dimensional case $\overline{R^2} = 6Dt$ $\tag{202}$

The cases most often considered in practice have been those of diffusion in cubic crystals, *i.e.* the isotropic three-dimensional case. D may then be related to ℓ^2 by combining (202) with (197). n is related to the jump frequency w discussed in section 3.2.1. As in that account, w will be used for the frequency of jumps in one specified direction (w_v for vacancies and w_{ion} for ions). In a sub-lattice of ions of one sign in which a vacancy has z nearest neighbours, the total frequency of all jumps of the vacancy is zw_v. Hence

$$D_v = (\tfrac{1}{6})zw_v\ell^2 \tag{203}$$

An ion is mobile only when a vacancy is adjacent to it, and for a lattice of N ions and n vacancies, the probability of this is $z(n/N)$. The ion can then jump with the frequency w_v, provided that a correlation factor f as discussed above is inserted. Hence

$$D_{ion} = (\tfrac{1}{6})fz(n/N)w_v\ell^2 \tag{204}$$

in which the equilibrium situation embodied in (n/N) has been discussed in section 3.1.3, and the kinetic considerations determining w_v have been discussed in section 3.2.1. The calculation of f is complicated, and will not be described in detail here. Some useful values[100, 169] are:

cubic lattice	simple	body-centred	face-centred	diamond
f	0.6531	0.7272	0.7815	0.5

The analysis of interdiffusion problems can be made along similar lines to those indicated above, except that (fw_v) must be replaced by a more complicated expression involving w's for all different jumps of a vacancy possible in the mixture, and n/N must be replaced by an expression involving the concentrations of vacancies and the particular species which is diffusing. For example, Lidiard *et al.*[100,171,172] have discussed the case of diffusion of an impurity atom in a face-centred cubic lattice by a vacancy mechanism. They distinguish three jump rates for a vacancy originally on a site adjacent to an impurity: w_1 for the jump to another nearest neighbour site, which rotates the impurity–vacancy complex without either dissociating it or causing diffusion; w_2 for interchange of vacancy and impurity, the diffusion step; and w_3 for any jump which dissociates the complex (an approximation, since these moves are not all equivalent). The equilibrium concentration of impurity–vacancy complexes has been discussed in section 3.1.3e (*ii*) above. If the ratio of complexes to total impurity atoms, *i.e.* $(f_2 - f_1)/f_2$ in equation (174), is represented by F, the expression finally obtained is

$$D_{\text{impurity}} = \frac{F\ell^2 w_2[w_1 + (7w_3/2)]}{3[w_1 + w_2 + (7w_3/2)]} \tag{205}$$

3.2.3 Electrical conduction by ionic defects

In an electric field E, the barrier to migration of a vacancy in any direction with a component in the direction to which the field attracts it ("forward" direction, subscript F) will be slightly lowered; and the barrier to migration with a component in the opposite direction will be slightly raised ("backward" direction, subscript B). Let one particular jump of the vacancy be represented by a vector

$$l = ia + jb + kc \tag{206}$$

and let the field direction be represented by a unit vector

$$u = iu_i + ju_j + ku_k \tag{207}$$

where

$$u_i^2 + u_j^2 + u_k^2 = 1 \tag{208}$$

The projection of the jump on to the field direction is

$$\ell_E = u \cdot l = u_i a + u_j b + u_k c \tag{209}$$

Consider the NaCl lattice, with nearest neighbour spacing r_0. For the face-centred cubic cation sublattice, the possible vectors $l = (a, b, c)$ are $(\pm r_0, \pm r_0, 0)$ with any permutation of the $+$ and $-$ signs and the position of the zero to give the 12 nearest neighbour directions of the FCC lattice.

With these values of a, b and c, there is for every ℓ_E an equal and opposite value. Thus for any field direction the jumps can be divided into two sets, one with "forward" components (ℓ_E positive) and the other containing a backward jump to match each forward jump in the first set. For either the forward or the backward set

$$\sum_{(\ell_E \text{ positive})} \ell_E^2 = 4(u_i^2 + u_j^2 + u_k^2)r_0^2 = 4r_0^2 \tag{210}$$

Consider any one jump ℓ_E, and the corresponding backward jump $-\ell_E$. To a good approximation, the energy barrier is crossed at $\pm(\tfrac{1}{2})\ell_E$ for these two jumps. If the activation energy at zero field is U, then for the forward jump

$$U_F = U - (\tfrac{1}{2})eE\ell_E \tag{211}$$

and for the backward jump

$$U_B = U + (\tfrac{1}{2})eE\ell_E \tag{212}$$

For the jump rates w_F and w_B, from (186)

$$w_{F, B} = Ae^{-U_{F, B}/kT} \tag{213}$$

The contribution of this pair of jumps alone to the drift velocity of the ion in the field direction is

$$v_1 = (w_F - w_B)\ell_E \tag{214}$$

and the mobility of the vacancy is

$$\mu_1 = (w_F - w_B)\ell_E/E \tag{215}$$

If there are n_v vacancies per unit volume, the conductivity resulting from these jumps is

$$\sigma_1 = n_v e\mu_1 \tag{216}$$

Now if $eE\ell_E \ll U$, equation (213) can be approximated as

$$w_{F, B} = Ae^{-U/kT}[1 \pm (\tfrac{1}{2})eE\ell_E/kT] \tag{217}$$

Substitution in (215) and (216) yields

$$\sigma_1 = (n_v e^2 A\ell_E^2/kT)e^{-U/kT} \tag{218}$$

The total conductivity σ is the sum of similar contributions for all pairs of equal and opposite jumps, so that σ is obtained by replacing ℓ_E^2 in (218) by the summation in (210), *viz.*

$$\sigma = (4n_v e^2 A r_0^2 / kT) e^{-U/kT}. \tag{219}$$

Equation (218), for jumps in a single direction only, with or against the field, was given by Mott and Gurney[98], while equation (219) for the NaCl lattice was given by Lidiard[173]. For NaCl ($U = 0.79$ eV) the mobility μ of a cation vacancy is about 10^{-5} cm^2 . sec^{-1} . volt^{-1} at 600 °C, as calculated from the above equations.

Comparison of (219) and (204) yields a simple relation between diffusion and conductivity (when both refer to motion of the same species of defect), *viz.*

$$\sigma / D_{ion} = N_v e^2 / fkT \tag{220}$$

in which n/N in (204) has been replaced by n_v/N_v, where the subscript specifies unit volume, and z has been given the value 12 appropriate to the FCC lattice. Equation (220) is often called the Einstein, or Nernst–Einstein equation. In its original form[174], developed in relation to colloid chemistry, it was an expression relating mobility and diffusivity, and with the addition of the factor f it is often needed in that form, *viz.*

$$\mu / D = e / fkT \tag{221}$$

For interdiffusion, it has often been assumed that the Einstein equation should be obeyed in the form (221) without modification, and results such as those of Chemla[175], which showed that the mobility of divalent ions in NaCl is much less than that given by (221) have been regarded as anomalous. Howard and Lidiard[100] have given a treatment of the problem of interdiffusion and mobility in the formalisms of irreversible thermodynamics. A full exposition of this is beyond the scope of this account. Briefly, the equations for transport parameters are set up in terms of linear coefficients L_{ij} relating the flux of species i to the thermodynamic force for movement of species j. (The thermodynamic force, for isothermal systems with no external forces, is the gradient of the chemical potential, with a negative sign). The L coefficients are an intermediate stage in relating the jump frequencies w to the transport parameters D and σ. The manipulations of irreversible thermodynamics are carried out almost entirely in terms of the L coefficients, and converted to expressions involving w and D only in the final stages. For a lattice of ions a (charge e_a), with a very small concentration of an impurity b (charge e_b), Howard and Lidiard derive a more general expression to replace (221)

$$\mu_b / D_b = |e_a L_{ba} + e_b L_{bb}| / kT L_{bb} \tag{222}$$

The L coefficients being converted into terms of the jump frequencies w_1, w_2 and w_3 as defined in relation to equation (206), they obtain finally

$$\mu_b/D_b = (e/kT)10w_3/[w_1+7(w_3/2)] \tag{223}$$

for the diffusion and mobility in an electric field of a divalent impurity ($e_a = e$, $e_b = 2e$) in the NaCl lattice. If $w_1 \gg w_3$, i.e. if impurity–vacancy complexes rotate much more rapidly than they dissociate, this equation explains the Chemla result with no need to postulate different mechanisms for diffusion and migration in an electric field.

REFERENCES

1 W. E. GARNER, P. W. M. JACOBS AND F. C. TOMPKINS, Chemistry of the Solid State, ed. W. E. GARNER, Butterworths, London, 1955, Chapters 7, 8, 9.
2 N. F. H. BRIGHT AND W. E. GARNER, J. Chem. Soc., (1934) 1872.
3 G. P. ACOCK, W. E. GARNER, J. MILSTED AND H. J. WILLAVOYS, Proc. Roy. Soc. (London), Ser. A, 189 (1946) 508.
4 R. C. CATTON, B. Sc. thesis, University of British Columbia, 1964 (unpublished).
5 J. HUME AND J. COLVIN, Proc. Roy. Soc. (London), Ser. A, 125 (1929) 635.
6 J. HUME AND J. COLVIN, Proc. Roy. Soc. (London), Ser. A, 132 (1931) 548.
7 B. TOPLEY AND J. HUME, Proc. Roy. Soc. (London), Ser. A, 120 (1928) 211.
8 W. D. SPENCER AND B. TOPLEY, J. Chem. Soc., (1929) 2633.
9 S. J. GREGG AND R. I. RAZOUK, J. Chem. Soc., (1949) 536.
10 J. TAYLOR, Solid Propellant and Explosive Compositions, Newnes, London, 1959, Plate 8.
11 W. E. GARNER AND E. W. HAYCOCK, Proc. Roy. Soc. (London), Ser. A, 211 (1952) 335.
12 W. E. GARNER AND L. W. REEVES, Trans. Faraday Soc., 51 (1955) 694.
13 W. E. GARNER AND L. W. REEVES, Trans. Faraday Soc., 50 (1954) 254.
14 L. G. HARRISON, M. D. BAIJAL AND D. J. BIRD, Trans. Faraday Soc., 60 (1964) 1099.
15 L. G. HARRISON, R. J. ADAMS, M. D. BAIJAL AND D. J. BIRD, Reactivity of Solids (Proc. 5th Internat. Symp., Munich, 1964), ed. G.-M. SCHWAB, Elsevier, Amsterdam, 1965, p. 279.
16 KH. S. BAGDASSARIAN, Acta Physicochim. U.R.S.S., 20 (1945) 441.
17 J. G. N. THOMAS AND F. C. TOMPKINS, Proc. Roy. Soc. (London), Ser. A, 209 (1951) 550; 210 (1951) 111.
18 J. M. HEDGES AND J. W. MITCHELL, Phil. Mag., 44 (1953) 357.
19 W. E. GARNER AND W. R. SOUTHON, J. Chem. Soc., (1935) 1705.
20 A. WISCHIN, Proc. Roy. Soc. (London), Ser. A, 172 (1939) 314.
21 F. C. TOMPKINS AND D. A. YOUNG, Trans. Faraday Soc., 52 (1956) 1245.
22 W. E. GARNER AND H. R. HAILES, Proc. Roy. Soc. (London), Ser. A, 139 (1933) 576.
23 J. A. MORRISON AND K. NAKAYAMA, Trans. Faraday Soc., 59 (1963) 2560.
24 R. W. BURTON, B. Sc. thesis, University of British Columbia, 1965 (unpublished).
25 A. FINCH, P. W. M. JACOBS AND F. C. TOMPKINS, J. Chem. Soc., (1954) 2053.
26 R. S. BRADLEY, J. COLVIN AND J. HUME, Phil. Mag., 14 (1932) 1102; Proc. Roy. Soc. (London), Ser. A, 137 (1932) 531.
27 M. AVRAMI, J. Chem. Phys., 7 (1939) 1103; 8 (1940) 212; 9 (1941) 177.
28 B. V. EROFEEV, Compt. Rend. Acad. Sci. U.R.S.S., 52 (1946) 511.
29 K. L. MAMPEL, Z. Physik. Chem., A187 (1940) 43, 235.
29a D. A. YOUNG, Decomposition of Solids, Pergamon, London, 1966.
30 A. R. ALLNATT AND P. W. M. JACOBS, Can. J. Chem., 46 (1968) 111.
30a M. KAHLWEIT, Reactivity of Solids (Proc. 6th Internat. Symp., Schenectady, 1968), to be published; M. KAHLWEIT AND L. KAMPMANN, Ber. Bunsenges. Phys. Chem., 71 (1967) 78.

31 E. G. PROUT AND F. C. TOMPKINS, *Trans. Faraday Soc.*, 40 (1944) 488.

32 F. BOOTH, *Trans. Faraday Soc.*, 49 (1953) 272.

33 J. E. SPICE AND L. A. K. STAVELEY, *J. Soc. Chem. Ind. (London)*, 68 (1949) 313, 348.

34 C. HUGGETT, C. E. BARTLEY AND M. M. MILLS, *Solid Propellant Rockets*, Princeton University Press, 1960, p. 32–33.

35 M. A. COOK, *The Science of High Explosives*, Reinhold, New York, 1958, especially Chapters 3 and 4 and Appendix II.

36 S. FORDHAM, *High Explosives and Propellants*, Pergamon, London, 1966, Chapter 2.

37 A. R. UBBELOHDE, *Chemistry of the Solid State*, ed. W. E. GARNER, Butterworths, London, 1955, Chapter 11.

38 S. PATERSON, *Discussions Faraday Soc.*, 22 (1956) 155.

39 D. L. CHAPMAN, *Phil. Mag.*, 47 (1899) 90.

40 E. JOUGUET, *J. Math.*, (1905) 347; (1906) 6; *Mécanique des Explosifs*, O. Doin et fils, Paris, 1917.

41 *Bulletin* 84, Report of the Committee on Hydrodynamics, Natl. Research Council, U.S.A., (1931) 551.

42 R. L. SCORAH, *J. Chem. Phys.*, 3 (1935) 425.

43 G. B. KISTIAKOWSKY AND E. B. WILSON, O.S.R.D. report No. 114 (1941); S. R. BRINKLEY AND E. B. WILSON, N. D. R. C., Div. 8, O. S.R.D. report No. 1707 (1943); R. O. COWAN AND W. FICKETT, *J. Chem. Phys.*, 24 (1956) 932.

44 S. PATERSON, *Research (London)*, 1 (1948) 221.

45 H. LAWTON AND I. C. SKIDMORE, *Discussions Faraday Soc.*, 22 (1956) 188.

46 R. S. BRADLEY, *Phil. Mag.*, 12 (1931) 290.

47 B. TOPLEY, *Proc. Roy. Soc. (London)*, Ser. A, 136 (1932) 413.

48 C. WAGNER, *Z. Physik. Chem.*, B34 (1936) 309.

49 C. WAGNER, *Z. Anorg. Chem.*, 236 (1938) 320.

50 C. WAGNER, *Z. Physik. Chem.*, B21 (1933) 25; B32 (1936) 447.

51 H. S. CARSLAW AND J. C. JAEGER, *Conduction of Heat in Solids*, Oxford University Press, Oxford, 1947.

52 R. M. BARRER, *Diffusion in and through Solids*, Cambridge University Press, Cambridge, 1941.

53 W. JOST, *Diffusion*, Academic Press, New York, 1952; revised edition 1960 has identical contents and pagination except for an extensive addendum.

54 J. CRANK, *The Mathematics of Diffusion*, Oxford University Press, Oxford, 1956.

55 P. H. S. HENRY, *Proc. Roy. Soc. (London)*, Ser. A, 171 (1939) 215.

56 A. T. McKAY, *Proc. Phys. Soc.*, 42 (1930) 547.

57 F. C. CARMAN AND R. A. W. HAUL, *Proc. Roy. Soc. (London)*, Ser. A, 222 (1954) 109.

58 R. S. BARNES, *Nature*, 166 (1950) 1032.

59 R. E. HOFFMAN AND D. TURNBULL, *J. Appl. Phys.*, 22 (1951) 634; *Acta Met.*, 2 (1954) 419.

60 J. F. LAURENT AND J. BÉNARD, *J. Phys. Chem. Solids*, 7 (1958) 218; 3 (1957) 7.

61 L. G. HARRISON, *Trans. Faraday Soc.*, 57 (1961) 1191.

62 E. W. HART, *Acta Met.*, 5 (1957) 597.

63 A. B. LIDIARD AND K. THARMALINGHAM, *Discussions Faraday Soc.*, 28 (1959) 64.

64 R. T. P. WHIPPLE, *Phil. Mag.*, 45 (1954) 1225.

65 D. PATTERSON, G. S. ROSE AND J. A. MORRISON, *Phil. Mag.*, Ser. 8, 1 (1956) 393; L. G. HARRISON, J. A. MORRISON AND R. RUDHAM, *Trans. Faraday Soc.*, 54 (1958) 106; L. W. BARR, I. M. HOODLESS, J. A. MORRISON AND R. RUDHAM, *Trans. Faraday Soc.*, 56 (1960) 697.

66 P. V. DANCKWERTS, *Trans. Faraday Soc.*, 47 (1951) 1014.

67 A. H. WILSON, *Phil. Mag.*, 39 (1948) 48.

68 J. CRANK, *Phil. Mag.*, 39 (1948) 362.

69 E. O. KIRKENDALL, L. THOMASSEN AND C. UPTHEGROVE, *Trans. A. I. M. E.*, 133 (1939) 186; E. O. KIRKENDALL, *Trans. A. I. M. E.*, 147 (1942) 104; A. D. SMIGELSKAS AND E. O. KIRKENDALL, *Metals Technology*, 13 (1946), Techn. Pub. 2071.

70 F. S. STONE AND R. J. D. TILLEY, *Reactivity of Solids (Proc. 5th Internat. Symp., Munich*, 1964), ed. G.-M. SCHWAB, Elsevier, Amsterdam, 1965, p. 583.

71 R. E. CARTER, *J. Am. Ceram. Soc.*, 44 (1961) 116.

72 C. Kooy, *Pure Appl. Chem.*, 9 (1964) 441.
73 P. V. Danckwerts, *Trans. Faraday Soc.*, 46 (1950) 701.
74 F. Booth, *Trans. Faraday Soc.*, 44 (1948) 796.
75 T. B. Grimley, *Chemistry of the Solid State*, ed. W. E. Garner, Butterworths, London, 1955, Chapter 14.
76 K. Lehovec, *J. Chem. Phys.*, 21 (1953) 1123.
77 J. D. Eshelby, C. W. A. Newey, P. C. Pratt and A. B. Lidiard, *Phil. Mag.*, 3 (1958) 75.
78 S. K. Deb, *Trans. Faraday Soc.*, 62 (1966) 3032.
79 E. A. Giess, *J. Am. Ceram. Soc.*, 46 (1963) 374.
80 W. Jander, *Z. Anorg. Allgem. Chem.*, 163 (1927) 1.
81 A. M. Ginstling and B. I. Brounshtein, *J. Appl. Chem. U.S.S.R.*, 23 (1950) 1327.
82 H. Dünwald and C. Wagner, *Z. Physik. Chem.*, B24 (1934) 53.
83 B. Serin and R. T. Ellickson, *J. Phys. Chem.*, 9 (1941) 742.
84 L. G. Harrison, R. C. Catton and A. K. Rantamaa, *Reactivity of Solids (Proc. 6th Internat. Symp., Schenectady, New York*, 1968), Wiley, New York (to be published 1969).
85 K. Hauffe and H. Pfeiffer, *Z. Elektrochem.*, 56 (1952) 390; *Z. Metallk.*, 44 (1953) 37.
86 J. W. Mitchell, *Repts. Progr. in Phys.*, 20 (1957) 433.
87 C. A. Wert and R. M. Thomson, *Physics of Solids*, McGraw-Hill, New York, 1964.
88 J. P. McKelvey, *Solid-State and Semiconductor Physics*, Harper and Row, New York, 1966.
88a K. Hauffe, *Reactivity of Solids (Proc. 6th Internat. Symp., Schenectady*, 1968), to be published.
89 F. C. Frank, *Discussions Faraday Soc.*, 5 (1949) 48, 66, 72, 76.
90 A. H. Cottrell, *Dislocations and Plastic Flow in Crystals*, Clarendon Press, Oxford, 1953.
91 G. I. Taylor, *Proc. Roy. Soc. (London), Ser. A*, 145 (1934) 362; E. Orowan, *Z. Physik*, 89 (1934) 604, 634; M. Polanyi, *Z. Physik*, 89 (1934) 660.
92 G. C. Benson, *Can. J. Phys.*, 34 (1956) 888.
93 G. C. Benson and F. van Zeggeren, *J. Chem. Phys.*, 26 (1957) 1083.
94 G. C. Benson: with F. van Zeggeren, *J. Chem. Phys.*, 26 (1957) 1077; with P. Balk and P. White, *J. Chem. Phys.*, 31 (1959) 109; *J. Chem. Phys.*, 35 (1961) 2113; with P. I. Freeman and E. Dempsey, *Advances in Chem. Ser.*, 33 (1961) 26; with P. I. Freeman and E. Dempsey, *J. Chem. Phys.*, 39 (1963) 302.
95 W. K. Burton, N. Cabrera and F. C. Frank, *Phil. Trans. Roy. Soc. (London), Ser. A*, 243 (1951) 299.
96 N. Cabrera, *Z. Elektrochem.*, 56 (1952) 294.
97 W. Schottky, *Z. Physik. Chem.*, B29 (1935) 335.
98 N. F. Mott and R. W. Gurney, *Electronic Processes in Ionic Crystals*, 2nd edition, Clarendon Press, Oxford, 1948, especially Chapter 2.
99 F. S. Stone, *Chemistry of the Solid State*, ed. W. E. Garner, Butterworths, London, 1955, Chapter 2.
100 R. E. Howard and A. B. Lidiard, *Repts. Progr. in Phys.*, 27 (1964) 161.
101 F. A. Kröger, *The Chemistry of Imperfect Crystals*, North-Holland, Amsterdam and Wiley, New York, 1964, especially pages 246, 277–279, 423–430 and 799–800.
102 G. H. Vineyard and G. J. Dienes, *Phys. Rev.*, 93 (1954) 265.
103 J. Frenkel, *Kinetic Theory of Liquids*, Clarendon Press, Oxford, 1946.
104 Y. Haven and J. H. van Santen, *Philips Research Repts.*, 7 (1952) 474.
105 H. W. Etzel and R. J. Maurer, *J. Chem. Phys.*, 18 (1950) 1003.
106 W. Biermann, *Z. Physik. Chem. (Frankfurt)*, 25 (1960) 90.
107 R. W. Dreyfus and A. S. Nowick, *J. Appl. Phys., Suppl.*, 33 (1962) 473.
108 O. Theimer, *Phys. Rev.*, 109 (1958) 1095.
109 O. Theimer, *Phys. Rev.*, 112 (1958) 1857.
110 F. G. Fumi and M. P. Tosi, *Discussions Faraday Soc.*, 23 (1957) 92.
111 H. B. Huntington, G. A. Shirn and E. S. Wajda, *Phys. Rev.*, 99 (1955) 1085.
112 W. Jost, *J. Chem. Phys.*, 1 (1933) 466.
113 N. F. Mott and M. J. Littleton, *Trans. Faraday Soc.*, 34 (1938) 485.
114 R. W. Dreyfus and A. S. Nowick, *Phys. Rev.*, 126 (1962) 1367.
115 C. Tubandt, *Z. Anorg. Chem.*, 110 (1920) 196; 115 (1921) 105; with E. Rindtorff and

W. Jost, 165 (1927) 195; with H. Reinhold and G. Liebold, 197 (1931) 225; Landolt-Bornstein, *Physikalisch-chemische Tabellen*, Springer, Berlin, 1931.

116 L. W. Barr, I. M. Hoodless, J. A. Morrison and R. Rudham, *Trans. Faraday Soc.*, 56 (1960) 697.

117 K. Tharmalingham and A. B. Lidiard, *Phil. Mag.*, 6 (1961) 1157.

118 F. Bassani and F. G. Fumi, *Nuovo Cimento*, 11 (1954) 274.

119 J. R. Reitz and J. L. Gammel, *J. Chem. Phys.*, 19 (1951) 894.

120 F. Seitz, *Rev. Mod. Phys.*, 18 (1946) 384.

121 G. J. Dienes, *J. Chem. Phys.*, 16 (1948) 620.

122 A. B. Lidiard, *Phys. Rev.*, 94 (1954) 29; 112 (1958) 56; *Phil. Mag.*, 46 (1955) 1218.

123 W. Lehfeldt, *Z. Physik*, 85 (1933) 717.

124 E. Koch and C. Wagner, *Z. Physik. Chem.*, B38 (1937) 295.

125 J. Frenkel, *Z. Physik*, 35 (1926) 652.

126 W. Jost, *Trans. Faraday Soc.*, 34 (1938) 860; C. Wagner and J. Beyer, *Z. Physik. Chem.*, B32 (1936) 113.

127 K. H. Lieser, *Z. Physik. Chem. (Frankfurt)*, 2 (1954) 238; 9 (1956) 216.

128 G. M. Hood and J. A. Morrison, *J. Appl. Phys.*, 38 (1967) 4796.

129 E. Barsis and A. Taylor, *J. Chem. Phys.*, 45 (1966) 1154.

130 R. W. Ure, *J. Chem. Phys.*, 26 (1957) 1363.

131 A. D. Franklin, *J. Phys. Chem. Solids*, 26 (1965) 933.

132 W. D. Compton and J. Maurer, *J. Phys. Chem. Solids*, 1 (1956) 191; A. S. Miller and J. Maurer, *J. Phys. Chem. Solids*, 4 (1958) 196; R. J. Friauf, *J. Phys. Chem.*, 66 (1962) 2380.

133 E. A. Secco, *Discussions Faraday Soc.*, 28 (1959) 94; *J. Chem. Phys.*, 29 (1958) 406.

134 H. R. Paneth, *Phys. Rev.*, 80 (1950) 708.

135 W. Känzig and T. O. Woodruff, *J. Phys. Chem. Solids*, 9 (1958) 70.

136 L. G. Harrison, R. J. Adams and R. C. Catton, *J. Chem. Phys.*, 45 (1966) 4023.

137 F. Seitz, *Phys. Rev.*, 80 (1950) 239.

137a C. H. Bamford, G. C. Eastmond and J. C. Ward, *Proc. Roy. Soc. (London), Ser. A*, 271 (1963) 357; J. M. Thomas and E. L. Evans, *Nature*, 214 (1967) 167; J. M. Thomas and J. O. Williams, *Trans. Faraday Soc.*, 63 (1967) 1922; J. M. Thomas and G. Roscoe, *Proc. Roy. Soc. (London), Ser. A*, 297 (1967) 397.

138 A. H. Cottrell, *Dislocations and Plastic Flow in Crystals*, Clarendon Press, Oxford, 1953.

139 F. Seitz, *Rev. Mod. Phys.*, 18 (1946) 384.

140 F. Seitz, *Rev. Mod. Phys.*, 26 (1954) 7.

141 J. H. Schulman and W. D. Compton, *Color Centers in Solids*, Pergamon, Oxford, 1963.

142 R. W. Pohl, *Proc. Phys. Soc. (London)*, 49 (extra part) (1937) 3.

143 J. W. Mitchell, *Photographic Sensitivity*, Butterworths, London, 1951, p. 242.

144 P. W. M. Jacobs, J. G. Sheppard and F. C. Tompkins, *Reactivity of Solids (Proc. 5th Internat. Symp., Munich, 1964)*, ed. G.-M. Schwab, Elsevier, Amsterdam, 1965, p. 509.

145 G. J. King, B. S. Miller, F. F. Carlson and R. C. McMillan, *J. Chem. Phys.*, 35 (1961) 1442.

146 F. E. Theisen and A. B. Scott, *J. Chem. Phys.*, 20 (1952) 529.

147 P. W. M. Jacobs and F. C. Tompkins, *Chemistry of the Solid State*, ed. W. E. Garner, Butterworths, London, 1955, Chapter 3.

148 E. E. Schneider, *Photographic Sensitivity*, Butterworths, London, 1951, p. 13.

149 E. Mollwo, *Nachr. Akad. Wiss. Göttingen*, 1 (1935) 215; *Ann. Physik*, 29 (1937) 394.

150 J. H. de Boer, *Rec. Trav. Chim.*, 56 (1937) 301.

151 T. G. Castner and W. Känzig, *J. Phys. Chem. Solids*, 3 (1957) 178.

152 M. H. Cohen, W. Känzig and T. O. Woodruff, *J. Phys. Chem. Solids*, 11 (1959) 120.

153 W. Känzig, *J. Phys. Chem. Solids*, 17 (1960) 80.

154 J. H. O. Varley, *J. Nucl. Energy*, 1 (1954) 130.

154a B. S. H. Royce. *Progr. Solid State Chem.*, 4 (1967) 213.

155 R. J. Adams and L. G. Harrison, *Trans. Faraday Soc.*, 60 (1964) 1792.

156 L. G. Harrison, I. M. Hoodless and J. A. Morrison, *Discussions Faraday Soc.*, 28 (1959) 103.

157 L. G. Harrison, *J. Chem. Phys.*, 38 (1963) 3039.

158 A. L. G. REES, *Chemistry of the Defect Solid State*, Methuen, London, 1954.

159 N. B. SLATER, *The Theory of Unimolecular Reactions*, Cornell University Press, 1959.

160 S. A. RICE, *Phys. Rev.*, 112 (1958) 804.

161 A. R. ALLNATT AND S. A. RICE, *J. Chem. Phys.*, 33 (1960) 573.

162 S. A. RICE AND N. H. NACHTRIEB, *J. Chem. Phys.*, 31 (1959) 139.

163 S. GLASSTONE, K. J. LAIDLER AND H. EYRING, *The Theory of Rate Processes*, McGraw-Hill, New York, 1941.

164 G. H. VINEYARD, *J. Phys. Chem. Solids*, 3 (1957) 121.

165 C. WERT, *Phys. Rev.*, 79 (1950) 601.

166 R. GUCCIONE, M. P. TOSI AND M. ASDENTE, *J. Phys. Chem. Solids*, 10 (1959) 162.

167 W. M. LOMER, *Progr. Metal Phys.*, 8 (1959) 255.

168 J. BARDEEN AND C. HERRING, *Imperfections in Nearly Perfect Crystals*, Wiley, New York, 1952.

169 K. COMPAAN AND Y. HAVEN, *Trans. Faraday Soc.*, 52 (1956) 786; 54 (1958) 1498.

170 J. R. MANNING, *Phys. Rev.*, 116 (1959) 819; J. G. MULLEN, *Phys. Rev.*, 124 (1961) 1723.

171 A. B. LIDIARD, *Phil. Mag.*, 46 (1955) 1218.

172 R. E. HOWARD AND A. B. LIDIARD, *J. Phys. Soc. Japan*, Suppl. II, 18 (1963) 197.

173 A. B. LIDIARD, *Handbuch der Physik*, Springer, Berlin, 1957, p. 246.

174 A. EINSTEIN, *Ann. Physik (Leipzig)*, 17 (1905) 549.

175 M. CHEMLA, *Ann. Phys. (Paris)*, 13 (1956) 959.

Index